Commonly Asked Questions in Thermodynamics

Commonly Asked Questions in Thermodynamics

Second Edition

Marc J. Assael, Geoffrey C. Maitland,
Thomas Maskow, Urs von Stockar,
William A. Wakeham, and Stefan Will

CRC Press
Taylor & Francis Group
Boca Raton London New York

CRC Press is an imprint of the
Taylor & Francis Group, an informa business

Second edition published 2023
by CRC Press
6000 Broken Sound Parkway NW, Suite 300, Boca Raton, FL 33487-2742

and by CRC Press
4 Park Square, Milton Park, Abingdon, Oxon, OX14 4RN

CRC Press is an imprint of Taylor & Francis Group, LLC

First edition published by CRC Press 2011

Library of Congress Cataloguing-in-Publication Data
Names: Assael, Marc J., author.
Title: Commonly asked questions in thermodynamics / Marc J. Assael [and five others].
Description: Second edition. | Boca Raton : CRC Press, 2022. | Includes bibliographical references and index. |
Summary: "Designed for a wide audience, from students and researchers to practicing professionals in related areas, this book is organized in a user friendly Question and Answer format. Presented questions become increasingly specific throughout the book, with clear and concise answers, as well as illustrations, diagrams, and tables are incorporated wherever helpful. Thermodynamics is a core discipline associated with the theoretical principles and practical applications underlying almost every area of science, from nanoscale biochemical engineering to astrophysics. Highlighting chemical thermodynamics in particular, this book is written in an easy-to-understand style and provides a wealth of fundamental information, simple illustrations, and extensive references for further research and collection of specific data. Designed for an audience that ranges from undergraduate students to scientists and engineers at the forefront of research, this indispensable guide presents clear explanations for topics with wide applicability. It reflects the fact that, very often, the most common questions are also the most profound"-- Provided by publisher.
Identifiers: LCCN 2021060049 (print) | LCCN 2021060050 (ebook) | ISBN 9780367338916 (hardback) | ISBN 9781032275895 (paperback) | ISBN 9780429329524 (ebook)
Subjects: LCSH: Thermodynamics--Miscellanea.
Classification: LCC QC319 .C66 2022 (print) | LCC QC319 (ebook) | DDC 536/.7--dc23/eng20220412
LC record available at https://lccn.loc.gov/2021060049
LC ebook record available at https://lccn.loc.gov/2021060050

ISBN: 978-0-367-33891-6 (hbk)
ISBN: 978-1-032-27589-5 (pbk)
ISBN: 978-0-429-32952-4 (ebk)

DOI: 10.1201/9780429329524

Typeset in Palatino
by MPS Limited, Dehradun

The authors wish to dedicate this second edition of Commonly Asked Questions in Thermodynamics *to **Dr. Anthony H.R.C. Goodwin**, who was the driving force behind the completion of the first edition. Tony died suddenly on 12th September 2014, at the age of 53. We have been conscious of his legacy as we have prepared this edition, mindful of his entreaties to be rigorous, innovative and forever to employ the guidelines established by the International Union of Pure and Applied Chemistry, with whom he worked assiduously. We have done our best to remain true to that aim but, in broadening the material in the text, this has proved a considerable challenge. We feel that he would forgive us for occasionally straying from rigid adherence to this principle, given the wider aims we have in this new edition.*

Contents

Preface

As we observed in the preface to the first edition of this book, the subject of thermodynamics plays at least some role in almost every other discipline of science, from molecules and the nanoscale to the cosmos and astrophysics, with biology and the life sciences not excepted. Some aspects of thermodynamics underpin the very fundamentals of these subjects while it has an impact on almost every application in engineering. It follows that the subject features in a wide range of undergraduate courses in some form or other and also in disparate research activities. For these reasons, the task facing authors of books on thermodynamics is what approach to adopt: whether to be axiomatic or not; whether to consider fundamentals or applications; which topics to select for detailed treatment and which to omit. These choices also faced the authors of the current volume, but the concept and format of texts entitled *Commonly Asked Questions in* … does at least provide a different rationale for the selection of material.

Accordingly, our selection of topics has been guided in part by our own experience and, as was the choice in the first edition, by questions posed by those we have taught or by colleagues from other disciplines who have sought advice. However, in this edition, we also wanted to recognize that questions arise in the minds of the public and governments in the twenty-first century to which thermodynamics can provide interesting partial answers. The most pressing of these questions now arise with respect to the steps that should be taken to mitigate climate change. Specifically, because thermodynamics is predominantly concerned with energy, its conservation, conversion and exploitation, we address the choices that face us about energy generation, storage and utilization as well as the environment. While it is not possible to argue that thermodynamics alone provides definitive answers to these questions or to identify specific solutions, it does point to the general directions that should be followed and provides some rules against which potential solutions should be evaluated. At the same time the growth of life science as a scientific discipline and the basis of an industrial biotechnology sector suggests that it is also worthwhile examining the extent to which thermodynamics can be helpful in this field.

Naturally, the fundamentals with which we deal in the first five chapters of this book remain the same as in the first edition, because they are universally valid. However, the material has been re-ordered and re-written to support the new applications we wish to treat and to correct the inevitable errors in the first edition. At the same time, we have taken the opportunity to introduce the very latest version of the SI system of units, which is now based on defined values of the fundamental constants of nature rather than on artifacts held in secure locations. In particular, this has had a consequence for the temperature scale so fundamental to thermodynamics. Our approach to the fundamental material has been pragmatic, with the intention of providing explanations that are complementary to those in other texts, perhaps providing a different approach in some cases which can aid understanding. We have not always been able to be deductive and adopt a rigorous pedagogical approach. Thus, the reader will find frequent reference to other texts that are devoted entirely to topics that we treat briefly. The concept of the book also means that readers may seek answers to a particular question rather than wish to read the book from the beginning to the end. For that reason, the material is, in its detail, not ordered to anticipate a reader methodically working their way through it. Hence, the contents list is detailed and the use of cross-references in the text ubiquitous.

In Chapter 1 we answer questions about the definitions of relevant quantities in thermodynamics, the First Law of Thermodynamics and some simple applications. Chapter 2 briefly expounds the basis of statistical mechanics, which links the macroscopically observable properties of materials in equilibrium with the properties of the molecules of which they are composed. Chapter 3 considers the Second Law of Thermodynamics, a range of additional thermodynamic functions and their use. In Chapter 4, we deal with phase equilibrium and the thermodynamics of fluid mixtures in some detail because of their importance to chemists and chemical engineers. In Chapter 5, we treat chemical reactions and electrochemical systems. In the first of the chapters concerned with practical applications, Chapter 6 deals with heat engines and power generation from fossil fuels, with attention paid to carbon capture and sequestration, and also covers refrigeration. Questions are posed and answered using the fundamental material of earlier chapters. Chapter 7 then explicitly considers the application of thermodynamics to the problems of energy generation, conversion, storage and usage in the context of a need to reduce our reliance on fossil fuels, as well as a brief discussion of environmental pollution. Again, questions are posed and answered that illustrate how thermodynamics can guide the solutions for which mankind must search. Recognizing the importance of the growth of biochemical engineering, Chapter 8 explores the topic of biothermodynamics by means of some fundamental and applied questions that are often raised about the field, using the earlier basic material. Finally, in Chapter 9, we provide an updated guide to the sources of thermodynamic data about materials that are essential for the practical implementation of the subject. The growth of electronic databases now means that this is the preferred means of acquiring the necessary thermodynamic data and we provide signposts to reliable data sources and their use.

The authors wish to acknowledge individually and collectively the many generations of students whom they have taught in many countries, whose inquisitiveness has provided both the inspiration for and the challenge of preparing this volume. They also wish to thank their families for their understanding, support and forbearance throughout this endeavor. Finally, the authors are grateful to Ms Katerina Tasidou for her careful and informed scrutiny of the text and her considerable help with the preparation of the final typescript.

About the Authors

Professor Marc J. Assael, FIChemE, FITCc, is emeritus professor in thermophysical properties in the Chemical Engineering Department of the Aristotle University of Thessaloniki, Greece, and the editor-in-chief of the *International Journal of Thermophysics* (Springer Nature). He is a Fellow of the Institute of Chemical Engineers UK, and a Fellow of the International Thermal Conductivity conferences. During the years 1995–1997, he served as the head of the Department of Chemical Engineering. In 1998, Marc J. Assael was TEPCO Chair Visiting Professor in Keio University, Tokyo, Japan, and during 2007–2010 he has also held the position of adjunct professor in Jiaotong University, Xi'an, China.

His research interests include the accurate measurements of the thermal conductivity of solids and fluids, accurate measurements of the viscosity of fluids and the preparation of wide-ranging reference correlations for these two properties. He has published more than 200 papers in international journals, 180 in conference proceedings, 28 chapters in books and 11 books.

He acts as a referee for most journals in the area of thermophysical properties, while he is also a member of the editorial board of the following scientific journals: *High Temperatures – High Pressures, IChemE Transactions Part D: Education for Chemical Engineers* and *International Review of Chemical Engineering*.

Marc J. Assael is also the secretary of the International Association of Transport Properties and the secretary of the International Organizing Committee of the European Conferences on Thermophysical Properties.

Recent information, as well as a full publication list, is available at https://ltpep.com/.

Professor Geoffrey C. Maitland CBE FREng FIChemE FEI FRSC is professor of energy engineering at Imperial College London and a past President of the Institution of Chemical Engineers (2014–2015). His career has spanned academia and industry, spending 20 years in the oil and gas industry and over 20 years at Imperial. He studied chemistry at Oxford University, where he also obtained his doctorate in physical chemistry. After a period as an ICI Research Fellow at Bristol University, he was appointed to a lectureship in chemical engineering at Imperial College in 1974. In 1986, he moved to the oilfield services company Schlumberger, where he carried out research in oilfield fluids engineering. He held a number of senior technical and research management positions in Cambridge and Paris, including as a research Director. He rejoined Imperial College in September 2005 as professor of energy engineering and his current research is centered on how we can continue to use fossil fuels where necessary without causing catastrophic climate change, particularly through carbon capture, use and storage (CCUS). He has chaired several CCUS public reports and was a member of the 2018 UK Government CCUS Cost Challenge Taskforce. He has published over 200 papers in international journals, been granted 10 patents and co-authored four books.

He is a Fellow of the Institution of Chemical Engineers, the Royal Society of Chemistry and the Energy Institute. In 2006, he was elected a Fellow of the Royal Academy of Engineering. He served as President of the British Society of Rheology from 2002–2005, as a Trustee of IChemE from 2013–2016, as a Trustee of RAEng from 2017–2020, chairing their Audit and Risk Committee and as an RSC Trustee and chair of their Publications Board from 2016–2020. He was awarded the Hutchison Medal by the Institution of Chemical Engineers in 1998, received the IChemE Chemical Engineering Envoy Award for 2010 for his media work explaining the engineering issues involved in the Gulf of Mexico oil spill and their Ambassador Award in 2021. In 2011, he chaired the government independent review of the UK Offshore Oil and Gas Regulatory Regime, received the Rideal Lecture Award from the Royal Society of Chemistry in 2012 and was awarded the Leverhulme Medal of the Society of Chemical Industry in 2017. In 2019, he was appointed a Commander of the Order of the British Empire (CBE) for services to chemical engineering.

PD Dr. Thomas Maskow studied physical and theoretical chemistry at the Technical University Carl Schorlemmer Leuna-Merseburg and received his Ph.D. from the Martin-Luther University, Halle Wittenberg on the calculation and measurement of distillation processes of complex multicomponent mixtures. In 2005, he obtained his habilitation at the Technical University Dresden (TUD) for the field of biothermodynamics. Now, he is leading a working group "Ecothermodynamics/Biocalorimetry" at the Helmholtz Center for Environmental Research – UFZ in Leipzig (Germany). Currently, he teaches biophysics/physical chemistry at the TUD, white biotechnology/systems biotechnology at the University of Leipzig and bioprocess technology in Riesa at the BA Sachsen University of Cooperative Education. He was elected as chairperson of the International Society for Biological Calorimetry (ISBC) between 2010 and 2012 and is a guest professor at the University of Geosciences in Wuhan (China).

His research aims at a quantitative understanding of bioconversions in natural and technical systems using thermodynamics and calorimetry. The research ranges from the application of single biocatalysts (i.e. enzymes, bacteria, fungus, yeasts and microalgae) over the analysis of networks in microbial ecology as well as in systems biology to the energetics of photosynthesis and microbial bioelectrocatalysis. The results of this work are published in more than 93 peer-reviewed articles in international journals, in 8 book chapters or books and in 5 patents. In addition to his work as referee for different journals, he is also associate editor/member of the editorial board/guest editor of several scientific journals (e.g. *Engineering in Life Sciences; Frontiers in Microbiology; Energy, Sustainability and Society; Journal of Bioenergetics*). These efforts were acknowledged in 2018 with the award of the Lavoisier-Medal by the ISBC and 2021 with an award by the Central and Eastern European Conference on Thermal Analysis and Calorimetry (CEEC-TAC) and the Mediterranean Conference on Calorimetry and Thermal Analysis (MEDICTA). Recent information, as well as a full publication list, is available at http://www.ufz.de/index.php?en=39077.

Professor Dr. Urs von Stockar is currently an honorary professor (prof. emeritus) at the Swiss Federal Institute of Technology Lausanne (Ecole polytechnique fédérale de Lausanne, EPFL). He studied engineering chemistry at the Swiss Federal Institute of Technology Zurich (ETHZ) and wrote an award-winning Ph.D. thesis at the same institution in chemical engineering under the supervision of Professor J. Bourne. He then spent more than three years at the Department of Chemical Engineering at the University of California, Berkeley, as a postdoctoral researcher and finally as a lecturer. During this period, he familiarized himself with biochemical engineering while contributing in the research group of Professor Charles Wilke towards the development of a process for hydrolyzing the cellulose in waste newspaper in order to sustainably produce ethanol as a fuel. After returning to Switzerland, he worked for a year for Ciba-Geigy in Schweizerhalle before he was appointed as associate professor of chemical engineering at the EPFL in 1977. In 1982, he was promoted to full professor and, in addition, the University of Geneva hired him from 1990 to 1996 as a part-time associate professor. After his retirement in 2007, he served another seven years as an adjunct professor at Dublin City University.

Urs von Stockar directed, together with his senior staff member Dr. I.W. Marison, research on mass transfer and mixing in packed industrial gas-liquid contacting columns and in bioreactors, advanced online monitoring and control of bioreactors using online mid-range IR and Raman spectroscopy, scanning dielectric spectroscopy and online biocalorimetry, integrated bioprocessing by using hydrophobic membranes, transmembrane distillation, cell and solvent encapsulation, reactive capsular perstraction and developed processes for the production and application of natural and recombinant proteins. Special emphasis was on the use of biocalorimetry for bioenergetic research and for thermodynamic analysis of chemotrophic and phototrophic growth. In order to promote the young field of biothermodynamics, he developed and organized, together with international colleagues, a course in biothermodynamics for biochemical engineers, which was held six times in various locations in Europe. The contributions of the many teachers of that course were collected in a book in 2013, for which Urs von Stockar served as an editor.

He has published around 230 peer-reviewed papers and supervised approximately 50 Ph.D. students. He won the Dubrunfault Award and the Lavoisier Medal of the International Society of Biological Calorimetry. Urs van Stockar served as department chairman and chair of the Conference of Department Chairmen at the EPFL, was the chairman of the Swiss Coordination Committee, was member of the Scientific Committee and of the Board of the Swiss Academy of Technical Sciences (SATW), is an individual member of this academy, served as vice-chairman and chairman of the European Federation of Biotechnology and organized the 11th European Congress of Biotechnology in Basel, which attracted over 1,000 delegates.

Professor Sir William A. Wakeham is emeritus professor of engineering at the University of Southampton, UK. He retired as vice-chancellor of that university in September 2009 after eight years in that position. He began his career with a training in physics at Exeter University, UK, at both the undergraduate and doctoral level and after a postdoctoral period in the United States at Brown University, he took up a lectureship in the Chemical Engineering Department at Imperial College London. He became a professor there in 1983 and was also head of the department, from 1988–1996 and prorector and deputy rector from 1996–2001. His academic publications include 7 books and about 400 peer-reviewed papers.

He is a Fellow of the Royal Academy of Engineering, and has been its Senior Vice-President and international secretary and holds the Academy's Silver Medal for Engineering Excellence. He is a Fellow of the Institution of Chemical Engineers, the Institution of Engineering and Technology and the Institute of Physics. He has been awarded honorary degrees in the United Kingdom and Portugal and has lifetime achievement awards from both the European and Asian Conferences on Thermophysical Properties as well as the Touloukian Award of the American Society of Mechanical Engineering.

He has conducted reviews for the U.K. government of U.K. Physics, of the Effectiveness of Full Economic Costing of Research and of the Employability of UK STEM graduates and was, until 2019, a Trustee of the Royal Anniversary Trust. He continues to be chair of the South-East of England Physics Network (SEPnet) and a Trustee of the Rank Prizes Fund. He was knighted in 2009 for services to chemical engineering and higher education.

Professor Dr.-Ing. Stefan Will has been a full professor in engineering thermodynamics at the Department of Chemical and Bioengineering of Friedrich-Alexander University Erlangen-Nürnberg (FAU), Germany, since 2012. After graduation in physics, Stefan Will received a doctoral degree in engineering from the Technical Faculty of FAU in 1995 for a thesis on "Viscosity Measurement by Dynamic Light Scattering."

After holding several academic positions at various universities, he became a full professor at the University of Bremen between 2002 and 2012. During the years 2003–2009, he served as deputy dean and dean respectively, of the Faculty of Production Engineering.

Stefan Will's research interests include the development and application of optical techniques in energy and process engineering, in particular for particle and combustion diagnostics, the measurement of thermophysical properties and the modeling of complex energy systems. In these fields he has authored and co-authored more than 250 publications in international journals, conference proceedings and books. He is an active member and delegate in several national and international organizations in thermodynamics and process engineering. Since 2014, he has also been Director and co-coordinator of the Erlangen Graduate School in Advanced Optical Technologies (SAOT), established in the framework of the German Universities Excellence Initiative. Further information about his research is available at http://ltt.tf.fau.de.

Nomenclature

English Symbols

A	area (m^2)
	Helmholtz function, Helmholtz free energy (J)
$\mathbf{A}$	affinity of chemical reaction (J)
a	relative activity
B	anergy (J)
	second virial coefficient ($m^3\ mol^{-1}$)
C	number of components
	third virial coefficient ($m^6\ mol^{-2}$)
C_p	heat capacity at constant pressure ($J\ K^{-1}$)
C_p, $C_{p,m}$	molar heat capacity at constant pressure ($J\ mol^{-1}\ K^{-1}$)
C_V	heat capacity at constant volume ($J\ K^{-1}$)
C_V, $C_{V,m}$	molar heat capacity at constant volume ($J\ mol^{-1}\ K^{-1}$)
c	molarity, concentration ($mol\ m^{-3}$)
c_p	specific heat capacity at constant pressure ($J\ kg^{-1}\ K^{-1}$)
c_V	specific heat capacity at constant volume ($J\ kg^{-1}\ K^{-1}$)
d	mean ion diameter (m)
E	energy (J)
	electrochemical cell electromotive force (V)
	exergy (J)
	isothermal Young's modulus (Pa)
E_P	electrode potential (V)
E_S	isentropic Young's modulus (Pa)
e	charge on a proton (positive) or electron (negative) (C)
	specific exergy ($J\ kg^{-1}$)
F	degrees of freedom
	force (N)
	Faraday's constant ($C\ mol^{-1}$)
F_B	Poynting factor
f	frequency (Hz)
	Mayer function
G	Gibbs energy (J)
	shear modulus (Pa)
g	acceleration of free fall ($m\ s^{-2}$)
	degeneracy
H	enthalpy (J)
H_m	Henry's law constant when concentration is molarity ($Pa\ mol\ kg^{-1}$)
H_x	Henry's law constant when concentration is mole fraction (Pa)
h	Planck constant (J s)
	specific enthalpy ($J\ kg^{-1}$)
I	ionic strength ($mol\ kg^{-1}$)
	moment of inertia ($kg\ m^2$)
i	current (A)

i	unit vector
j	quantum number
$j_{R,B}$	degree of reduction of compound B
K	equilibrium constant
	isothermal bulk modulus (Pa)
K_S	isentropic bulk modulus (Pa)
$K_{XW,i}$	phase X-Water partition coefficient of species i
	– (X is O for oil; B for biota; S for soil)
k_A	first-order rate constant (s^{-1})
k_{AB}	second-order rate constant ($m^3\ mol^{-1}\ s^{-1}$)
k_B	Boltzmann's constant ($J\ K^{-1}$)
k_{ij}	binary interaction parameter
L or N_A	Avogadro's constant (mol^{-1})
l	length (m)
M	molar mass ($kg\ mol^{-1}$)
m	mass (kg)
$\dot{m}$	mass flow rate ($kg\ s^{-1}$)
m_B	molality of B ($mol\ kg^{-1}$)
N	number of molecules
N_A or L	Avogadro constant (mol^{-1})
N_j	number of molecules in quantum state j
	number of moles in stream j
n	amount of substance (mol)
	polytropic index
n_x	quantum number
P	power (W)
	number of phases
P_w	power coefficient (wind turbine)
p	pressure (Pa)
$\tilde{p}$	fugacity (Pa)
Q	canonical partition function
	– (translational Q^T, rotational Q^R, vibrational Q^v)
	heat (J)
	reaction quotient
$\dot{Q}$	rate of heat flow (W)
Q_G	proportionality coefficients between the combustion Gibbs energy and the degree of reduction ($J\ mol^{-1}$)
Q_H	proportionality coefficients between the combustion enthalpy and the degree of reduction ($J\ mol^{-1}$)
q	charge (C)
	heat supplied or withdrawn per unit mass ($J\ kg^{-1}$)
	partition function
	– (translational q_T, rotational q_R, vibrational q_v, electronic q_E, nuclear q_N, lowest state q_0)
R	universal gas constant ($J\ mol^{-1}\ K^{-1}$)
	resistance (ohm)
R_s	specific gas constant = universal gas constant per unit mass ($J\ kg^{-1}\ K^{-1}$) = R/M
r	radius, distance (m)

r_B	reaction rate of compound B (mol s^{-1} m^{-3})
r_j	reaction rate for jth reaction (mol s^{-1} m^{-3})
S	entropy (J K^{-1})
	symmetry number
$\dot{S}$	rate of entropy generation in a system owing to irreversible processes (J K^{-1} s^{-1})
s	solubility (mol m^{-3})
	specific entropy (J kg^{-1} K^{-1})
T	absolute temperature (K)
T_0	low temperature level (K)
t	time (s)
t_i	transport number of ion i
U	internal energy (J)
	potential energy (J)
u	specific internal energy (J kg^{-1})
	speed of sound (m s^{-1})
	velocity (m s^{-1})
V	volume (m^3)
V_{TN}	thermo-neutral voltage (V)
v	specific volume (m^3 kg^{-1})
W	work (J)
	– (electric W_{el}, flow W_f, other W_o, shaft W_s, dissipative W_{diss}, volume W_v)
$\dot{W}$	work rate (W)
w	specific work (J kg^{-1})
X	biomass
X	extensive property
X_B	property X per mole of compound B
	partial molar property X of compound B
x	vapor quality
x_B	liquid phase mole fraction of compound B
Y	force (N)
$Y_{B/X}$	yield coefficient, expressing how much of species B is produced or consumed per "molecule" of biomass
y_B	gas/vapor phase mole fraction of compound B
Z	compression factor
	number of surface adsorption sites
	state variable (various)
z	force vector (N)
z_B	fraction of biota mass in lipid
z_i, z_B	number of charges of species i or B

Greek Symbols

α	isobaric thermal expansion coefficient (K^{-1})
	parameter in the Debye-Hückel limiting law (mol$^{1/2}$ kg$^{-1/2}$)
	ratio of heat extracted to loss of electrical energy generation
β	parameter in the Debye-Hückel limiting law (m^{-1} kg$^{1/2}$ mol$^{-1/2}$)
	vapor fraction

Γ	independent variables
γ	activity coefficient
	heat capacity ratio
$\gamma_{\pm}^{\infty,m}$	mean ion activity coefficient
$\Delta_c H$	enthalpy of combustion (J)
ΔG^*	activation energy (J)
$\Delta_l^g H$	enthalpy of evaporation (J)
$\Delta_r H$	enthalpy of reaction (J)
$\Delta_s^l H$	enthalpy of fusion (J)
ΔX	finite change in the quantity X
δX	infinitesimal change in the quantity X
ε	compression ratio
	energy (J)
	– (translational ε_T, rotational ε_R, vibrational ε_v, electronic ε_E, nuclear ε_N, lowest state ε_0)
	Lennard-Jones energy parameter (J)
ε_o	vacuum permittivity ($J^{-1}C^2{}_m{}^{-1}$)
ε_r	relative permittivity ($J^{-1}C^2{}_m{}^{-1}$)
ζ	fractional recovery
η	efficiency
θ	fractional surface coverage
	temperature (°C)
κ_S	isentropic compressibility (Pa^{-1})
κ_T	isothermal compressibility (Pa^{-1})
λ	absolute activity
	first Lamé constant (Pa)
	mean free path (m)·
μ	chemical potential (J mol^{-1})
	second Lamé constant (Pa)
$\tilde{\mu}_B^{\alpha}$	electrochemical potential of species B in the phase α (J mol^{-1})
μ_{JT}	Joule-Thomson coefficient (K Pa^{-1})
$\nu_{B,j}$	stoichiometric number of compound B in the *j*th reaction
ξ	extent of chemical reaction (mol)
ξ_j	extent of chemical reaction for the *j*th reaction (mol)
$\dot{\xi}_j$	rate of advancement of the *j*th chemical reaction, reaction rate of reaction *j* for the whole system (mol s^{-1})
$\dot{\xi}_X$	rate of synthesis of fresh dry biomass by the cell (C-mol s^{-1})
Π	pressure ratio
	probability
	osmotic pressure (Pa)
ρ	density (kg m^{-3})
ρ_n	amount-of-substance density (mol m^{-3})
σ	Lennard-Jones separation parameter (m)
	Poisson's ratio
υ	frequency of vibration (Hz)
Φ	relative humidity
φ	cutoff ratio
ϕ	electrostatic potential (J)
	fugacity coefficient (–)

$\phi(r)$	intermolecular potential (J)
ϕ_{JT}	isothermal Joule-Thomson coefficient (K Pa^{-1})
ϕ_{m}	osmotic coefficient
ψ_B	order of chemical reaction with respect to species B
Ω	configurational integral (m^{3N})
ω	acentric factor
	humidity ratio or moisture content (kg_{water} / $kg_{dry\ air}$)

Superscripts

.	time derivative or rate average
$\ominus$	standard state
*	pure component property
′	related to saturated liquid
′′	related to saturated vapor
o	reference state
∞	infinite dilution
α, β	phases
adiab	adiabatic
assoc	associative
b	boiling
c	molarity basis
cal	calcium looping capture
disp	dispersive
E	excess
eq	equilibrium
hc	hard chain
id	ideal
ig	ideal gas
isoth	isothermal
m	mean
	molality basis
mix	mixing
pg	perfect gas
ref	reference value
res	residual
s	isentropic
sat	saturation
surf	surface

Subscripts

0	dead environment state
	lower temperature level
∞	infinity
a	ambient
	absorber
abs	absorbed
ad	adsorption

an	anabolic
B	Brayton
	chemical species label
bp	bubble point
C	Carnot
	chemical species label for all chemical species in mixture
c	combined
	combustion
	compressor
	cooling
	critical
cap	capture
cat	catabolic
cc	combined cycle
D	Diesel
d	dewpoint
des	desorption
dil	dilution
dis	discharge
diss	dissipative
e	external
el	electrical
eq	equilibrium
ex	exergetic
ext	extracted
FS	fluid stream
f	final
	flow
	formation
g	gaseous
gen	generated, generator
h	heating
i	initial
id	ideal
in	input
irrev	irreversible
JT	Joule-Thomson
j	chemical reaction label
kin	kinetic
L	longitudinal
l	liquid
	lost
m	molar quantity (maybe omitted when there is no risk of confusion)
max	maximum
min	minimum
mix	mixing
ms	molten salt
OP	operating
o	other

out	output
ox	oxidizer
P	product
p	pressure
	pump
pot	potential
R	Rankine
	reactant
	regeneration
RT	round trip
r	reaction
	rotor
red	reducer
ref	refrigerant
rev	reversible
s	shaft
	shear
	solar
	solid
soln	solution
solv	solvent
st	stirrer
	stored, storage
strip	stripping
sys	system
sur	surroundings
t	turbine
th	theoretical
	thermal
vap	vaporized
w	water

Examples of Use

$\Delta_{r,j}H^{\ominus}$	enthalpy of the jth reaction in the standard state (J mol^{-1})
$\tilde{p}_{B,l}^{\infty,m}$	fugacity of species B in the liquid phase at infinite dilution on a molality basis (Pa)
$\dot{n}_{B,k}$	molar flow of species B entering through the kth mass exchange port (mol s^{-1})
μ_B^{α}	chemical potential of species B in a mixture in phase α

Abbreviations

ASHP	Air-Source Heat Pump
ASHRAE	American Society of Heating, Refrigerating and Air-Conditioning Engineers
ASME	American Society of Mechanical Engineers
ASOG	Analytical Solution of Groups
BECCS	Bioenergy with Carbon Capture and Storage
BIPM	Bureau International des Poids et Mesures
BWR	Benedict-Webb-Rubin
CaC	Calcium Carbonate
CAES	Compressed Air Energy Storage
CCS	Carbon Capture and Storage
CCU	Carbon Capture and Utilization
CCGT	Combined Cycle Gas Turbine (Power Plant)
CFC	Constant Flux Calorimetry
CHP	Combined Heat and Power
CLC	Chemical Looping Combustion
COP	Coefficient of Performance
CP	Critical Point
CSP	Concentrated Solar Power
DAC	Direct Air Capture
DDT	DichloroDiphenylTrichloroethane
DIPPR	Design Institute for Physical Property Data
DME	Dimethyl Ether
DMFC	Direct Methanol Fuel Cell
DMR	Dry Methane Reforming
DSC	Differential Scanning Calorimeter
EOE	Ethoxyethane
EOS	Equation of State
EOO	1-Ethoxyoctane
EP	Energy Penalty
ePC-SAFT	Electrolyte Perturbated-Chain Statistical Associating Fluid Theory
FT	Fischer-Tropsch (Process)
GHG	Greenhouse Gas
GSHP	Ground-Source Heat Pump
GWP	Global Warming Potential
HBC	Heat Balance Calorimeter
HFC	Heat Flow Calorimeter
HHV	Higher Heating Value
HRSG	Heat Recovery Steam Generator
IACT	International Association for Chemical Thermodynamics
IAPS	International Association for the Properties of Water and Steam
IATP	International Association for Transport Properties
ICE	Internal Combustion Engine
IEAGHG	International Energy Agency Greenhouse Gas (R&D Program)
IGCC	Integrated Gasification Combined Cycle (Power Plant)

IMC	Isothermal Micro-Calorimeter
IUPAC	International Union of Pure and Applied Chemistry
KEGG	Kyoto Encyclopedia of Genes and Genomes
LC_{50}	Lethal Concentration 50%
LHV	Lower Heating Value
LNG	Liquified Natural Gas
LOHC	Liquid Organic Hydrogen Carrier
LPG	Liquified Petroleum Gas (mainly C_3H_8 + C_4H_{10})
MBWR	Modified Benedict-Webb-Rubin
MCFC	Molten Carbonate Fuel Cell
MEA	Monoethanolamine
MOSCED	Modified Separation of Cohesive Energy Density
MPP	Maximum Power Point
MSA	Moisture Swing Adsorption
NG	Natural Gas
NGCC	Natural Gas Combined Cycle (Power Plant)
NIST	National Institute for Standards and Technology
PAFC	Phosphoric Acid Fuel Cell
PAH	Polycyclic Aromatic Hydrocarbon
PC	Propylene Carbonate
PCB	Polychlorinated Biphenyl
PCC	Pulverized Coal Combustion (Power Plant)
PCC	Power Compensation Calorimetry
PC–SAFT	Perturbed Chain – Statistical Associating Fluid Theory
PEM	Polymer Exchange Membrane
PEMFC	Polymer Exchange Membrane Fuel Cell
PHCA	Pumped Hydro-Compressed Air (Energy Storage)
PHES	Pumped Hydro Energy Storage
PO	Propylene Oxide
POM	Partial Oxidation of Methane
PPC	Poly(Propylene Carbonate)
PSA	Pressure Swing Adsorption
PTC	Poly(Trimethylene Carbonate)
PV	Photovoltaic
RC	Reaction Calorimeter
REFPROP	Reference Fluid Thermodynamic and Transport Properties Database
SAFT	Statistical Association Fluid Theory
SGSS	Steam Generator Subsystem
SI	Le System International d' Unites
SMR	Steam Methane Reforming
SOFC	Solid Oxide Fuel Cell
SRD	Specific Reboiler Duty
STH	Solar-to-Hydrogen
STP	Standard Temperature and Pressure
Syngas	Synthesis Gas (CO + H_2)
TECRDB	Thermodynamics of Enzyme-Catalyzed Reactions Database
TDE	Thermo Data Engine
TFA	Thermodynamic Feasibility Analysis
TDE	Thermo Data Engine

TMC	Trimethylene Carbonate
TMO	Trimethylene Oxide
TN	Thermo-Neutral
TRC	Thermodynamics Research Center
UNIFAC	Universal Functional Group Activity Coefficients
WGS	Water Gas Shift

1

Definitions and the First Law of Thermodynamics

1.1 Introduction

The subjects of thermodynamics, statistical mechanics, kinetic theory and transport phenomena are almost universal within university courses in physical, chemical and biological sciences and engineering. The intensity with which these topics are studied as well as the balance between them varies considerably by discipline. However, to some extent, the development and indeed ultimate practice of these disciplines requires thermodynamics as a foundation. It is therefore rather more than unfortunate that for many studying courses in one or more of these topics thermodynamics present a very great challenge. It is often argued by students that the topics are particularly difficult and abstract with a large amount of complicated mathematics and rather few practical examples that arise in everyday life. Probably for this reason, surveys of students reveal that most strive simply to learn enough to pass the requisite examination but do not attempt serious understanding. However, our lives use and require energy, its conversion in a variety of forms and understanding these processes is intimately connected to thermodynamics and transport phenomena (the transport of mass, momentum or energy). For example, whether a proposed new source of energy or a new product is genuinely renewable and/or carbon-neutral, depends greatly on a global energy balance, on the processes of its production and its interaction with the environment. This analysis is necessarily based on the laws of thermodynamics, which makes it even more important now for all scientists and engineers to have a full appreciation of these subjects as they seek to grapple with increasingly complex and interconnected problems.

This book sets out to provide answers to some of the questions that undergraduate students and new researchers raise about thermodynamics. The list of topics is therefore rather eclectic and, perhaps in some sense, not entirely coherent. It is certainly true that the reader of any level should not expect to "learn" any of these subjects from this book alone. It is, instead, intended to complement existing texts, dealing in greater detail and in a different way with "some" of the topics deemed least straightforward by our own students over many years. If you do not find the question you have treated in this text, then we apologize. Alternative sources of information include Sandler (1989), Levenspiel (1996), Cengel and Boles (2014), Sonntag et al. (2003), Smith (2013), Smith et al. (2017), Moran et al. (2018) and Grossman (2020).

This chapter provides definitions that are required in all chapters of this book, along with the definition of standard states. In addition, we consider how some of the quantities important to thermodynamics may be measured.

DOI: 10.1201/9780429329524-1

1.2 What Is Thermodynamics?

Thermodynamics provides a rigorous mathematical formulation of the interrelationships among measurable physical quantities that are used to describe the energy and equilibria of macroscopic systems as well as the experimental methods used to determine those quantities. These formulations include contributions from pressure, volume, chemical potential and electrical work; but there can also be significant energy contributions arising from electromagnetic sources, gravitation and relativity. The contributions that are important change with the discipline in which the problem arises. For example, for the majority of chemists the inclusion of gravitational and relativistic contributions is unimportant because of their dominant requirement to understand chemical reactions and equilibrium; whereas for physicists the same contributions may be dominant and chemical and mechanical engineers may need to include electromagnetic forces but will also need to account for phenomena associated with nonequilibrium states such as the processes that describe the movement of energy, momentum and matter.

The fact that thermodynamics relates measurable physical quantities implies that measurements of those properties must be carried out for useful work to be done in the field. Generally speaking, the properties of interest are called *thermophysical properties*, a subset that pertains to equilibrium states being referred to as *thermodynamic properties*, and a further subset that refers to dynamic processes in nonequilibrium states being called *transport properties*. Thermodynamics is an exacting experimental science because it has turned out to be quite difficult and time consuming to make very accurate measurements of properties over a range of conditions (temperature, pressure and composition) for the wide range of materials of interest in the modern world. Given the exact relationship between properties that follows from thermodynamics the lack of accuracy has proved problematic. Thus, very considerable efforts have been made over many decades to refine experimental measurements, using methods for which complete working equations are available in the series *Experimental Thermodynamics* (Vol. I 1968, Vol. II 1975, Vol. III 1991, Vol. IV 1994, Vol. V 2000, Vol. VI 2003, Vol. VII 2005, Vol. VIII 2010, Vol. IX 2014). It has been important that any such measurements have a quantifiable uncertainty because properties derived from them, for example, are required to design an effective and efficient air conditioning system. In this paragraph itself, several terms have been used, such as "system" which, in the field of thermodynamics, have a particular meaning and require definition; we have provided these definitions in the following text.

1.3 What Vocabulary Is Needed to Understand Thermodynamics?

The A–Z of thermodynamics has been prepared by Perrot in 1998; hence, we do not provide a comprehensive dictionary of thermodynamics here, but instead give some clear definitions of commonly encountered terms.

1.3.1 What Is a System?

A system is the part of the universe chosen for study while everything else is part of the surroundings. The system must be defined in order that one can analyze a particular problem but can be chosen for convenience to make the analysis simpler. Typically, in practical applications, the system is macroscopic and of tangible dimensions, such as a bucket of water; however, a single molecule is an acceptable microscopic system. A system is characterized both by its contents and the system boundary; the latter in the end is always virtual. For example, if one considers a container with a rigid enclosure, the boundary of the system is set in a way to include all the material inside, but to exclude the walls. Especially in engineering applications a careful and advantageous choice of the system boundary is of enormous importance; defining the right system boundary may considerably ease setting up energy and mass balances for example.

Systems are deemed *open* if material is allowed to cross its boundary. Systems are deemed *closed* if material is not allowed to cross the system boundary. Provided that any chemical reactions in the system have ceased, the state of a closed system is unchanging unless work or heat are transferred across the system boundary. When the system is thermally insulated, so that heat cannot cross the system boundary, it is called *adiabatically enclosed.* A Dewar flask with a stopper approximates an adiabatic enclosure. A system with thermally conducting walls, such as those made of a metal, is called *diathermic*. When a closed system is adiabatic and when no work can be done on it the system is termed *isolated*.

1.3.2 What Is a State?

The state of a system is defined by specifying a number of thermodynamic variables for the system under study. In principle, these could be any of or all the measurable physical properties of a system. Fortunately, not all of the variables or properties need to be specified to define the state of the system because only a few can be varied independently; the exact number of independent variables depends on the system but rarely exceeds five. The exact choice of the independent variables for a system is a matter of convenience but pressure and temperature are often included within them. As an illustration of the point, if the temperature and pressure of a pure gas are specified then the density of the gas takes a value (dependent variable) that is determined. The general rule for calculating the number of independent variables for a system at equilibrium is given by the phase rule that will be introduced and discussed in Question 4.1.2.

1.3.3 What Are the Types of Property: Extensive and Intensive?

For a system that can be divided into parts any property of the system that is the sum of the property of the parts is *extensive*. For example, the mass of the system is the sum of the mass of any parts into which it is divided. Volume and amount of substance (see Question 1.3.10) are all extensive properties as are energy, enthalpy, Gibbs function, Helmholtz function and entropy, all of which are discussed later. A system property that can have the same value for each of the parts is an *intensive* property. The most familiar intensive properties are temperature and pressure. It is also worth remembering that the quotient of two extensive properties gives an intensive property. For example, the mass of a system (extensive) divided by its volume (extensive) yields its density which is intensive.

1.3.4 What Is a Phase?

If a system has the same temperature and pressure, and so on throughout, and if none of these variables change with time, the system is said to be in equilibrium. If, in addition, the system has the same composition and density throughout it is said to be *homogeneous* and is defined as a *phase*. When the system contains one or more phases so that the density and composition may vary but the system is still at equilibrium it is termed *heterogeneous*. Water contained in a closed metallic vessel near ambient conditions will have a layer of liquid water at the lowest level (liquid phase) and a vapor phase above it consists of a mixture of air and water vapor. Necessarily this picture implies an interface exists between the liquid and the vapor. The properties of the system are therefore discontinuous at this interface and generally interfacial forces will be present at the interface that are not present in the two phases on either side.

A phase that can exchange material with other phases or surroundings, depending on how the system boundary is defined, is termed *open* while a *closed* phase is one that does not exchange material with other phases or surroundings. This is entirely consistent with the terminology adopted above for systems. In the example given previously, the closed metallic vessel contained liquid water and water vapor. If we define the system to include the two phases then the system is closed, but it contains two open phases exchanging material within it.

1.3.5 What Is a Thermodynamic Process?

A thermodynamic process has taken place when, at two different times, there is a difference in any macroscopic property of the system. A change in the macroscopic property is infinitesimal if it has occurred through an infinitesimal process. Processes can occur spontaneously and generally proceed towards equilibrium or can be brought about by external intervention in which case they may move the system away from equilibrium. Processes may also be *reversible* or *irreversible*. The latter topics are discussed in Question 1.3.8. But, as examples of spontaneous processes, we consider a closed system of substance B held at a constant temperature and a pressure, p, where the vapor pressure is p_B^{sat}. For the case that $p < p_B^{sat}$ liquid will evaporate in a process that occurs spontaneously until the pressure rises to p_B^{sat}. Equally when $p > p_B^{sat}$ vapor will condense, and the process will see the pressure fall to p_B^{sat}.

The term *process* can have a variety of other implications for mechanical and chemical engineers, because they involve external intervention, and while some are discussed in this chapter others are not.

1.3.6 What Is Work?

There are two ways by which not only open, but also closed, systems can exchange energy with their surroundings: (i) by work and (ii) by heat. A differential amount of work δW is usually specified as the scalar product $\boldsymbol{z}\cdot\mathrm{d}\boldsymbol{Y}$ where $\boldsymbol{z}$ represents a force vector acting on the system and $\mathrm{d}\boldsymbol{Y}$ a vector showing a differential amount of displacement; if the two vectors are collinear, i.e. having the same direction, the scalar product simplifies to $z\mathrm{d}Y$. The most important form of work is *volume work*, or *boundary work*. The quantitative description may be understood by imagining a compressible fluid as the system in a cylindrical container closed at one end by a mobile piston acting on the system with a force F (Figure 1.1). This force can result from a mechanical action, such as in compressors or heat engines, or from a pressure existing in the surrounding, such as in reactors operating under constant pressure. Assuming that this force is in equilibrium with the force acting on the piston from within owing to the system pressure, the magnitude of the force

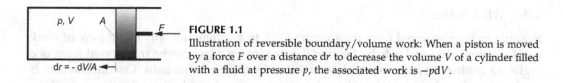

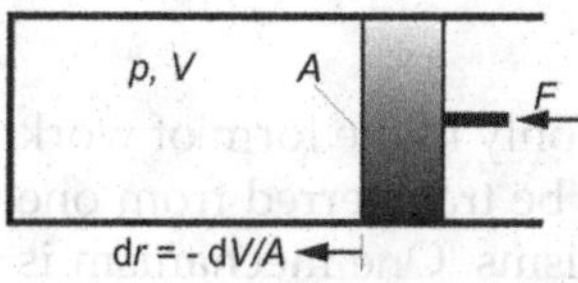

FIGURE 1.1
Illustration of reversible boundary/volume work: When a piston is moved by a force F over a distance dr to decrease the volume V of a cylinder filled with a fluid at pressure p, the associated work is $-pdV$.

acting on the system equals the system pressure p times A, the area of the piston. The displacement of the piston is dr, which, when multiplied with A yields dV. The amount of work done on the system is therefore

$$\delta W = -pdV. \tag{1.1}$$

Work is counted positively when it is done *on* the system, that is when the system gains energy, and negatively when it does work on the surrounding and thus loses energy. The minus sign indicates that δW is positive when the volume diminishes.

Equation 1.1 expresses only the *reversible* part of the volume work; for it to be valid, a process must allow the system to stay in equilibrium all the time as any departure from equilibrium will generate entropy and thus *irreversible work*. This happens also when friction is involved, as is the case for any real system. As a result, the amount of work that has to be provided in order to compress the system increases and the amount of work that can be gained from an expansion of the fluid is reduced. This additional work is categorized as part of W_{diss} (see below).

Other forms of work are designated as W_o. In this group, there are a number of other forms of *reversible work* that can be described as zdY an extensive property, Y, ("work coordinate/displacement"). Most important among them is electric work, for which the electrostatic potential at electrodes represents the force and the flow of charge (current) the displacement. Other examples include surface tension work, where the surface tension is the force and the creation of interfacial area the displacement (see Question 4.12), and magnetic work for which the force is the magnetic field and the displacement is the change of magnetic moment.

A special form of work is shaft work W_s, which is typical of open systems and occurs in engineering equipment such as compressors or turbines. Besides heat, it is a typical form of energy that changes the enthalpy of an open system, see Question 1.5.5, Equation 1.76. Often, the reversible part of shaft work, in differential notation, may be written as $\delta W_s = Vdp$. Evidently this term does not fit into the general scheme zdY; the reason is that shaft work in these cases derives from a combination of volume work, $-pdV$, and so-called *flow work* $\delta W_f = d(pV)$, with $Vdp = -pdV + d(pV)$. These relations are further detailed in Question 1.6.3. It should be noted that shaft work, W_s, is related to the differences between quantities of state at the inlet and outlet ports of open systems. Accordingly, the pressure change dp is always regarded as the difference between the pressure at two points in space and not in time.

A further subgroup in W_o is dissipative work, W_{diss}. This includes, on the one hand, the additional work required owing to the irreversible processing mentioned before, such as friction work, and, on the other hand, forms of irreversible work done on the system that are directly dissipated into the system and thus cannot be recovered any more. Examples of the latter are the electric work provided by an electric resistance placed in a fluid (W_{el}) and stirrer work (W_{st}); for example, the mechanical work dissipated into a reaction mixture by a stirrer.

1.3.7 What Is Heat?

Energy can be transferred from the surroundings to a system not only in the form of work (see Question 1.3.6), but also by heating the system up. Heat can be transferred from one region to another or from one body to another by three mechanisms. One mechanism is *conduction*, where heat transfer takes place from one part of a body to another part of the same body, or between two bodies in physical contact through the combination of molecular motion that transports the kinetic energy of the molecules or through collisions between the molecules that allow transfer of energy from one molecule to another. A second mechanism is *convection*, where heat transfer takes place from a point to another within a fluid, or between a fluid and a solid or another fluid, by virtue of the bulk motion of the fluid as a continuum that transports warmer fluid from one location to another. A typical example of convective heat flow would be the removal of energy from a reacting mixture (the system) in a chemical reactor by the way of cooling coils through which water is circulated. Energy would then be removed from the system in the form of heat, which would result in a temperature increase of the cooling water. Evidently, convection is not a mechanism of heat transfer that has any meaning for the transfer of heat within solids. A third mechanism of heat transfer is by the exchange of electromagnetic *radiation*. The radiation can be emitted by one region of a material and absorbed and/or reflected by other regions of the material or by surfaces. Heating oneself up by standing in the sun demonstrates that heat can be transferred by radiation.

It is important to note that the terms "heat" and "work" are, strictly speaking, only used in connection with a transfer of energy to and from the system. To say that a hot reaction mixture contains "a lot of heat" is in principle a misuse of the term and should be replaced by "a lot of thermal energy" (see Question 1.5.7).

1.3.8 What Is a Reversible Process or Reversible Change?

In Question 1.3.5 an example was used to illustrate spontaneous processes and this will be used again for the topic of reversibility. We again define a system of substance B in both liquid and gaseous phases of vapor pressure $p_{\mathrm{B}}^{\mathrm{sat}}$ where the phases are at a pressure p. If $p = p_{\mathrm{B}}^{\mathrm{sat}}$, both evaporation and condensation can occur for any infinitesimal decrease or increase in p, respectively, and the process is reversible. That is, for $p = p_{\mathrm{B}}^{\mathrm{sat}} - \delta p$, where $\delta p > 0$ the process conforms to the first spontaneous process of Question 1.3.5 and when $p = \lim_{\delta p \to 0} p_{\mathrm{B}}^{\mathrm{sat}}$ the process is reversible; it can be considered to be a passage through a continuous series of equilibrium states between the system and the surroundings.

Another, albeit difficult to comprehend but more important, example of a reversible process concerns the work done on a phase α by the surroundings. In this case, if the work on α is restricted to an external pressure p_{e}^{α} acting on the phase α, which is at a pressure p^{α} then the change in volume of α is $\mathrm{d}V^{\alpha}$ and in the absence of friction given by

$$\delta W = -p_{\mathrm{e}}^{\alpha}\mathrm{d}V^{\alpha}, \tag{1.2}$$

where $p_{\mathrm{e}}^{\alpha} = p^{\alpha}$ the change in volume is said to be reversible. That is, if $p_{\mathrm{e}}^{\alpha} = p^{\alpha} + \delta p$, where δp is an infinitesimal change in pressure, then $\mathrm{d}V^{\alpha} < 0$ and the phase α contracts. When $p_{\mathrm{e}}^{\alpha} = p^{\alpha} - \delta p$, then $\mathrm{d}V^{\alpha} > 0$ and the phase α expands. In both cases the change can be reversed by a change in p_e^α equal in magnitude to δp but of opposite sign; $\delta p_e^\alpha = -\delta p$. When the pressure of the phase $p_e^\alpha \neq p^\alpha$, the change in volume is not reversible.

However, when we refer to the passage of the system through a sequence of internal equilibrium states without the establishment of equilibrium with the surroundings this is referred to as a reversible change or process. We define a process as *reversible* if the initial state of the system can be reached again during the process without leaving any effect upon the environment. Conversely, then an *irreversible process* is one where the initial state of the system cannot be recovered without having an effect on the environment.

This concept is best illustrated first by a mechanical example where a perfectly elastic ball is dropped under gravity in a vacuum. The ball falls, losing potential energy and gaining kinetic energy until it reaches the ground when it is deformed by the contact and the energy is converted to elastic energy which is then released to cause the ball to rebound and eventually reach the height from which it was dropped with the same potential energy. Here we see that although there are many changes in the form of energy in the system (the ball), the initial state is recovered and there is no effect upon the environment. It is a reversible process. If the ball in this example is **not** perfectly elastic, some energy is dissipated as thermal energy when the ball strikes the ground and therefore the ball will not recover its initial state (height) unless there is an effect left on the environment such as would be the case if an observer raises the ball to its initial height.

An example that combines the concept of reversible change and reversible process will now be considered. We define the system as a liquid and a vapor of a substance in equilibrium contained within a cylinder that on one circular end has a rigid immovable wall and on the other has a piston exerting a pressure equal to the vapor pressure of the fluid at the system temperature. Energy in the form of heat is now applied to the outer surface of the metallic cylinder and the heat flows through the cylinder (owing to the relatively high thermal conductivity) increasing the liquid temperature. This results in further evaporation of the liquid and an increase in the vapor pressure. Work must be done on the piston at constant temperature to maintain the pressure. This change in the system is termed a reversible change. It can only be called a reversible process if the temperature of the substance surrounding the cylinder is at the same temperature as that of the liquid and vapor within the cylinder. This requirement arises because if the temperatures were not equal the heat flow through the walls would not be reversible and thus the whole process would not be reversible. If the system is only the liquid and the gas within the cylinder the process is reversible. Another example is provided by considering two systems both in complete equilibrium and in which the heat flows from one to the other. Each system undergoes a reversible change provided each remains at equilibrium. The heat flow is not a reversible process unless the temperature of both systems is equal.

The obvious difficulties of realizing any of these examples in practice suggests that most practical processes are irreversible; however, the importance of reversible processes and changes along with the content of Question 1.3.5 will become apparent first in Questions 1.6.3, 1.6.4 and 1.6.5, as well as in Chapter 7.

1.3.9 What Is Thermal Equilibrium?

If an adiabatically enclosed system is separated into two parts by a diathermic wall, then the two parts will be in thermal equilibrium with each other. This implies that the states of the two subsystems that are at thermal equilibrium are dependent on each other. In other words, there is a relationship between the independent variables that define the states of the two subsystems. Mathematically, for a system consisting of two parts A and

B with independent variables Γ_A and Γ_B at thermal equilibrium there is a function f that relates the two sets of variables

$$f(\Gamma_A, \Gamma_B) = 0. \tag{1.3}$$

For three systems, A, B and C, that are all adiabatically enclosed, if A is in thermal equilibrium with B which is also in equilibrium with C then A must be in thermal equilibrium with C. This is often referred to as the Zeroth *law of thermodynamics*. This of course assumes that sufficient time has elapsed to permit attainment of internal thermal equilibrium.

1.3.10 What Is the Amount of Substance?

The amount of substance n_B of a chemical entity B in a system is a physical quantity defined by its proportionality to the number of entities N_B in the system that is given by $N_B = Ln_B$, where L is the Avogadro constant (BIPM 2019). For example, if the chemical entity B is an atom of argon, then N_B is the number of atoms of argon in the system. The SI unit for the amount of substance is the mole defined currently by *Le Système International d'unités (SI)* (BIPM 2019).

The *mole* is defined to contain 6.022 140 76 × 10^{23} elementary entities. When the mole is used, the elementary entities must be specified and may be atoms, molecules, ions, electrons, other particles or specified groups of such particles.

The SI symbol for mole is mol. The specified groups need not be confined to independent entities or groups containing integral numbers of atoms. For example, it is quite correct to state an amount of substance of 0.5 mol of H_2O or of $(H_2 + 0.5O_2)$ or of $0.2MnO_4^-$. We set out the most recent (2019) definitions of all the basic SI units in Question 1.7.1 and link them there to exact values of fundamental constants.

1.3.11 What Are Molar and Mass or Specific Quantities?

The molar volume of a phase is the quotient of the volume and the total amount of substance of the phase. Generally, any extensive quantity X divided by the total amount of substance $\sum_B n_B$ is, by definition, an intensive quantity called the *molar quantity* X_m

$$X_m = X/\sum_B n_B. \tag{1.4}$$

In Equation 1.4, the subscript m designates a molar quantity and can be replaced by the chemical symbol for the substance in this example subscript B; when no ambiguity can result the subscripts m and B may be omitted entirely.

In engineering applications, quantities are very often related to the mass instead of the amount of substance. The specific volume of a phase is the quotient of the volume and the total mass of substance of the phase. By analogy with molar quantities, any extensive quantity X divided by the total substance mass $\sum_B m_B$ is an intensive variable called the specific quantity x

$$x = X/\sum_B m_B. \tag{1.5}$$

Specific quantities are normally designated by lowercase letters.

To elucidate the differences between molar and mass quantities, a few examples are provided. The volume of a phase is given the symbol V and when this refers to a molar volume, the symbol V_m is used, and the quantity given by $V_m = M/\rho = \rho_n^{-1}$ where M is the molar mass and ρ is the mass density, which is given by $\rho = m/V$, where m is the mass and ρ_n the amount-of-substance density, which is related to the mass density by $\rho_n = \rho/M$. The specific volume v is given by $v = V/m = \rho^{-1}$ and defines the volume of a mass of material.

In the remainder of this book we make use of both molar and mass notation. The choice depends on whether the focus of the discussion is on chemistry and the (fundamental) properties of matter, or upon engineering applications where mass or specific quantities are usually adopted. We may occasionally switch between molar and mass quantities without explicit mention. Throughout the text we have defined each symbol when it has either been first introduced or when it is used for a different purpose.

1.3.12 What Is Chemical Composition?

The properties of a system consisting of a mixture of chemical components depend on the composition of the phase, which is specified by a measure of the amount of each chemical component present. The composition of a phase can change by virtue of the extent of a chemical reaction, or by the gain or loss of one or more components. To study the variation of the properties of a mixture it is convenient to define other, non-thermodynamic quantities. The purpose of the following questions is to introduce these parameters.

The *mole fraction* y of a substance B in a phase is given by y_B, which is an intensive quantity

$$y_B = n_B / \sum_B n_B, \tag{1.6}$$

the sum of the mole fractions in a phase must then equal unity. An analogous definition is, of course, possible for mass fraction.

1.3.13 What Are Partial Molar Quantities?

When a homogeneous mixture of several chemicals A, B, C … is formed there are some mixtures for which the thermodynamic quantities, such as volume and internal energy behave additively. Such mixtures are termed *ideal.* That means that the resulting total thermodynamic quantity X can be evaluated as the sum, $\Sigma_B n_B X_B^*$, where X_B^* is the molar quantity X of *pure* B.

In general, however, mixtures are *non-ideal,* because of the specific interactions of the different substances at the molecular level. The molar quantity X of pure B then must be replaced by what is called *the partial molar quantity* X_B. In a sense, this represents the molar property of B *within the mixture* of a certain composition as opposed to B in the pure state; it is therefore defined as the effect of adding a mole of B to a mixture of otherwise unchanged composition.

Thus, the partial molar quantity X_B (which is an intensive quantity) of substance B in a mixture is defined by

$$X_B = (\partial X/\partial n_B)_{T,p,n_{A\neq B}}, \tag{1.7}$$

where $n_A \neq n_B$ means all the n's except n_B are held constant; for a pure substance B, $X_B = X/n_B = X_m$. The total differential of an extensive quantity X of a mixture can thus be written as

$$dX = (\partial X/\partial T)_{p,n_B} dT + (\partial X/\partial p)_{T,n_B} dp + \sum_B X_B dn_B, \tag{1.8}$$

and, by the use of Euler's theorem (see Question 1.10.2), as

$$X = \sum_B n_B X_B, \tag{1.9}$$

or recast as

$$X_m = \sum_B x_B X_B, \tag{1.10}$$

on division on both sides by $\sum_B n_B$. Differentiation of Equation 1.9 and combination with Equation 1.8 gives

$$0 = -(\partial X/\partial T)_{p,n_B} dT - (\partial X/\partial p)_{T,n_B} dp + \sum_B n_B dX_B, \tag{1.11}$$

so that at constant temperature and pressure we have

$$0 = \sum_B n_B dX_B, \tag{1.12}$$

which, when substituted into the total derivative of Equation 1.9, gives

$$dX = \sum_B X_B dn_B, \tag{1.13}$$

a result which also follows directly from Equation 1.8 when $dT = dp = 0$.

It can also be shown that

$$dX_m = \sum_B X_B dx_B. \tag{1.14}$$

Equations 1.10 and 1.14 can be used to determine all partial molar quantities of a mixture as a function of composition.

For a binary mixture of chemical species $\{xA + (1 - x)B\}$ Equations 1.10 and 1.14 are

$$X_m = (1 - x)X_A + xX_B \tag{1.15}$$

and

$$dX_m = (X_B - X_A)dx. \tag{1.16}$$

When Equations 1.15 and 1.16 are solved for X_A and X_B, they give

$$X_A = X_m - x(\partial X_m/\partial x)_{T,p} \quad (1.17)$$

and

$$X_B = X_m + (1-x)(\partial X_m/\partial x)_{T,p}. \quad (1.18)$$

The partial molar quantities X_A and X_B for a particular composition can be obtained from measurements of X_m and the variation of X_m with x provided that the latter is nearly linear. When this is not so, as is often the case for example for the volume, then an alternative approach must be sought and this is provided by the molar quantity of mixing.

1.3.14 What Are Molar Quantities of Mixing?

The molar quantity of mixing $\Delta_{mix}X_m$ of quantity X is the difference between the molar quantity X_m of a mixture of specified composition and the value one would expect if X_m were evaluated for an hypothetical *ideal* mixture of the same species. For a binary mixture $\{(1-x)A + xB\}$, the molar quantity of mixing at a given temperature and pressure $\Delta_{mix}X_m$ is given by

$$\Delta_{mix} X_m = X_m - (1-x)X_A^* - xX_B^*, \quad (1.19)$$

where X_A^* and X_B^* are the appropriate molar quantities of pure A and B. For example, the molar volume of mixing can be determined from measurements of the density, ρ, of the mixture the densities of the pure materials and a knowledge of the molar masses M of A and B from

$$\Delta_{mix} V_m = \frac{(1-x)M_A + xM_B}{\rho} - \frac{(1-x)M_A}{\rho_A^*} - \frac{xM_B}{\rho_B^*}. \quad (1.20)$$

1.3.15 What Are Mixtures, Solutions and Molality?

Mixture is the word reserved for systems (whether they be gases, liquids and solids) containing more than one substance; all components in the mixture are treated equally. On the other hand, the term *solution* is reserved for liquids or solids containing more than one substance where one substance is deemed to be a solvent and the others are solutes; these entities are not treated in the same way. If the sum of the mole fractions of the solutes is small compared with unity the solution is termed *dilute*.

The composition of a solution is usually expressed in terms of the molalities of the solutes. The *molality* of a solute B m_B in a solvent A of molar mass M_A is defined by

$$m_B = n_B/(n_A M_A), \quad (1.21)$$

and is related to the mole fraction x_B by

$$x_B = m_B \frac{M_A}{1 + \sum_B m_B}, \quad (1.22)$$

or

$$m_{\mathrm{B}} = \frac{x_{\mathrm{B}}}{M_A(1 - \Sigma_{\mathrm{B}} x_{\mathrm{B}})}. \tag{1.23}$$

1.3.16 What Are Dilution and Infinite Dilution?

For a mixture of species A and B containing amounts of substance n_{A} and n_{B}, the change in a quantity X on dilution by the addition of an amount of substance Δn_{A} is $\Delta_{\mathrm{dil}}X$, which is given by

$$\Delta_{\mathrm{dil}}\,X = \Delta_{\mathrm{mix}}\,X\{(n_{\mathrm{A}} + \Delta n_{\mathrm{A}})\mathrm{A} + n_{\mathrm{B}}\mathrm{B}\} - \Delta_{\mathrm{mix}}\,X\,(n_{\mathrm{A}}\mathrm{A} + n_{\mathrm{B}}\mathrm{B}), \tag{1.24}$$

or when divided by n_{B}

$$\Delta_{\mathrm{dil}}\,X/n_{\mathrm{B}} = \Delta_{\mathrm{mix}}X\,(x_{\mathrm{f}})/x_{\mathrm{f}} - \Delta_{\mathrm{mix}}X\,(x_{\mathrm{i}})/x_{\mathrm{i}}, \tag{1.25}$$

where the subscripts f and i indicate the final and initial mole fractions of B. As $x_{\mathrm{f}} \to 0$ one speaks of infinite dilution of species B in solvent A and the quantity is given a superscript ∞ so that Equation 1.25 becomes

$$(\Delta_{\mathrm{dil}}X/n_{\mathrm{B}})^{\infty} = \{(1 - x)/x\}\{X_{\mathrm{A}}\,(x) - X_{\mathrm{A}}^{*}\} + \{X_{\mathrm{B}}^{\infty} - X_{\mathrm{B}}\,(x)\}. \tag{1.26}$$

In Equation 1.26 the subscripts f and i were removed because at infinite dilution $x_{\mathrm{f}} \approx x_{\mathrm{i}}$.

When a solid B dissolves in a liquid solvent A to give a solution, the change in X is denoted by $\Delta_{\mathrm{soln}}X$, which is given by

$$\Delta_{\mathrm{soln}}\,X/n_{\mathrm{B}} = \{(1 - x)/x\}\{X_{\mathrm{A}}\,(\mathrm{l}, x) - X_{\mathrm{A}}^{*}\,(\mathrm{l})\} + \{X_{\mathrm{B}}\,(\mathrm{l}, x) - X_{\mathrm{B}}^{*}\,(\mathrm{s})\}, \tag{1.27}$$

in which l denotes the liquid state and s denotes the solid.

At infinite dilution of the solid in the solvent Equation 1.27 becomes

$$(\Delta_{\mathrm{soln}}X/n_{\mathrm{B}})^{\infty} = X_{\mathrm{B}}^{\infty}\,(\mathrm{l}) - X_{\mathrm{B}}^{*}\,(\mathrm{s}), \tag{1.28}$$

and

$$(\Delta_{\mathrm{soln}}X/n_{\mathrm{B}})^{\infty} - (\Delta_{\mathrm{dil}}X/n_{\mathrm{B}})^{\infty} = \Delta_{\mathrm{soln}}\,X/n_{\mathrm{B}}. \tag{1.29}$$

Equation 1.27 is also used when the solute is a gas with an appropriate designation of the phase of the solute.

1.3.17 What Is the Extent of Chemical Reaction?

A chemical reaction from reagents R to products P can be written as

$$\sum_{\mathrm{R}} (-\nu_{\mathrm{R}})\mathrm{R} = \sum_{\mathrm{P}} \nu_{\mathrm{P}}\mathrm{P}, \tag{1.30}$$

where ν is the stoichiometric number and is, by convention, negative for reactants and positive for products. The extent of a chemical reaction, ξ, (an extensive property) for a substance B that reacts according to Equation 1.30 is defined by

$$n_B(\xi) = n_B(\xi = 0) + \nu_B \xi, \tag{1.31}$$

where $n_B(\xi = 0)$ is the amount of substance present when the extent of reaction is zero; for example, before the reaction commenced.

1.4 How Do We Formulate Balances?

1.4.1 What Is the Use and What Is the General Structure of Balances?

Balances are one of the most important intellectual tools for analyzing physical, chemical, biological and many other systems and for predicting their evolution with time. Balances can be written for any extensive property of a system. Writing a balance begins with the careful choice of the extensive property to be balanced. The next highly important point is the definition of the system boundaries, or the control volume. By this definition we distinguish clearly what we consider to be part of the system we analyze from whatever is "outside" or in the environment. The appropriate choice of the system boundaries may dramatically simplify the solution of some problems, but requires a bit of practice.

Balances always express the fact that whatever flows into the system minus what leaves the system plus whatever is produced inside, accumulates in the system (Figure 1.2).

Mathematically the balance may be formulated showing the differences before and after a given time interval Δt or for an infinitesimally small-time interval as a differential equation.

1.4.2 What Is a Mass Balance?

A mass balance could be formulated in differential form as

$$dm = dm_{in} - dm_{out}, \tag{1.32}$$

where dm (with $[m]$ = kg) represents a differential accumulation of mass inside the system and dm_{in} and dm_{out} designate differential amounts of mass that have been added and removed from the system, respectively. Mass balances do not contain source terms since mass can neither be created nor destroyed.

In cases with more than one mass feed point and/or several ports through which the material may enter or leave the system, one will have to sum up the amounts through the various ports

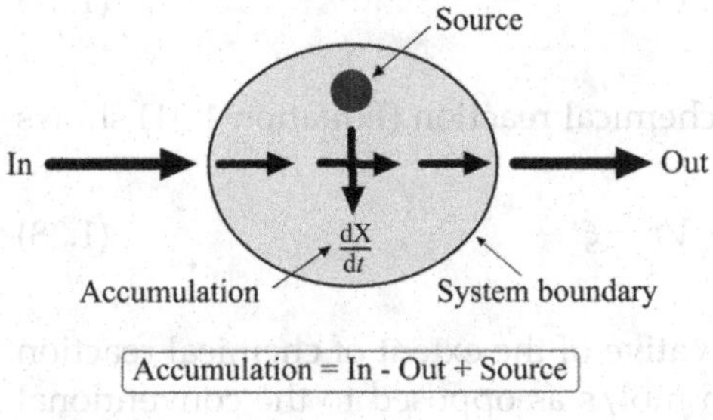

FIGURE 1.2
The general structure of balances.

$$dm = \sum_k dm_k, \tag{1.33}$$

where dm_k designates the differential amount of mass that has been added or removed from the system through the k-th exchange port. Mass leaving the system counts negatively.

A formulation that is closer to engineering applications is (in units of kg/s)

$$\frac{dm}{dt} = \sum_k \dot{m}_k, \tag{1.34}$$

where $\dot{m}_k$ stands for the mass flow rate entering the system through the k-th exchange port. The dot above the symbol means "per unit time" and is generally used to avoid the need for defining new symbols for rates.

An integrated form of the balance for a given time interval Δt would read

$$\Delta m = \sum_k m_k, \tag{1.35}$$

where Δm represents the change of mass within the system and m_k the amount of mass exchanged through the k-th exchange port through the system boundary.

1.4.3 What Is a Molar Balance on the Amount of Substance?

A differential molar balance for the amount of substance B in a *closed system* of volume, V, can be formulated as

$$dn_B = \nu_B r V dt, \tag{1.36}$$

where dn_B is the molar accumulation of B in the system in moles. The right-hand side of the equation does not account for B exchanged with the environment, because the system is assumed to be closed. But in contrast to the mass balance, a molar balance for a given substance contains a *source term,* because B could be synthesized or consumed inside the system through a chemical reaction. This term appears on the right side and expresses the infinitesimal number of moles of B produced or consumed during an infinitesimal time span dt. ν_B represents the stoichiometric coefficient of B in the chemical reaction and r the reaction rate in mol s^{-1} m^{-3}. As explained in connection with Equation 1.31, the stoichiometric coefficient is positive if B is a product but negative if it is a *reactant*. In the latter case, B is consumed by the chemical reaction and one would speak of a sink rather than of a source.

The integrated balance can be formulated as

$$\Delta n_B = \nu_B \int_0^t rV dt. \tag{1.37}$$

Comparing this to the definition of the extent of chemical reaction (Equation 1.31) shows

$$\int_0^t rV dt = \xi \quad \text{or} \quad Vr = \dot{\xi}. \tag{1.38}$$

The last equation demonstrates that the time derivative of the extent of chemical reaction really is the reaction rate for the whole system in mol/s as opposed to the conventional reaction rate r, which is expressed per unit volume in mol s^{-1} m^{-3}. The source term has been formulated here, assuming a system of uniform composition.

In **open systems**, the balance must allow for both the substance B exchanged with the surroundings and the amount produced or consumed by a chemical reaction

$$\mathrm{d}n_\mathrm{B} = \sum_k \mathrm{d}n_{\mathrm{B},k} + \nu_\mathrm{B} V r \,\mathrm{d}t, \tag{1.39}$$

or

$$\frac{\mathrm{d}n_\mathrm{B}}{\mathrm{d}t} = \sum_k \dot{n}_{\mathrm{B},k} + \nu_\mathrm{B} V r, \tag{1.40}$$

where $\dot{n}_{\mathrm{B},k}$ denotes the molar exchange flow rate of B through the k-th exchange port through the system boundary (mol/s). Leaving streams count negatively.

In many real cases, the compound B is produced and consumed by several chemical reactions simultaneously. The source term must then sum up the effects of all these reactions (in mol/s) as

$$\frac{\mathrm{d}n_\mathrm{B}}{\mathrm{d}t} = \sum_k \dot{n}_{\mathrm{B},k} + \sum_j \nu_{\mathrm{B},j} V r_j, \tag{1.41}$$

where r_j designates the rate of the j-th chemical reaction and $\nu_{\mathrm{B},j}$ the stoichiometric coefficient of substance B in the j-th reaction.

1.4.4 What Is Internal, or Thermodynamic, Energy? What Is an Energy Balance for a Closed System?

In mechanical systems, energy is stored primarily in the form of potential and kinetic energy. In thermodynamic systems, however, the most important form of energy is often the so-called *internal* or *thermodynamic energy*. This comprises all forms of energy stored at the molecular level in force fields such as attraction and repulsion forces, the force required to create a surface, the kinetic energy of the molecules themselves moving around at a certain temperature and so on (see Question 2.5). It is called thermodynamic or internal energy U and quantified in joules.

The internal energy of a *closed system* can be increased by adding energy to it in the form of heat (δQ), volume work ($-p\mathrm{d}V$) or other work (δW_o) (see Questions 1.3.6 and 1.3.7). Hence, the energy balance of a closed system is

$$\mathrm{d}U = \delta Q - p\mathrm{d}V + \delta W_\mathrm{o} \tag{1.42}$$

or

$$\Delta U = Q - \int p\mathrm{d}V + W_\mathrm{o}. \tag{1.43}$$

The internal energy U is a *state function*. This signifies that the difference ΔU between two states of the system that are unequivocally defined in terms of e.g., temperature, pressure and composition will always have the same value, independently of the method used to bring the system from state 1 to state 2. This is not the case for Q and W, which only quantify the amount of energy transferred to the system or removed from it. The amount of energy ΔU required to change the system from state 1 to state 2 can be brought about by different combinations of Q and W. These values can thus not be found by integrating δQ or δW, whence the notation δ is used to indicate an infinitesimal amount rather than dQ and dW. Q and W are thus not state functions.

1.4.5 What Is Enthalpy and Why Was It Introduced?

It is obvious from the previous equations that changes in internal energy, ΔU, directly predict the heat evolved or taken up if the state change is effected under conditions in which no work can be exchanged with the environment ($dV = 0$ and $\delta W_o = 0$ in Equation 1.42). On the other hand, it is possible to quantify the internal energy change by measuring the heat exchanged under such conditions, or else by measuring the work exchanged in the form of electrical work if the heat exchange is suppressed by an adiabatic insulation and the volume work is eliminated by working at constant volume (see Question 1.6.1).

In such experiments, suppressing the volume work despite the tendency of the system to expand or contract owing, for example, to changes in temperature, requires working in an autoclave, which is obviously very expensive and cumbersome (see Question 1.8.2). The vast majority of actions involving thermodynamic state changes, such as analytical measurements, separation processes and industrial chemical reactions are therefore carried out at constant pressure rather than at constant volume.

It would thus be extremely useful to identify a state function like the internal energy, but which predicted the heat exchange at constant pressure rather than at constant volume. This is precisely what enthalpy does. It is defined as

$$H = U + pV \tag{1.44}$$

or

$$dH = dU + pdV + Vdp. \tag{1.45}$$

Solving the second expression for dU and substituting into the differential energy balance for a *closed system* above yields

$$dH - pdV - Vdp = \delta Q + (-pdV) + \delta W_o \tag{1.46}$$

or

$$dH = \delta Q + Vdp + \delta W_o \tag{1.47}$$

or

$$\Delta H = Q + \int Vdp + W_o. \tag{1.48}$$

It is obvious that in a *closed system at constant pressure*, changes in enthalpy correspond directly to the heat exchanged with the environment if there is no other work involved.

The PV-term that is added to the internal energy in the definition of enthalpy (Equation 1.44) may be interpreted as the energy that was needed to push aside the environment in order to make room for a system of volume V at a given pressure p. If the volume of the system increases, more of such PV-energy is needed and must be supplied from somewhere. Enthalpy accounts for this added PV-energy at constant pressure.

1.5 What is the First Law of Thermodynamics?

1.5.1 What Is the Statement of the First Law?

The First Law says that energy balances do not contain a source term. For a *closed* system without the influence of potential or kinetic energy, the First Law is thus written as shown in Equations 1.43 and 1.48. (These equations are restricted to *closed* systems but remain valid for reactive systems.) This means that energy cannot be created inside the system nor destroyed. It is therefore also impossible to invent a perpetual motion machine, because that would need to constantly create energy from nothing. A very clear example of the application of the First Law in the form of Equation 1.43 is given in Question 1.6.2.

1.5.2 How Are Changes in Thermodynamic Energy Linked to Measurable State Variables?

In many cases, the state functions internal energy and enthalpy need to be linked to the variables defining the state of the system to make them useful. The number of variables that must be specified is given by the phase rule (see Question 4.1.2) and amounts to two plus the number of compounds in a homogeneous mixture without chemical reactions. A reasonable choice of variables might be: Temperature, volume or pressure, and the number of moles of each constituent. The changes of these variables may then be linked to the change of thermodynamic energy by a total differential (Question 1.10.1)

$$dU = \left(\frac{\partial U}{\partial T}\right)_{V,n_B} dT + \left(\frac{\partial U}{\partial V}\right)_{T,n_B} dV + \sum_B \left(\frac{\partial U}{\partial n_B}\right)_{T,V,n_{i\neq B}} dn_B. \tag{1.49}$$

The partial derivatives of internal energy with respect to temperature and volume and with respect to the number of moles of the different constituents of the mixture must be determined experimentally by introducing changes in temperature, volume and mole number and measuring the change in U based on heat or electrical work exchanged.

It should be emphasized that this type of equation **is not a balance.** It is just a purely mathematical formalism to relate the change of a state function to the changes of the state variables it depends on.

A similar total differential may be written for enthalpy

$$dH = \left(\frac{\partial H}{\partial T}\right)_{p,n_B} dT + \left(\frac{\partial H}{\partial p}\right)_{T,n_B} dp + \sum_B H_B \, dn_B, \tag{1.50}$$

where H_B stands for the partial molar enthalpies of the various compounds B.

The coefficients $\left(\frac{\partial U}{\partial T}\right)_{V,n_B}$ and $\left(\frac{\partial H}{\partial T}\right)_{p,n_B}$ are called the heat capacity at constant volume (C_V) and at constant pressure (C_p), respectively (in $\mathrm{J\,K^{-1}}$). They are often expressed as specific heat capacities, i.e. per unit mass of a given material ($\mathrm{J\,K^{-1}\,kg^{-1}}$) for which we use lowercase letters

$$\left(\frac{\partial U}{\partial T}\right)_{V,n_B} = mc_V \text{ and } \left(\frac{\partial H}{\partial T}\right)_{p,n_B} = mc_p, \tag{1.51}$$

where m is the mass of material.

If these coefficients are known as a function of the state of the system, the above expressions can in principle be substituted for the accumulation term in the differential balance, yielding equations that link the amounts of energy added or removed from the system in the form of heat or work to the change of the state variables obtained.

1.5.3 How Can the First Law Be Used to Calculate the Energy Required for a State Change in Closed Systems without Chemical Reactions?

One of the simplest examples for illustrating this question might be a system consisting of a non-reacting liquid mixture of several chemical compounds placed in a stirred tank reactor that is being heated up by the action of a heating jacket (see Figure 1.3).

Based on Equation 1.42, the energy balance for this system may be formulated as

$$dU = \delta Q - pdV + \delta W_{st}, \tag{1.52}$$

where the term δW_o has been replaced by the stirrer work, δW_{st}, i.e. the energy introduced by the stirrer.

The internal energy change, dU, can be linked to the change in the measurable state variables by the total differential dU given in Equation 1.49. Using the specific heat capacity c_V (see Question 1.5.2) in the total differential, this reads for our case,

$$dU = mc_V dT + \left(\frac{\partial U}{\partial V}\right)_{T,n_B} dV, \tag{1.53}$$

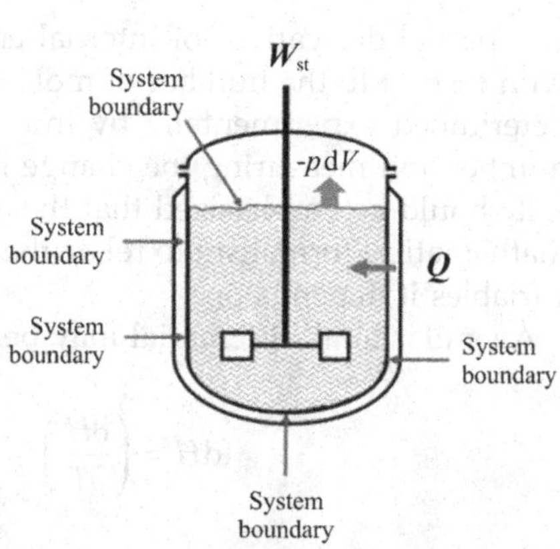

FIGURE 1.3
Example of a closed system.

where c_V stands for the specific heat capacity of the mixture (in J K^{-1} kg^{-1}) and m is the mass of the system. The last term in Equation 1.49 plays no role because dn_B is zero in a closed non-reactive system.

Combining Equations 1.52 and 1.53 by eliminating dU yields a relationship between the changes of measurable state variables of the system and the amounts of energy exchanged with the environment

$$mc_V \mathrm{d}T + \left(\frac{\partial U}{\partial V}\right)_{T,n_B} \mathrm{d}V = \delta Q - p\mathrm{d}V + \delta W_{st}. \tag{1.54}$$

The same treatment for enthalpy results in the following equation

$$mc_p \mathrm{d}T + \left(\frac{\partial H}{\partial p}\right)_{T,n_B} \mathrm{d}p = \delta Q + V\mathrm{d}p + \delta W_{st}. \tag{1.55}$$

Fortunately, integration of these two equations can be simplified for several common cases. First, for any system, the specific heat capacities may often be assumed to remain constant in the temperature interval of interest. Secondly, for ideal gases the volume dependence of internal energy and the pressure dependence of enthalpy can be shown to be zero (see Equations 3.107 and 3.60 for the volume dependence of U and Equation 3.70 for the pressure dependence of H). Thus, for ideal gases the integrated left-hand side of Equations 1.54 and 1.55 reduce to $mc_V\Delta T$ and $mc_p\Delta T$, respectively.

Thirdly, if the system consists of an incompressible liquid or of a solid, it can be shown that the partial derivatives of each equation cancel with the $-p\mathrm{d}V$ term in the internal energy balance and with the $V\mathrm{d}p$ term in the case of enthalpy. As a result, these two terms also disappear from each of the balances (see Question 1.5.6). Thus, if the constancy of heat capacities is again assumed, the two integrated balances turn out to be

$$mc_V \Delta T = Q + W_{st} \tag{1.56}$$

$$mc_p \Delta T = Q + W_{st}. \tag{1.57}$$

These equations show that in the case of an incompressible liquid or a solid c_V and c_p must be identical and that integrating Equations 1.54 and 1.55 reveals that ΔU and ΔH must also be identical. These equations permit prediction of the energy required in terms of heat and in terms of stirrer energy to heat the fluid up from one temperature to another.

1.5.4 How Can the First Law Be Applied to a Closed, Stirred Chemical Reactor at Constant Pressure?

At constant pressure, the energy balance ought to be formulated in terms of enthalpy (see Equation 1.47)

$$\mathrm{d}H = \delta Q + \delta W_{st}, \tag{1.58}$$

where no $V\mathrm{d}p$ work is implied because of the constant pressure. According to Equation 1.50, dH is linked to changes of temperature and of the number of moles as follows (p = const)

$$dH = mc_p dT + \sum_B H_B dn_B. \quad (1.59)$$

Assuming one single chemical reaction

$$0 = \sum_B \nu_B B, \quad (1.60)$$

where ν_B is the stoichiometric number, and substituting the differential molar balance for dn_B (Equation 1.36) links all the changes of the mole numbers dn_B in Equation 1.59 to the reaction rate, r,

$$dH = mc_p dT + \sum_B \nu_B H_B Vr\, dt. \quad (1.61)$$

Here it has been assumed that c_p can be approximated by that of the solvent and that it is nearly constant despite the chemical reaction. This is a valid assumption if the reactants are sufficiently diluted in the solvent. Eliminating dH from Equations 1.58 and 1.61 yields

$$mc_p dT + \sum_B H_B \nu_B Vr\, dt = \delta Q + \delta W_{st}, \quad (1.62)$$

where $\Sigma_B \nu_B H_B$ is known as the *enthalpy of reaction*

$$\sum_B \nu_B H_B \equiv \Delta_r H. \quad (1.63)$$

The balance (Equation 1.50) makes no reference to pressure because pressure was assumed to stay constant. If the specific heat capacity and the enthalpy of reaction may be assumed approximately constant over a given temperature range, the balance may be integrated to yield

$$mc_p (T_2 - T_1) = Q + W_{st} - \Delta_r H \int_1^2 Vr\, dt. \quad (1.64)$$

This equation permits predicting the temperature increase or the heat evolution when performing a chemical reaction. For an isothermal reaction without stirring work, the heat evolution would be

$$Q = \Delta_r H \int_1^2 Vr\, dt. \quad (1.65)$$

The integral is equal to the extent of reaction ξ (see Question 1.4.3). Therefore, the enthalpy of reaction represents the heat evolution per mole of reaction advancement, so that it is also called *"heat of reaction"*.

By analogy with the enthalpy of reaction, similar "properties of reaction" also exist for other state functions. Of particular importance is the *Gibbs energy of reaction* $\Delta_r G$, defined in Question 3.3.4.

The energy balance (Equation 1.62), together with the definition of Equation 1.63, can readily be generalized to include the effect of many simultaneous reactions

$$mc_p\Delta T = Q + W_{st} - \sum_j \left\{\Delta_{rj}H \int_1^2 Vr_j \, dt\right\}, \tag{1.66}$$

where $\Delta_{rj}H$ and r_j stand for the heat of reaction and for the rate of the j-th reaction, respectively.

In evaluating the enthalpies of reaction, $\Delta_{rj}H$, the question arises what to substitute for the partial molar enthalpies H_B in using the definition Equation 1.61. They must be defined based on a reference state common to all participating chemicals. Two such reference states are in use: The constituent elements in their most stable states at a given temperature, in which case the H_B are substituted by the enthalpies of formation of B ($\Delta_f H_B$), and the completely oxidized state of the chemicals at a given temperature, in which case the H_B are substituted by the negative enthalpies of combustion of B ($-\Delta_c H_B$), (see Chapter 5).

1.5.5 How Can the First Law Be Applied to Open, Non-Reactive Flow Systems at Steady State?

In *open systems*, the energy balance must account for the fact that entering and leaving matter can import and export considerable amounts of internal energy convectively (by flow). In addition, work (energy) is done on the system by pushing the incoming matter into the system, whereas the system does work on the environment by pushing the outgoing streams into it. This is called *flow work* $w_f = p_{out}v_{out} - p_{in}v_{in}$. In many open systems, e.g. for power generation, a characteristic form of work is shaft work, W_s, denoting mechanical work delivered or withdrawn by a rotating shaft as in pumps, compressors or turbines. In these cases, shaft work derives from the pressure difference between the entrance port and exit port of an open system. The internal energy balance (Equation 1.42) must, therefore, now be modified as (Figure 1.4)

$$dU = \delta Q - p dV + \delta W_s + (u_{in} + p_{in}v_{in})dm_{in} - (u_{out} + p_{out}v_{out})dm_{out}, \tag{1.67}$$

which may be written as

$$dU = \delta Q - p dV + \delta W_s + h_{in}dm_{in} - h_{out}dm_{out}. \tag{1.68}$$

When adding $d(pV)$ on either side of this equation, we obtain the enthalpy balance (cf. Equation 1.45)

$$dH = \delta Q + V dp + \delta W_s + h_{in}dm_{in} - h_{out}dm_{out}. \tag{1.69}$$

In these equations, u_{out} and u_{in} indicate the specific internal energy of the outflowing and the inflowing material, respectively, p_{in} and p_{out}, v_{in} and v_{out}, h_{in} and h_{out} denote the

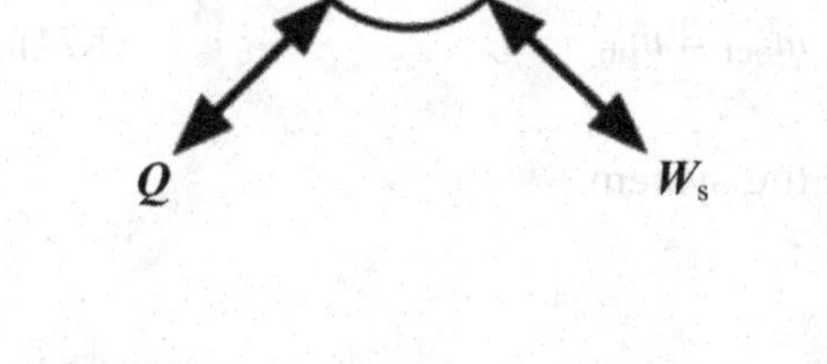

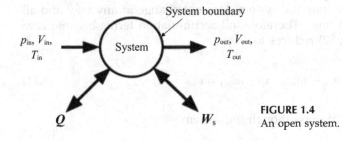

FIGURE 1.4
An open system.

pressure, the specific volume and the specific enthalpy in the feed and in the discharge, respectively.

In these equations it has been assumed that no reversible work other than volume work is implied. Also, most forms of dissipative work, W_{diss}, have not been considered. The only other work that is done on or by the system would be shaft work W_s, denoting mechanical work delivered by a rotating shaft such as in pumps, compressors or turbines. It should also be remembered that the p-V expressions above denote only the reversible parts of work. The shaft work computed through Equations 1.67–1.69 must thus be corrected by applying a correction factor that accounts for the additional work required to overcome irreversible volume work caused, for instance, by friction.

The classical notation used so far is quite confusing for open systems, owing to the many free-standing operators such as d, δ, Δ and the like, which all have different meanings. The situation becomes even worse for open systems in transients because of the need to distinguish between infinitesimal amounts accumulating inside the system (e.g. dm, dn_B) from infinitesimal amounts added or removed from the system (e.g. dm_{in}, $dn_{B,in}$).

As a result, it is preferable to reformulate the differential balances for an infinitesimal time interval dt rather than for an infinitesimal amount energy accumulated (dU, dH). In this form, the balances expressed by Equations 1.67 and 1.69 become (in watts)

$$\frac{dU}{dt} = \dot{Q} - p\frac{dV}{dt} + \dot{W}_s + (u_{in} + p_{in}v_{in})\,\dot{m}_{in} - (u_{out} + p_{out}v_{out})\,\dot{m}_{out} \tag{1.70}$$

and

$$\frac{dH}{dt} = \dot{Q} + V\frac{dp}{dt} + \dot{W}_s + h_{in}\dot{m}_{in} - h_{out}\dot{m}_{out}, \tag{1.71}$$

where $\dot{Q}$ and $\dot{W}_s$ represent the rate of heat and shaft work introduced into the system in watts, and $\dot{m}_{in}$ and $\dot{m}_{out}$ the mass flow rate in and out, respectively, of the system in kg/s (for an explanation of the symbol with dots, see Equation 1.34).

The total differentials that link dU and dH to dT, dV, dp and dn_B (Equations 1.49 and 1.50) must also be reformulated in terms of time derivatives

$$\frac{dU}{dt} = mc_V\frac{dT}{dt} + \left(\frac{\partial U}{\partial V}\right)_{T,n_B}\frac{dV}{dt} + \sum_B\left(\frac{\partial U}{\partial n_B}\right)_{T,V,n_{i\neq B}}\frac{dn_B}{dt} \tag{1.72}$$

$$\frac{dH}{dt} = mc_p\frac{dT}{dt} + V\frac{dp}{dt} + \sum_B H_B\frac{dn_B}{dt}. \tag{1.73}$$

But as the present case concerns a system of mass m at *steady state*, the equations can be dramatically simplified. At steady state the system does not change in any way and all state variables stay constant with time. Therefore, all accumulation terms become zero and the mass balance (Equation 1.32) reduces to

$$\frac{dm}{dt} = 0 = \dot{m}_{in} - \dot{m}_{out} \quad \text{or} \quad \dot{m}_{out} = \dot{m}_{in} = \dot{m}, \tag{1.74}$$

where $\dot{m}$ represents the flow rate of mass though the system.

In the energy and enthalpy balances (Equations 1.70 and 1.71), and in the Equations 1.72 and 1.73 the time derivative of U and H may also be set to zero at steady state. The terms $-p\frac{dV}{dt}$ and $V\frac{dp}{dt}$ on the right hand sides of the balances 1.70 and 1.71, respectively, also disappear since the volume, V, and pressure, p, must necessarily stay constant over time at steady state. Equations 1.70 and 1.71 thus reduce to

$$(u_{out} + p_{out}v_{out} - u_{in} - p_{in}v_{in})\dot{m} = \dot{Q} + \dot{W}_s, \tag{1.75}$$

or, because of $h = u + pv$, according to the definition of enthalpy

$$(h_{out} - h_{in})\dot{m} = \dot{Q} + \dot{W}_s. \tag{1.76}$$

Often the In and Out streams also carry *kinetic and potential energy*. Accounting for these yields

$$(u_{out} + p_{out}v_{out} + e_{kin,out} + e_{pot,out})\,\dot{m} - (u_{in} + p_{in}v_{in} + e_{kin,in} + e_{pot,in})\dot{m} = \dot{Q} + \dot{W}_s, \tag{1.77}$$

where $e_{pot} = gz$, the potential energy per unit mass, with z = height and $e_{kin} = u_f^2/2$, the kinetic energy per unit mass, with u_f = fluid velocity.

In many cases, these equations are only useful if the internal energy and the enthalpy in the feeding and discharge streams can be linked to their respective state variables in those streams. Noting that $h = u + pv$ and assuming that the specific heat capacities are independent of temperature, Equation 1.77 can be written

$$\dot{m}c_p(T_{out} - T_{in}) + \frac{\dot{m}}{2}\left(u_{f,out}^2 - u_{f,in}^2\right) + \dot{m}g\,(z_{out} - z_{in}) = \dot{Q} + \dot{W}_s. \tag{1.78}$$

Many technical devices such as compressors, turbines or pumps can be analyzed with energy balances such as Equation 1.78 (cf Questions 1.6.3–1.6.5).

A special case of the energy balance (Equation 1.77) is the *Bernoulli equation*. It describes the changes of velocity, pressure and height in fluid flow when the fluid can be assumed incompressible (v, V constant) and the process being simultaneously isothermal and adiabatic, and without any involvement of shaft work. In this case Equation 1.77 simplifies to

$$v\,(p_{out} - p_{in}) + \frac{1}{2}\left(u_{f,out}^2 - u_{f,in}^2\right) + g\,(z_{out} - z_{in}) = 0, \tag{1.79}$$

where v stands for the specific volume so that $1/v = \rho$. Dividing throughout this equation by v yields the famous Bernoulli equation in a familiar form

$$\Delta p + \frac{1}{2}\rho\Delta u_f^2 + \rho g\Delta z = 0. \tag{1.80}$$

1.5.6 How Can the First Law Be Applied to Transient Open Systems with Several Feed and Discharge Points and Multiple Chemical Reactions?

A real example of this kind is provided by a chemical reactor operated in fed-batch mode ("semi-batch" in chemical engineering parlance). As a specific case, we consider a

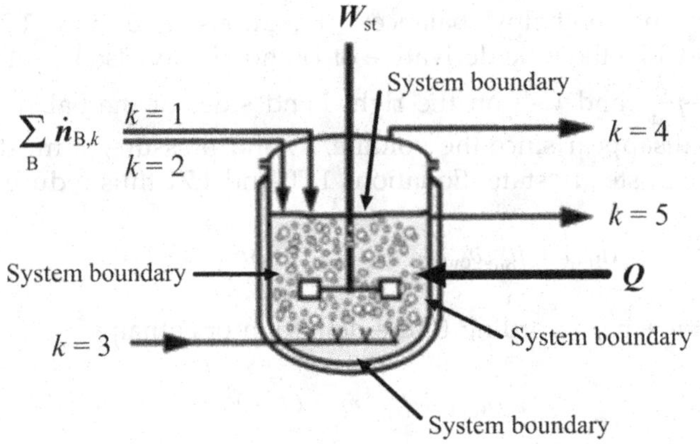

FIGURE 1.5
An open bioreactor in transient state with several feeding and discharge ports, and with a multitude of simultaneous chemical and biological reactions.

bioreactor containing live cells. There is a constant feed of a medium filling the reactor tank up slowly, but there are also feeds and drains for pH control, aeration and online analysis (see Figure 1.5). In order to simplify the equation, the development assumes *constant pressure.*

The system is best defined as the liquid phase with the live cells suspended in it but excluding the air bubbles.

The enthalpy balance for this case must allow for varying system compositions, and also for not only two, but several mass-exchange ports with the environment. As a consequence, the balance for open systems (Equation 1.69) must be rewritten as a molar balance and must contain a sum over all feed and discharge ports

$$\frac{dH}{dt} = \dot{Q} + \dot{W}_{st} + \sum_k \sum_B H_{B,k}\, \dot{n}_{B,k}, \tag{1.81}$$

where $H_{B,k}$ stands for the partial molar enthalpy of B at the *k*-th mass exchange port and $\dot{n}_{B,k}$ represents the molar flow rate of B through the *k*-th mass exchange port. Substituting Equation 1.73 for dH/dt and accounting for a constant pressure

$$mc_p \frac{dT}{dt} + \sum_B H_B \frac{dn_B}{dt} = \dot{Q} + \dot{W}_{st} + \sum_k \sum_B H_{B,k} \dot{n}_{B,k}. \tag{1.82}$$

Whereas the left-hand terms express the enthalpy accumulation rate in the system, the right-hand terms enumerate all enthalpy flows into and out of the system. The c_p has been assumed constant despite the chemical reactions. This simplification is justified if the reactants are sufficiently diluted in a dominating solvent.

The changes in the number of moles n_B in the second left-hand term of the equation are linked through the stoichiometries of the various reactions (see Question 1.4.3). Expressing the time derivatives of n_B by the molar balances for B in a reactive system (Equation 1.39) yields

$$mc_p\frac{dT}{dt} + \sum_k\sum_B H_B\,\dot{n}_{B,k} + \sum_j\sum_B\{H_B\nu_{B,j}V\,r_j\} = \dot{Q} + \dot{W}_{st} + \sum_k\sum_B H_{B,k}\,\dot{n}_{B,k}. \tag{1.83}$$

The sum $\sum_B\{\nu_{B,j}H_B\}$ is the enthalpy of reaction $\Delta_{rj}H$ of the reaction j. The balance thus can be written as

$$mc_p\frac{dT}{dt} + \sum_k\sum_B H_B\dot{n}_{B,k} + \sum_j\{\Delta_{rj}H\;\;Vr_j\} = \dot{Q} + \dot{W}_{st} + \sum_k\sum_B H_{B,k}\dot{n}_{B,k}, \tag{1.84}$$

where H_B denotes the partial molar enthalpy of B inside the reactor and $H_{B,k}$ the partial molar enthalpy at the k-th matter exchange port. This equation can also be written as

$$mc_p\frac{dT}{dt} + \sum_j\{\Delta_{rj}H\;\;Vr_j\} = \dot{Q} + \dot{W}_{st} + \sum_k\sum_B (H_{B,k} - H_B)\,\dot{n}_{B,k}. \tag{1.85}$$

The difference between the partial molar enthalpies in the streams entering or leaving B and the partial molar enthalpy of B inside the reactor is often determined only by the temperature difference and may be calculated by $C_{p,B}(T_{in} - T_{out})$. Therefore, the balance turns out to read

$$mc_p\frac{dT}{dt} + \sum_j\{\Delta_{rj}H\;\;Vr_j\} = \dot{Q} + \dot{W}_{st} + \sum_k\sum_B C_{pB}(T_k - T)\dot{n}_{B,k}. \tag{1.86}$$

In practice it is common to use the mass flow rate at each port rather than the molar flowrate of each species and to assume that in place of heat capacities of all chemicals in a stream we may use an average specific heat capacity, $\bar{c}_{p,k}$ so that one obtains

$$m\bar{c}_p\frac{dT}{dt} = \dot{Q} + \dot{W}_{st} + \sum_k \bar{c}_{p,k}(T_k - T)\,\dot{m}_k - \sum_j\{\Delta_{rj}H\;\;Vr_j\}, \tag{1.87}$$

where $\bar{c}_p$ is the average specific heat capacity of material in the reactor.

This derivation could readily be extended for variable pressures by including the pressure dependence of enthalpy $\{(\partial H/\partial p)_{T,n_B}(dp/dt)\}$ on the left-hand side of the enthalpy balance Equation 1.82 and the work $V(dp/dt)$ on the right-hand side. The term $(\partial H/\partial p)_{T,n_B}$ can be expressed by Equation 3.70. If the system is composed of an ideal gas, this is equal to zero. If the system is composed of an incompressible liquid or solid the newly inserted term $\{(\partial H/\partial p)_{T,n_B}(dp/dt)\}$ on the left-hand side of Equation 1.82 can be combined with the $V(dp/dt)$ term on the right and thus becomes $\{-T(\partial V/\partial T)_{p,n_B}(dp/dt)\}$. This expression is zero in incompressible fluids and may thus be disregarded, such that the balance reverts again to Equation 1.82, finally again yielding Equation 1.87.

The balance can be used to predict the temperature evolution of an adiabatically insulated system. Alternatively, it indicates the cooling rate ($\dot{Q}$) that would be required in order to keep the temperature stable. If that cooling rate can be measured online in an isothermal system, it can serve as an online signal for the chemical or biological activity in the system. This may be of practical value if there is only one major chemical reaction, for which the rate of reaction r could then be monitored based on the heat evolution rate $\dot{Q}$ (see for example Chapter 8).

1.5.7 Can One Write Heat Balances?

Equations such as the last equation of the previous question and Equations 1.64 and 1.66 are sometimes indeed viewed as a "heat balances." The left-hand terms are then interpreted as the "accumulation of sensible heat," the first right-hand term of Equations 1.64, 1.66 and 1.87 as "sensible heat flows" exchanged with the environment, and the last terms as "source of heat" from chemical reactions.

In a strict thermodynamic sense, however, only $\dot{Q}$ counts as heat, the other terms are enthalpic ones. Heat may be transferred from the environment to a system, but as soon as it is inside the system it becomes internal energy, just as rain feeding a lake ceases to be rain once it is in the lake.

1.6 Examples of the Application of the First Law of Thermodynamics

In this section we seek to pose and answer several practical and realistic problems using the notions and laws of thermodynamics we have covered so far. As is the case throughout this book, the examples are chosen to illustrate features of the subjects that are often found difficult by students; the list of topics is not exhaustive but is intended to be illustrative.

1.6.1 How Does a Dewar Flask Work?

A *Dewar* flask, shown schematically in Figure 1.6, is a vessel used for maintaining materials at temperatures other than those of the surroundings for a finite duration. This is accomplished by slowing down the heat transfer between the object in the vessel and the surroundings by all of the mechanisms discussed in Question 1.3.7.

First, as can be seen in Figure 1.6, it has a double wall and the space between the walls is evacuated to a very low pressure (less than 1 Pa). At such low pressures, the mean free path, λ, of the gas molecules (diameter σ) that remain is very long $\lambda \approx k_B T/(\pi p \sigma^2)$ and, for the pressure quoted, is greater than the distance between the walls of the vessel (Atkins 2018). As a result, the only mechanism for the transport of molecular energy between one

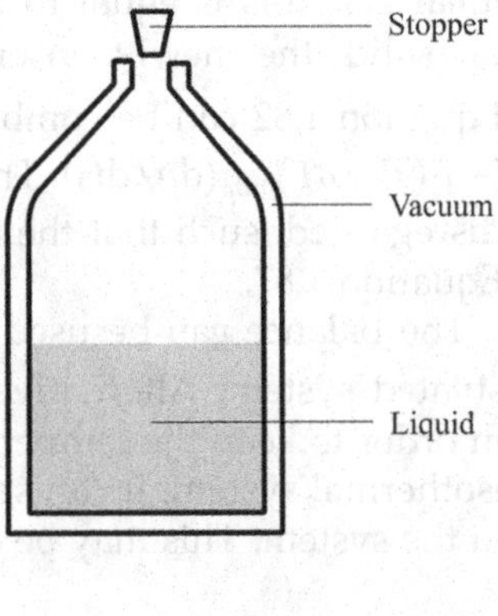

FIGURE 1.6
Dewar flask.

wall and another is associated with the kinetic energy of the molecules that collide with the inner wall and then next collide at the outer wall. This kinetic energy is very small and the number of molecules making the trip per second is also very small so that the heat conducted between the two walls is very small indeed. The heat transported by convection (bulk fluid motion) is similarly reduced. The magnitude of the heat transported by bulk motion must depend upon the heat capacity of the fluid per unit mass, the mass per unit volume and the velocity of the motion. The fact that we have a very low pressure in the evacuated space ensures that the density of the gas is very low and in itself this reduces the convective effects to a very small level irrespective of whether the remaining gas has a significant heat capacity per unit mass or there are convective currents.

Finally, the surfaces of the walls of the vessel inside the evacuated space are coated with silver, which is a weak emitter of radiation (it has a low emissivity) and highly reflective. Thus, neither surface emits very much radiation, according to the Stefan-Boltzmann Law. What it does emit is largely reflected from the opposing surface. Thus, the amount of heat transported by radiation between the object on the inside of the Dewar flask to the surroundings is very small. This inhibition of all three heat transfer processes results in a long delay of approach to thermal equilibrium between the contents of the flask and the surroundings. Thus, the contents of a Dewar flask will remain either hot or cold for a long time.

In laboratories and industry, vacuum flasks are often used to store liquids which become gaseous at well below ambient temperature, such as O_2, which has a normal boiling temperature of 90.2 K at a pressure of 0.1 MPa and N_2 (normal boiling temperature of 77.3 K). It is possible to maintain such materials in the liquid state for several days without the need for expensive refrigeration equipment.

1.6.2 In a Thermally Isolated Room, Why Does the Temperature Go up When a Refrigerator Powered by a Compressor Is Placed Within?

The reader will recall from the earlier discussion that to answer any thermodynamic question the first thing that must be done is to define the system considered. In this case, we define the system to include all the entities and masses contained within the walls of the isolated room (including the air and the refrigerator itself). Anything outside the walls (the boundary of the system) is defined as the *surroundings*. If no mass enters or leaves the system through the walls (including the doors and the windows), the mass in the system remains fixed (does not change with time) and the system is *closed*, as we defined it in Question 1.3.1. In a closed system, only *energy* may be transferred in or out of the system through the boundaries. The First Law of Thermodynamics (or the law of conservation of energy) states that for a time interval Δt, the energy accumulated in the system is equal to the net energy transfer through the system boundaries.

We assume, in our example, that the room is stationary in some reference frame so that the kinetic energy of the system itself is zero and its potential energy is constant and that there are no magnetic or other external forces. The internal energy of the system U is given by

$$U = \sum_i m_i u_i, \tag{1.88}$$

where m_i is the mass of one part of the room or refrigerator of specific material and u_i is the specific internal energy of component i. The sum extends over all components within the system.

We remind the reader now that *heat* is the energy that is transferred between the system and its surroundings and *work* is the energy of interaction between a system and its surroundings as a result of force acting. Thus, a piston compressing a gas, a rotating shaft and an electric wire heated by a current within the system are all examples of *work*. This gives Equation 1.43.

In the room there is no heat transfer through the boundaries because it is thermally isolated and so $Q = 0$. The only work crossing the system boundary is the electrical work, W_{el}, done by the electric current in the wire entering to move the compressor of the refrigerator, which must come from outside in the surroundings. The first law for the isolated room is

$$\Delta U = W_{el}. \tag{1.89}$$

The addition of electric work to the system causes the internal energy to increase and, because of the constant mass, the temperature must also increase with time. This description is neither dependent on the position of the refrigerator door nor on the water vapor content of the air within the room that condenses on the cold refrigerator nor anything else that occurs inside the room. Indeed, it has not been necessary to consider these aspects of the problem essentially because of where we placed the system boundary.

1.6.3 What Is the Best Mode of Operation for a Gas Compressor?

In order to answer the question, we must, of course, first define what "best" means in this context. Most often in engineering the "best" way to compress gas (in most circumstances used to increase the pressure of a flowing gas) is that which requires the minimum work be done on the system. While there are of course many means to compress a gas, all of which could be subject to a similar analysis, we consider here only a reciprocating compressor, illustrated in Figure 1.7. This example has the advantage that the relationship between the forms of work can be connected with both closed and open systems.

In the reciprocating compressor, shown in Figure 1.7, the gas flows into a compression cylinder through an inlet valve. After closure of the valves the gas is compressed by a piston and then discharged after opening of the outlet valve. While the whole process may be regarded as an open system, the actual compression takes place within a closed system. Indeed, this is rather a good illustration of the difference. The total shaft work per unit mass required for the whole process, w_{s12}, consists of the volume work exerted to achieve the compression of the gas, w_{v12}, and the flow work, w_{f12}, for moving the piston when the gas enters the cylinder and again when it is discharged,

$$w_{s12} = w_{v12} + w_{f12}. \tag{1.90}$$

For an ideal (that is reversible) process where

$$(w_{v12})_{rev} = -\int_1^2 p dv, \tag{1.91}$$

we may now calculate the total work required

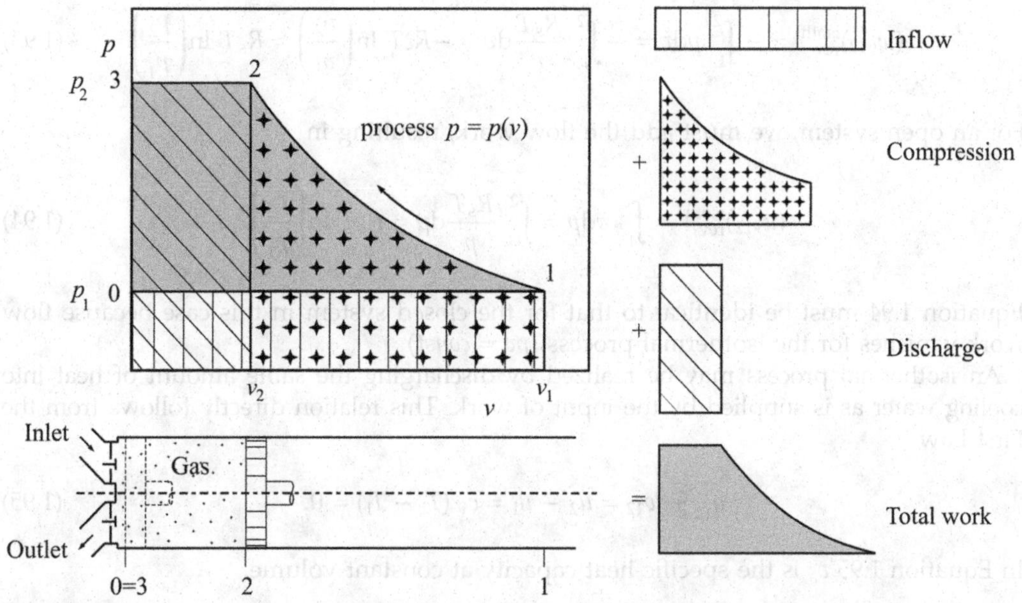

FIGURE 1.7
Scheme of a reciprocating compressor for the illustration of the total work required for the compression of a gas stream: The gas enters into a compression cylinder (0→1, associated with flow work for the displacement of the piston, negative sign), after closure of the valves the gas is compressed (1→2, volume work in a closed system) and then discharged after opening of the outlet valve (2→3, flow work). The total shaft work required for the whole process is the sum of these contributions.

$$(w_{s12})_{rev} = -\int_1^2 p dv + p_2 \cdot v_2 - p_1 \cdot v_1 = \int_1^2 [-p dv + d(pv)] = \int_1^2 (-p dv + v dp + p dv) = \int_1^2 v dp. \quad (1.92)$$

In a (p, v) diagram, the total work may now be illustrated as the area between the ordinate and the line describing the chosen process, as shown in Figure 1.7.

We now turn to the question as to what modes of operation are possible for such a compressor, that is, what are reasonable thermodynamic processes for gas compression. It is obvious that the process can neither be performed in an isobaric nor in an isochoric (constant volume) manner. The processes under question are therefore either an isothermal or an adiabatic process.

1.6.4 What Is the Work Required for an Isothermal Compression?

If, for simplicity, but without detriment to the argument, we restrict ourselves to an ideal gas, the ideal gas law provides the simple relation $pv = RT/M = R_s T$ where R_s is called the specific gas constant and M is the molar mass. For an isothermal process (at constant temperature T), the product of pressure and volume is a constant ($pv = const$). For the ideal, reversible case the volume work for compression of the gas in a closed system is given by

$$(w_{v12})_{rev}^{isoth.} = -\int_1^2 p dv = -\int_1^2 \frac{R_s T}{v} dv = -R_s T \ln\left(\frac{v_2}{v_1}\right) = R_s T \ln\left(\frac{p_2}{p_1}\right). \quad (1.93)$$

For an open system, we must add the flow work, resulting in

$$(w_{s12})_{rev}^{isoth.} = \int_1^2 v dp = \int_1^2 \frac{R_s T}{p} dp = R_s T \ln\left(\frac{p_2}{p_1}\right). \quad (1.94)$$

Equation 1.94 must be identical to that for the closed system in this case because flow work vanishes for the isothermal process ($pv = const$).

An isothermal process may be realized by discharging the same amount of heat into cooling water as is supplied by the input of work. This relation directly follows from the First Law

$$q_{12} + w_{12} = u_2 - u_1 = c_V (T_2 - T_1) = 0. \quad (1.95)$$

In Equation 1.95 c_v is the specific heat capacity at constant volume.

1.6.5 What Is the Work Required for an Adiabatic Compression?

Again, we consider the reversible work in connection with the compression of an ideal gas. The process is reversible and adiabatic and is characterized by a property called the isentropic (expansion) exponent κ, defined by $\kappa = -vp^{-1}(\partial p/\partial v)_s$ (cf with Questions 3.5.2 and 3.5.6).

For an ideal gas κ is equal to γ, the ratio of the isobaric and isochoric heat capacities, $\gamma = c_p/c_V$. For our derivation, we additionally assume that γ is constant with temperature. This statement strictly only holds for monatomic gases but is a good approximation also for polyatomic gases because of the similar dependence of c_p and c_V on temperature. From the first law in a differential form,

$$\delta q + (\delta w)_{rev} = du, \quad (1.96)$$

and with $\delta q = 0$ and $(\delta w)_{rev} = -pdv$, we obtain

$$du + pdv = 0. \quad (1.97)$$

For the ideal gas, $du = c_V dT, p = RT/(Mv), c_p - c_V = R/M$, so that Equation 1.97 becomes

$$\frac{dT}{T} + (\gamma - 1)\frac{dv}{v} = 0. \quad (1.98)$$

Integration from state 1 to state 2 yields

$$\ln\left(\frac{T_2}{T_1}\right) + (\gamma - 1)\ln\left(\frac{v_2}{v_1}\right) = 0, \quad (1.99)$$

or

$$T_2\, v_2^{\gamma-1} = T_1\, v_1^{\gamma-1}, \tag{1.100}$$

and with $T = M\, p\, v/R = p\, v/R_s$, we obtain

$$p_1\, v_1{}^{\gamma} = p_2\, v_2{}^{\gamma} = p\, v^{\gamma}. \tag{1.101}$$

This equation generally describes the relation between p and v for a reversible and adiabatic process of an ideal gas. We can finally combine the preceding equations to find a relation between p and T for such a process

$$T_1\, p_1{}^{(\gamma-1)/\gamma} = T_2\, p_2{}^{(\gamma-1)/\gamma}. \tag{1.102}$$

From the first law for an adiabatic process, we then obtain the work for the ideal gas in a closed system of

$$(w_{v12})_{\text{rev}}^{\text{adiab.}} = u_2 - u_1 = c_V\,(T_2 - T_1) = c_V\, T_1\left(\frac{T_2}{T_1} - 1\right), \tag{1.103}$$

with $c_V = R/M\,(\gamma - 1)$ and the ideal gas law, this expression can be rewritten in the form

$$(w_{v12})_{\text{rev}}^{\text{adiab.}} = \frac{p_1 v_1}{\gamma - 1}\left(\frac{T_2}{T_1} - 1\right), \tag{1.104}$$

or, using Equation 1.101 into

$$(w_{v12})_{\text{rev}}^{\text{adiab.}} = \frac{p_1 v_1}{\gamma - 1}\left[\left(\frac{p_2}{p_1}\right)^{(\gamma-1)/\gamma} - 1\right]. \tag{1.105}$$

Similarly, the work for an open system with an ideal gas is

$$(w_{s12})_{\text{rev}}^{\text{adiab.}} = h_2 - h_1 = c_p\,(T_2 - T_1) = c_p\, T_1\left(\frac{T_2}{T_1} - 1\right). \tag{1.106}$$

Equation 1.106 can, with, $c_p = R_s\gamma/(\gamma - 1)$, be cast as

$$(w_{s12})_{\text{rev}}^{\text{adiab.}} = \frac{\gamma\, p_1 v_1}{\gamma - 1}\left(\frac{T_2}{T_1} - 1\right) = \frac{\gamma\, p_1 v_1}{\gamma - 1}\left[\left(\frac{p_2}{p_1}\right)^{(\gamma-1)/\gamma} - 1\right]. \tag{1.107}$$

Having answered the questions for the work required in the most relevant processes (isothermal and adiabatic), we can now generalize it to the case of a polytropic process, which is characterized by the relation $pv^n = const.$ with the polytropic exponent n, with little extra effort.

A compression is really performed in a manner that lies between the idealized cases of an isothermal (with $n = 1$) and an adiabatic (with $n = \gamma$) process. Rather than comparing the individual equations for the work required, we consider the relations for the shaft work with reversible processes,

$$(w_{v12})_{rev} = -\int_1^2 p dv \tag{1.108}$$

and

$$(w_{s12})_{rev} = \int_1^2 v dp, \tag{1.109}$$

respectively, and view the (p, v) diagram for the cases of interest.

The most relevant and familiar case is that of the compression of a flowing gas stream from pressure p_1 to pressure p_2. For such an open system, the total (shaft) work is the quantity of interest, which for an ideal (i.e. reversible) process may be obtained as the area between the respective process (path) and the ordinate (p axis) from a (p, v) diagram, shown in Figure 1.8a. Because $(\gamma > 1)$, the magnitude of $|dp/dv|$ is greater for an adiabatic than for an isothermal process, as shown in Figure 1.8, and more work is required in the adiabatic case. For practical purposes, this result implies that a gas compressor should indeed be operated as close as possible to the isothermal case and therefore requires efficient heat exchange.

If the task, however, is to compress a gas in a closed system such as a cylinder, the answer to our initial question for the best mode of operation may be different. The result depends upon whether the gas is to be brought to a defined pressure or to a defined specific volume (the latter case does not make sense for an open system). In this case, we have to consider the volume work for the two processes, which may be identified in a (p,v) diagram as the area between the respective process and the abscissa (v axis, Figure 1.8b and 1.8c). Whereas for the compression to a defined pressure the adiabatic process is the right choice, things reverse when it comes to reaching a defined smaller specific volume, where the isothermal process is to be preferred.

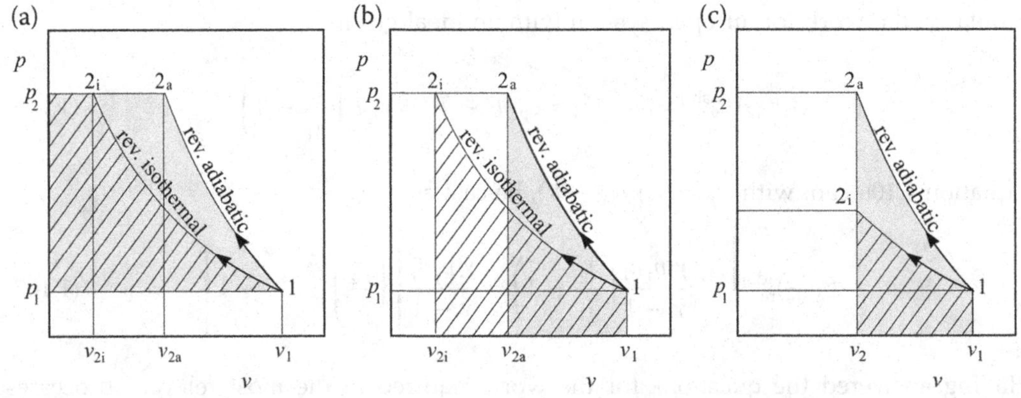

FIGURE 1.8
Illustration of reversible work required for gas compression as a function of boundary conditions for reversible adiabatic and isothermal compression: (a), for an open system where the total work is defined by the area between the p axis and the respective process. (b) and (c), for a closed system where the relevant quantity is volume work obtained from the area between the process and the v axis; the compression method, be it adiabatic or isothermal, depends on if a specific pressure is to be achieved as shown in (b) or to be reached as shown in (c).

1.7 How Are Thermodynamic Properties Measured?

Thermodynamics is an experimental science; almost all the properties that we have discussed so far and that occur in thermodynamics and transport phenomena must be measured by experimental means. Molecular simulation or theoretical calculation based on molecular physics can provide estimates of thermophysical properties that are usually significantly less accurate than the measured values. For very few systems such as helium, quantum mechanical calculations and the fundamental constants have been used, *ab initio*, to determine the pair interaction potential energy. When combined with the methods of statistical mechanics (Chapter 2), this potential provided estimates of the thermophysical properties of helium in the low-density gas phase with an estimated uncertainty less than that obtained from measurements. Thermophysical properties calculated by this approach have been used to provide data for instruments used for measurements on other materials and to form a standard for pressure. However, for most other molecules and atoms even at low density, these calculations have yet to be done with sufficient precision to be able to replace measurement; at higher densities in the gas and in the liquid phase there is little likelihood that such calculations will be performed in the near future. For that reason, the current reliance on experimental determination of properties will persist for a considerable time.

Techniques used to measure thermophysical properties can be found in the series *Experimental Thermodynamics* (Vol. I 1968, Vol. II 1975, Vol. III 1991, Vol. IV 1994, Vol. V 2000, Vol. VI 2003, Vol. VII 2005, Vol. VIII 2010 and Vol IX 2014), which also includes two volumes concerned with equations of state for fluids and fluid mixtures that are discussed in Chapter 4. The first of these two volumes (Vol. V 2000) has been updated in 2010 and (Volume VIII 2010) places a greater emphasis on the application of theory. The latter volume specifically includes theoretical and practical information regarding equations of state for chemically reacting fluids and methods applicable to nonequilibrium thermodynamics than hitherto provided in Volume V. However, computer simulations for the calculation of thermodynamic properties were omitted from Volume VIII because the subject requires an in-depth coverage such as that given in a special issue of *Fluid Phase Equilibria* (Case et al. 2008; Eckl et al. 2008; Ketko et al. 2008; Li et al. 2008; Müller et al. 2008; Olson and Wilson 2008) or the comprehensive monograph by Rapaport (2004). The problem of evaluating the thermodynamic properties for industrial use by means of calculation and simulation is treated in other publications (Case et al. 2004; 2005; 2007 as well as, for example, Jimenez-Serratos et al. 2019 and Jin and Firoozabadi 2016).

The monographs in the series *Experimental Thermodynamics* (Vol. I, 1968; Vol. II, 1975; Vol. III 1991; Vol. IV 1994; Vol. V 2000; Vol. VI 2003; Vol. VII 2005; Vol. VIII 2010, Vol IX 2014) were published under the auspices of the International Union of Pure and Applied Chemistry (IUPAC) and since 2004 in association with the International Association of Chemical Thermodynamics (IACT) that is an affiliate of IUPAC. Throughout this text we have adopted the quantities, units and symbols of physical chemistry defined by IUPAC in the text commonly known as the *Green Book*. We have also adopted where possible the ISO guidelines for the expression of uncertainty (Evaluation of Measurement – Guide to the Expression of Uncertainty in Measurement 2008) and vocabulary in metrology (International Vocabulary of Basic and General Terms in Metrology 2012). Values of the fundamental constants and atomic masses of the elements have been obtained from BIPM (2019).

The series *Experimental Thermodynamics* is complemented by other publications associated with IUPAC and IACT that have covered a range of diverse issues reporting applications of solubility data (Developments and Applications of Solubility 2007), to the topical issue of alternate sources of energy (Future Energy: Improved Sustainable and Clean Options for our Planet 2008) and the application of chemical thermodynamics to other matters of current industrial and scientific research including separation technology, biology, medicine and petroleum in one (Chemical Thermodynamics 1999), and heat capacity measurements in another (Heat Capacities: Liquids, Solutions and Vapours 2010).

1.7.1 What Is the SI System of Units?

There is a globally agreed system of units that has been known as the International System of Units (SI) since 1960. It covers every conceivable form of measurement so that the number of units in practical use is very large. However, there are just seven base units. They are the kilogram (kg) for mass, the meter (m) for length, the second (s) for time, the ampere (A) for electric current, the kelvin (K) for temperature, the mole (mol) for the amount of substance and the candela (cd) for luminous intensity. In the past, some of these units were defined relative to physical objects or properties of materials, established by international agreement and held securely in a specific location. For example, the internationally agreed standard for the kilogram was realized by the International Prototype Kilogram manufactured from a platinum-iridium alloy and held in France.

Over long periods of time, the mass of such a block of material will change (albeit very slightly) so that there has always been a desire to move away from a system of units based upon artifacts. In fact, the kilogram was the last of the units realized in this way. Science has now advanced so much that it has been possible to measure some of the fundamental constants of nature so accurately that they can now provide the basis for the definition of the base units of the SI system. An excellent overview on the development of the new SI system is given by Göbel and Siegner (2019).

From 20 May 2018, the SI units have been defined in terms of constants of nature:

- The unperturbed ground state hyperfine transition frequency of the caesium-133 atom, Δv, is exactly 9,192,631,770 hertz
- The speed of light in vacuum, c, is exactly 299,792,458 meters per second
- The Planck constant h is exactly 6.626,070 15 × 10^{-34} joule second
- The elementary charge e is exactly 1.602,176,634 × 10^{-19} coulomb
- The Boltzmann constant k_B is exactly 1.380,649 × 10^{-23} joule per kelvin
- The Avogadro constant L or N_A is exactly 6.022,140,76 × 10^{23} reciprocal mole
- The luminous efficacy of monochromatic radiation of frequency 540 × 10^{12} hertz K_{cd}, is exactly 683 lumen per watt

So far as thermodynamics is concerned, the most often used of these units are the mole, the kilogram, the meter, the second and the kelvin. The formal definitions of each unit are as follows:

One **mole** is defined to contain 6.022 140 76 × 10^{23} entities, such as atoms or molecules.

The **kilogram** is defined by taking the fixed numerical value of the Planck constant, h, to be 6.626 070 15 × 10^{-34} when expressed in the unit J s, which is equal to kg m^2 s^{-1}, where the meter and the second are defined in terms of the speed of light, *c*, and the hyperfine transition frequency of caesium-133, $\Delta\nu$, respectively.

The **meter** is defined by taking the fixed numerical value of the speed of light in vacuum, *c*, to be 299,792,458 when expressed in the unit m s^{-1}, where the second is defined in terms of the caesium-133 atom frequency $\Delta\nu$.

The **second** is defined by taking the fixed numerical value of the caesium frequency $\Delta\nu$, the unperturbed ground-state hyperfine transition frequency of the caesium-133 atom, to be 9,192,631,770 when expressed in the unit Hz, which is equal to s^{-1}.

The **kelvin** is defined by taking the fixed numerical value of the Boltzmann constant k_B to be $1.380\ 649 \times 10^{-23}$ when expressed in the unit J K^{-1}, which is equal to kg m^2 s^{-2} K^{-1}, where the kilogram, meter and second are defined in terms of *h*, *c* and $\Delta\nu$.

The kelvin is, in many senses, the most important unit for thermodynamics because all substances are composed of atoms and, fundamentally, temperature is a measure of the average energy of the motion of the atoms within an object. We see then that the kelvin is defined in terms of this microscopic motion and is based on the fundamental constant known as the Boltzmann constant that measures how much energy of motion corresponds to one kelvin. This definition has the advantage that any fixed point can be used as a standard temperature, and any appropriate method for temperature measurement could be used. This allows for the possibility of improved uncertainty of temperature measurement at extremely high and extremely low temperatures.

1.7.2 How Is Temperature Measured?

The equations of thermodynamics are written using for the temperature what is known as the *thermodynamic temperature* along with a thermodynamic temperature scale. The thermodynamic temperature is defined (Kestin 1965, McGovern 2004, MIT 2021) using the properties of the reversible Carnot cycle discussed in Chapter 3. There it is shown that the efficiency of a reversible heat engine operating on a Carnot cycle is independent of any property of the working fluid in the heat engine or how exactly the engine is constructed. This universal feature allows the identification of a universal function which relates the heat absorbed and rejected by a reversible Carnot cycle (see Equation 3.3), that function is known as the *thermodynamic temperature*. It can further be shown that this thermodynamic temperature function is exactly the same temperature that occurs in the perfect gas equation of state (Kestin 1965, MIT 2021). It is then a result of the statistical mechanical treatment of the perfect gas model (see Chapter 2) that relates the energy of one molecule to the product $k_B T$ and thus to the perfect gas equation of state. The identification of the numerical value of the Boltzmann constant, k_B, in the previous question then determines the thermodynamic temperature uniquely from among the possible functions. A unique thermodynamic temperature scale then results. Until very recently, the practical use of the thermodynamic temperature has been limited by the difficulty of realizing simple instruments for its measurement. This situation is now changing, and the complete story of the evolution of the most recent definition and realization of thermodynamic temperature is summarized in a special issue of the Philosophical Transactions of the Royal Society (Machin et al. 2016), which provides access to the vast amount of work that lies behind this development.

In principle, because many properties of substances change with changing temperature, any of them and any substance can be used as a thermometer. Two primary thermometers have been the gas thermometer, exploiting the changes of pressure and volume of a perfect gas with temperature using the equation of state of the perfect gas and the identity between the thermodynamic temperature and that of the perfect gas,

$$T_2/T_1 = \lim_{p \to 0}\{p_2 V_2/(p_1 V_1)\}, \tag{1.110}$$

and acoustic thermometers exploiting the speed of sound, c, in a gas through the equation

$$T_2/T_1 = \lim_{p \to 0}(c_2^2/c_1^2). \tag{1.111}$$

However, more practical (secondary) thermometers are more commonly encountered. They include mercury-in-glass thermometers, which exploit the thermal expansion of mercury, platinum resistance thermometers, which exploit the resistivity change of platinum with temperature, thermocouples or thermistors. The last three require the measurement of a resistance or voltage rather than direct visual observation. All of these devices require calibration against a primary thermometer or a set of accepted standard reference points so that the temperature they report can be related to the thermodynamic temperature. The choice of instrument is determined by the precision required in the temperature and the range in which it is required. The interpretation of the resistivity of platinum in terms of thermodynamic temperature is achieved by the use of the International Temperature Scale of 1990 (ITS-90) (Preston-Thomas 1990; Nicholas 2003, BIPM 2019).

In materials science (e.g. for solids and nanoparticles) but also in biology and medicine (e.g. for micelles, cells, tissues and biofilms), one is interested in non-contact temperature measurements with a high spatial resolution in the nanometer range. Optical temperature sensors often called "molecular thermometers" are currently being developed for this purpose. Thereby, the temperature dependence of fluorescence or luminescence is exploited. For example, one considers soluble organic dyes that are excited with light of a certain wavelength and then emit light of lower energy (e.g. in the infrared range). The ratio of the emission taken within two different wavelength bands is temperature-dependent and, after calibration, can be used for temperature measurement with a spatial resolution of the same order of magnitude as the applied "molecular thermometer." Open questions for many applications are still the low sensitivity, biocompatibility and suitable nano-carriers (see for example Otto et al. 2017).

1.7.3 How Is Pressure Measured?

Pressure can be measured with a piston or "dead-weight" gauge. In this instrument, the pressure to be measured is applied to the base of a piston of known effective cross-sectional area contained within a close-fitting cylinder. Masses are added to a carrier connected to the piston so as to balance the pressure. The force exerted by the masses is determined from the local acceleration owing to gravity and the pressure is determined from the known cross-sectional area of the piston. This experiment is far from routine and is very time-consuming and delicate. Thus, the majority of pressure measurements are obtained from transducers that have been calibrated against deadweight gauges. These transducers usually determine the mechanical strain induced by the applied pressure with an appropriately located resistive strain gauge and a Wheatstone bridge or from variations in the resonance frequency of a quartz object. All methods of pressure measurement have been extensively reviewed elsewhere (Suski et al. 2003).

1.7.4 What Is the Difference between Uncertainty and Accuracy?

It is common in the literature for the words *accuracy* and *uncertainty* to be used interchangeably but there is a difference between them that is significant and vital. The term "accuracy of measurement" has the internationally agreed definition that, paraphrased, states it is the difference between the measured and the true values and is a hypothetical term because in most circumstances the true value is not known. The phrase "uncertainty of measurement" defines the range of values of the result within which it is reasonable with a cited statistical confidence the value will lie. This is achieved without recourse to the assumption of a true value. On the basis of these definitions, the vast majority of measurements are uncertain and not accurate. Those interested in this topic should also refer to the NIST Technical Note 1297 (Taylor and Kuyatt 1994) and the Evaluation of Measurement – Guide to the Expression of Uncertainty in Measurement (2008).

1.8 Calorimetry

Calorimetry is the science of measuring heat associated with physical, chemical or biological processes. These processes can either release (exothermic) or absorb heat (endothermic). We distinguish between direct calorimetry, which is explained in the following questions, and indirect calorimetry. In indirect calorimetry, the amount of heat is calculated indirectly from the measured oxygen consumption. For this purpose, a relationship between heat evolution and oxygen consumption, called the oxycaloric equivalent (–430 to –480 kJ mol^{-1}), is exploited (Gnaiger and Kemp 1990). This is particular useful for large organisms such as humans! Direct calorimetry is performed with a calorimeter. The word *calorimetry* is derived from the Latin word *calor*, meaning "heat" and the Greek word *μέτρον* (metron), meaning measure.

1.8.1 How Are Energy and Enthalpy Differences Measured?

Unfortunately, the absolute value of the energy U cannot be measured directly; only the difference between two states $\Delta U = U_2 - U_1$ can be determined with a calorimeter. A calorimeter for measuring differences of energy and enthalpy of a system as a function of state variables such as T, V and p is an enclosed container in which work is done to change the state of a material and heat as well as work is exchanged according to the first law (Equation 1.41). The total work includes electrical work, any volume work and other forms of work, W_o, such as stirring the contents of the calorimeter. The electrical work, $W_{el} = I^2Rt$, is usually obtained by passing a constant current I through an electrical resistance R within the system for a measured time t. Consequently, the working equation for the calorimeter becomes

$$\Delta U = W_{el} - \int p\mathrm{d}V + W_o + Q. \tag{1.112}$$

If the calorimeter volume is held constant by rigid walls, then $\mathrm{d}V = 0$ and, if the system is adiabatically enclosed, then $Q = 0$ so that Equation 1.112 becomes

$$\Delta U = W_{el} + W_o. \tag{1.113}$$

This means that the internal energy difference between two states can be determined via electrical work, taking into account that the work of, for example, the stirrer can be neglected or is measured.

If the pressure in the calorimeter is maintained equal to that of the surroundings and the calorimeter walls are still adiabatic but not rigid (so that the volume of the system changes from V_1 to V_2), then Equation 1.112 becomes

$$\Delta U = U_2 - U_1 = W_{el} - pV_2 + pV_1 + W_o. \tag{1.114}$$

On rearrangement, Equation 1.114 is

$$\Delta(U + pV) = (U_2 + pV_2) - (U_1 + pV_1) = W_{el} + W_o, \tag{1.115}$$

or in terms of the enthalpy, H, Equation 1.115 becomes

$$\Delta H = W_{el} + W_o, \tag{1.116}$$

and we see that the enthalpy difference between two states can also be determined with a calorimeter.

1.8.2 How Is the Energy or Enthalpy Change of a Chemical Reaction Measured?

For an adiabatically enclosed calorimeter of constant volume ($Q = 0$, $\Delta V = 0$, $W = 0$) that contains reactants at initial temperature, T_i and extent of reaction, ξ_i, and continues to temperature T_f and extent of reaction, ξ_f, the energy change can be written as

$$\Delta U_1 = U(T_f, V, \xi_f) - U(T_i, V, \xi_i) \approx 0. \tag{1.117}$$

The internal energy has not changed because the effect of the chemical reaction has been compensated by an increase (or a reduction) in temperature. In order to separate the effect of the chemical reaction from the (compensating) effect of the temperature change, the experiment is repeated with the chemical conversion blocked at ξ_f, but by heating the calorimeter up electrically. (In case of an exothermic reaction, the reaction mixture must first be cooled down to T_i.) As a result,

$$\Delta U_2 = U(T_f, V, \xi_f) - U(T_i, V, \xi_f) \approx W_{el}. \tag{1.118}$$

Subtracting Equation 1.118 from Equation 1.117 yields the internal energy change

$$\Delta U = U(T_i, V, \xi_f) - U(T_i, V, \xi_i) + 0 \approx -W_{el}. \tag{1.119}$$

From the measured internal energy change, the internal reaction energy, $\Delta_r U$, may be determined as

$$\Delta_r U = \Delta U / (\xi_f - \xi_i) \tag{1.120}$$

provided $\Delta_r U$ may be assumed constant.

A similar treatment for enthalpy yields

$$\Delta_r H = \Delta H/(\xi_f - \xi_i). \tag{1.121}$$

1.8.3 How Are Heat Capacity and Phase Transitions Measured?

Consider a system comprising a substance contained within an adiabatic calorimeter (so that $Q = 0$) when the temperature is changed from T_1 to T_2 while the sample volume is held constant (so that $\int p\mathrm{d}V = 0$) and only electrical work, W_{el}, is done so that no other external work is done. The experiment is performed first with the substance in the calorimeter to determine the electrical work necessary to achieve the prescribed change of temperature and then again to determine the electrical work required to change the temperature of solely the calorimeter from T_1 to T_2 without the substance.

The electrical work required to increase the temperature of the substance from T_1 to T_2 is therefore

$$\Delta U = W_{el}(\text{sample + calorimeter}) - W_{el}(\text{calorimeter}). \tag{1.122}$$

Because ΔU can be written as

$$\Delta U = \int_{T_1}^{T_2} (\partial U/\partial T)_V \mathrm{d}T, \tag{1.123}$$

then with the definition

$$C_V = (\partial U/\partial T)_V \tag{1.124}$$

for the heat capacity at constant volume, Equation 1.124 becomes

$$\int_{T_1}^{T_2} C_V \mathrm{d}T = \Delta U = W_{el}(\text{sample + calorimeter}) - W_{el}(\text{calorimeter}). \tag{1.125}$$

If the heat capacity is independent of temperature, then

$$C_V = \{W_{el}(\text{sample + calorimeter}) - W_{el}(\text{calorimeter})\}/(T_2 - T_1), \tag{1.126}$$

which shows how the heat capacity can be measured.

However, in practice, the pressure required to maintain the volume of a sample constant when it is either a solid or a liquid sample under a temperature change requires a container constructed from a material that has a volume independent of temperature and one that is also rigid. The container would require a linear thermal expansion of zero (which is impractical if not impossible except over a limited temperature range) and, for its construction, either a material of unrealizable elastic properties or with very thick walls. The latter implies a mass much greater than the sample so that most of the heat capacity and work done would be that of the container and the effect of the sample would be rather lost in the experiment.

To see the problem clearly, we give a comparison. The pressure increase at constant volume is given by $(\partial p/\partial T)_V$ and, for a liquid hydrocarbon, it is about 1 MPa K^{-1} so that a 10 K temperature increase gives rise to a pressure increase of 10 MPa. For a gas, $(\partial p/\partial T)_V$ is about 0.001 MPa K^{-1} and a 10 K temperature increase results in a pressure change of only 0.01 MPa which is more easily contained. Thus, the mass of container required to maintain a zero-volume change is much less for a gas, however, the heat capacity of a gas is correspondingly lower than that for a liquid and so the mass of the container is still about 100 times greater than that of the sample. These experimental difficulties, which result in an unacceptable uncertainty, require the measurements to be done at constant pressure rather than at constant volume and, therefore, to be of enthalpy differences rather than energy differences. Thus, we have

$$\Delta H = \int_{T_1}^{T_2} (\partial H/\partial T)_p \mathrm{d}T = W_{\mathrm{el}}(\text{sample} + \text{calorimeter}) - W_{\mathrm{el}}(\text{calorimeter}) \quad (1.127)$$

which, in view of the definition of the heat capacity at constant pressure, C_p, of

$$C_p = (\partial H/\partial T)_p, \quad (1.128)$$

can be written as

$$C_p = \{W_{\mathrm{el}}(\text{sample} + \text{calorimeter}) - W_{\mathrm{el}}(\text{calorimeter})\}/(T_2 - T_1), \quad (1.129)$$

if it is assumed that C_p is independent of temperature over the range. The heat capacity at constant volume, C_V, can then be obtained from C_p with Equation 1.159, that is equivalent to Equation 3.102. However, for a gas, the experiment defined by Equation 1.129 is a very complex and demanding one and yields the heat capacity only with a high uncertainty. Thus, for gases, an alternative method is required, and one is described in Question 1.8.6 as a means of demonstrating an application of the First Law of Thermodynamics.

The heat capacity of a gas can also be determined from measurements of the speed of sound, as outlined in Chapter 3. This method has the special advantage that is independent of the amount of substance in the sample, but it is outside the scope of this text and the reader is referred to Goodwin and Trusler (2003 and 2010) for the details.

Often, however, the heat required to increase the temperature of a sample is not constant over a range of temperature. In material sciences or even in biology, the type and temperature of phase transitions influencing the measurement of the heat absorption or release are important. Figure 1.9 shows the pattern of heat absorption in some cases. Examples are melting/crystallization processes, glass transition (especially for plastics), folding and unfolding of proteins and decomposition of sensitive material.

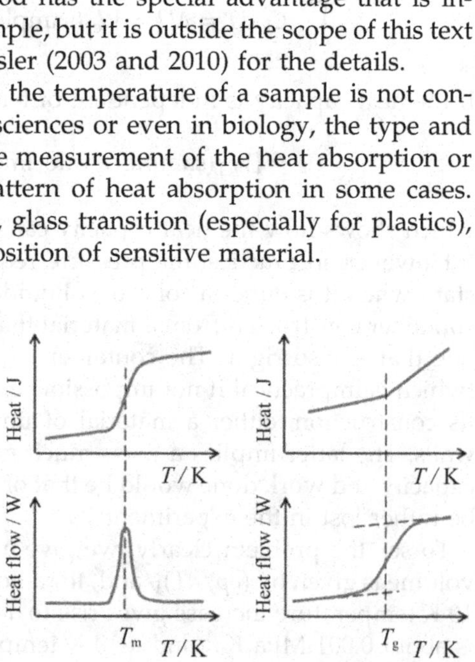

FIGURE 1.9
Influence of melting (left) and glass transition (right) on the heat and heat flow as observed with a DSC.

To study such processes a differential scanning calorimeter **(DSC)** is applied, which measures differentially the heat flow into a sample and compares it with a reference material as a function of temperature.

1.8.4 How Do I Measure the Energy in a Food Substance?

Here we describe the methods used to determine the energy in a food substance and this value is reported on the container of processed food. This energy can be measured by completely burning the substance in the presence of an excess of oxygen. The evolved heat is measured in an adiabatic bomb calorimeter (Figure 1.10). In this reactor, the heat evolved during the combustion is absorbed by a known mass of water that surrounds the calorimeter, resulting in an increase of the water temperature that can be measured. A description of this apparatus is given below for the emotive example of a chocolate bar.

Consider a chocolate bar consisting of peanut butter, nougat topped with roasted peanuts and caramel covered with milk chocolate. It is well known that such a bar contains substantial food energy. *Food energy* is the energy in the food available through digestion. The values for food energy are found on all commercially available processed food. The material within the food comprises large organic molecules that are broken down into smaller molecules by digestion. Some of these molecules are used by the body to build complicated molecules necessary for the body's function. Others are metabolized (burned) with the oxygen we breathe in from air. The products of complete combustion are CO_2 and H_2O and the energy of combustion $\Delta_c U$. It is the $\Delta_c U$ that is used to power the body including both physical and mental activity. The amount of energy in a substance (including food) can be measured by completely burning the substance in the presence of excess oxygen within a bomb calorimeter shown in Figure 1.10. It is nothing more than a plausible assumption that the heat of complete combustion of a substance can be equated to the *food* energy.

The "bomb" is actually a high-pressure vessel, usually made of steel, immersed in a water bath. The bomb is designed to change its volume by a negligible amount when the pressure inside it changes. The temperature of this water is continuously monitored with a high precision thermometer. The water is itself contained in a Dewar flask (Question 1.3.8) that prevents heat flow from the water to the surroundings.

The sample of the chocolate bar is dried and then ground into a powder and placed on a sample container inside the constant volume "bomb". The bomb is then charged with a supply of oxygen up to a pressure of about 2.5 MPa so that there is adequate oxygen for complete combustion of the sample. The sample is then ignited electrically. The heat evolved is transferred to the bomb and the water surrounding it leading to a temperature

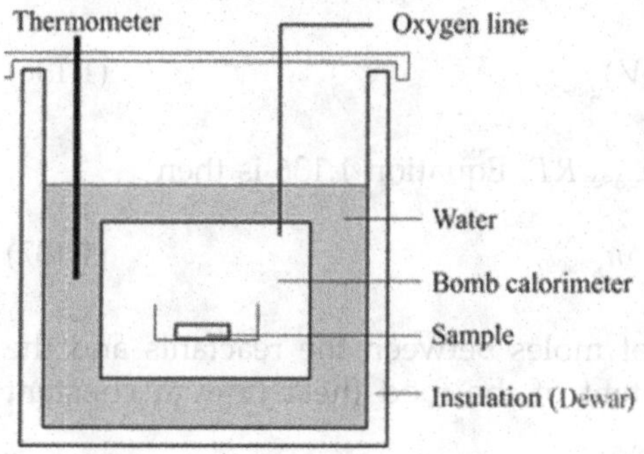

FIGURE 1.10
Bomb calorimeter.

increase of both. Owing to the thermal isolation provided by the Dewar flask, the food sample and the oxygen can be taken as a closed system and the bomb and the water as the surroundings and assuming that there is no thermal exchange out of the Dewar flask so that

$$\Delta U_{\text{tot}} = \Delta U_{\text{sys}} + \Delta U_{\text{sur}} = 0, \tag{1.130}$$

and

$$\Delta U_{\text{sys}} = -\Delta U_{\text{sur}}. \tag{1.131}$$

The First Law of Thermodynamics for closed systems is given by Equation 1.41 and a process conducted at constant volume in the closed system has $dV = 0$ and the work done, $W = \int p dV = 0$, so that

$$Q = \Delta U_{\text{sys}}. \tag{1.132}$$

The heat transferred from the system to the surroundings consisting of the bomb container and the water is given by

$$\Delta U_{\text{sur}} = (mc)_{\text{bomb}}\Delta T + (mc)_{\text{water}}\Delta T, \tag{1.133}$$

where c is the specific heat of each component, m is its mass and ΔT is the temperature increase of the bomb wall and that of the water. The energy change of the system is obtained from

$$\Delta U_{\text{sys}} = (mc)_{\text{bomb}}\Delta T + (mc)_{\text{water}}\Delta T, \tag{1.134}$$

so that the change of the internal energy of the closed system can be determined from the temperature change measured in the water and knowledge of the masses and heat capacities of the bomb and the water. The heat capacity of the calorimeter can be determined from measurements with a substance for which the heat capacity is known precisely for example benzoic acid. However, we are interested in the change in enthalpy of the food in the process; it is given by

$$\Delta H = \Delta U + \Delta(pV). \tag{1.135}$$

The volumes of the solids and liquids and their changes are small compared to those of the gases in the bomb, so we assume $\Delta(pv) \approx 0$ for both and Equation 1.135 becomes

$$\Delta H = \Delta U + \Delta(pV)_{\text{gases}}, \tag{1.136}$$

and assuming the gases are ideal so that $pV = n_{\text{gases}}RT$, Equation 1.136 is then

$$\Delta H = \Delta U + RT\Delta n_{\text{gases}}, \tag{1.137}$$

where Δn_{gases} is the difference of the number of moles between the reactants and the products in the gas. Thus, ΔH, the energy content of the food (heat flow at constant pressure), can be determined from ΔU (heat flow under constant volume) plus the pV work done under constant pressure conditions.

1.8.5 How Do I Measure the Energy Transitions during Metabolic Processes?

Metabolic processes are always associated with heat exchange with the environment. They are usually exothermic but can also be endothermic (Liu et al. 2001). The calorimetric monitoring of this heat exchange provides real-time information on the amount of biomass, its status and reproductive capacity and on the metabolic reaction sequence (Maskow and Paufler 2015). This is important, for example, in medicine, pharmacy and the food industry, where very small bacterial contaminations must be detected at an early stage, or in biotechnology where, conversely, high cell concentrations should convert educts into products with high space-time yields. The first application requires calorimeters which accurately sense very small amounts of heat. Since the metabolism is often strongly dependent on temperature, isothermal micro-calorimeters (**IMCs**) or Calvet calorimeters are preferentially used for such applications. In IMC or Calvet-type calorimeters, the thermoelectric effect is exploited, which correlates a heat flux $\dot{Q}$ detected by a thermoelectric device called a Peltier element with a Seebeck coefficient π_{AB} yielding a measurable current, I, or voltage. An equation describing this type of calorimeter written for a measured current is obtained if Equation 1.85 is simplified for isothermal operation in a closed system and assuming a fixed metabolic stoichiometry,

$$\Delta_r HVr = \dot{Q} \approx -\dot{W}_{el} = \pi_{AB} I. \tag{1.138}$$

The second application requires a calorimeter that can enclose a (bio-)reactor. In such a (bio-)reaction calorimeter, the temperature is kept constant by the continuous removal of the reaction heat using a cooling device with temperature T_j or by balancing the heat removal. The first type of reaction calorimeter is called a heat flow calorimeter (**HFC**) and the second a heat balance calorimeter (**HBC**).

In the first case for HFC, we have

$$\Delta_r HVr = |\dot{Q}| = h_{HT} A (T_i - T_j), \tag{1.139}$$

where h_{HT}, A, T_i and T_j are the overall heat transfer coefficient, the heat transfer area, the temperature of the reaction mixture and the temperature of the cooling jacket, respectively. It is essential that the cooling jacket of the HFC is able to respond rapidly in order that it retains the temperature of the contents of the calorimeter and thus the internal energy constant. The flow rate of the cooling fluid must be high enough to avoid a significant temperature gradient inside the jacket.

In the second case for HBC

$$\Delta_r HVr = |\dot{Q}| = \dot{m} c_p (T_{in} - T_{out}), \tag{1.140}$$

where $\dot{m}$, c_p, T_{in}, T_{out}, the mass flow of the transfer fluid, the specific heat capacity of the transfer fluid and the temperature of the transfer fluid in the inflow and the outflow of the cooling device. The HBC must be able to vary T_{in} rapidly so as to maintain the temperature of the contents of the calorimeter and thus the internal energy constant.

Variations of the heat flow techniques are power compensation calorimetry (**PCC**) and constant flux calorimetry (**CFC**). Temperature constancy is then ensured by controlling

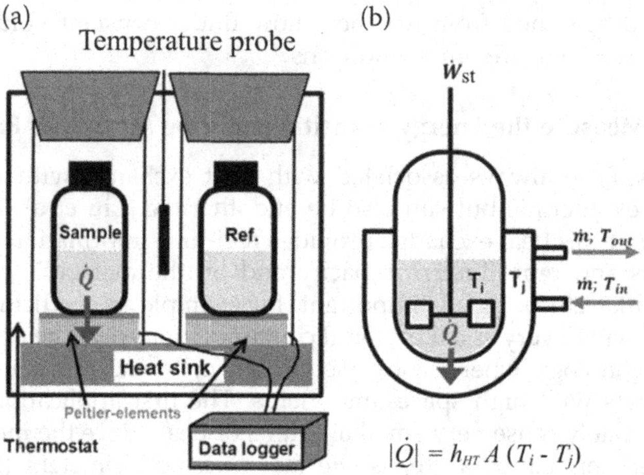

$|\dot{Q}| = \pi_{AB}\,(I_{Sample} - I_{Ref})$

$|\dot{Q}| = h_{HT}\,A\,(T_i - T_j)$

$|\dot{Q}| = \dot{m}\,c_p\,(T_{in} - T_{out})$

FIGURE 1.11
Principles of isothermal microcalorimeter (IMC) (a) and of reaction calorimeter (b).

the power in an electrical heater in the case of PCC or the heat transfer area h_{HT} in the case of CFC. A sketch of the most popular measuring principles (e.g. (bio-)reaction calorimeter and IMC) is given in Figure 1.11.

The advantages of (bio-)reaction calorimeters are (i) they allow measurement under operational conditions (e.g. stirring, reactant addition, pH control, additional sensors), (ii) a large reaction volume (in the range of liters) to allow extensive sampling and (iii) a short response time (a few seconds to minutes), although they have the disadvantage of poor signal resolution (5–200 mW L^{-1}). The advantage of an IMC is the resolution of the heat signal which is often less than 0.1 mW L^{-1}; but this is paid for by the poor possibilities for homogenization, sampling and dosing, and the accommodation of additional sensors.

1.8.6 How Do I Measure Joule-Thomson Coefficients?

Heat, or more correctly, temperature, effects that are triggered by pressure changes are designated as Joule-Kelvin, Joule-Thomson or Kelvin-Joule effects. The Joule-Thomson effect is explained in Question 3.5.8. Here we consider only how the coefficient can be measured. When a gas expands through a valve or porous plug, the process is described quantitatively by the Joule-Thomson coefficient μ_{JT} (Equation 3.118)

$$\mu_{JT} = \left(\frac{\partial T}{\partial p}\right)_H = \frac{V}{C_p}(\alpha\, T - 1). \tag{1.141}$$

The first of these two equations is the formal definition of the Joule-Thomson coefficient. The symbol α represents the coefficient of thermal expansion of the gas. Especially for the design of cooling processes such as air conditioners, heat pumps and liquefiers, μ_{JT} is a key parameter.

The coefficient can be measured using an adiabatic flow calorimeter, shown in Figure 1.12. It consists of a thermally isolated tube with a throttle (a constriction), through

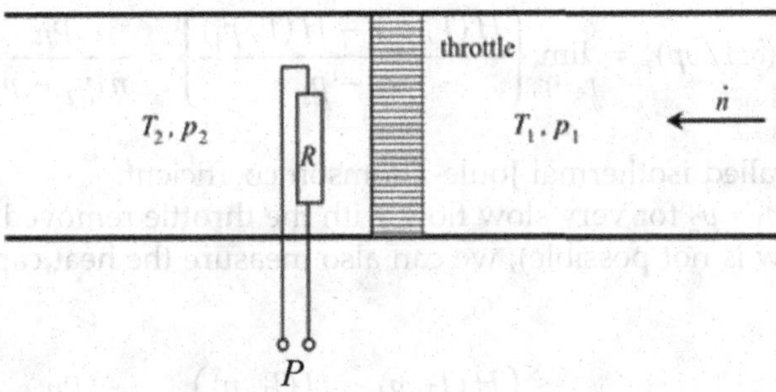

FIGURE 1.12
Schematic cross-section through an adiabatic flow calorimeter with gas flowing at an amount of substance rate $\dot{n}$ fitted with a throttle and resistor R through which a power P can be dissipated.

which gas flows leading to a pressure drop across it, a resistance heater and a measured source of power P downstream of the throttle, and a means of measuring the temperature and pressure before and after the throttle.

Material present upstream of the throttle at temperature T_1 and pressure p_1 passes at a rate $\dot{n}$ through the throttle where it emerges at temperature T_2 and pressure p_2 in an *adiabatic* enclosure where a power P is applied to the resistor, R. For an amount of substance n, the First Law of Thermodynamics for an open system for this situation (Equation 1.75) becomes

$$\Delta U = U_2 - U_1 = \frac{Pn}{\dot{n}} + p_1 V(T_1, p_1) - p_2 V(T_2, p_2), \tag{1.142}$$

where we assume that the tube is horizontal and that the kinetic energy of the gas is negligible as is often the case. In terms of the enthalpy, then

$$H(T_2, p_2) - H(T_1, p_1) = Pn/\dot{n}. \tag{1.143}$$

Here we note that when $P = 0$,

$$H(T_2, p_2) = H(T_1, p_1), \tag{1.144}$$

i.e. the process is isenthalpic.

For a gas, if we can measure the temperatures of the material on both sides of the throttle and if the process is carried out at low pressures, then it is possible to measure the Joule-Thomson coefficient because

$$\mu_{\mathrm{JT}} = (\partial T/\partial p)_T = \lim_{p_2 \to p_1} \left\{ \frac{T(H_1, p_2) - T(H_1, p_1)}{p_2 - p_1} \right\}. \tag{1.145}$$

This quantity has proven of importance in understanding the forces between simple molecules.

Alternatively, it is possible to operate the equipment so as to adjust P and to maintain $T_2 = T = T_1$, and Equation 1.145 can be written as

$$\phi_{JT} = (\partial H/\partial p)_T = \lim_{p_2 \to p_1} \left\{ \frac{H(T, p_2) - H(T, p_1)}{p_2 - p_1} \right\} = \frac{Pn}{\dot{n}(p_2 - p_1)}, \tag{1.146}$$

where ϕ_{JT} is the so-called isothermal Joule-Thomson coefficient.

In the case where $p_1 \approx p_2$ for very slow flow with the throttle removed (since an exactly zero rate of fluid flow is not possible), we can also measure the heat capacity at constant pressure

$$C_p = (\partial H/\partial T)_p = \lim_{T_2 \to T_1} \left\{ \frac{H(T_2, p) - H(T_1, p)}{T_2 - T_1} \right\} = \frac{Pn}{\dot{n}(T_2 - T_1)}. \tag{1.147}$$

The three quantities, μ_{JT}, ϕ_{JT} and C_p, are related by Equation 1.153, the –1 rule, as follows

$$(\partial T/\partial p)_H (\partial p/\partial H)_T (\partial H/\partial T)_p = -1, \tag{1.148}$$

that can be written as

$$\mu_{JT} = -\phi_{JT}/C_p. \tag{1.149}$$

1.9 What Are Standard Quantities?

By convention, the height of a mountain peak on Earth is expressed as the height above mean sea level; indeed, this is the convention adopted for all points on Earth's surface. It is a convention that leads to just one entry for the height of each location on Earth. The same convention means that the difference in height between any two locations from among N requires just N listed values to describe the $N(N-1)/2$ differences. By analogy, the same efficiency can be secured for the tabulation of thermodynamic quantities by reference to a standard state that is independent of pressure and composition. This is very important because, in most applications, it is not the absolute value of the thermodynamic quantity that matters but the changes that may be brought about in these quantities by variations in the temperature, pressure or composition.

It is therefore important to define a reference value for each thermodynamic quantity to which these changes can be referred. Suitable reference values are known as *standard thermodynamic quantities*. Generally, the specification of a standard thermodynamic quantity will require a statement of a state of aggregation (whether it is gas, g; liquid, l; or solid, s, together with a definition of a standard pressure and composition). By convention, since 1982, the standard pressure denoted by $p^{\ominus}$, has been taken as 10^5 Pa (= 1 bar). The conventional means of stating the standard composition is through the molality. The standard molality is denoted by $m^{\ominus}$ and for standard thermodynamic quantities the recommended value, which is used almost exclusively, is 1 mol kg^{-1}. Nevertheless, the value of $m^{\ominus}$ adopted should always be stated when the numerical value of any standard thermodynamic quantity of a solute is given. Finally, there are several temperatures in common use as a reference temperature, including $T = 298.15$ K and $T = 273.15$ K so that the reference temperature always needs to be specified in any tabulations.

We shall return to a discussion of standard thermodynamic quantities in Chapters 3 and 5 when we have introduced more of the thermodynamic quantities themselves. We shall also discuss reference states in Question 5.2.1, noting here that the standard state is but one form of the more general term reference states.

1.10 What Mathematical Relationships Are Useful in Thermodynamics?

Much of this book seeks to provide explanations of the fundamental laws of thermodynamics and thermophysics. Many of the derivations of the results we quote in this book are omitted in the interest of brevity and because they are not actually the intent of this book. However, we recognize that some readers will want to attempt the derivations themselves for further understanding and for that reason we provide here a few useful mathematical relationships. In any case, some of the difficulty in understanding thermodynamics arises because of the long, nonintuitive manipulation of thermodynamic relationships particularly through partial derivatives. Thus, in the last question of this chapter, we provide a statement of several important relationships among partial derivatives that will help students of thermodynamics to keep them at their fingertips while attempting to understand other texts that do provide (or expect!) full derivations.

1.10.1 What Is Partial Differentiation?

For a function of $x, y, z, \cdots, u = u(x, y, z, \cdots)$ the total derivative of u is

$$du = (\partial u/\partial x)_{y,z,\ldots}\mathrm{d}x + (\partial u/\partial y)_{x,z,\ldots}\mathrm{d}y + (\partial u/\partial z)_{y,x,\ldots}\mathrm{d}z + \cdots. \quad (1.150)$$

For a function $u(x, y)$, Equation 1.150 becomes

$$du = (\partial u/\partial x)_y dx + (\partial u/\partial y)_x dy. \quad (1.151)$$

Equation 1.151 can be used to illustrate three theorems. The first is for the change of variable held constant

$$(\partial u/\partial x)_z = (\partial u/\partial x)_y + (\partial u/\partial y)_x(\partial y/\partial x)_z. \quad (1.152)$$

The second is the −1 rule that is

$$(\partial u/\partial x)_y(\partial x/\partial y)_u(\partial y/\partial u)_x = -1. \quad (1.153)$$

The third theorem is for cross differentiation

$$\{\partial(\partial u/\partial x)_y/\partial y\}_x = \{\partial(\partial u/\partial y)_x/\partial x\}_y. \quad (1.154)$$

An expression for the difference $(C_p - C_V)$ can be found by applying the first two rules and using the definitions of C_p, C_V and $H = U + pV$. First, we note from Equations 1.124 and 1.128

$$C_p - C_V = (\partial H/\partial T)_p - (\partial U/\partial T)_V \\ = (\partial H/\partial T)_p - (\partial H/\partial T)_V + V(\partial p/\partial T)_V. \quad (1.155)$$

Use of the rule for change of variable held constant on $(\partial H/\partial T)_V$ gives

$$(\partial H/\partial T)_V = (\partial H/\partial T)_p + (\partial H/\partial p)_T(\partial p/\partial T)_V \quad (1.156)$$

so that Equation 1.155 can be written as

$$C_p - C_V = \{V - (\partial H/\partial p)_T\}(\partial p/\partial T)_V. \quad (1.157)$$

From $H = G + TS$ (Chapter 3), it can be shown that

$$(\partial H/\partial p)_T = V - T(\partial V/\partial T)_p, \quad (1.158)$$

when combined with Equation 1.157, this gives

$$C_p - C_V = T\{(\partial V/\partial T)_p\}(\partial p/\partial T)_V. \quad (1.159)$$

When the −1 rule is used on $(\partial p/\partial T)_V$, we find

$$(\partial p/\partial T)_V(\partial T/\partial V)_p(\partial V/\partial p)_T = -1, \quad (1.160)$$

and rearrangement gives

$$(\partial p/\partial T)_V = -(\partial V/\partial T)_p/(\partial V/\partial p)_T. \quad (1.161)$$

Substitution of Equation 1.161 into Equation 1.159 yields

$$C_p - C_V = -T\{(\partial V/\partial T)_p\}^2/(\partial V/\partial p)_T. \quad (1.162)$$

This is an important result because the derivatives $(\partial V/\partial T)_p$ and $(\partial V/\partial p)_T$ can be measured directly. An example of the application of the cross-differentiation rule can be found in a method of evaluating the change in entropy of a material arising from a pressure change at constant temperature. For a phase of fixed composition when the variables are T and p, Equation 3.39 leads to

$$\mathrm{d}G = -S\mathrm{d}T + V\mathrm{d}p \quad (1.163)$$

and directly from Equation 1.163, it follows that

$$(\partial G/\partial T)_p = -S, \quad (1.164)$$

and

$$(\partial G/\partial p)_T = V. \quad (1.165)$$

Differentiating Equation 1.165 with respect to p at constant T gives

$$\{\partial(\partial G/\partial T)_p/\partial p\}_T = -(\partial S/\partial p)_T \tag{1.166}$$

and Equation 1.165 with respect to T at constant p

$$\{\partial(\partial G/\partial p)_T/\partial T\}_p = (\partial V/\partial T)_p. \tag{1.167}$$

Comparison of Equations 1.166 and 1.167 gives

$$(\partial S/\partial p)_T = -(\partial V/\partial T)_p, \tag{1.168}$$

that is called a Maxwell equation and by integration provides a means of determining the entropy difference

$$\{S(T, p_2) - S(T, p_1)\} = -\int_{p_1}^{p_2} (\partial V/\partial T)_p \mathrm{d}p. \tag{1.169}$$

1.10.2 What Is Euler's Theorem?

When u is a homogeneous function of the *nth* degree in the variables $x, y, z, \cdots$, then, by definition

$$u(ax, ay, az, \cdots) = a^n u(x, y, z, \cdots) \tag{1.170}$$

and Euler's theorem states that then

$$x(\partial u/\partial x)_{y,z,\ldots} + y(\partial u/\partial y)_{x,z,\cdots} + z(\partial u/\partial z)_{x,y,\cdots} + \cdots = nu. \tag{1.171}$$

For a mixture (A + B) containing amounts of substance n_{A} and n_{B}, respectively, at constant temperature and pressure the volume V is a homogeneous function of n_{A} and n_{B} of the first degree and from Equations 1.7 and 1.9 it is

$$V = n_{\mathrm{A}}(\partial V/\partial n_{\mathrm{A}})_{T,p,n_{\mathrm{B}}} + n_{\mathrm{B}}(\partial V/\partial n_{\mathrm{B}})_{T,p,n_{\mathrm{A}}}. \tag{1.172}$$

We have already made use of Euler's theorem in this chapter.

1.10.3 What Is Taylor's Theorem?

For an analytic function $f(x)$, the Taylor's expansion about $x = a$ is

$$\begin{aligned} f(x) = f(a) + (x-a)\left\{\frac{\partial f(x)}{\partial x}\right\}_{x=a} + (x-a)^2 \frac{\{\partial^2 f(x)/\partial x^2\}_{x=a}}{2!} \\ + (x-a)^3 \frac{\{\partial^3 f(x)/\partial x^3\}_{x=a}}{3!} + \cdots. \end{aligned} \tag{1.173}$$

1.10.4 What Is the Euler-MacLaurin Theorem?

The sum of a function $f(n)$ over all integral values of n from 0 to ∞ is given by

$$\sum_{n=0}^{\infty} f(n) = \int_0^{\infty} f(n)\mathrm{d}n + \frac{1}{2}f(0) - \frac{1}{12}\left(\frac{\partial f}{\partial n}\right)_{n=0} + \frac{1}{720}\left(\frac{\partial^3 f}{\partial n^3}\right)_{n=0} - \cdots. \quad (1.174)$$

References

Atkins, P. W., de Paula, J., and Keeler, J., 2018, *Atkins' Physical Chemistry*, 11th ed., Oxford University Press, Oxford.

BPM, Bureau International des Poids et Mesures, *Le Système international d'unités (SI)*, 2006, 8th ed. , STEDI Media, France.

Case F., Chaka A., Friend D.G., Frurip D., Golab J., Johnson R., Moore J., Mountain R.D., Olson J., Schiller M., and Storer J., 2004, "The first industrial fluid properties simulation challenge", *Fluid Phase Equilib.* **217**:1–10.

Case F.H., Chaka A., Friend D.G., Frurip D., Golab J., Gordon P., Johnson R., Kolar P., Moore J., Mountain R.D., Olson J., Ross R., and Schiller M., 2005, "The second industrial fluid properties simulation challenge", *Fluid Phase Equilib.* **236**:1–14.

Case F.H., Brennan J., Chaka A., Dobbs K.D., Friend D.G., Frurip D., Gordon P.A., Moore J., Mountain R.D., Olson J., Ross R.B., Schiller M., and Shen V.K., 2007, "The third industrial fluid properties simulation challenge", *Fluid Phase Equilib.* **260**:153–163.

Case F.H., Brennan J., Chaka A., Dobbs K.D., Friend D.G., Gordon P.A., Moore J.D., Mountain R.D., Olson J.D., Ross D.B., Schiller M., Shen V.K., and Stahlberg E.A., 2008, "The fourth industrial fluid properties simulation challenge", *Fluid Phase Equilib.* **274**:2–9.

Cengel Y.A., and Boles M.A., 2014, *Thermodynamics—an Engineering Approach*, 8th ed., McGraw-Hill, New York.

Chemical Thermodynamics, 1999, ed. Letcher T.M., for IUPAC, Blackwells Scientific Publications, Oxford.

Developments and Applications in Solubility, 2007, ed. Letcher T.M., for IUPAC, Royal Society of Chemistry, Cambridge.

Eckl B., Vrabec J., and Hasse H., 2008, "On the application of force fields for predicting a wide variety of properties: Ethylene oxide as an example", *Fluid Phase Equilib.* **274**:16–26.

Experimental Thermodynamics, Volume I, Calorimetry of Non-Reacting Systems, 1968, eds. McCullough J.P., and Scott D.W., for IUPAC, Butterworths, London.

Experimental Thermodynamics, Volume II, Experimental Thermodynamics of Non-Reacting Fluids, 1975, eds. Le Neindre B., and Vodar B., for IUPAC, Butterworths, London.

Experimental Thermodynamics, Volume III, Measurement of the Transport Properties of Fluids, 1991, eds. Wakeham W.A., Nagashima A., and Sengers J.V., for IUPAC, Blackwell Scientific Publications, Oxford.

Experimental Thermodynamics, Volume IV, Solution Calorimetry, 1994, eds. Marsh K.N., and O'Hare P.A.G., for IUPAC, Blackwell Scientific Publications, Oxford.

Experimental Thermodynamics, Volume V, Equations of State for Fluids and Fluid Mixtures, Parts I and II, 2000, eds. Sengers J.V., Kayser R.F., Peters C.J., and White Jr. H.J., for IUPAC, Elsevier, Amsterdam.

Experimental Thermodynamics, Volume VI, Measurement of the Thermodynamic Properties of Single Phases, 2003, eds. Goodwin A.R.H., Marsh K.N., and Wakeham W.A., for IUPAC, Elsevier, Amsterdam.

Experimental Thermodynamics, Volume VII, Measurement of the Thermodynamic Properties of Multiple Phases, 2005, eds. Weir R.D., and de Loos T.W., for IUPAC, Elsevier, Amsterdam.

Experimental Thermodynamics, Volume VIII, Applied Thermodynamics of Fluids, 2010, eds. Goodwin A.R.H., Sengers J.V., and Peters C.J., for IUPAC, RSC Publishing, Cambridge.

Experimental Thermodynamics Volume IX, Advances in Transport Properties of Fluids, 2014, eds. Assael M.J., Goodwin A.R.H.C., Vesovic V., and Wakeham W.A. for IUPAC, RSC Publishing, Cambridge.

Evaluation of measurement - Guide to the Expression of Uncertainty in Measurement, 2008, by Joint Committee for Guides in Methodology (JCGM), Geneva, Switzerland. https://www.bipm.org/utils/common/documents/jcgm/JCGM_100_2008_E.pdf

Future Energy: Improved, Sustainable and Clean Options for our Planet, 2008, ed. Letcher T.M., for IUPAC, Elsevier, Amsterdam.

Gnaiger, E., and Kemp, R.B., 1990, "Anaerobic metabolism in aerobic mammalian cells: Information from the ratio of calorimetric heat flux and respirometric oxygen flux", *Biochim. Biophys. Acta*, **1016**:328–332.

Göbel E.O., and Siegner, U., 2019, *The New International System of Units (SI)*, Wiley-VCH, Berlin.

Goodwin A.R.H., and Trusler, J.P.M., 2003, *Speed of Sound*, Chapter 6, in *Experimental Thermodynamics, Volume VI, Measurement of the Thermodynamic Properties of Single Phases*, eds. Goodwin A.R.H., Marsh K.N., and Wakeham W.A., for IUPAC, Elsevier, Amsterdam.

Goodwin A.R.H., and Trusler J.P.M., 2010, *Speed of Sound Measurements and Heat Capacities of Gases*, Chapter 9, in *Heat Capacities: Liquids, Solutions and Vapours*, eds. Letcher T.M., and Willhelm E., for IUPAC, RSC Publishing, Cambridge.

Grossman J.C., 2020, *Thermodynamics: 4 Laws that Move the Universe*, The Great Courses, https://www.thegreatcourses.com/courses/thermodynamics-four-laws-that-move-the-universe.html

Heat Capacities: Liquids, Solutions and Vapours, 2010, eds. Willhelm E. and Letcher T.M., for IUPAC, RSC Publishing, Cambridge.

International Vocabulary of Metrology - Basic and General Concepts and Associated Terms, 2012, by Joint Committee for Guides in Methodology (JCGM), Geneva, Switzerland. https://www.bipm.org/utils/common/documents/jcgm/JCGM_200_2012.pdf

Jimenez-Serratos G., Totton T.S., Jackson G., and Müller E.A., 2019, "Aggregation behavior of model asphaltenes revealed from large-scale coarse-grained molecular simulations", *J. Phys. Chem. B*. **123**:2380–2396.

Jin Z., and Firoozabadi A., 2016, "Phase behavior and flow in shale nanopores from molecular simulations", *Fluid Phase Equilib*. **430**:156–168.

Kestin, J., 1965, *A Course in Thermodynamics*, Blaisdell, Mass, USA.

Ketko M.H., Rafferty J., Siepmann J.I., and Potoff J.J., 2008, "Development of the TraPPE-UA force field for ethylene oxide", *Fluid Phase Equilib*. **274**:44–49.

Levenspiel O., 1996, *Understanding Engineering Thermodynamics*, Prentice Hall PTR, NJ, USA.

Liu J.S., Marison I.W., and von Stockar U., 2001, "Microbial growth by a net heat up-take: A calorimetric and thermodynamic study on acetotrophic methanogenesis by *Methanosarcina barkeri*", *Biotechnol. Bioeng*. **75**:170–180.

Li X., Zhao L., Cheng T., Liu L., and Sun H., 2008, "One force field for predicting multiple thermodynamic properties of liquid and vapor ethylene oxide", *Fluid Phase Equilib*. **274**:36–43.

McGovern J., 2004, "Thermodynamic temperature", https://theory.physics.manchester.ac.uk/~judith/stat_therm/node27.html

Machin, G.F., Hänggi, J., and Trusler, J.P.M., 2016, "Towards implementing the new kelvin," *Phil. Trans. Roy. Soc. A*. **374**, 2064.

Maskow T., and Paufler S., 2015, "What does calorimetry and thermodynamics of living cells tell us?", *Methods* **76**:3–10.

MIT, 2021, https://web.mit.edu/16.unified/www/FALL/thermodynamics/notes/node45.html

Moran M.J., Shapiro H.N., Boettner D.D., and Bailey M.B., 2018, *Fundamentals of Engineering Thermodynamics*, 9th ed., John Wiley & Sons, New York.

Müller T.J., Roy S., Zhao W., Maaß A., and Reith D., 2008, "Economic simplex optimization for broad range property prediction: Strengths and weaknesses of an automated approach for tailoring of parameters", *Fluid Phase Equilib.* **274**:27–35.

Nicholas J.W., 2003 *Temperature*, Chapter 2, in *Experimental Thermodynamics, Volume VI, Measurement of the Thermodynamic Properties of Single Phases*, eds. Goodwin A.R.H., Marsh K.N., and Wakeham W.A., for IUPAC, Elsevier, Amsterdam.

Olson J.D., and Wilson L.C., 2008, "Benchmarks for the fourth industrial fluid properties simulation challenge", *Fluid Phase Equilib.* **274**:10–15.

Otto S., Scholz N., Behnke T., Resch-Genger U., and Heinze K., 2017, "Thermo-Chromium: A contactless optical molecular thermometer", *Chem. Eur. J.* **23**:12131–12135.

Perrot P., 1998, *A to Z of Thermodynamics*, Oxford University Press, Oxford.

Preston-Thomas H., 1990, "The International Temperature Scale of 1990 (ITS-90)", *Metrologia* **27**:3–10.

Rapaport, D., 2004, *The Art of Molecular Dynamics Simulation*, 2nd ed., Cambridge University Press. Cambridge

Sandler S.I., 1989, *Chemical and Engineering Thermodynamics*, 3rd ed., John Wiley & Sons Inc., New York.

Smith E.B., 2013, *Basic Chemical Thermodynamics*, 6th ed., Imperial College Press, London.

Smith J.M., van Ness H.C., Abbott M.M., and Swihart M.T., 2017, *Introduction to Chemical Engineering Thermodynamics*, 8th ed., McGraw-Hill, New York.

Sonntag R.E., Borgnakke C., and van Wylen G.J., 2003, *Fundamentals of Thermodynamics*, 6th ed., John Wiley & Sons, New York.

Suski J., Girard M., Ehrlich C.D., Schmidt J.W., Abramson E., and Sutton C., 2003, *Pressure*, Chapter 3, in *Experimental Thermodynamics, Volume VI, Measurement of the Thermodynamic Properties of Single Phases*, eds. Goodwin A.R.H., Marsh K.N., and Wakeham W.A., for IUPAC, Elsevier, Amsterdam.

Taylor B.N., and Kuyatt C.E., 1994, *Guidelines for Evaluating and Expressing the Uncertainty of NIST Measurement Results*, NIST Technical Note 1297.

The International System of Units (SI), 2019, https://www.bipm.org/documents/20126/41483022/SI-Brochure-9-EN.pdf

2

Statistical Mechanics

2.1 Introduction

Chapter 1 dealt with the definitions of many of the quantities required for the macroscopic description of the thermodynamic behavior of systems viewed as continua, including the definition of a system. However, we are familiar with the notion that all matter is made up of atomic or molecular entities and it is the purpose of statistical mechanics to provide a microscopic description of the behavior of a thermodynamic system in terms of the properties, interactions and motions of the atoms or molecules that make up the system. Because macroscopic thermodynamic systems contain very large numbers of molecules, the task of statistical mechanics is not to describe exactly what happens to every single molecule, but rather to derive results that pertain to the complete assembly of molecules that comprise the system in a probabilistic manner. The atoms and molecules that comprise the system are best described using quantum mechanics rather than classical mechanics so that is the basis for the development of the theory of statistical mechanics.

The solution of Schrödinger's equation of quantum mechanics is a wave that describes the probable state of the system that includes a description of the quantum states (eigenstates) and energy levels (energy eigenvalues) an individual molecule and the system can attain; in quantum theory the energy levels are discrete. It is much easier to solve Schrödinger's equation for a single molecule (or realistically for a single atom) than for a system of N molecules or atoms to obtain the quantum states or energy levels so we begin with that problem. For a single, relatively simple molecule (such as nitrogen), the problem of solving Schrödinger's equation is made tractable by separation of the modes of motion of the molecule (translation of the center of mass, rotation and vibration) so that each is handled independently. This is legitimate provided that certain conditions are met and a number of texts on quantum mechanics and/or statistical physics will provide you with the means of deducing the allowed energy states for each of these modes of motion.

The question then arises as to how are the molecules distributed between the energy levels available to a single molecule? It is reasonable to anticipate that a system of N molecules will be arranged so that in each quantum state of discrete energy there will be a number of molecules. When the energy of the molecules is very much higher than the difference between the energies of the various quantum levels, which happens for the translational kinetic energy of the molecules of a gas in a macroscopic system at moderate temperature, then the relationship between the number of molecules with a specified energy and the energy of that state, is given by Boltzmann's distribution (see Question 2.2). For some other types of systems, other distributions of energy are possible. For these other cases, the spin of the molecular system matters. For a set of entities with an integral spin the system will obey Bose-Einstein statistics while, if it is a half-integral spin system, Fermi-Dirac statistics are used. When the energy of the system is sufficiently high, both reduce to Boltzmann's

DOI: 10.1201/9780429329524-2

distribution for the molecules and conditions of interest to chemists and engineers and Boltzmann statistics are likely to be appropriate. On the other hand, for physicists, particularly at low temperatures, the other types of distribution are often appropriate. The distinction between low and high temperatures will be quantified as we proceed.

We now consider a system of N molecules (where $N > 10^{15}$). If we suppose that we have a distribution of the molecules so that N_i of the molecules is in the ith quantum state each with energy ε_i then the total number of molecules is

$$N = \sum_i N_i, \tag{2.1}$$

and the internal energy U is

$$U = \sum_i N_i \varepsilon_i. \tag{2.2}$$

It can also be shown (McQuarrie 2000) that the thermodynamic pressure may be written as

$$p = \sum_i N_i(-d\varepsilon_i/dV), \tag{2.3}$$

where V is the volume of the system.

Indeed, building on these methods it is now possible to evaluate intermolecular pair potentials (see Question 2.5.1) for molecules as complicated as hydrogen sulfide, nitrogen, propane or ethylene oxide (Hellmann et al. 2012; Hellmann 2013; 2014; 2017), and obtain their potential energy surface. Therefrom, macroscopic properties such as dilute-gas viscosity and thermal conductivity of these pure gases or binary mixtures containing these molecules (Hellmann 2020), can be obtained from statistical mechanics. In some cases the uncertainty of such calculated values can be less than that available by direct experimentation (Hurly and Moldover 2000; Hurly and Mehl 2007) and in others, such as hydrogen sulfide, such calculations provide the only source of properties (Hellmann et al. 2012). More generally, the principles of statistical mechanics are used within molecular simulation or computational chemistry to provide estimates of the thermophysical properties of materials but for most systems the calculations are significantly less precise than those available through direct measurement.

2.2 What Is Boltzmann's Distribution?

According to Boltzmann's distribution, the number of molecules N_i in the ith quantum state of energy, ε_i, is given by

$$N_i = \lambda \exp(-\varepsilon_i/k_B T). \tag{2.4}$$

Here, λ is called the absolute activity of a substance. It is defined in terms of another important thermodynamic quantity called the chemical potential, μ, by

$$\lambda = \exp(\mu/k_B L T). \tag{2.5}$$

where k_B is Boltzmann's constant, exactly equal to 1.380649×10^{-23} J K^{-1} (Question 1.7.1), and is the proportionality between statistical and classical thermodynamics. In addition, L is Avogadro's constant,[1] equal to $6.02214076 \times 10^{23}$ mol^{-1} (Question 1.7.1) and T is the thermodynamic temperature. The chemical potential and the absolute activity are central to various aspects of classical thermodynamics, including the treatment of the equilibrium between the various phases of materials and chemical reaction equilibrium. The two quantities will be developed in those contexts in Chapters 3, 4 and 5 and the reader is referred to those chapters for a detailed exposition.

The use of Equation 2.4 in Equation 2.1 gives

$$N = \lambda \sum_i \exp(-\varepsilon_i / k_B T), \tag{2.6}$$

where the sum on the right-hand side is defined by

$$\sum_i \exp(-\varepsilon_i / k_B T) = q \tag{2.7}$$

and called the molecular partition function. The combination, Lk_B, is also special

$$R = Lk_B, \tag{2.8}$$

and is known as the universal gas constant, R.

From Equations 2.6 and 2.7, we then have

$$N = \lambda q, \tag{2.9}$$

so that

$$\mu = RT \ln N - RT \ln q \tag{2.10}$$

using Equation 2.5.5.

The fraction of molecules in particular states, N_i/N, is then given by

$$N_i/N = \exp(-\varepsilon_i/k_B T)/\sum_i \exp(-\varepsilon_i/k_B T) = \exp(-\varepsilon_i/k_B T)/q, \tag{2.11}$$

and, as $\varepsilon_i/k_B T$ increases, N_i/N decreases, so that fewer molecules are found at higher energies, in line with intuition. Employing this result in Equations 2.2 and 2.3 gives

$$U = N \frac{\sum_i \varepsilon_i \exp(-\varepsilon_i/k_B T)}{\sum_i \exp(-\varepsilon_i/k_B T)} = Nk_B T^2 \left(\frac{\partial \ln q}{\partial T}\right)_V, \tag{2.12}$$

and

[1] For Avogadro's constant ($6.02214076 \times 10^{23}$ mol^{-1}) the SI symbol is N_A, but to avoid confusion in certain cases, the symbol L (accepted by SI), is also used.

$$p = N\frac{\sum_i (-\mathrm{d}\varepsilon_i/\mathrm{d}V)\exp(-\varepsilon_i/k_\mathrm{B}T)}{\sum_i \exp(-\varepsilon_i/k_\mathrm{B}T)} = Nk_\mathrm{B}T\left(\frac{\partial \ln q}{\partial V}\right)_T. \tag{2.13}$$

In relation to Equation 2.12, looking at its first part, the numerator can be obtained by the differentiation the denominator with respect to $\beta = 1/k_\mathrm{B}T$. Thus, $U = N(1/q)(\mathrm{d}q/\mathrm{d}\beta)_V = N\mathrm{d}\ln q/\mathrm{d}\beta = -Nk_\mathrm{B}T^2\ (\mathrm{d}\ln q/\mathrm{d}T)_V$.

In the case when a number of independent quantum states, g_i, have the same energy, it is a matter of convenience to write Equation 2.7 as

$$q = \sum_i g_i \exp\left(-\frac{\varepsilon_i}{k_\mathrm{B}T}\right), \tag{2.14}$$

where g_i is the degeneracy of the energy level, ε_i.

2.3 How do I Evaluate the Partition Function Q for Non-Interacting Molecules?

The concept of solving Schrödinger's equation and thus evaluating q from quantum theory was alluded to in Question 2.1. Here we consider the evaluation of the partition function for a system of molecules that act independently of each other, so that they exert no forces upon each other and there is no intermolecular energy. We separate the modes of motion of the molecule, so that the energy ε of a molecule in an eigenstate can be written as

$$\varepsilon = \varepsilon_\mathrm{T} + \varepsilon_\mathrm{R} + \varepsilon_\mathrm{V} + \varepsilon_\mathrm{E} + \varepsilon_\mathrm{N} + \varepsilon_0, \tag{2.15}$$

that is, as the sum of the energy eigenvalues for translational ε_T, rotational ε_R, vibrational ε_V, electronic ε_E, nuclear ε_N and the lowest energy state ε_0, often called the 'ground state'. As was indicated earlier, this separation is valid only under certain circumstances.

It follows from the existence of these various forms of energy and their summation to the whole that the partition function for a molecule can be written as

$$q = q_\mathrm{T} q_\mathrm{R} q_\mathrm{V} q_\mathrm{E} q_\mathrm{N} q_0. \tag{2.16}$$

The translational partition function q_T can be separated into three parts, one for each of the Cartesian coordinates. For a molecule of n atoms that is linear, there are two rotational modes corresponding to rotation about the two axes perpendicular to the axis of the linear molecule and $(3n - 5)$ vibrational modes, while for a nonlinear molecule there are three rotational modes and $(3n - 6)$ vibrational modes. This ideal separation fails when the molecule is in a high vibrational state because the rotational energy levels depend on the moment of inertia of the molecule which can change in practice as molecules come near to dissociation; then there is some interaction between the modes. However, for many molecules of interest to chemists, biologists and engineers, only the lowest vibrational energy levels are accessible so that the intramolecular potential nearly approximates to that of a simple harmonic oscillator that does not permit dissociation, so that the separation of rotational and vibrational modes is very often valid.

For the likely readership of this text, the remaining modes of energy of a molecule are of small interest. For example, electronic modes at room temperatures are unimportant for chemists and engineers because only the lowest $\varepsilon_{0,E}$, and at most the first excited states, $\varepsilon_{1,E}$, of atomic or molecular orbitals are populated so that the electronic partition function can be written as

$$q_E = g(\varepsilon_{0,E})\exp\left(-\frac{\varepsilon_{0,E}}{k_B T}\right)\left[1 + \frac{g(\varepsilon_{1,E})}{g(\varepsilon_{0,E})}\exp\left(-\frac{(\varepsilon_{1,E} - \varepsilon_{0,E})}{k_B T}\right)\right]. \tag{2.17}$$

The nuclei are all in the ground state for molecular gases of interest with a mass greater than hydrogen and so $q_N = 1$. For homonuclear molecular gases with a nuclear spin and low mass (e.g. hydrogen) at low temperature (i.e. <300 K), the effect of nuclear spin contributes to the thermodynamic properties but this very special topic will not be considered further here.

Assuming the preceding approximations of mode separation are valid and the nuclei are in the ground state the molecular partition function can then be written as

$$q = q_x q_y q_z q_R q_V g(\varepsilon_0) g(\varepsilon_{0,E})\exp\left(-\frac{\varepsilon_{0,E}}{k_B T}\right)\left[1 + \frac{g(\varepsilon_{1,E})}{g(\varepsilon_{0,E})}\exp\left(-\frac{(\varepsilon_{1,E} - \varepsilon_{0,E})}{k_B T}\right)\right]\exp\left(-\frac{\varepsilon_0}{k_B T}\right). \tag{2.18}$$

When, as is often the case $(\varepsilon_{1,E} - \varepsilon_{0,E}) >> k_B T$, Equation 2.17 becomes

$$q_E = g(\varepsilon_{0,E})\exp\left(-\frac{\varepsilon_{0,E}}{k_B T}\right) \tag{2.19}$$

and is electronically unexcited so that Equation 2.18 reduces to

$$q = q_x q_y q_z q_R q_V g(\varepsilon_0)\exp\left(-\frac{\varepsilon_0}{k_B T}\right), \tag{2.20}$$

where the $g(\varepsilon_{0,E})$ and $\exp(-\varepsilon_{0,E}/k_B T)$ of Equation 2.17 are represented by $g(\varepsilon_0)$ and $\exp(-\varepsilon_0/k_B T)$, respectively.

For *translational motion*, the solution of Schrödinger's equation for a particle of mass m moving in the x-direction within a box of length l_x (Sandler 2010; McQuarrie 2000; Reed and Gubbins 1973) gives the energy

$$\varepsilon_x = \frac{n_x^2 h^2}{8 l_x^2 m}, \tag{2.21}$$

where n_x is the quantum number, of value 1, 2, 3, ⋯, and h is Planck's constant so that the partition function, q_x, is

$$q_x = \sum_{n_x=1}^{\infty} \exp\left(-\frac{n_x^2 h^2}{8 l_x^2 m k_B T}\right). \tag{2.22}$$

Because the separation of the energy levels in translational motion is small compared with $k_B T$ (as a simple calculation using Equation 2.21 will illustrate), the sum in

Equation 2.22 can be replaced by an integral, which is tantamount to the assumption that the energy is a continuous variable. Upon integration with the Euler-Maclaurin theorem (Question 1.10.4), we find that

$$q_x = \left(\frac{2\pi m k_B T}{h^2}\right)^{1/2} l_x - \frac{1}{2}. \tag{2.23}$$

The first term on the right-hand side of Equation 2.23 is much larger than 10^6; the second term (½) can therefore be ignored and the motion is termed *classical* (because the energy levels have been assumed continuous) so that for the three independent translational directions, recognizing that the system volume $V = l_x l_y l_z$

$$q_T = q_x q_y q_z = \left(\frac{2\pi m_B k T}{h^2}\right)^{3/2} V. \tag{2.24}$$

The form of the rotational partition function is different depending on whether the molecule is linear or nonlinear: Both are considered here. For a *linear* molecule, the solution of Schrödinger's equation gives the rotational energy ε_R as (McQuarrie 2000; Reed and Gubbins 1973)

$$\varepsilon_R = \frac{j(j+1)h^2}{8\pi^2 I} \tag{2.25}$$

where I is the moment of inertia and j is the quantum number equal to 0, 1, 2, 3, ···. Each of the energy levels is $(2j + 1)$ degenerate so that the partition function for rotation is

$$q_R = \sum_{j=0}^{\infty} (2j+1)\exp\left(-\frac{j(j+1)h^2}{8\pi^2 I k_B T}\right). \tag{2.26}$$

Again, because the separation of the energy levels in rotation is usually small compared with $k_B T$, the sum can be replaced by an integral (assuming the energy levels are continuous) and on integration with the Euler-Maclaurin theorem (Question 1.10.4), we find that

$$q_R = \frac{8\pi^2 I k_B T}{h^2} + 0.42. \tag{2.27}$$

The numerical value of $(8\pi^2 I k_B T/h^2)$ is usually (but not always) large compared to 0.42, as a simple calculation reveals. For the extreme case of hydrogen $(8\pi^2 I k_B/h^2) \approx 0.01\ \mathrm{K}^{-1}$ and at a temperature of 300 K $(8\pi^2 I k_B T/h^2) \approx 3$; so that ignoring 0.42 gives rise to a fractional uncertainty of 0.1 and the treatment of hydrogen as a classical rotator fails. However, for iodine at a temperature of 300 K, the fractional error is $<10^{-4}$ and it is thus an acceptable approximation to assume that rotational motion is classical. For molecules with a molar mass greater than hydrogen and for temperatures on the order of 100 K or greater, Equation 2.27 can be approximated by

$$q_R \approx \frac{8\pi^2 I k_B T}{s h^2}, \tag{2.28}$$

where s is the symmetry number. The symmetry number is $s = 1$ for linear molecules with no center of symmetry such as hydrogen fluoride and 2 for molecules with a center of symmetry, for example, oxygen. The difference arises because for symmetrical molecules each distinguishable orientation has been counted twice (Reed and Gubbins 1973).

The rotational partition function for a *nonlinear* molecule is (Sandler 2010)

$$q_R = \frac{\pi^{1/2}}{s}\left(\frac{8\pi^2 (I_x I_y I_z)^{1/3} k_B T}{h^2}\right)^{3/2}, \tag{2.29}$$

where the I's are the moments of inertia for each direction of the coordinate system selected and s the symmetry number; which has a similar interpretation as that given above, $s = 1$ for molecules with no symmetry axis, 2 for water, 3 for ammonia, 4 for ethane and 12 for methane and ethane.

If the vibrational motion is assumed to be simple harmonic, which means the molecule can never dissociate, but allows the modes to be treated as separable, the solution to Schrödinger's equation gives the energy

$$\varepsilon_V = \left(\nu + \frac{1}{2}\right) h\upsilon \tag{2.30}$$

where the quantum number $\nu = 0, 1, 2, 4, \cdots$, and υ is the frequency of vibration so that the partition function is

$$q_V = \sum_{\upsilon=0}^{\infty} \exp\left(-\left(\nu + \frac{1}{2}\right)\frac{h\upsilon}{k_B T}\right) \tag{2.31}$$

Even at room temperature, the molecular vibrations of most molecules do not behave classically and, indeed may not be excited at all, so the sum cannot be replaced by an integral. However, the summation is a geometrical progression; hence, we can write the vibrational partition function as (Olver et al. 2010)

$$q_V = \sum_{\upsilon=0}^{\infty} \exp\left(-\left(\nu + \frac{1}{2}\right)\frac{h\upsilon}{k_B T}\right) = 1 + e^{-h\upsilon/k_B T} + e^{-2h\upsilon/k_B T} + \ldots. \tag{2.32}$$

and

$$q_V = \left(1 - \exp\left(-\frac{h\upsilon}{k_B T}\right)\right)^{-1}. \tag{2.33}$$

2.4 What Can Be Calculated Using the Molecular Partition Function?

Now that we have the molecular partition function, it is possible to evaluate a number of thermodynamic properties of several systems and we provide five examples. They are all in some way or another results for idealized systems and are important because they illustrate the power of the methodology of statistical thermodynamics. For many systems encountered in engineering practice, the results derived for these idealized systems are useful limiting values for calculations that are generally much more complex and where approximations must be used to obtain meaningful results.

2.4.1 What Is the Heat Capacity of an Ideal Diatomic Gas?

We first consider the pressure exerted by a gas composed of the non-interacting atoms of a perfect gas for which the only energy mode is translational energy. In this case, the only component of the partition function is the translational component given by Equation 2.24. Evaluation of the pressure according to Equation 2.13 leads to

$$p = nk_BT/V \tag{2.34}$$

which is, of course, the perfect gas equation of state.

For an electronically unexcited, diatomic molecule, when the molecules do not interact with each other, the molecular partition function is given by

$$q = q_T q_R q_V = V\left(\frac{2\pi mk_BT}{h^2}\right)^{3/2}\left(\frac{8\pi^2 Ik_BT}{sh^2}\right)\left(1-\exp\left(-\frac{h\,\upsilon}{k_BT}\right)\right)^{-1} g\,(\varepsilon_0)\exp\left(-\frac{\varepsilon_0}{k_BT}\right). \tag{2.35}$$

A gas of such particles will obey the ideal gas equation of state, which is equivalent to Equation 2.34. The use of Equation 2.35 in Equation 2.12 gives the internal energy of N molecules of a gas of *diatomic* molecules as

$$U = Nk_BT^2\left(\frac{\partial \ln q}{\partial T}\right)_V = \frac{3}{2}Nk_BT + Nk_BT + \frac{Nh\upsilon}{\{\exp(h\upsilon/k_BT) - 1\}} + N\varepsilon_0. \tag{2.36}$$

The molar heat capacity at constant volume $C_{V,\mathrm{m}}$, which is the molar form of Equation 1.124, can be obtained from Equation 2.36, cast in terms of the molar internal energy, U_m, using

$$\frac{(\partial U_\mathrm{m}/\partial T)_V}{R} = \frac{C_{V,\mathrm{m}}}{R} = \frac{5}{2} + \left(\frac{h\upsilon}{k_BT}\right)^2\frac{\exp(h\upsilon/k_BT)}{\{1-\exp(h\upsilon/k_BT)\}^2}, \tag{2.37}$$

which is the formal result for the heat capacity of a diatomic molecule. Evidently, a knowledge of the single vibrational frequency allows calculation of the heat capacity quite simply. The single frequency required can be obtained spectroscopically. Equations 2.36 and 2.37 also show that for a symmetric diatomic gas, each of the 3 translational directions contributes $R/2$ to the molar heat capacity as do each of the two rotational modes (one around the molecular axis and the other perpendicular to it), giving a total of $5R/2$. This is an example of the equipartition of energy, which applies at temperatures sufficiently high for classical mechanics to apply to different modes of molecular motion (McQuarrie 2000). At very high temperatures when $h\upsilon/k_BT \ll 1$, the second vibrational term in Equation 2.37 simplifies to 1, showing that each molecular vibration contributes a quantity R to the molar heat capacity when fully excited.

For a *monatomic* gas, both the rotational and vibrational partition functions are omitted from Equation 2.35 and so Equation 2.37 becomes

$$\frac{C_{V,\mathrm{m}}}{R} = \frac{3}{2}, \tag{2.38}$$

which is independent of temperature.

For a polyatomic molecule with n atoms there are $(3n - 5)$ vibrational modes for a linear molecule and $(3n{-}6)$ for a nonlinear molecule. Each of these will have a characteristic vibrational frequency ν_m from which a vibrational partition function $q_V(m)$ may be evaluated; the vibrational partition coefficient of the whole molecule is then $\Pi_m\, q_V(m)$, $m = 1$ to $(3n - 5)$ or $(3n - 6)$. If the temperature is sufficiently high that all the vibrational modes are fully excited, then the vibrational contribution to $C_{V,\mathrm{m}}$ is $(3n - 5)R$ or $(3n - 6)R$, depending on whether the molecule is linear or not. The appropriate rotational partition function (Equation 2.29) should also be used and if electronic excitation is significant it must also be included.

2.4.2 What Is the Heat Capacity of a Crystal?

A perfect crystal formed from N identical atoms has no modes of motion except for the N three-dimensional vibrations of each atom about the occupied lattice sites. If this motion is considered simple harmonic, the partition function q of an atom is then

$$\begin{aligned} q_V &= \sum_{\nu=0}^{\infty} \exp\left(-\left(\nu + \frac{1}{2}\right)\frac{h\upsilon}{k_B T}\right) = \exp\left(\frac{-h\upsilon}{2k_B T}\right)\sum_{\nu=0}^{\infty}\exp\left(\frac{-\nu h\upsilon}{k_B T}\right) \\ &= \exp\left(\frac{-h\upsilon}{2k_B T}\right)\sum_{\nu=0}^{\infty}\left[\exp\left(\frac{-h\upsilon}{k_B T}\right)\right]^{\nu}. \end{aligned} \tag{2.39}$$

For a one-dimensional harmonic oscillator, employing the series developed in Equation 2.32, we obtain

$$q_V = \frac{\exp(-h\upsilon/2k_B T)}{1 - \exp(-h\upsilon/k_B T)}. \tag{2.40}$$

The partition function for vibration in each of the other two dimensions is the same. Hence, for the three-dimensional harmonic oscillator

$$q_V = \left[\frac{\exp(-h\upsilon/2k_B T)}{1 - \exp(-h\upsilon/k_B T)}\right]^3. \tag{2.41}$$

Employing Equation 2.12, the energy of a crystal of N, independent, distinguishable atoms is

$$U = N k_B T^2\left(\frac{\partial \ln q_V}{\partial T}\right)_V = \frac{3Nh\upsilon}{2} + \frac{3Nh\upsilon\exp(-h\upsilon/k_B T)}{1 - \exp(-h\upsilon/k_B T)}. \tag{2.42}$$

The molar heat capacity at constant volume $C_{V,\mathrm{m}}$, which is the molar form of Equation 1.124, can be obtained from Equation 2.36, cast in terms of the molar internal energy, U_{m}, using

$$\frac{C_{V,\mathrm{m}}}{R} = \frac{(\partial U_{\mathrm{m}}/\partial T)_V}{R} = 3\left(\frac{h\upsilon}{k_{\mathrm{B}}T}\right)^2 \frac{\exp(h\upsilon/k_{\mathrm{B}}T)}{\{1 - \exp(h\upsilon/k_{\mathrm{B}}T)\}^2}. \tag{2.43}$$

At high temperatures, $h\upsilon/k_{\mathrm{B}}T$ tends to zero, and $\exp(h\upsilon/k_{\mathrm{B}}T) \approx 1 + (h\upsilon/k_{\mathrm{B}}T)$. Using this result in Equation 2.43,

$$\frac{C_{V,\mathrm{m}}}{R} \approx 3\left(\frac{h\upsilon}{k_{\mathrm{B}}T}\right)^2 \frac{1 + (h\upsilon/k_{\mathrm{B}}T)}{\{1 - (h\upsilon/k_{\mathrm{B}}T) - 1\}^2} = 3(1 + (h\upsilon/k_{\mathrm{B}}T)) \approx 3. \tag{2.44}$$

This result is essentially equivalent to the law stated by Dulong and Petit (Petit and Dulong 1819). In 1819, they found experimentally that the molar heat capacity at constant volume for many solid elements was approximately equal to 25 J mol^{-1} K^{-1}, which is about $3R$ – at that time, the universal gas constant R had not yet been defined.

2.4.3 What Is the Change of Gibbs Function Associated with the Formation of a Mixture of Gases?

In subsequent material, the significance of the Gibbs function will become apparent and so the reader is invited to look forward to Chapter 3 to explore the definition of the Gibbs function $G = U + pV - TS = H - TS$. It is also worth saying that the Gibbs function, which is a measure of energy, is the appropriate thermodynamic variable for conditions where we specify temperature and pressure, that are themselves the most easily controlled experimental variables. Here we use the Gibbs function to consider the formation of a perfect gas mixture at a specified pressure from the two separate pure gases at the same pressure.

For any species, j, in a mixture of substances we have from Equation 2.9

$$\lambda_j = \frac{N_j}{q_j}. \tag{2.45}$$

We also have from Question 3.4,

$$G = \sum_j N_j \mu_j \tag{2.46}$$

and in view of the definition (Equation 2.5)

$$\mu_j \overset{\mathrm{def}}{=} k_{\mathrm{B}}T \ln \lambda_j, \tag{2.47}$$

we can cast Equation 2.46 as

$$G = k_{\mathrm{B}}T \sum_j N_j \ln(N_j/q_j). \tag{2.48}$$

For a perfect gas, there are only translational modes of energy so that Equation 2.24 provides the partition function, q_j, for each species, which we can write as

$$q_j = \left(\frac{2\pi m_j k_B T}{h^2}\right)^{3/2} V = Y_j V. \tag{2.49}$$

Furthermore, for a perfect gas mixture at a pressure, p, the volume is

$$V = \frac{k_B T}{p}\sum_j N_j \tag{2.50}$$

and the j species act entirely independently of each other. For the pure components at the same pressure, the volume is

$$V_j = \frac{k_B T}{p} N_j. \tag{2.51}$$

Substitution of Equations 2.49 and 2.50 into Equation 2.48 yields, for the Gibbs energy of the mixture after its formation,

$$\frac{G_{\text{final}}}{k_B T} = \sum_j N_j \ln\left(\frac{N_j p}{\sum_j N_j Y_j k_B T}\right). \tag{2.52}$$

The total Gibbs energy of the j gases held separately at the same pressure is

$$\frac{G_{\text{initial}}}{k_B T} = \sum_j N_j \ln\left(\frac{p}{Y_j k_B T}\right). \tag{2.53}$$

Thus, from the definition of the molar quantity of mixing (Equation 1.19), the Gibbs function for mixing of a perfect gas at constant pressure is then given by

$$\frac{\Delta_{\text{mix}} G}{k_B T} = \frac{\Delta G_{\text{final}} - \Delta G_{\text{initial}}}{k_B T}, \tag{2.54}$$

so that

$$\frac{\Delta_{\text{mix}} G}{k_B T} = \sum_j N_j \ln\left(\frac{N_j}{\sum_j N_j}\right). \tag{2.55}$$

Dividing both sides of this equation by $\sum_j N_j$, setting that sum equal to L, Avogadro's number, and also recognizing the definition of the mole fraction (Equation 1.6), the molar Gibbs energy of mixing is

$$\Delta_{\text{mix}} G_{\text{m}} = RT\sum_j x_j \ln x_j. \tag{2.56}$$

The molar entropy of mixing can be shown by a similar method to be

$$\Delta_{\text{mix}} S_{\text{m}} = -R\sum_j x_j \ln x_j. \tag{2.57}$$

Both Equations 2.56 and 2.57 are important because they can also be derived from thermodynamic assumptions (see also Section 3.8).

The Helmholtz function, A, also defined in Chapter 3, can be written as

$$A = U - TS = G - pV = G - k_{\text{B}}T\sum_j N_j. \tag{2.58}$$

Employing Equation 2.48

$$A = -k_{\text{B}}T\sum_j N_j - k_{\text{B}}T\sum_j N_j \ln(q_j/N_j), \tag{2.59}$$

or

$$A = -k_{\text{B}}T \ln\left\{\prod_j \left(\frac{q_j^{N_j}}{N_j^{N_j}\, e^{-N_j}}\right)\right\}. \tag{2.60}$$

Defining a quantity Q called the canonical partition function of the system, we can cast Equation 2.60 as

$$A = -k_{\text{B}}T \ln Q. \tag{2.61}$$

In Question 2.5, we will discuss Q for interacting particles and how it may be used to calculate the properties of substances from statistical mechanics. However, for now, we are considering a system of independent particles when Q is given by

$$Q = \prod_j \left(\frac{q_j^{N_j}}{N_j^{N_j}\, e^{-N_j}}\right). \tag{2.62}$$

Using Stirling's approximation, which states that for large N

$$N! = N^N \mathrm{e}^{-N}, \tag{2.63}$$

Equation 2.62 can be cast as

$$Q = \prod_j \left(\frac{q_j^{N_j}}{N_j!}\right), \tag{2.64}$$

which gives a simple means to calculate A for the perfect gas mixture. Here we note that

Equation 2.63 is nearly exact already for $N > 10^3$, which is still small compared with the number of particles in a mole ($\approx 10^{23}$), in which we are usually interested.

2.4.4 What Is the Equilibrium Constant for a Chemical Reaction in a Gas?

Another important quantity to be determined is the equilibrium constant of a chemical reaction and, in certain conditions, statistical mechanics enables us to calculate this quantity. For a chemical reaction from reagents R to products P, as introduced in Question 1.3.17, and Equation 1.30, we write

$$\sum_{\mathrm{R}} (-\nu_{\mathrm{R}})\mathrm{R} = \sum_{\mathrm{P}} \nu_{\mathrm{P}}\mathrm{P}, \tag{2.65}$$

where ν is the stoichiometric number and is, by convention, negative for reactants and positive for products. Thermodynamic equilibrium in the chemical reaction is achieved when (Equation 5.23) $\sum_j \nu_j \mu_j = 0$, which means that, using Equation 2.47,

$$\prod_j (\lambda_j)^{\nu_j} = 1. \tag{2.66}$$

Using Equation 2.51 for N_j at the standard pressure $p^{\ominus}$ (Question 1.9), the standard absolute activity can be written for one mole of species j, using Equation 2.45, as

$$\lambda_j^{\ominus} = \left(\frac{x_j p}{p^{\ominus}}\right)^{-1} = \frac{N_j}{q_j} = \left(\frac{p^{\ominus} V}{k_{\mathrm{B}} T}\right)\frac{1}{q_j}. \tag{2.67}$$

In the treatment of chemical reactions in Chapter 5, one quantity which is very important is the standard equilibrium constant. From Equation 5.25, the definition of the standard equilibrium constant is given by

$$K^{\ominus}(T) = \prod_j \left\{\lambda_j^{\ominus}(T)\right\}^{-\nu_j}, \tag{2.68}$$

and the standard absolute activity of substance j is related to the standard chemical potential by the definition

$$\lambda_j^{\ominus}(T) = \exp\left(\frac{\mu_j^{\ominus}}{RT}\right). \tag{2.69}$$

For a gaseous phase with mole fractions y_j Equation 2.68 can, in light of Equation 2.67, be written as

$$K^{\ominus}(T) = \prod_j \left\{\frac{y_j p}{p^{\ominus}}\right\}^{\nu_j} = \prod_j \left\{\frac{k_{\mathrm{B}} T q_j}{V p^{\ominus}}\right\}^{\nu_j}. \tag{2.70}$$

For a reaction between diatomic gases for which the electronic modes are unexcited at a temperature T, substitution of q_j/V from Equation 2.35, yields the following

$$\ln\{K^{\ominus}(T)\} = \sum_j \nu_j \ln\left\{\left(\frac{k_B T}{p^{\ominus}}\right)\left(\frac{2\pi M_j RT}{L^2h^2}\right)^{3/2}\right\} + \sum_j \nu_j \ln\left(\frac{8\pi^2 I_j k_B T}{s_j h^2}\right) - \sum_j \nu_j \ln\left\{1 - \exp\left(-\frac{h\upsilon_j}{k_B T}\right)\right\} + \sum_j \nu_j \ln\{g(\varepsilon_{0,j})\} - \sum_j \nu_j \frac{\varepsilon_{0,j}}{k_B T}. \quad (2.71)$$

The standard equilibrium constant for the reaction posed can be evaluated from Equation 2.71 at a temperature T, if we have values of the molar mass M_j ($=Lm_j$); vibrational frequency, υ_j; the moment of inertia, I_j; the degeneracy, g_j and $\nu_j\varepsilon_{0,j}$ for each reacting substance j as well as the symmetry number, s_j. Aside from the molar mass and $\sum_j \nu_j\varepsilon_{0,j}$, all quantities can be determined spectroscopically. In order to make Equation 2.71 useful, therefore the term $\sum_i \nu_i\varepsilon_{0,i}$ in Equation 2.71 must be eliminated. To do so requires use of van't Hoff's equation (Equation 5.33)

$$\Delta H_m^{\ominus} = RT^2 \frac{d(\ln K^{\ominus})}{dT}. \quad (2.72)$$

Differentiation of Equation 2.71, and the use of Equation 2.72, gives

$$\frac{\Delta H_m^{\ominus}(T^{\ominus})}{RT} = \frac{7T^{\ominus}}{2T}\sum_j \nu_j + \frac{T^{\ominus}}{T}\sum_j \nu_j \left[\frac{h\upsilon_j/(k_B T^{\ominus})}{\{\exp(-h\upsilon/k_B T^{\ominus}) - 1\}}\right] + \sum_j \nu_j \frac{\varepsilon_{0,j}}{k_B T}, \quad (2.73)$$

where $T^{\ominus}$ is a temperature for which $\Delta H_m^{\ominus}(T^{\ominus})$ is known; as discussed in Chapter 5, $T^{\ominus}$ is usually chosen to be 298.15 K. The addition of Equations 2.71 and 2.73 gives

$$\ln\{K^{\ominus}(T)\} = \sum_j \nu_j \ln\left\{\left(\frac{k_B T}{p^{\ominus}}\right)\left(\frac{2\pi M_j RT}{L^2h^2}\right)^{3/2}\right\} + \sum_j \nu_j \ln\left(\frac{8\pi^2 I_j k_B T}{s_j h^2}\right) - \sum_j \nu_j \ln\left\{1 - \exp\left(-\frac{h\upsilon_j}{k_B T}\right)\right\} + \sum_j \nu_j \ln\{g(\varepsilon_{0,j})\} - \frac{\Delta H_m^{\ominus}(T^{\ominus})}{RT} + \frac{7T^{\ominus}}{2T}\sum_j \nu_j + \frac{T^{\ominus}}{T}\sum_j \nu_j \left[\frac{h\upsilon_j/(k_B T^{\ominus})}{\left\{\exp\left(-\frac{h\upsilon_j}{k_B T^{\ominus}}\right) - 1\right\}}\right], \quad (2.74)$$

and we see that we have replaced the requirement to obtain $\sum_j \nu_j\varepsilon_{0,j}$ from spectroscopic measurements with the need for the calorimetric determination of $\Delta H_m^{\ominus}(T^{\ominus})$. This is one reason for the effort expended in measuring precise values of $\Delta H_m^{\ominus}(T^{\ominus})$ at a temperature of 298.15 K, which the reader will find decorates the literature of chemical thermodynamics.

2.4.5 What Is the Entropy of a Perfect Gas?

From the definition of the standard molar entropy for species j (Equation 3.150), Equation 2.69 enables us to write

$$S_j^{\ominus} = -\left(\frac{\partial \mu_j^{\ominus}}{\partial T}\right)_p = -R\ln \lambda_j^{\ominus} - RT\frac{\mathrm{d}\lambda_j^{\ominus}}{\mathrm{d}T}. \tag{2.75}$$

For a perfect pure gas (see Equation 2.67), it follows that

$$S_j^{\ominus}(T) = R\ln\left\{\left(\frac{k_B T}{p^{\ominus}}\right)\left(\frac{q_j}{V}\right)\right\} + RT\left(\frac{\partial \ln q_j}{\partial T}\right)_V, \tag{2.76}$$

so that the evaluation of the entropy requires the molecular partition function for the pure gas.

For an electronically unexcited monatomic gas, Equations 2.16, 2.20 and 2.24 enable us to use Equation 2.76 to yield the entropy at the standard pressure as

$$S_j^{\ominus}(T) - R\ln\{g(\varepsilon_{0,j})\} = R\ln\left\{\frac{(2\pi M_j)^{3/2}(RT)^{5/2}}{p^{\ominus}L^4h^3}\right\} + 2.5R. \tag{2.77}$$

For an electronically unexcited diatomic gas, using Equations 2.16 and 2.35, Equation 2.76 yields for the entropy at the standard pressure

$$\begin{aligned} S_j^{\ominus}(T) - R\ln\{g(\varepsilon_{0,j})\} &= R\ln\left\{\frac{(2\pi M_j)^{3/2}(RT)^{5/2}}{p^{\ominus}L^4h^3}\right\} + R\ln\left(\frac{8\pi^2 I_j k_B T}{s_j h^2}\right) \\ &\quad - R\ln\left\{1 - \exp\left(-\frac{h\upsilon_j}{k_B T}\right)\right\} + 3.5R + R\left[\frac{h\upsilon_j/(k_B T^{\ominus})}{\left\{\exp\left(-\frac{h\upsilon_j}{k_B T^{\ominus}}\right) - 1\right\}}\right]. \end{aligned} \tag{2.78}$$

2.5 What Are Intermolecular Forces and How Do We Know They Exist?

The fact that liquids and solids exist at all means that there must exist forces that bind molecules together under some conditions so that individual molecules do not simply evaporate into the gas phase. On the other hand, we know that it is extremely hard (taking considerable energy) to compress solids and liquids so as to reduce their volume. This implies that as we try to push atoms and molecules even closer together a force acts to keep them apart. Thus, we conceive a model of intermolecular forces between two molecules that are highly repulsive at small intermolecular distances but attractive at longer distances. In this question, we develop this concept to explore the origins of these forces, how they are modeled and some other direct demonstrations of their existence.

2.5.1 What Is the Intermolecular Potential Energy?

Consider first the interaction of two spherical neutral atoms a and b. The total energy $E_{tot}(r)$ of the pair of atoms at a separation r is written as

$$E_{tot}(r) = E_a + E_b + \phi(r). \tag{2.79}$$

Here, E_a and E_b are the energies of the isolated atoms, and $\phi(r)$ is the contribution to the total energy arising from interactions between them. We call $\phi(r)$ the *intermolecular pair potential energy function* and, in the present example, it depends only on the separation of the two atoms. Since this energy is equal to the work done in bringing the two atoms from infinite separation to the separation r, it is given in terms of the intermolecular force $F(r)$ by

$$F(r) = \frac{d\phi(r)}{dr}. \tag{2.80}$$

By convention, the force F is positive when repulsive and negative when attractive.

The general forms of $\phi(r)$ and $F(r)$ are illustrated in Figure 2.1 (Maitland et al. 1981). We see as foreshadowed previously that, at short range, a strong repulsion acts between the molecules while, at longer range, there is an attractive force that decays to zero as $r \to \infty$. Consequently, the potential energy, $\phi(r)$, is large and positive at small separations but negative at longer range. It is known that, for neutral atoms at least, there is only one minimum and no maximum in either $F(r)$ or $\phi(r)$. The parameters σ, r_o and ε usually employed to characterize the intermolecular pair potential energy are defined in Figure 2.1. σ is the separation at which the potential energy crosses zero, r_o is the separation at which $\phi(r)$ is minimum and $-\varepsilon$ is the minimum energy.

For molecules that are not spherically symmetric, the situation is more complex because the force between the molecules, or equivalently the intermolecular potential energy, depends not just upon the separation of the center of the molecules but also upon the

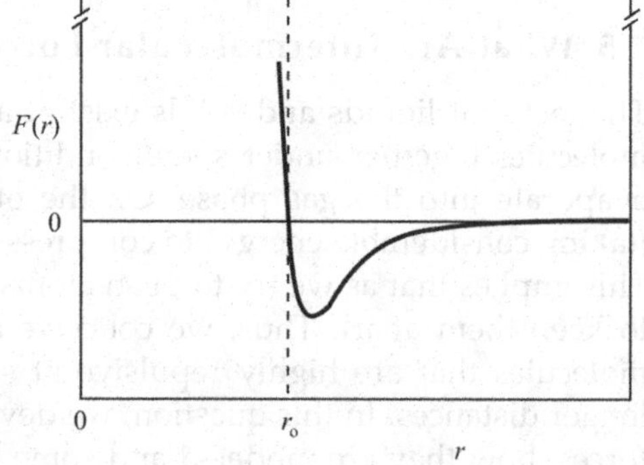

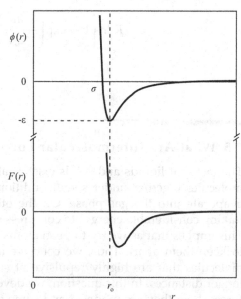

FIGURE 2.1
The intermolecular pair potential energy $\phi(r)$ and force $F(r)$ as a function of r about the equilibrium separation r_0.

orientation of the two molecules with respect to each other. Thus, the intermolecular potential is not spherically symmetric. We shall consider this in a little more detail later.

In general, the potential energy, U, of a cluster of molecules is a function of the intermolecular interactions, which in turn depend upon the type and number of molecules under consideration, the separation between each molecule and their mutual orientation. The term *configuration* is used to define the set of coordinates that describe the relative position and orientation of the molecules in a cluster.

To estimate the potential energy of a configuration it is usual, and often necessary, to make some or all of the following simplifications

1. The term *intermolecular pair-potential energy* is used to describe the potential energy involved in the interaction of an isolated pair of molecules. It is very convenient to express the total potential energy U of a cluster of molecules in terms of this pair potential ϕ. This leads to a very important assumption, the *pair additivity approximation,* according to which the total potential energy of a system of molecules is equal to the summation of all possible pair interaction energies. This implies that the interaction between a pair of molecules is unaffected by the proximity of other molecules.
2. The second important assumption is that the pair-potential energy depends only on the separation of the two molecules. As we have argued, this assumption is strictly valid only for monatomic species where, owing to the spherical symmetry, the centers of molecular interaction coincide with the centers of mass.
3. Finally, since the intermolecular potential is known accurately for only a few simple systems, model functions need to be adopted in most cases. Typically, such models give U as a function only of the separation between molecules, but nevertheless the main qualitative features of molecular interactions are incorporated.

For a system of N spherical molecules, the general form of the potential energy U may be written as

$$U(\mathbf{r}_1, \mathbf{r}_2, \ldots, \mathbf{r}_N) = \sum_{i=1}^{N-1} \sum_{j=i+1}^{N} \phi_{ij} + \Delta\phi_N, \tag{2.81}$$

where ϕ_{ij} is the potential energy of the isolated pair of molecules i and j, and ϕ_N is an increment to the potential energy, characteristic of the whole system, over and above the strictly pairwise additive interactions. According to the pair-additivity approximation, this reduces to

$$U(\mathbf{r}_1, \mathbf{r}_2, \ldots, \mathbf{r}_N) = \sum_{i=1}^{N-1} \sum_{j=i+1}^{N} \phi_{ij} = \sum_{i<j} \phi_{ij}. \tag{2.82}$$

The approximation of Equation 2.82 implies that the N-body interactions (with $N > 2$) are negligible compared with the pairwise interactions. In fact, many-body forces are known to make a small but significant contribution to the total potential energy when $N \geq 3$ and, for systems at higher density, the pair-additivity approximation can lead to significant errors. However, it is often possible to employ an effective pair potential that gives satisfactory results for the dense fluid while still providing a reasonable description of dilute-gas properties.

2.5.2 What Is the Origin of Intermolecular Forces?

Intermolecular forces are known to have an electromagnetic origin (Maitland et al. 1981) and the main contributions are well established. The strong repulsion that arises at small separations is associated with overlap of the electron clouds. When this happens, there is a reduction in the electron density in the overlap region leaving the positively charged nuclei incompletely shielded from each other. The resulting electrostatic repulsion is referred to as an *overlap force*. At greater separations, where attractive forces predominate, there is little overlap of electron clouds and the interaction arises in a different manner. Here, the attractive forces are associated with electrostatic interactions between the essentially undistorted charge distributions that exist in the molecules.

There are, in fact, three distinct contributions to the attractive forces that will be discussed here only briefly; for a more detailed description, the reader is referred to the specialized literature (Maitland et al. 1981). For polar molecules, such as HCl, the charge distribution in each molecule gives rise to a permanent electric dipole and, when two such molecules are close, there is an *electrostatic force* between them that depends upon both separation and orientation. The force between any two molecules may be either positive or negative, depending upon the mutual orientation of the dipoles, but the averaged net effect on the bulk properties of the fluid is that of an attractive force.

Such electrostatic interactions are not associated exclusively with dipole moments. Molecules, such as CO, which have no dipole moment, but a quadrupole moment also have electrostatic interactions of a similar nature. These interactions exist in general when both molecules have one or more nonzero multipole moments.

There is a second contribution to the attractive force that exists when at least one of the two molecules possess a permanent multipole moment. This is known as the *induction force* and it arises from the fact that molecules are polarizable so that a multipole moment is induced in a molecule when it is placed in any electric field, including that of another molecule. Thus, a permanent dipole moment in one molecule will induce a dipole moment in an adjacent molecule. The permanent and induced moments interact to give a force that is always attractive and, at long range, proportional to r^{-6}.

The third contribution to the attractive force, and the only one present when both molecules are non-polar, is known as the *dispersion force*. This arises from the fact that even nonpolar molecules generate fluctuating electric fields associated with the motion of the electrons. These fluctuating fields around one molecule give rise to an induced dipole moment in a second nearby molecule and a corresponding energy of interaction. Like induction forces, dispersion forces are always attractive and, at long range, vary as r^{-6} to leading order.

2.5.3 What Are Model Pair-Potentials and Why Do We Need Them?

The difficulties encountered in the evaluation of the intermolecular pair-potential energy from an *ab initio* basis have led to the adoption of the following heuristic approach. We use the spherically symmetric potential as an example. The evaluation procedure starts with the assumption of an analytical form for the relationship between the potential energy, ϕ, and the distance, r, between molecules. Subsequently, macroscopic properties are calculated using the appropriate molecular theory. Comparisons between calculated and experimental values of these macroscopic properties provide a basis for the determination of the parameters in the assumed intermolecular potential energy function. Finally, predictions may be made of thermodynamic properties of the substance in regions where experimental information is unavailable.

In the following questions, we present some of the most widely used model potential-energy functions. For a more comprehensive discussion the reader is referred to specialized literature (Maitland et al. 1981).

2.5.3.1 What Is the Hard-Sphere Potential?

In this model, the molecules are assumed to behave as hard spheres of diameter σ. It is apparent that the minimum possible distance between the molecules is then equal to σ and that the energy needed to bring the molecules closer together than $r = \sigma$ is infinite, as shown in Figure 2.2. For separations $r > \sigma$, there is no interaction between the molecules. The mathematical form of the potential is given by the following discontinuous function

$$\begin{aligned} \phi(r) &= \infty \quad r < \sigma \\ \phi(r) &= 0 \quad r \geq \sigma. \end{aligned} \tag{2.83}$$

Although this model is not very realistic, it does incorporate the basic idea that the molecules themselves occupy some of the system volume. The hard-sphere model is especially important in the theory of the transport properties of dense fluids.

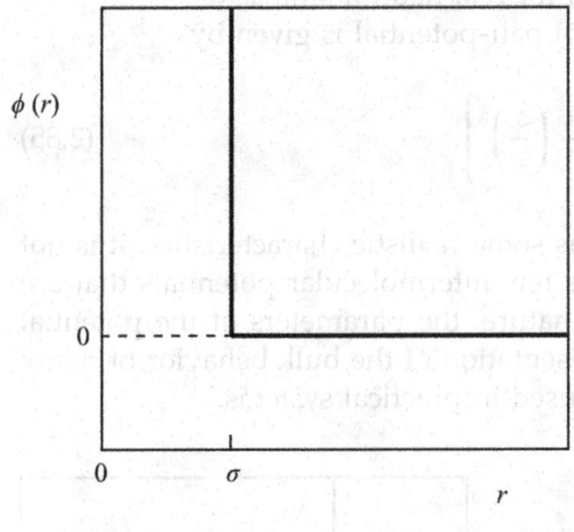

FIGURE 2.2
Hard-sphere potential $\phi(r)$ as a function of r.

2.5.3.2 What Is the Square Well Potential?

This potential function is a more realistic one in the sense that it includes an attractive potential field, of depth ε and range $g\sigma$, surrounding the spherical hard core shown in Figure 2.3. Commonly used values of g are between 1.5 and 2.0. The mathematical form of the model is

$$\begin{aligned} \phi(r) &= \infty \quad r < \sigma \\ \phi(r) &= -\varepsilon \quad \sigma \leq r < g\sigma \\ \phi(r) &= 0 \quad r \geq g\sigma. \end{aligned} \tag{2.84}$$

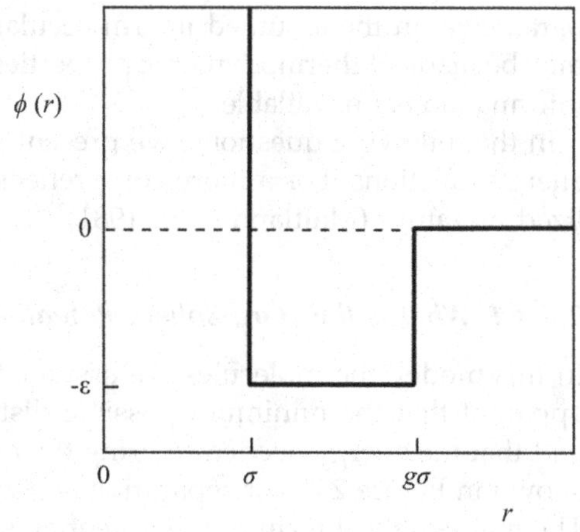

FIGURE 2.3
Square-well potential $\phi(r)$ as a function of r.

2.5.3.3 What Is the Lennard-Jones (12-6) Potential?

The Lennard-Jones (12-6) potential, illustrated in Figure 2.4, accounts for both attractive and repulsive energies and assumes that the interaction between the molecules occurs along the line joining their centers of mass. It is one of the most commonly used models owing to its mathematical simplicity and the fact that it embodies the most important features of many real interactions, especially because its attractive component conforms to the leading term for the dispersion interaction for real neutral atoms.

The functional form of the Lennard-Jones (12-6) pair-potential is given by

$$\phi(r) = 4\varepsilon\left[\left(\frac{\sigma}{r}\right)^{12} - \left(\frac{\sigma}{r}\right)^{6}\right]. \tag{2.85}$$

Although the model (given by Equation 2.85) has some realistic characteristics, it is not actually an accurate representation of any of the few intermolecular potentials that are well known. However, despite its approximate nature, the parameters of the potential model can be chosen so as to give a useful representation of the bulk behavior of many real systems and for that reason it is very often used in practical systems.

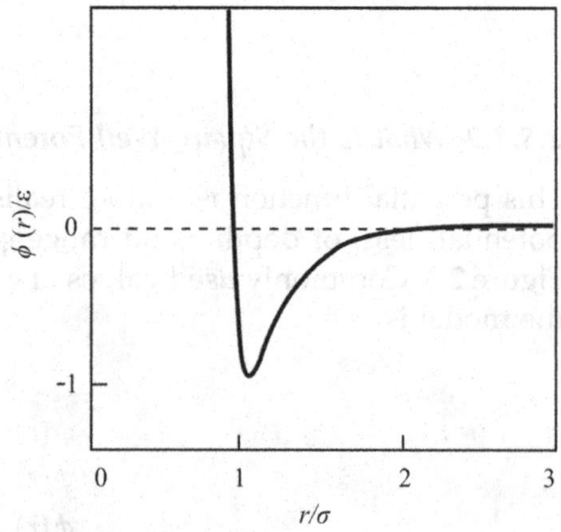

FIGURE 2.4
Lennard-Jones (12-6) potential $\phi(r)$ as a function of r.

2.5.3.4 What Is the Potential for Nonspherical Systems?

In the more general case of the interaction of polyatomic molecules, the angular dependence of the potential must be considered as we have illustrated. It may be necessary to include up to five angular variables to describe the relative orientation of a pair of molecules explicitly. However, should we wish to do so, we can still think in terms of a one-dimensional function for any fixed orientation of the molecules. As an example, Figure 2.5 shows two cross-sections through a model potential which has been proposed for the system Ar+CO_2. In this case, the potential is quite strongly anisotropic and the parameters σ and ε characterizing the interaction along different paths of fixed orientation show marked differences.

Clearly, the exact mathematical description of such potentials is very complicated. The key features of nonspherical molecules that give rise to anisotropic forces are

1. the nonspherical "core" geometry which dominates the anisotropy of the repulsive part of the potential; and
2. the presence of electric multipoles, especially dipole or quadrupole moments, which give rise to anisotropic electrostatic forces that may be dominant at longer range.

This last point is of considerable importance and dipolar forces are often included in model intermolecular pair potentials where appropriate. The most common model that includes such forces is the *Stockmayer potential* which consists of a central Lennard-Jones (12-6) potential plus the energy of interaction of two dipole moments

$$\phi(r, \theta_1, \theta_2, \psi) = 4\varepsilon\left[\left(\frac{\sigma}{r}\right)^{12} - \left(\frac{\sigma}{r}\right)^{6}\right] - \frac{\mu^2}{4\pi\varepsilon_0 r^3}(2\cos\theta_1\cos\theta_2 - \sin\theta_1\sin\theta_2\cos\psi). \quad (2.86)$$

In Equation 2.86, the angles θ_1, θ_2 and ψ define the mutual orientation of the dipole moments. θ_i is the angle made between the dipole moment on molecule i and the intermolecular axis, while ψ is the relative azimuthal angle between the two dipoles about the same axis.

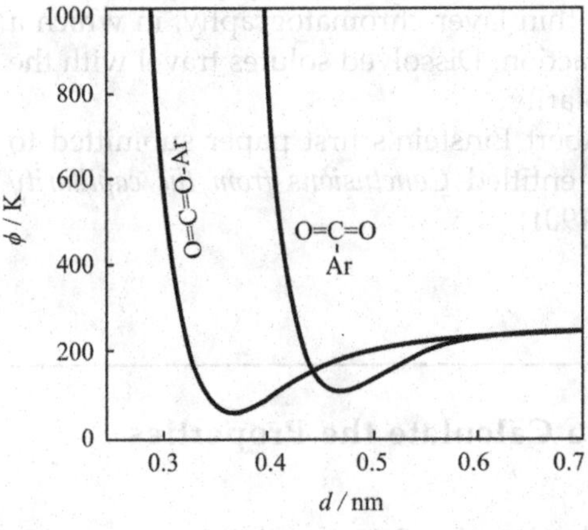

FIGURE 2.5
Cross-sections through a pair potential for the system Ar + CO_2 for a "T" shaped and a linear configuration as a function of the separation d between the centers of mass of Ar and CO_2.

2.5.4 Is There Direct Evidence of the Existence of Intermolecular Forces?

Capillary action or wicking is the ability of a porous substance to draw another liquid substance into it. A common example is the tubes in the stems of plants, but this can also be seen readily with porous paper. Capillary action occurs when the attractive intermolecular forces between the liquid at a surface and (usually) a solid substance are stronger than the cohesive intermolecular forces in the bulk of the liquid. If the solid surface is vertical, the liquid "climbs" the wall made by the solid and a concave meniscus forms on the liquid surface.

A common apparatus used to demonstrate capillary action is the *capillary tube*. When the lower end of a vertical glass tube is placed in a liquid such as water, a concave meniscus is formed. Surface tension pulls the liquid column up until there is a sufficient mass of liquid for gravitational forces to balance the intermolecular forces. The weight of the liquid column is proportional to the square of the tube's diameter, but the contact length (around the edge) between the liquid and the tube is proportional only to the diameter of the tube, so a narrow tube will draw a liquid column higher than a wide tube. For example, a glass capillary tube 0.5 mm in diameter will lift a column of water approximately 2.8 mm.

With some pairs of materials, such as mercury and glass, a convex meniscus forms and capillary action works in reverse so that the liquid is depressed in the tube relative to that in the absence of interfacial forces. These forces are known generally as surface tension or, more properly, as interfacial tension since they may arise at any interface between different materials. The inclusion of these forces in thermodynamics is briefly discussed in Chapter 4.

There are many areas where capillary action is important. In hydrology, capillary action describes the attraction of water molecules to soil particles. Capillary action is responsible for moving groundwater from wet areas of the soil to dry areas. It is also essential for the drainage of constantly produced tear fluid from the eye: two canaliculi of tiny diameter are present in the inner corner of the eyelid, also called the lachrymal ducts; their openings can be seen with the naked eye within the lachrymal sacs when the eyelids are turned inside-out. Paper towels absorb liquid through capillary action, allowing a fluid to be transferred from a surface to the towel. The small pores of a sponge act as narrow capillaries, causing it to adsorb a comparatively large amount of fluid. Some modern sport and exercise fabrics use capillary action to "wick" sweat away from the skin. These are often referred to as wicking fabrics, presumably after the capillary properties of a candle wick. Chemists utilize capillary action in thin layer chromatography, in which a solvent moves vertically up a plate via capillary action. Dissolved solutes travel with the solvent at various speeds depending on their polarity.

Maybe it is, finally, worth mentioning that Albert Einstein's first paper submitted to Annalen der Physik was on capillarity. It was entitled *Conclusions from the capillarity phenomena* and was published in 1901 (Einstein 1901).

2.6 Can Statistical Mechanics Be Used to Calculate the Properties of Real Fluids?

The idealized systems that have been examined in Question 2.4 are of immense value as limiting cases approached occasionally by real systems. The analysis presented is necessarily simplified in a number of ways compared to that which needs to be applied to real materials. The majority of the differences between real systems and the idealized models we have considered lie in the fact that the non-interacting particles of the idealized system must be replaced by particles that interact. In the case of molecular entities, they interact through intermolecular forces which can affect the total energy of the ensemble of molecules because the total internal energy is not simply the sum of that of individual molecules. It is this difference which is the subject of this question where we illustrate the use of statistical mechanics for the evaluation of the thermodynamic properties of fluids. We are not attempting to be comprehensive in this question, and the reader is referred to specialized texts for greater detail and breadth (e.g. McQuarrie 2000).

2.6.1 What Is the Canonical Partition Function?

As has been explained previously, the role of statistical mechanics is that of a bridge between the microscopic and macroscopic descriptions of the system. The statistical mechanics of systems at equilibrium, from which the thermodynamic properties may be obtained, is based upon two postulates. The first postulate, concerning the probability distribution of molecules occupying available energy microstates and how this relates to bulk thermodynamic properties, was introduced earlier in Question 2.1, and has enabled us to evaluate some of the properties of some idealized systems. To try to calculate the properties of more complex systems that are less than ideal in some way, in particular where the molecules interact with each other, we need to move away from the single molecular partition function discussed earlier to the canonical partition function, Q. To introduce this concept, we first consider a real system in a thermodynamic state defined by the macroscopic variables of thermodynamics and consisting of N molecules. The individual molecules in this system are in an unknown quantum state, but we know that a very large number of systems must exist in which individual molecules are in different states, but the overall thermodynamic state is the same. The collection of all of these possible systems is consistent with the real system, each of which is a unique quantum state of the system, called the canonical ensemble.

The second postulate of statistical mechanics states that the only dynamic variable upon which the quantum states of the entire canonical ensemble depend is the total ensemble energy. From this postulate, we deduce that all states of the ensemble having the same energy are equally probable. It can then be shown (Sandler 2010; Reed and Gubbins 1973; Hill 1988) that the probability Π_i that a system selected at random from the ensemble will be found in quantum state i varies exponentially with the energy E_i of that state. That is

$$\Pi_i(E_i) \propto \exp\left(-\frac{E_i}{k_B T}\right). \tag{2.87}$$

Since, however, there is unit probability that the system resides in *some* state, we have that $\sum_i \Pi_i(E_i) = 1$ and

$$\Pi_i(E_i) = \frac{\exp\{-E_i/(k_B T)\}}{Q}, \tag{2.88}$$

where

$$Q(N, V, T) = \sum_i \exp\left(-\frac{E_i}{k_B T}\right). \tag{2.89}$$

Equation 2.89 defines the quantity Q, known as the *canonical partition function*, which plays a central role in statistical thermodynamics. It does not have a well-defined physical meaning, but it serves as a useful statistical device in terms of which all of the thermodynamic properties of a system may be expressed. We now examine the relation between the thermodynamic properties and the canonical partition function for the most general case.

The internal energy of the system is just the ensemble average system energy. Following the first postulate of statistical mechanics, the ensemble average of the energy is defined as

$$\langle E \rangle = \sum_i E_i \Pi_i, \tag{2.90}$$

where Π_i is the probability that a system chosen at random from the ensemble will be found in the quantum state i with energy E_i. According to the first postulate, this ensemble average will approach the thermodynamic internal energy U of the real system as $N \to \infty$,

$$U = \lim_{N\to\infty} \sum_i E_i \Pi_i. \tag{2.91}$$

Combining Equation 2.89 and Equation 2.91 with Equation 2.88, we obtain the expression

$$U = \frac{1}{Q}\sum_i E_i \exp\left(-\frac{E_i}{k_B T}\right), \tag{2.92}$$

which, in view of the definition of Q, may be written as

$$U = k_B T^2 \left(\frac{\partial \ln Q}{\partial T}\right)_{N,V}. \tag{2.93}$$

Equation 2.93 provides a direct relation between the internal energy and the canonical partition function.

To obtain an expression for the entropy in terms of the partition function, we compare the relation between internal energy and the probability function with the Second Law of Thermodynamics (see Question 3.2.1 and Equation 3.7). According to macroscopic thermodynamics, the fundamental equation for a change in the state of a system of fixed composition is Equation 3.31

$$\mathrm{d}U = T\mathrm{d}S - p\mathrm{d}V. \tag{2.94}$$

Now, according to our statistical-mechanical arguments, when N is constant, a change in the internal energy of the system can occur only if either the probability function or the energy levels change. Thus,

$$\mathrm{d}U = \sum_i E_i \mathrm{d}\Pi_i + \sum_i \Pi_i \mathrm{d}E_i. \tag{2.95}$$

Let us start with the second term of Equation 2.95. With N constant, the energy levels may change only if the volume changes and, hence, $\mathrm{d}E_i = (\partial E_i/\partial V)\,\mathrm{d}V$. Thus, comparing Equations 2.94 and 2.95, we see that

$$p\mathrm{d}V = -\sum_i \Pi_i \mathrm{d}E_i \tag{2.96}$$

and thus

$$T\mathrm{d}S = -\sum_i E_i \mathrm{d}\Pi_i. \tag{2.97}$$

To obtain the entropy, we eliminate the energy levels E_i from Equation 2.97 in favor of the partition function Q. We do this by obtaining an expression for E_i from the logarithm of Equation 2.88, with the result

$$T\mathrm{d}S = -k_B T\left(\sum_i \ln \Pi_i \mathrm{d}\Pi_i + \ln Q \sum_i \mathrm{d}\Pi_i\right) = -k_B T \sum_i \ln \Pi_i \mathrm{d}\Pi_i, \tag{2.98}$$

where we have used the fact that $\Sigma_i\ \Pi_i = 1$ and hence $\Sigma_i\ \mathrm{d}\Pi_i = 0$.

Equation 2.98 can be also written as

$$\mathrm{d}S = -k_B \mathrm{d}\left(\sum \Pi_i \ln \Pi_i\right) \tag{2.99}$$

and, since $\mathrm{d}S$ is an exact differential, we see that the right-hand side of this equation is the product of a constant and an exact differential. We may, therefore, integrate Equation 2.99 directly with the result

$$S = -k_B \sum_i \Pi_i \ln \Pi_i. \tag{2.100}$$

Finally, using Equations 2.88 and 2.89 to eliminate Π_i in favor of Q, we obtain

$$S = k_B T\left(\frac{\partial \ln Q}{\partial T}\right)_{N,V} + k_B \ln Q, \tag{2.101}$$

which is the desired relation between S and Q.

We now have expressions for both U and S in terms of Q, from which the Helmholtz free energy, A, can readily be obtained through the relation

$$A = U - TS. \tag{2.102}$$

Combining Equations 2.91, 2.92 and 2.102, we find that A is given by the simple relation

$$A = -k_B T \ln Q. \tag{2.103}$$

Since A is the characteristic state function for the choice of N, V and T as the independent variables, all of the other thermodynamic properties follow from this quantity. For example, from Equation 2.102, we have

$$\mathrm{d}A = \mathrm{d}U - T\mathrm{d}S - S\mathrm{d}T. \tag{2.104}$$

We shall see in Chapter 3 that for a closed phase of fixed composition the second law requires that

$$dU - TdS = -pdV, \tag{2.105}$$

so that the total differential of Equation 2.104 is

$$dA = -pdV - SdT, \tag{2.106}$$

and

$$dA = \left(\frac{\partial A}{\partial V}\right)_T dV + \left(\frac{\partial A}{\partial T}\right)_V dT, \tag{2.107}$$

from which we can deduce that

$$p = -\left(\frac{\partial A}{\partial V}\right)_T, \tag{2.108}$$

so that

$$p = k_B T \left(\frac{\partial \ln Q}{\partial V}\right)_{N,T}. \tag{2.109}$$

2.6.2 Why Is the Calculation So Difficult for Real Systems?

The difficulty of applying statistical mechanics to the evaluation of all the thermodynamic properties of real systems is twofold. First, the energy of the system of molecules in a real system arises not just from the energies of individual isolated molecules but the energies arising from their interactions with each other in pairs or other many-body configurations. Those interactions as a function of the distance between the atoms or molecules are not, in general, available. It has been pointed out (Question 2.1) that the potential energy that characterizes the forces between just two atoms or molecules at a time has been evaluated theoretically for only a small number of systems, such as helium, hydrogen, nitrogen and water. For other systems, the forces have been deduced empirically (Stone 2013; Israelachvili 2011; Maitland et al. 1981).

Even if the interaction energies were known with great precision, to compute the thermodynamic properties of such systems exactly for the large number of molecular interactions that would be involved is evidently a very large problem that is beyond even the fastest computers today. Consequently, a means of sampling the ensemble has been introduced, followed by various means of averaging through techniques known as equilibrium molecular simulation. This subject is beyond the scope of this book and an interested reader is referred for example to (McQuarrie 2000).

Because of the difficulties of exact calculation of the properties, while possible in principle, a series of methods have been developed which rely on models of systems and they have provided the basis of much of the development of the engineering application of the properties of fluids. To introduce these approaches, we first characterize a number of limiting models. We then sketch the development of statistical mechanics for real systems and quote results derived elsewhere in the interests of brevity. In this question, we are more interested in the practical application of the methods than their derivation. A reader wishing to know more than we can include here is invited to consult a number of suitable texts such as van Ness (2015); Poling et al. (2001); Prausnitz et al. (1998) and Assael et al. (1996).

2.7 What Are Real, Ideal and Perfect Gases and Fluids?

At very low pressures, every gas conforms to the very simple, ideal equation of state for n moles

$$pV = nRT, \tag{2.110}$$

which is also the equation of state for the perfect gas, composed of infinitesimal particles that exert no forces on each other.

The behavior of a real material is shown in Figure 2.6, alongside that for the perfect gas in a general (p, V, T) diagram. The diagram reveals the liquid and solid phases as well as the vapor phase of a real substance. The behavior of even the vapor phase of this real system departs considerably from that embodied in Equation 2.110. The very existence of the liquid and solid phases is a result of the attractive forces that hold the molecules together, while their incompressibility reveals the strong repulsive forces that must exist between the same molecules at small separations.

The transition between vapor and liquid phases received systematic attention in 1823 from Faraday, but it was not until the work of Andrews on carbon dioxide in 1869 that the volumetric and phase behavior of a pure fluid was established over appreciable ranges of temperature and density. This behavior is illustrated by the three-dimensional phase diagram shown, together with its projections on to the p-V and p-T planes, in Figure 2.6. The pioneering experimental work of Andrews and others paved the way for the modern view of the equation of state and led van der Waals to postulate in his dissertation in 1873 "On the continuity of the gas and liquid states" the famous equation of state that now bears his name, and which described for the first time gas and liquid phases.

The general equation describing the behavior of a real fluid is usually written in terms of the compressibility factor, Z, as

$$Z = \frac{pV}{nRT}, \tag{2.111}$$

where p the pressure, V the volume, n the number of moles, R the universal gas constant and T the thermodynamic temperature.

When the compressibility factor equals unity, then Equation 2.111 reduces to Equation 2.110, which can be written as

$$p = \rho_{\mathrm{n}} RT, \tag{2.112}$$

where ρ_{n} is the amount-of-substance density.

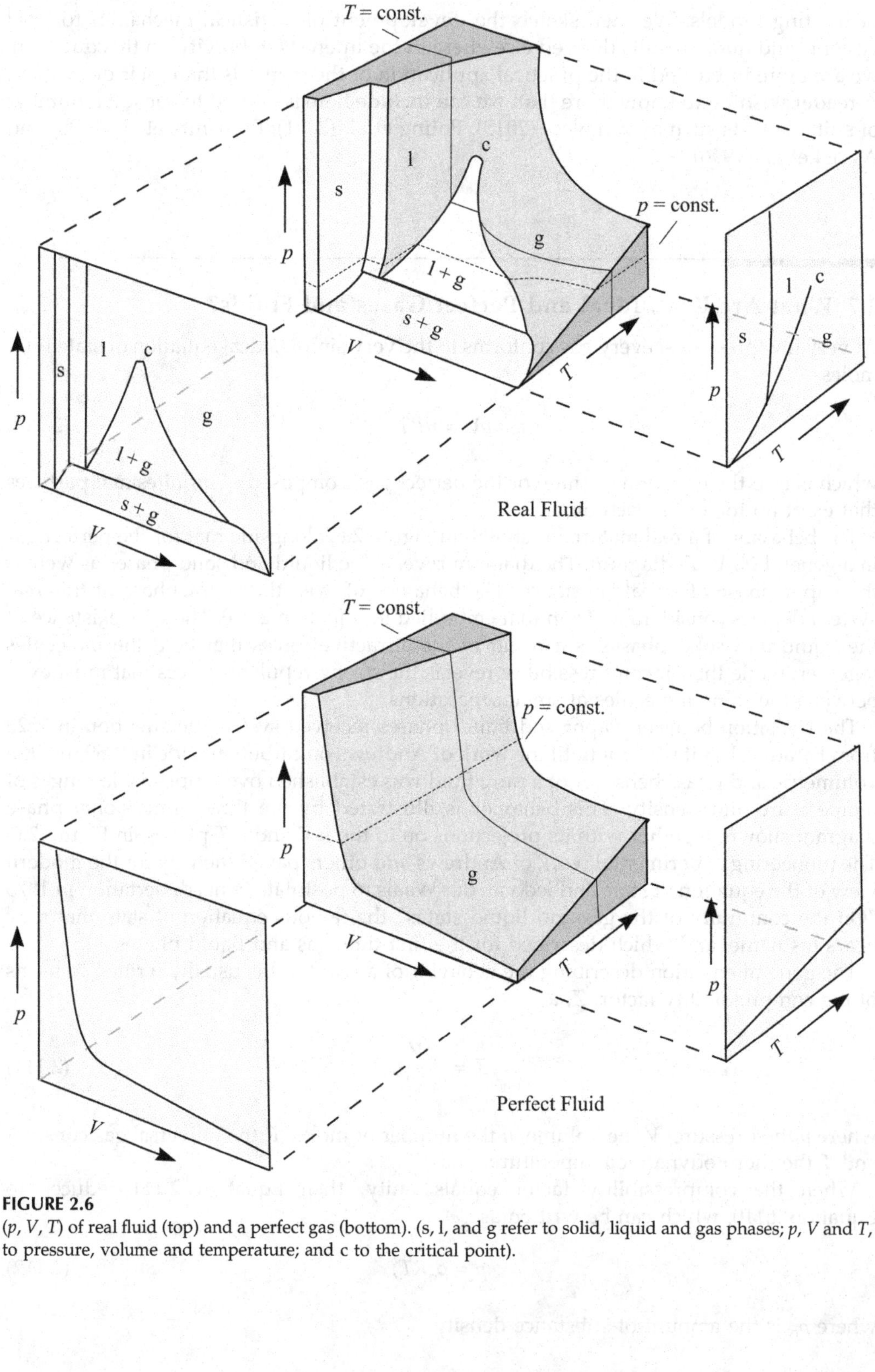

FIGURE 2.6
(p, V, T) of real fluid (top) and a perfect gas (bottom). (s, l, and g refer to solid, liquid and gas phases; p, V and T, to pressure, volume and temperature; and c to the critical point).

A real gas that obeys Equation 2.110 under some conditions is then called an "ideal gas", or is said to be acting as an "ideal gas"; Equation 2.110 is referred to as the ideal gas equation of state. Of course, from a different perspective, one can say that the compressibility factor, Z, is used to modify the ideal gas equation so that it can account for the real gas behavior.

As implied earlier, a "perfect gas" is the model of a material in which there are supposed to be point particles that make up the gas, which have no volume and do not interact. Hence, the perfect gas is a hypothetical substance for which the total potential energy is zero. This definition of the perfect gas implies that $p(V, T)$ properties conform exactly to Equation 2.110, which can be derived from statistical mechanical methods or from kinetic theory.

Any thermodynamic property, X, of a real fluid is usually separated into contributions arising from a perfect-gas, X^{pg}, and a residual part, X^{res}, by

$$X = X^{pg} + X^{res}. \tag{2.113}$$

In Equation 2.113, the X^{res} arises from the interactions between molecules. The calculation of the residual part from first principles would require calculation of the canonical partition function and this is, in general, an impossible task. We consider several techniques to obviate the need for this calculation in a later question. Here, we first concentrate on the perfect gas contribution.

The perfect-gas contribution can be obtained in many different ways that follow from what has been discussed in Question 2.4. We have already seen in this chapter how some of the properties of a gas, treated as a perfect gas containing molecules with translational, rotation and vibrational degrees of freedom, can be calculated. It was made clear earlier that to perform these calculations it is essential to know some properties of the molecules so that the energy levels of its quantum states (or at least molecular constants) that relate to them, such as the molecular moment of inertia or the vibrational constant, are known. In practice, the various molecular constants required for the calculation of perfect-gas properties are obtained from spectroscopic measurements of rotational, vibrational and electronic energy levels. Such data are readily available for a wide variety of molecules (Moore 1949–1958; Landolt-Bornstein 1951; Sutton 1965; Janz 1967; Herzberg 1989, 1991) and, where they are not, bond-contribution methods exist for their estimation (Howerton 1962). Tables of perfect-gas properties, based on a combination of theoretical and experimental work, are available in the literature (Selected Values of Properties of Hydrocarbons and Related Compounds, 1977, 1978), but it is now much more convenient to make use of computer programs from which the properties may be evaluated routinely (e.g. see NIST TRC Ideal Gas Database). Because of the difficulties with internal rotations and, to a lesser extent, vibration-rotation interaction, it is pragmatic to adopt empirical representations for some of the properties rather than to calculate everything directly from the partition function. One common approach (Assael et al. 1996) is to base perfect-gas property calculation on correlations of the perfect-gas specific heat capacity.

We also note that real gases are composed of molecules between which the interactions fall off rapidly with increasing separation. When such a gas is very "dilute" (i.e. the density is low), the average molecular separation becomes large and the condition $U = 0$ is fulfilled if no external fields are present. Thus, all real gases exhibit ideal gas behavior in the limit of zero density and can sometimes be modeled by the perfect gas model with sufficient accuracy at non-zero but low densities.

One technique that makes use of this limiting behavior leads to the virial equation of state that we consider in the next question.

2.8 What Is the Virial Equation and Why Is It Useful?

2.8.1 What Is the Virial Equation for a Pure Substance?

As we have seen thermodynamic properties can be expressed as a function of the canonical partition function. This partition function is itself a product of two terms. The first term, called the *molecular part,* includes information about isolated molecules and therefore depends only on the molecular properties of the system such as mass, moment of inertia, and so on. The second term, known as the *configuration integral,* contains information about the interactions between the molecules of the system, and it is the only part that depends upon the density. Intermolecular forces in real systems enter calculations through the configuration integral. Unfortunately, even with the assumption that the total energy of interaction of a set of molecules can be described solely by the sum of interactions between pairs of molecules (pair-additivity), evaluation of the configuration integral is very difficult for real systems. An alternative treatment that leads to results of great utility is to expand the configuration integral as a power series in the density about the zero-density limit. Then the *m*th coefficient of this series is rigorously related to molecular interactions in clusters of *m* molecules. Hence, provided that the series converges satisfactorily, the intractable *N*-body problem is transformed into a soluble series of 1-body, 2-body, 3-body, ⋯ problems.

The canonical partition function Q can be expressed as a product of a number of factors just as we did for the single particle partition function (Question 2.3) so that we can write

$$Q = Q^{\mathrm{T}}(N, T, V)\ Q^{\mathrm{V}}(N, T)\ Q^{\mathrm{R}}(N, T), \tag{2.114}$$

where Q^{T} is the translational partition function with obvious meanings for R and V as superscripts. Strictly, this product is an approximation because it assumes that the rotational motion of a molecule is unaffected by the density and not connected to the motion of the center of gravity of a molecule. While this may be true for small, nearly symmetric molecules, it is unlikely to be true for more complex asymmetric molecules since Q^{T} involves the energy connected with the motion of the molecules which, for a real fluid is of two kinds: the kinetic energy and the potential energy associated with the interactions between the molecules. This is conventionally expressed by splitting Q^{T} into two parts, one of which is the kinetic component identical to that of a perfect gas (see Equation 2.24) and the other is known as the configurational integral, Ω:

$$Q^{\mathrm{T}} = \frac{1}{N!}\left(\frac{2\pi m\, k_{\mathrm{B}} T}{h^2}\right)^{3N/2} \Omega(N, V, T), \tag{2.115}$$

where

$$\Omega(N, V, T) = \int_V \cdots \int \exp\left(\frac{-U(\mathbf{r}_1 \ldots \mathbf{r}_N)}{k_{\mathrm{B}} T}\right) \mathrm{d}\mathbf{r}_1 \ldots \mathrm{d}\mathbf{r}_N. \tag{2.116}$$

In Equation 2.115, the factor related to kinetic energy is easily seen to be the canonical analog of the same component for the single particle partition function (Equation 2.24). The complete derivation of Equation 2.115 is beyond the scope of this text and the reader is referred to (McQuarrie 2000).

The configurational integral is the only part that depends upon the volume, V (or the density). It is an integral involving the potential energy $U(\mathbf{r}_1 \ldots \mathbf{r}_N)$ for the entire N molecules whose positions are described by vectorial positions $(\mathbf{r}_1 \ldots \mathbf{r}_N)$; thus, Equation 2.109 for the pressure (from Equation 2.13) may be written as

$$p = k_B T \left(\frac{\partial \ln \Omega}{\partial V} \right)_{N,T}. \tag{2.117}$$

Given the earlier comments about the expansion of the configurational integral in density about the zero density limit we see from Equation 2.117 that the pressure will also be a power series of density

$$\frac{p}{\rho_n RT} = 1 + B\rho_n + C\rho_n^2 + D\rho_n^3 + \cdots . \tag{2.118}$$

Equation 2.118 is known as the virial equation of state and the coefficients of the virial series, $B, C, D, \ldots$, known as *virial coefficients*, are functions of temperature and composition but not of density. The importance of the virial equation of state lies in its rigorous theoretical foundation by which the virial coefficients appear not merely as empirical constants but with a precise relation to the intermolecular potential energy of groups of molecules. Specifically, the second virial coefficient, B, arises from the interaction between a pair of molecules; the third virial coefficient, C, depends upon interactions in a cluster of three molecules; D involves a cluster of four molecules and so on. Consequently, experimental values of the virial coefficients can be used to obtain information about intermolecular forces or, conversely, virial coefficients may be calculated from known, or assumed, intermolecular potential energy function. Moreover, exact relations can be derived for the virial coefficients of a gaseous mixture in terms of like- and unlike-molecular interactions.

The virial series converges only for sufficiently low densities. The radius of convergence is not well established theoretically, except for hard spheres, for which it encompasses all fluid densities. In real systems, the empirical evidence suggests that the series converges up to approximately the critical density. It certainly does not converge either for the liquid phase or in the neighborhood of the critical point. Furthermore, since not all of the coefficients of the virial series are known from theory or experiment, the series is usually limited in practice to densities much below the critical.

In the case when the potential energy of the system of N molecules is the sum of the interaction between all possible pairs, we can express the configuration integral, Equation 2.116, in terms of the Mayer function as (Assael et al. 1996)

$$\Omega = \int_V \cdots \int \left\{ \prod_{i=1}^{N-1} \prod_{j=i+1}^{N} (1 + f_{ij}) \mathrm{d}\mathbf{r}_1 \cdots \mathrm{d}\mathbf{r}_N \right\} \tag{2.119}$$

and expanding

$$\Omega = \int_V \cdots \int \left\{ 1 + \sum_{i=1}^{N-1} \sum_{j=i+1}^{N} f_{ij} + \cdots \right\} \mathrm{d}\mathbf{r}_1 \cdots \mathrm{d}\mathbf{r}_N, \tag{2.120}$$

where the Mayer function, f_{ij}, is

$$f_{ij} = \exp\left\{-\frac{\phi_{ij}(\mathbf{r})}{k_B T}\right\} - 1, \tag{2.121}$$

in which $\phi_{ij}(\mathbf{r})$ represents the intermolecular pair potential between molecules i and j. One can show that the third term in the expansion involves summations over the product of two f_{ij}'s, the fourth term summations over the product of three f_{ij}'s and so on. Since these higher terms involve the interaction of more than two molecules, the assumption of the pair-additivity is an especially significant approximation. The full derivation of Equation 2.118 can be found elsewhere (Reed and Gubbins 1973; McQuarrie 2000; Parthria and Beale 2011).

We now integrate Equation 2.120 term by term. The first term is readily evaluated as V^N. The second term involves interactions between all distinct pairs of molecules in the system and there are $N(N-1)/2$ such terms. However, since all the molecules in a pure material interact with each other according to the same function ϕ, we can replace f_{ij} with, say, f_{12} and integrate over the coordinates $\mathbf{r}_3 \ldots \mathbf{r}_N$ one by one. Each such integration results in the factor V so that, approximating $N(N-1)/2$ by $N^2/2$ (for large N), and integrating over the coordinates $\mathbf{r}_1$, we finally obtain the configurational integral as

$$\Omega = V^N\left\{1 + 2\pi(N^2/V)\int_0^\infty f_{12}\, r_{12}^2\, dr_{12} + \cdots\right\}. \tag{2.122}$$

The thermodynamic properties of the system all depend upon the logarithm of Ω and it is therefore useful to develop $\ln \Omega$ as a power series in $(1/V)$. This may be accomplished by noting that, at sufficiently low densities, the second and higher terms between brackets in Equation 2.122 are small so that

$$\ln \Omega = N \ln(V) + 2\pi(N^2/V)\int_0^\infty f_{12}\, r_{12}^2\, dr_{12} + \cdots. \tag{2.123}$$

Expressions for the virial coefficients can be obtained by then inserting Equation 2.123 in Equation 2.117. Then, carrying out the differentiation with respect to volume, we obtain

$$p = \frac{Nk_B T}{V}\left\{1 - 2\pi(N/V)\int_0^\infty f_{12}\, r_{12}^2\, dr_{12} + \cdots\right\}. \tag{2.124}$$

Comparison of Equation 2.124 with Equation 2.118, then shows that the second virial coefficient is given for a mole of substance by

$$B = 2\pi L\int_0^\infty \{1 - \exp(-\phi_{ij}/k_B T)\}r^2\, dr. \tag{2.125}$$

In Equation 2.125, we note that B has the dimensions of molar volume.

One can show (McQuarrie 2000; Reed and Gubbins 1973) that, in the pair additivity approximation, the third virial coefficient is given by

$$C = -\frac{8\pi^2}{3}L^2\iiint f_{12}f_{13}f_{23}\, r_{12}r_{13}r_{23}\, dr_{12}\, dr_{13}\, dr_{23}. \tag{2.126}$$

Corrections to Equation 2.126 that allow for the fact that the energy of interaction of three molecules may not be the sum of that of all pairs have been evaluated (Reed and Gubbins 1973). Expressions can also be obtained for the higher virial coefficients, although they rapidly become complicated by the increasing number of coordinates over which integrations must be performed.

For a pure gas, values for these coefficients can be obtained in the following ways:

a. Second virial coefficients may be represented rather well by one of the several model intermolecular potentials such as the square-well or Lennard-Jones models introduced in Chapter 1. Tables of reduced (dimensionless) second and third virial coefficients have been compiled for several model intermolecular potentials (Sherwood and Prausnitz 1964) and values of the scaling parameters σ and ε in the Lennard-Jones (12–6) potential are available for a large number of systems (Reid et al. 1988; Assael et al.1996). Corrections to C for the effects of nonadditivity of the intermolecular forces have also been tabulated (Sherwood and Prausnitz 1964; Poling et al. 2001).
b. It is possible to represent the first two coefficients empirically as a function of temperature by correlating values obtained from p-V-T measurements. This approach works well but it is obviously restricted to cases where measurements exist (Dymond and Smith 1980; Dymond et al. 2002; 2003).
c. Although experimental data on second virial coefficients are abundant (Dymond and Smith 1980; Dymond et al. 2002; 2003), it is often necessary to estimate values of B for substances that have not been studied in sufficient detail. Several correlations have been developed for this purpose. One of the most common for nonpolar gases is the extended corresponding-states method of Pitzer and Curl (1958) and Tsonopoulos and Prausnitz (1969) in which the virial coefficients are given as a function of the critical constants and the acentric factor (see Equation 2.140), which may be evaluated easily from vapor-pressure data (Assael et al. 1996). Third virial coefficients of nonpolar gases have also been correlated using a similar model by Orbey and Vera (1983).
d. In the special case of the hard-sphere potential, all of the virial coefficients are independent of temperature. The first eight virial coefficients have been evaluated (Maitland et al. 1981) for this system and the results are given in Table 2.1, wherein σ represents the diameter of the rigid sphere.

TABLE 2.1

Virial coefficients for the hard-sphere potential

$B = 2\pi N_A\sigma^3/3 = b_o$
$C = (5/8)\ b_o^2$
$D = 0.28695\ b_o^3$
$E = 0.11025\ b_o^4$
$F = 0.03888\ b_o^5$
$G = 0.01307\ b_o^6$
$H = 0.00432\ b_o^7$

2.8.2 What Happens to the Virial Series for Mixtures?

For gas mixtures, Equation 2.118 remains formally the same but the interpretation of the terms is different. In order that the density is the molar density of the mixture, the second and third virial coefficients of a multicomponent gas mixture are given exactly by a quadratic and a cubic expression in the mole fractions, respectively, as

$$B_{\text{mix}}(T) = \sum_{i=1}^{v} \sum_{j=1}^{v} x_i x_j B_{ij}(T) \tag{2.127}$$

and

$$C_{\text{mix}}(T) = \sum_{i=1}^{v} \sum_{j=1}^{v} \sum_{k=1}^{v} x_i x_j x_k C_{ijk}(T). \tag{2.128}$$

In Equations 2.127 and 2.128, x_i is the mole fraction of species i in the mixture of v components. In Equation 2.127, B_{ii} is the second virial coefficient of the pure species i, and B_{ij} is called the interaction second virial coefficient. B_{ij} is defined as the second virial coefficient corresponding to the potential energy function $\phi_{ij}(r)$ that describes the interaction of one molecule of species i with one of species j. B_{ij} is also referred to as the cross virial coefficient, the cross-term virial coefficient, or the mixed virial coefficient.

Depending upon the availability of experimental (p, V, T) data, one of two general approaches may be adopted when dealing with multicomponent mixtures. If the (p, V, T) data for each pure component and for some compositions of each binary and ternary mixtures have been studied in great detail, one can fit the experimental data to the virial equation truncated after, say, the third virial coefficient and derive each of the possible pure-component and interaction virial coefficients. The significant advantage provided by this approach is the use of the exact Equation 2.118 to generate the behavior of any composition. An excellent example of this approach is offered by the GERG virial equation (Kunz and Wagner 2012; Jaeschke et al. 1988) for natural-gas type mixtures. The work was supported by the EU and for the 13 specified components, a total of 297 virial coefficients were required (B_{ii}, B_{ij}, C_{iii}, C_{iij}, C_{ijj} and C_{ijk}). The resulting equation predicts the density of natural-gas mixtures of up to 13 components of arbitrary composition with an uncertainty of approximately 0.1% at pressures up to 12 MPa and at temperatures between 265 and 335 K. The ubiquity and importance of natural gas in the world has, in the past, justified the enormous effort represented by this program of measurement and analysis.

If, however, experimental measurements of second virial coefficients are not available, for example, for binary mixtures, then it is necessary to resort to predictive methods. To obtain interaction second virial coefficients, a wide range of empirical methods exist. Some apply combining rules to critical constants while others use combining rules for the parameters of simple potential models, most of which are based on the Lorentz-Berthelot combining rules (Assael et al. 1996), as well as the formulae that relate the virial coefficients to intermolecular forces that are the subject of the next question. Similarly, several methods have been proposed for the estimation of the interaction third virial coefficients C_{ijk}, for example Orbey and Vera (1983) who followed Chueh and Prausnitz (1967).

2.9 How Can I Estimate Thermodynamics Properties?

So far, we have discussed the properties of a perfect gas and a moderately dense gas and their mixtures, but many of the materials encountered in practice do not fall into those categories and yet their properties are often required in an engineering context. It is not practicable to seek to measure all of the properties of all materials that might be encountered in engineering practice over a wide range of conditions. There is, therefore, a very considerable body of literature devoted to the estimation of the properties of such materials that is well beyond the scope of this book. However, many of the better estimation methods, which have often been implemented in computer software for the prediction of properties, e.g. REFPROP (Lemmon et al. 2018), have their origins in statistical mechanical methods based around Equation 2.113. It is such methods we consider briefly here.

2.9.1 How Can the Principle of Corresponding States Be Used to Estimate Properties?

We first consider the principle of corresponding states because it underpins, in some form, the very many thermodynamic models used in engineering. Here, we will briefly describe its scientific basis for pure fluids and the extension to mixtures is discussed in Chapter 4 on the basis of very clear assumptions that provide a powerful predictive tool.

The principle of corresponding states establishes a connection between the configuration integrals of different substances and thereby allows each of the configurational thermodynamic properties of one fluid to be expressed in terms of those of another fluid. If one fluid can be selected as a reference fluid and the properties of all others related to it then the basis for powerful property prediction approaches can be created. Since configurational and residual thermodynamic properties are related in a very simple way, the same results apply also to the latter.

The theoretical basis of the two-parameter corresponding states principle is the assumption that the intermolecular potentials of two substances may be rendered identical by the suitable choice of two scaling parameters, one applied to the separation and the other to the energy. Thus, the intermolecular potential of a substance that conforms to the principle is taken to be

$$\phi(r) = \varepsilon F(r/\sigma), \tag{2.129}$$

where ε and σ are, respectively, scaling parameters for energy and distance, and F is a *universal* function among all relevant materials. Substances that obey Equation 2.129 are said to be *conformal*. One of the great strengths of the method is that the function F need not be known. Instead, a reference substance is introduced, identified by the subscript 0, for which the thermodynamic properties of interest are known, and this is used to eliminate F from the problem. The configurational (and hence, residual) properties of another conformal substance, identified by the subscript i, are thereby given in terms of those of the reference fluid. We shall also see that the parameters ε and σ may be eliminated in favor of measurable macroscopic quantities.

The consequences of conformality may be derived by means of the following thought experiment. Consider two conformal substances, one of which is designated as the reference fluid, contained in separate vessels of the same shape but different volumes, as illustrated in Figure 2.7. Let there be N molecules of type i contained in volume V at

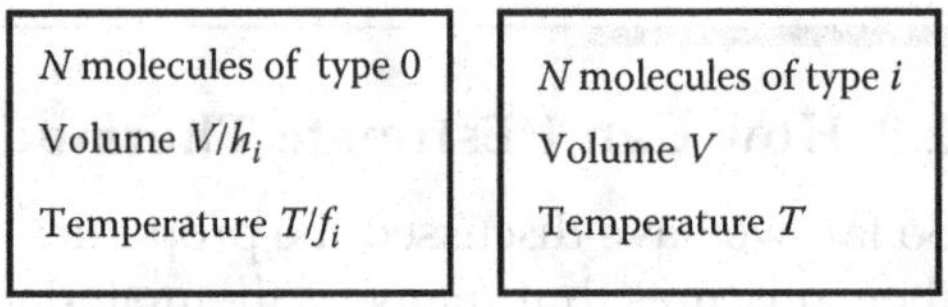

FIGURE 2.7
Corresponding states principle.

temperature T, while the N molecules of the reference fluid are contained in volume V/h_i at temperature T/f_i. Here, $h_i = (\sigma_i/\sigma_0)^3$ and $f_i = (\varepsilon_i/\varepsilon_0)$ are scaling ratios. We now suppose that the molecules are arranged in geometrical similar positions within their respective containers. Then, for each molecule in the system on the right with position vector $\mathbf{r}_i$ defined relative to the origin in that system, there is a corresponding molecule in the reference system with position vector $\mathbf{r}_0$ defined relative to the origin in that system and these position vectors are related by

$$\mathbf{r}_0 = \mathbf{r}_i/h_i^{1/3}. \tag{2.130}$$

Since it is assumed that the pair potentials are conformal and that either (i) the pair additivity approximation is obeyed, or (ii) that the N-body potentials are also conformal, the configurational energies of the two systems are related by

$$\Upsilon_0(\mathbf{r}_{0,1}, \mathbf{r}_{0,2}, \cdots, \mathbf{r}_{0,N}) = \Upsilon_i(\mathbf{r}_{i,1}, \mathbf{r}_{i,2}, \cdots, \mathbf{r}_{i,N})/f_i. \tag{2.131}$$

Equation 2.131 must apply to any configuration because for each configuration of the reference system a geometrically similar one exists for the second system.

The configuration integral Ω_0 for the reference system is given (Assael et al. 1996) by

$$\Omega_0(V/h_i, T/f_i) = \int \underset{V/h_i}{\cdots} \int \exp\left(-\frac{f_i \Upsilon_0}{k_B T}\right) d\mathbf{r}_0^N, \tag{2.132}$$

while that for the other system is

$$\Omega_i(V, T) = \int \underset{V}{\cdots} \int \exp\left(-\frac{\Upsilon_i}{k_B T}\right) d\mathbf{r}_i^N. \tag{2.133}$$

Upon changing the variables of integration from $\mathbf{r}_i$ to $\mathbf{r}_0$, in accordance with Equation 2.131, and making use of Equation 2.132, Ω_i becomes

$$\Omega_i(V, T) = \int \underset{V/h_i}{\cdots} \int \exp\left(-\frac{f_i \Upsilon_0}{k_B T}\right) h_i^N d\mathbf{r}_0^N. \tag{2.134}$$

Then, comparing Equations 2.132 and 2.134, we see that the configuration integrals of the two systems are related by the simple equation

$$\Omega_i(V, T) = h_i^N \Omega_0(V/h_i, T/f_i). \tag{2.135}$$

The compression factor is defined by Equation 2.111 and it follows from Equation 2.117 that

$$Z = \frac{V}{N}\left(\frac{\partial \ln \Omega}{\partial V}\right)_T, \tag{2.136}$$

so that it is a purely configurational property. It then follows from Equation 2.135 that

$$Z_i(V, T) = Z_0(V/h_i, T/f_i). \tag{2.137}$$

Thus, the compression factor of one conformal substance may be equated with that of another at a scaled volume and a scaled temperature. As this relation must hold also at the critical point, it follows that the scaling parameters are related to the critical constants by

$$f_i = T_{c,i}/T_{c,0}, \tag{2.138}$$

and

$$h_i = V_{c,i}/V_{c,0}, \tag{2.139}$$

and that the reduced pressure $p_r = p/p_c$ is the same function of the reduced volume $V_r = V/V_c$ and the reduced temperature $T_r = T/T_c$ in all conformal systems. Consequently, the compression factor is a universal function of T_r and V_r or, alternatively, of T_r and p_r.

Generalized equations are available (Assael et al. 1996, Lee and Kesler 1975), giving the compression factor Z, as well as the residual enthalpy and entropy in terms of the residual molar enthalpy (H_m^{res}/RT_c) and molar entropy (S_m^{res}/R), as a function of reduced temperature and pressure. The principle of corresponding states does not provide the perfect gas contribution to either the enthalpy or the entropy so that must be evaluated by alternative means, which we have already discussed in Question 2.4.

The simple treatment outlined previously will be applicable to substances that have conformal pair potentials. A group of substances for which it is nearly true is the monatomic gases Ar, Kr and Xe, for which it works remarkably well. But for He and, to some extent, Ne, deviations from the principle arise because at low temperatures quantum effects, which depend upon mass and not the potential, have to be considered. Several simple molecules, including N_2, CO and CH_4, deviate only slightly from the principle but most other molecules depart considerably.

The reasons for the conformality of the monatomic gases and the relative failure for other species rest on the fact that the former group of systems are spherically symmetric (in agreement with the assumptions of the model) while polyatomic molecules evidently do not have this symmetry. To apply the principle of corresponding states with any accuracy to molecular fluids, it is necessary to take into account the non-spherical nature of the molecules. The anisotropic nature of the intermolecular potential ϕ in these cases has already been briefly described in Question 2.5, and here we simply recall that ϕ is a function not only of the separation r but also of the relative orientation of the two molecules. Hence, in addition to the two scaling parameters described previously, others are necessary in principle. In the first attempt to deal with this problem from an engineering perspective, a third parameter was introduced, leading to a three-parameter corresponding states principle. A third parameter was first proposed by Pitzer (Pitzer 1955), who defined the acentric factor ω by

$$\omega = -1 - \log_{10}\left\{\frac{p^{sat}(T = 0.7\,T^c)}{p_c}\right\}. \tag{2.140}$$

where p_c is the critical pressure and p^{sat} the vapor pressure.

Pitzer (1955) proposed that a generalized thermodynamic property X can be written as a function of reduced temperature and pressure by

$$X(T_r, p_r) = X_0(T_r, p_r) + \omega X_1(T_r, p_r). \tag{2.141}$$

In Equation 2.141, X_0 is known as the simple fluid term and X_1 is known as the correction term. Equations representing the simple fluid and correction terms as functions of reduced temperature and pressure are available for the cases when $X = Z$, (H_m^{res}/RT_c) and (S_m^{res}/R).

Finally, to incorporate polar effects, four-parameter corresponding states models are usually employed (Wu and Stiel 1985). In this case, the extra parameter is usually obtained experimentally, and the reader is referred to the corresponding literature (Wu and Stiel 1985). We should also mention that in 2008, Xiang and Deiters proposed a new equation of state applicable to small, large, chain, hydrogen bonding and associating molecules covering wide ranges of temperature, density and pressure used in industry, where as the fourth corresponding-states parameter they defined an aspherical factor (Xiang and Deiters 2008).

To demonstrate the use of the principle of corresponding states, we show a few simple examples that are chosen to provide readers with an exposure to the estimation methods commonly employed in chemical engineering practice or in software routines that inform chemical engineering practice. The methods are all exact in some hypothetical limit but are approximate in any real case so that the results of the application of these methods should always be used with circumspection about their uncertainty.

First, if we suppose that all substances conform to the same reduced pair potential $\phi(r) = \varepsilon F(r/\sigma)$ as set out in Equation 2.129, then it is easily shown from Equation 2.125 that the second virial coefficient for all such substances obeys the simple two parameter law of corresponding states

$$B^* = \frac{B}{(2L\pi\sigma^3/3)} = \int_0^\infty \left[\frac{r}{\sigma}\right]^2 \left\{1 - \exp\left[-\frac{F(r/\sigma)}{T^*}\right]\right\} d(r/\sigma), \tag{2.142}$$

where $T^* = k_B T/\varepsilon$. For an assumed functional form of the pair potential tables of the reduced second virial coefficient can be calculated from which the real virial coefficient of any of the substances can be calculated using values for the two parameters σ and ε.

The power of the two-parameter principle of corresponding states can be demonstrated by estimating the density of argon from the density of krypton at some other temperature and pressure. In this example, the critical temperature and critical pressure are the scaling parameters. The density of krypton at T = 348.15 K and p = 2 MPa is 59.28 kg m^{-3} (Evers et al. 2002). The critical temperature, T_c, the critical pressure, p_c, and the critical mass density, ρ_c, of krypton are 209.48 K, 5.525 MPa and 909.21 kg m^{-3}, respectively (Lemmon et al. 2018). For Kr, the reduced temperature T_r, pressure p_r and mass density ρ_r are as follows

$$T_{r,Kr} = \frac{T}{T_{c,Kr}} = \frac{348.15 \text{ K}}{209.48 \text{ K}} = 1.662, \tag{2.143}$$

$$p_{r,Kr} = \frac{p}{p_{c,Kr}} = \frac{2 \text{ MPa}}{5.525 \text{ MPa}} = 0.362 \tag{2.144}$$

and

$$\rho_{\mathrm{r,Kr}} = \frac{\rho}{\rho_{\mathrm{c,Kr}}} = \frac{59.28\ \mathrm{kg.m^{-3}}}{909.21\ \mathrm{kg.m^{-3}}} = 0.0652. \tag{2.145}$$

For argon, $T_{\mathrm{c,Ar}} = 150.687$ K, $p_{\mathrm{c,\,Ar}} = 4.863$ MPa and $\rho_{\mathrm{c,\,Ar}} = 535.60$ kg m^{-3} (Lemmon et al. 2018) and when combined with Equations 2.143, 2.144 and 2.145, respectively, the temperature, pressure and density of argon are given by the following

$$T_{\mathrm{Ar}} = T_{\mathrm{r,Kr}} \cdot T_{\mathrm{c,Ar}} = 1.662 \times 150.687\ \mathrm{K} = 250.44\ \mathrm{K}. \tag{2.146}$$

$$p_{\mathrm{Ar}} = p_{\mathrm{r,Kr}} \cdot p_{\mathrm{c,Ar}} = 0.362 \times 4.863\ \mathrm{MPa} = 1.760\ \mathrm{MPa}. \tag{2.147}$$

$$\rho_{\mathrm{Ar}} = \rho_{\mathrm{r,Kr}} \cdot \rho_{\mathrm{c,Ar}} = 0.0652 \times 535.60\ \mathrm{kgm^{-3}} = 34.921\ \mathrm{kg\ m^{-3}}. \tag{2.148}$$

Hence the principle of corresponding states predicts the density for argon, from that of krypton, to be ρ_{Ar}(250.44 K, 1.760 MPa) = 34.921 kg m^{-3}, which lies 1.0% above the literature value of 34.555 kg m^{-3} (Lemmon et al. 2018). The comparison between the experiment and the corresponding-states method is obviously quite good in this case, but this is not surprising because it is known that the pair potentials of argon and krypton are almost conformal (Maitland et al. 1981). A more significant test is provided by calculating the properties of methane at the same corresponding state.

We will now use Equations 2.143, 2.144 and 2.145 to estimate the density of methane for which $T_{\mathrm{c,CH_4}} = 190.564$ K, $p_{\mathrm{c,CH_4}} = 4.5992$ MPa and $\rho_{\mathrm{c,CH_4}} = 162.658$ kg m^{-3} (Lemmon et al. 2018) to give

$$T_{\mathrm{CH_4}} = T_{\mathrm{r,Kr}} \cdot T_{\mathrm{c,CH_4}} = 1.662 \times 190.564\ \mathrm{K} = 316.717\ \mathrm{K}. \tag{2.149}$$

$$p_{\mathrm{CH_4}} = p_{\mathrm{r,Kr}} \cdot p_{\mathrm{c,CH_4}} = 0.362 \times 4.5992\ \mathrm{MPa} = 1.665\ \mathrm{MPa}. \tag{2.150}$$

$$\rho_{\mathrm{CH_4}} = \rho_{\mathrm{r,Kr}} \cdot \rho_{\mathrm{c,CH_4}} = 0.0652 \times 162.658\ \mathrm{kg\ m^{-3}} = 10.61\ \mathrm{kg\ m^{-3}}. \tag{2.151}$$

The principle of corresponding states thus predicts the density for methane, from that of krypton, to be $\rho_{\mathrm{CH_4}}$(316.717 K, 1.665 MPa) = 10.61 kg m^{-3} which lies 2.2% above the literature value of 10.377 kg m^{-3} (Lemmon et al. 2018). Methane is a nonspherical molecule while krypton is spherical, and the greater difference between the estimated and actual density is thus not surprising.

These two examples demonstrate the two-parameter principle of corresponding states, generally written as

$$X(T_r, p_r) = X_0(T_r, p_r), \tag{2.152}$$

where a property X can easily be related to the same property of another fluid X_0, at the same reduced conditions.

TABLE 2.2
The amount-of-substance density ρ_n of dodecane estimated from both Equation 2.152 and Equation 2.141 at the following temperature and pressure: (a) T = 298.15 K and p = 0.1 MPa; (b) T = 358.15 K and p = 13.8 MPa (Snyder and Winnick 1970)

	ρ_n/mol m^{-3}		
	Equation 2.152	**Equation 2.141**	**Ref**
298.15 K, 0.10 MPa	3354	4446	4375
358.15 K, 13.8 MPa	3257	4277	4193
ΔT = 60 K Δp = 13.7 MPa			

As the complexity of the molecule's structure increases and consequently the intermolecular potential is no longer purely spherical, the departure of the properties predicted with Equation 2.152 increases. Pitzer (1955) proposed a modification of Equation 2.152 that included the acentric factor ω, defined by Equation 2.140 (Lee and Kesler 1975). This approach produced a consistent scheme for the calculation of the density, enthalpy, entropy and fugacity of hydrocarbons based on the properties of octane.

We will now use Equations 2.141 and 2.152 to estimate (Assael et al. 1996) the density of dodecane at the following temperatures and pressures: (a) T = 298.15 K and p = 0.1 MPa (b) T = 358.15 K and p = 13.8 MPa. The procedures required for Equation 2.152 follow those described for Equations 2.149, 2.150 and 2.151, while those for Equation 2.141 are provided elsewhere (Assael et al. 1996). The results obtained are listed in Table 2.2, which also includes the accepted experimental values (Snyder and Winnick 1970) against which the predictions are compared. Clearly, Equation 2.141 provides the best estimates when compared with experiment. We also estimated from both Equations 2.141 and 2.152 the enthalpy change between conditions (a) and (b). Equation 2.152 provided ΔH_m = 22.5 kJ mol^{-1} while Equation 2.141 returned ΔH_m = 25.5 kJ mol^{-1}. The ΔH_m estimated with Equation 2.141 differs by less than 1.6% from the measured value, while that predicted with Equation 2.152 lies >10% below the measured (Snyder and Winnick 1970) value of ΔH_m = 25.1 kJ mol^{-1}.

For further examples of the application of two and three parameter corresponding states the reader is referred to Assael et al. (1996).

2.9.2 How Can the Statistical Association Fluid Theory (SAFT) Be Used to Estimate Properties?

Fluid mixtures of different molecular species often contain not only individual molecules of components, but also clusters of like and unlike molecules associated by hydrogen bonds or by donor-acceptor interactions. The properties of these clusters (size, energy, shape) are very different from those of the monomeric molecules and influence the bulk fluid properties of the mixture. One way to account for such association effects is to use statistical mechanical methods to quantify the relationship between well-defined site-site interactions and the bulk fluid behavior. The theory behind this approach is called Statistical Association Fluid Theory (SAFT) and the related equation of state (EOS) was developed by Chapman et al. (Chapman et al. 1989), based on the perturbation theory of Wertheim (Wertheim 1984a, 1984b). Over the years since its introduction, the SAFT EOS has been improved and developed, resulting in PC-SAFT (perturbed chain SAFT)

(Gross and Sadowski 2001) and, by adding a term for ionic compounds, in electrolyte PC-SAFT or ePC-SAFT (Cameretti and Sadowski 2005, Held et al. 2014). The reader is referred to the literature cited for full details. Here, we briefly start from the model that states that chains of molecules are linked chains of hard spheres that cannot penetrate each other but interact over distance through dispersion forces, association forces (such as hydrogen bonds) and ionic (Coulombic) forces. These interactions lead to terms in the ePC-SAFT EOS (Equation 2.153) that contribute to the residual Helmholtz energy A^{res}. In thermodynamics, a residual property is the difference between the real fluid property and an ideal gas property (see Question 2.7.).

$$\frac{A^{res}}{N} = a^{res} = a^{hc} + a^{disp} + a^{assoc} + a^{ion}. \tag{2.153}$$

Here, A^{res} is the extensive residual Helmholtz energy (Question 2.7) and N the total number of molecules. The terms a^{hc}, a^{disp}, a^{assoc}, a^{ion} account for the contribution of the hard chain (an extension of the hard-sphere potential described in Question 2.5.3.1), dispersive, associative and Coulomb interactions, respectively (see Figure 2.8).

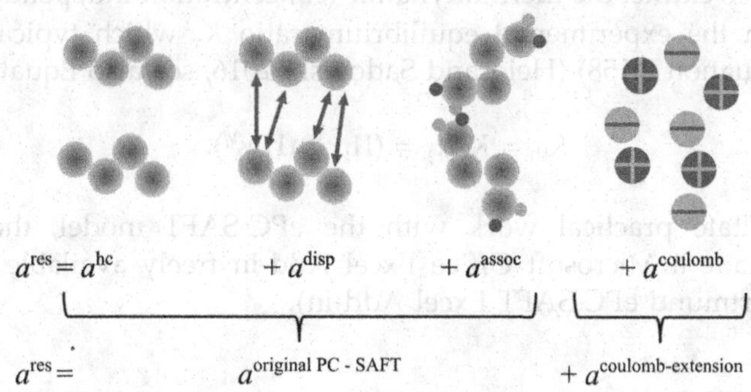

FIGURE 2.8
Contributions to the Helmholtz energy in the ePC-SAFT EOS.

For the expressions for the different contributions, the reader is referred to the paper of Held (Held et al. 2014). Important thermodynamic properties can be estimated via derivatives of the residual Helmholtz energy, a^{res}. For example, the fugacity coefficient ϕ_B (Chapter 4) of the species B in a multicomponent mixture may be written

$$\ln \phi_B = (Z - 1) - \ln Z + a^{res} + \frac{\partial a^{res}}{\partial x_B} - \sum_j x_j \left(\frac{\partial a^{res}}{\partial x_j} \right) \tag{2.154}$$

with

$$Z(\rho) = 1 + \rho \left(\frac{\partial a^{res}}{\partial \rho} \right)_{T,x}. \tag{2.155}$$

The symbols x_B, Z, ρ represent the molar fraction of the species B, the compression factor and the density, respectively. The activity coefficient of the species B then results from the fugacity coefficients so that (Question 4.6.1)

$$\gamma_B = \frac{\phi_B(T, p, x_B)}{\phi_B^*(T, p)}, \tag{2.156}$$

where $\phi_B^*(T, p)$ is the fugacity coefficient for pure B.

The activity coefficient is a key parameter for the prediction of the liquid-vapor and liquid-liquid equilibria required for the sizing of separation process equipment. How the activity coefficient can be measured and how it is used for equilibrium calculations is described in Question 4.6.

In the case of solid-liquid equilibria, the solubility of component B in the solvent is of particular interest, which results from Equation 2.157:

$$x_B^l = \frac{1}{\gamma_B}\exp\left\{-\frac{\Delta H_{0B}^{sl}}{RT}\left(1 - \frac{T}{T_{0B}^{sl}}\right)\right\}. \tag{2.157}$$

ΔH_{0B}^{sl}, T_{0B}^{sl} are the melting enthalpy and the melting temperature of the pure component B. In (bio-)chemical reactions, the ePC-SAFT concept can be used to calculate activity coefficients γ_i and so extract the thermodynamic (concentration-independent) equilibrium constant K_{th} from the experimental equilibrium ratio K_x which typically depends on concentration (Equation 2.158) (Held and Sadowski 2016, see also Equation 5.73);

$$K_{th} = K_x K_\gamma = (\Pi_i x_i^{\nu_i})(\Pi_i \gamma_i^{\nu_i}). \tag{2.158}$$

In order to facilitate practical work with the ePC-SAFT model, the University of Dortmund has made a Microsoft Office Excel Add-in freely available on the internet (University of Dortmund ePC-SAFT Excel Add-in).

2.10 What Is Entropy in Statistical Thermodynamics?

In this chapter we have repeatedly used the state variable entropy. Entropy is probably one of the properties that newcomers to thermodynamics have most problems with. These problems may be associated with the fact that – in contrast to terms such as energy or temperature – entropy is not a quantity experienced in everyday life. Moreover, there are different definitions of entropy that focus on different thermodynamically important aspects. And finally, one will find a multiplicity of introductions to entropy in various textbooks. Clausius was the first to employ the word entropy, taken from the Greek εντροπία whose translation is "*turning toward*"; Clausius preferred its interpretation as the energy of transformation (Clausius 1850a and 1850b). For the purpose of thermodynamics, the term might be taken to mean the energy lost to dissipation (Clausius 1865).

Clausius's now famous aphorism "Die Entropie der Welt strebt einem Maximum zu" that when translated yields "The entropy of the Universe tends toward a maximum" (Clausius 1865) might give rise to concepts of "mixed-upness" and indeed has been invoked to describe the fate of the universe. The latter has been argued because any process that takes place in the universe results in an increase of entropy and, it is implied, albeit wrongly, an increase of "mixed-upness." While Clausius coined the term *entropy*, the treatment of entropy in terms of statistical mechanics goes back to Boltzmann. We give the definition of entropy from the viewpoint of classical thermodynamics in Question 3.2.1 (Equation 3.5).

While entropy is, loosely speaking, often regarded as a measure for disorder, it is, more correctly, related to the probability with which a certain thermodynamic state is assumed. As an example (Meschede 2006), consider a volume that is divided by a wall with an orifice into two sub-volumes and which is – for simplicity – occupied by only four (identical) molecules (Figure 2.9).

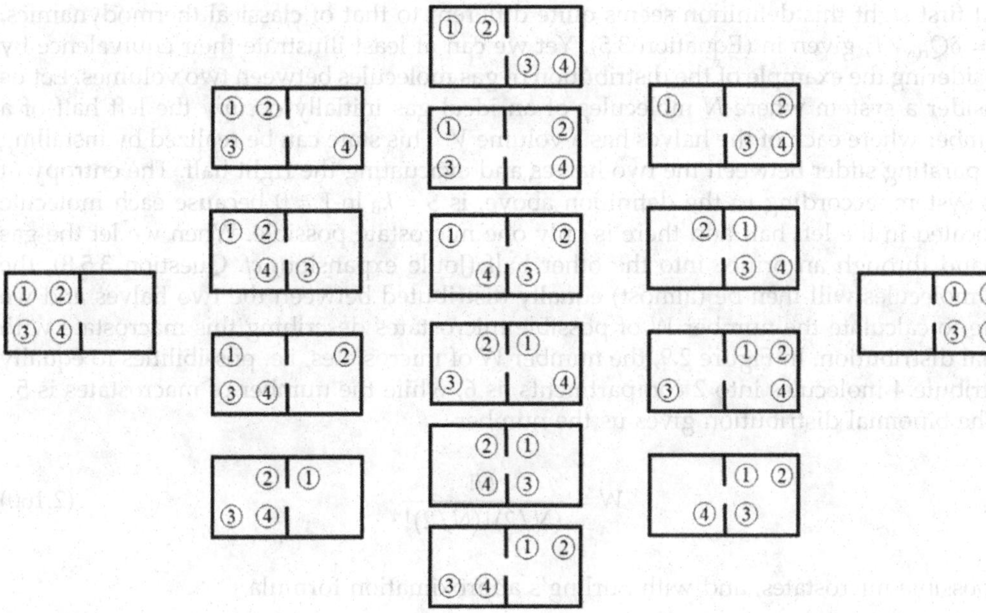

FIGURE 2.9
Possibilities of distributing four molecules into two sub-volumes.

Because there is an equal probability for each molecule to be found either in the left or in the right half of the volume, each of the 16 *microstates* shown is equally likely. If the molecules are regarded as indistinguishable and only the number of molecules present in each of the sub-volumes is considered, the probability with which each of the five possible *macrostates* is occupied, is obviously different. It is much more likely to find a condition where two molecules each are in each of the sub-volumes than a state where all molecules are either in the left or in the right half, respectively. This higher probability also tells us something about the direction a process will undergo. If the system is isolated, i.e. if there is no interaction with the surroundings, a process will tend towards an equilibrium state; in this example, this means towards an equal distribution of the number of molecules over both sub-volumes. An isolated system tends to transform away from a less probable macrostate to a state with higher probability, or – in other words – the entropy of an isolated system will increase (until it reaches an equilibrium state).

Entropy S may then be defined as

$$S = k_B \ln W \tag{2.159}$$

where k_B is Boltzmann's constant and W is the number of microstates contributing to a certain macrostate. This formula is engraved – with a little different symbolism – on Boltzmann's tombstone in Vienna's central cemetery. The (natural) logarithm is used because the number of molecules in a real system is extremely large; the use of the logarithm thus simplifies the calculation considerably. In addition, entropy-like internal energy is an extensive property, i.e. when two systems with entropies S_1 and S_2, respectively, are combined, the entropy of the entire system can be simply calculated as $S_{total} = S_1 + S_2$. By the calculation rules for the logarithm, the product for the combination of microstates becomes a sum in the entropies.

At first sight this definition seems quite different to that of classical thermodynamics, $dS = \delta Q_{rev}/T$, given in (Equation 3.5). Yet we can at least illustrate their equivalence by considering the example of the distribution of gas molecules between two volumes. Let us consider a system where N molecules of an ideal gas initially occupy the left half of a chamber where each of the halves has a volume V. This state can be realized by installing a separating slider between the two halves and evacuating the right half. The entropy of this system, according to the definition above, is $S = k_B \ln 1 = 0$ because each molecule is located in the left half and there is only one microstate possible. When we let the gas expand through an orifice into the other half (Joule expansion, *cf.* Question 3.5.8), the gas molecules will then be (almost) equally distributed between the two halves and we have to calculate the number W of possible microstates describing this macrostate with equal distribution. In Figure 2.9, the number W of microstates, i.e. possibilities to equally distribute 4 molecules into 2 compartments, is 6, while the number of macrostates is 5.

The binomial distribution gives us the number

$$W = \frac{N!}{(N/2)!(N/2)!} \tag{2.160}$$

of possible microstates, and with Stirling's approximation formula

$$\ln N! \approx N \ln N - N \approx N \ln N, \tag{2.161}$$

which holds for large N, we can calculate the entropy change for this process

$$\begin{aligned} \Delta S &= k_B \ln \frac{N!}{(N/2)!(N/2)!} - k_B \ln 1 = k_B[\ln(N!) - 2\ln((N/2)!)] - 0 \\ &\approx k_B\left[N \ln N - 2\frac{N}{2}\ln(N/2)\right] = k_B N \ln 2. \end{aligned} \tag{2.162}$$

When trying to calculate the entropy change on the basis of the classical definition $dS = \delta Q_{rev}/T$ (Equation 3.5), we imagine that the gas from the left volume does not expand directly into the right volume but displaces a movable piston so that the volume is increased in infinitesimal steps. We consider an expansion at constant temperature, T, reversing the isothermal compression process detailed in Question 1.6.4. This means we let a system of volume V expand reversibly and isothermally in an infinitely slow process so that the gas exerts work on the movable piston. For the process to be isothermal for an ideal gas, where the internal energy only depends on temperature, the volume work $W_{rev}^{isoth.}$ (*cf.* Equation 1.93) withdrawn from the system must be balanced by the same amount of heat added. Thus, the heat, Q, added when expanding the volume from V to $2V$ is

$$Q = -W_{rev}^{isoth.} = mR_sT \ln\left(\frac{2V}{V}\right) = mR_sT \ln 2 = nRT \ln 2, \tag{2.163}$$

where m is mass of the gas and R_s and R are the specific and universal gas constants, respectively. Taking into account that the number n of moles of the gas may be written as $n = N/L = k_B N/R$ with Avogadro's constant, L, since $\Delta S = Q_{rev}/T = k_B N \ln 2$, exactly the same entropy change as in Equation 2.162 results.

Interestingly, an equation similar in form to Equation 2.159 is used in the subject of information theory (Shannon 1948a, 1948b, Brillouin 1961). We consider a state space Ω with a cardinality, i.e. number of elements, $|\Omega|$. For example, a coin has the state space Ω = {heads, tails} with $|\Omega| = 2$ states, a die has the state space $\Omega = \{1, 2, 3, 4, 5, 6\}$ with $|\Omega| = 6$ states. We may then define: "The entropy H of a system is the minimum number of bits needed to define its condition." Mathematically, this translates into

$$H = \log_2 |\Omega|. \quad (2.164)$$

Entropy H may thus be seen as a number quantifying the information encoded in a system. The information contained becomes larger with an increasing number of possible states in the system, by analogy with the number of microstates in the definition for S above. Equation 2.164 yields $H = 1$ for a coin and $H \approx 2.585$ for a die. When adding a second coin or die, respectively, the amount of information encoded is squared, i.e. the entropy then becomes $H = \log_2 4 = 2$ in the first case and $H = \log_2 36 \approx 5.170$ in the latter. Finally, it should be noted that Equation 2.164 is a simplification of Shannon's original formula, where we assume here that each state is equally probable. For a detailed description of entropy in information theory the reader is referred to Gray (2011); the use of entropy in various fields of mathematics, science and information theory is explained by Frigg and Werndl (2011).

References

Assael M.J., Trusler J.P.M., and Tsolakis Th., 1996, *Thermophysical Properties of Fluids. An Introduction to their Prediction*, Imperial College Press, London.

Brillouin L., 1961, "Thermodynamics, statistics and information", *Am. J. Phys.* **29**:318–328.

Cameretti L.F., and Sadowski G., 2005, "Modelling of aqueous electrolyte solutions with perturbed-chain statistical association fluid theory", *Ind. Eng. Chem. Res.* **44**:3355–3362.

Chapman W.G., Gubbins K.E., Jackson G., and Radosz M., 1989, "SAFT: Equation-of-state solution model for associating fluids", *Fluid Phase Equilib.* **52**:31–38.

Chueh P.L., Prausnitz, J.M., 1967, "Vapor-liquid equilibria at high pressures. Vapor-phase fugacity coefficients in nonpolar and quantum-gas mixtures", *Ind. Eng. Chem. Fundam.* **6**:49–98.

Clausius R., 1850a, "Über die bewegende Kraft der Wärme und die Gesetze, welche sich daraus für die Wärmelehre selbst ableiten lassen, Part I", *Annalen der Physik* **155**:368–397 (also printed in 1851, "On the Moving Force of Heat, and the Laws regarding the Nature of Heat itself which are deducible therefrom. Part I," *Phil. Mag.* **2**:1–21).

Clausius R., 1850b, "Über die bewegende Kraft der Wärme und die Gesetze, welche sich daraus für die Wärmelehre selbst ableiten lassen, Part II", *Annalen der Physik* **155**:500–524 (also printed in 1851, "On the Moving Force of Heat, and the Laws regarding the Nature of Heat itself which are deducible therefrom. Part II," *Phil. Mag.* **2**:102–119).

Clausius R., 1865, *The Mechanical Theory of Heat—with Its Applications to the Steam Engine and to Physical Properties of Bodies*, ed. Hirst T.A., John van Voorst, London.

Dymond D.H., Marsh K.N., and Wilhoit R.C., 2003, *Virial Coefficients of Pure Gases and Mixtures (Landolt-Börnstein Numerical Data and Functional Relationships in Science and Technology) Group*

IV Physical Chemistry Vol. 21 Subvolume B Virial Coefficients of Mixtures, eds. Frenkel M., and Marsh K.N., Springer, New York.

Dymond J.H., and Smith E.B., 1980, *The Virial Coefficients of Pure Gases and Mixtures. A Critical Compilation*, Clarendon Press, Oxford.

Dymond J.H., Marsh K.N., Wilhoit R.C., and Wong K.C., 2002, *Virial Coefficients of Pure Gases and Mixtures (Landolt-Börnstein Numerical Data and Functional Relationships in Science and Technology) Group IV Physical Chemistry Vol. 21 Subvolume A Virial Coefficients of Pure Gases.*, eds. Frenkel M., and Marsh K.N., Springer, New York.

Einstein A., 1901, "Folgerungen aus den Capillaritätserscheinungen", *Ann. der Phys.* **4309**:513–523.

Evers C., Losch H.W., and Wagner W., 2002, "An absolute viscometer-densimeter and measurements of the viscosity of nitrogen, methane, helium, neon, argon, and krypton over a wide range of density and temperature", *Int. J. Thermophys.* **23**:1411–1439.

Frigg R., and Werndl C., 2011, *Entropy: A Guide for the Perplexed*, in *Probabilities in Physics*, eds. Beisbart, C., and Hartmann, S., Oxford University Press, Oxford.

Gray R. M., 2011, *Entropy and Information Theory*, Springer Science & Business Media, New York.

Gross J., and Sadowski G., 2001, "Pertubed-chain SAFT: An equation of state based on pertubation theory for chain molecules",*Ind. Eng. Chem. Res.* **40**:1244–1260.

Held C., Reschke T., Mohammad S., Luza A., and Sadowski G., 2014, "ePC-SAFT revised", *Chem. Eng. Res. Des.* **92**:2884–2897.

Held C., and Sadowski G., 2016, "Thermodynamics of bioreactions", *Ann. Rev. Chem. Biomol. Eng.* **7**: 395–414.

Hellmann R., Bich E., Vogel E., and Vesovic V., 2012, "Thermophysical properties of dilute hydrogen sulfide gas", *J. Chem. Eng. Data* **4**:1312–1317.

Hellmann R., 2013, "Ab-initio potential energy surface for the nitrogen molecule pair and thermophysical properties of nitrogen gas", *Mol. Phys.* **111**:387–401.

Hellmann R., 2014, "Intermolecular potential energy surface and thermophysical properties of ethylene oxide", *J. Chem. Phys.* **141**:164322.

Hellmann R., 2017, "Intermolecular potential energy surface and thermophysical properties of propane", *J. Chem. Phys.* **146**:114304.

Hellmann R., 2020, "Cross second virial coefficients and dilute gas transport properties of the systems (CO2 + C2H6) and (H2S + C2H6) from accurate intermolecular potential energy surfaces", *J. Chem. Eng. Data* **65**:968–979.

Herzberg G., 1989, *Molecular Spectra and Molecular Structure. Vol. 1-Spectra of Diatomic Molecules*, 2nd ed., Krieger Pub, USA.

Herzberg G., 1991, *Spectra and Molecular Structure. Vol. 2-Infrared and Raman Spectra of Polyatomic Molecules*, Krieger Pub, USA.

Hill T.L., 1988, *An Introduction to Statistical Thermodynamics*, Dover Publications, New York.

Howerton M.T., 1962, *Engineering Thermodynamics*, Van Nostrand, Princeton, NJ.

Hurly J.J., and Mehl J.B., 2007, "He-4 thermophysical properties: New ab initio calculations", *J. Res. Natl. Inst. Stand. Technol.* **112**:75–94.

Hurly J.J., and Moldover M.R., 2000, "Ab initio values of the thermophysical properties of helium as standards", *J. Res. Natl. Inst. Stand. Technol.* **105**:667–688.

Israelachvili J.N., 2011, *Intermolecular and Surface Forces*, 3rd ed., Academic Press.

Jaeschke M., Audibert S., van Caneghem P., Humphreys A.E., Janssen-Van R., Pellei Q., Michels J.P.J., Schouten J.A., and ten Seldam C.A., 1988, *High Accuracy Compressibility Factor Calculation for Natural Gases and Similar Mixtures by Use of a Truncated Virial Equation (GERG technical monograph)*, Verlag des Vereins Deutscher Ingenieure, Dusseldorf.

Janz G.J., 1967, *Thermodynamic Properties of Organic Compounds*, rev. ed., Academic Press, New York.

Kunz O., and Wagner W., 2012, "The GERG-2008 wide-range equation of state for natural gases and other mixtures: An expansion of GERG-2004", *J. Chem. Eng. Data* **57**:3032–3091.

Landolt-Bornstein, 1951, "Atom und Molekularphysik, Teil Molekeln I", *Z. Elektr. Chem.* **55**:666.

Lee B.I., and Kesler M.G., 1975, "A generalized thermodynamic correlation based on 3-parameter corresponding states", *AIChE J* **21**:510–527.

Lemmon E.W., Bell I.H., Huber M.L., and McLinden M.O., 2018, NIST Standard Reference Database 23, NIST Reference Fluid Thermodynamic and Transport Properties Database (REFPROP): Version 10.0.

Maitland G.C., Rigby M., Smith E.B., and Wakeham W.A., 1981, *Intermolecular Forces: Their Origin and Determination*, Clarendon Press, Oxford.

Meschede D., 2006, *Gerthsen Physik*, Springer, Berlin, Heidelberg, New York.

McQuarrie D.A., 2000, *Statistical Mechanics*, University Science Books, Sausalito, CA.

Moore G.E., 1949–1958, *Atomic Energy States, Nat. Bur. Stand. Circ.* 467, vols.1–3.

Olver F.W.J., Lozier D.W., Boisvert R.F., and Clark C.W. (Eds.), 2010, *NIST Handbook of Mathematical Functions*, 1st ed., Cambridge University Press., Cambridge

Orbey H., and Vera J.H., 1983, "Correlation for the 3rd virial coefficient using T_c, P_c and omega as parameters", *AIChE* **29**:107–113.

Parthria R.K., and Beale P.D., 2011, *Statistical Mechanics*, 3rd ed., Academic Press, USA.

Petit, A.-T., and Dulong, P.-L., 1819, "Recherches sur quelques points importants de la Théorie de la Chaleur", *Ann. Chim. Phys.* **10**:395–413.

Pitzer K.S., 1955, "The volumetric and thermodynamic properties of fluids. I. Theoretical basis and virial coefficients", *J. Am. Chem. Soc.* **77**:3427–3440.

Pitzer K.S., and Curl R.F., 1958, "Volumetric and thermodynamic properties of fluids — Enthalpy, free energy and entropy", *Ind. Eng. Chem.* **50**:265–274.

Pitzer K.S., Lippman D.Z., Curl R.F., Huggins C.M., and Petersen D.E., 1955, "The volumetric and thermodynamic properties of fluids 2. Compressibility factor, vapor pressure and entropy of vaporization", *J. Am. Chem. Soc.* **77**:3433.

Poling B., Prausnitz J.M., and O'Connell J.P., 2001, *The Properties of Gases and Liquids*, 5th ed., McGraw-Hill, New York.

Prausnitz J.M., Lichtenthaler R.N., and Gomes de Azevedo E., 1998, *Molecular Thermodynamics of Fluid-Phase Equilibria*, 3rd ed., Pearson Education, USA.

Reed T.M., and Gubbins K.E., 1973, *Applied Statistical Mechanics*, McGraw-Hill, Kogakusha.

Reid R.C., Prausnitz J.M., and Poling B.E., 1988, *The Properties of Gases and Liquids*, 4th ed., McGraw-Hill, New York.

Sandler S.I., 2010, *An Introduction to Applied Statistical Thermodynamics*, John Wiley & Sons Inc., USA.

Selected Values of Properties of Hydrocarbons and Related Compounds, 1977, 1978, Thermodynamic Research Center, Texas A&M University.

Shannon C.E., 1948a, "A mathematical theory of communication", *Bell Sys. Tech. J.* **27**:379–423.

Shannon C.E., 1948b, "A mathematical theory of communication", *Bell Sys. Tech. J.* **27**:623–656.

Sherwood A.E., and Prausnitz J.M., 1964, "Third virial coefficient for the Kihara Exp-6 and Square well potentials", *J. Chem. Phys.* **41**:413–428.

Snyder P.S., and Winnick J., 1970, *Proc. of 5th Symp. Thermophys. Prop.*, ASME, Boston, p. 115.

Stone A., 2013, *The Theory of Intermolecular Forces*, 2nd ed., Oxford University Press, Oxford.

Sutton L.E., 1965, *Tables of Interatomic Distances and Configuration in Molecules and Ions, Supplement*, The Chemical Society, London.

Tsonopoulos C., and Prausnitz J., 1969, "A review for engineering applications", *Cryogenics* **9**: 315–327.

University of Dortmund ePC-SAFT Excel Add-in, (https://www.th.bci.tu-dortmund.de/cms/de/Aktuelles/ePC-SAFT/index.html last accessed December 1st, 2020).

van Ness H.C., 2015, *Classical Thermodynamics of Nonelectrolyte Solutions*, McGraw-Hill, New York.

Wertheim M.S., 1984a, "Fluids with highly directional attractive forces. I. statistical thermodynamics", *J. Stat. Phys.* **35**: 19–34.

Wertheim, M.S., 1984b, "Fluids with highly directional attractive forces. II. Thermodynamic pertubation theory and integral equations", *J. Stat. Phys.* **35**:35–47.

Wu G.Z.A., and Stiel L.I., 1985, "A generalized equation of state for the thermodynamic properties of polar fluids", *AIChEJ.* **31**:1632–1644.

Xiang H.W., and Deiters U.K., 2008, "A new generalized corresponding-states equation of state for the extension of the Lee–Kesler equation to fluids consisting of polar and larger nonpolar molecules", *Chem. Engng. Sci.* **63**:1490–1496

3

Second Law of Thermodynamics, Thermodynamic Functions and Relationships

3.1 Introduction

In Chapter 1 of this book we argued that thermodynamics is an experimental science consisting of a collection of axioms, derivable from statistical mechanics and in many circumstances from Boltzmann's distribution. So far, we have introduced the First Law of Thermodynamics that interrelates physical quantities, some of which are far more easily measured than others. We have also emphasized the applications of the First Law of Thermodynamics to various systems such as closed, open and reactive systems in steady states and in transients. We are also armed with two types of "thermodynamic-meter": (1) a thermometer to measure temperature (cf. Question 1.7.2) and (2) a calorimeter used to measure differences in energy and enthalpy (cf. Question 1.8.1). Both of these will be put to good use in this chapter, which considers the Second Law of Thermodynamics. We will also introduce a third "meter": A chemical potentiometer used to measure differences in chemical potential. We shall introduce some additional physical quantities that will prove of value in treating systems containing more than one phase as well as systems in which chemical reactions take place. A more detailed treatment of these two important topics is found in Chapters 4 and 5, respectively. The applications of the Second Law of Thermodynamics we consider in this chapter do not rely upon these topics.

Clausius provided the first broad statements of the Second Law of Thermodynamics (1850a and 1850b) and these were refined by Thomson,[1] and those readers interested in the history of the formulation of the laws of thermodynamics should consider consulting the work of Rowlinson (2003 and 2005) and Atkins (2007).

3.2 What Is the Second Law of Thermodynamics?

The concepts of the Second Law of Thermodynamics and of entropy have their roots in the early nineteenth century, when engineers started to develop and to operate heat engines. One of the problems was that not all forms of energy are of equal "utility". For example, electrical or mechanical work (from a resistance heater or a stirrer) can – without restriction – be used to increase the internal energy of a water reservoir. It is not possible, however, to transform all the energy stored within a warm water reservoir into useful work. This statement holds for all processes. In a Clausius-Rankine process, as an example of a heat

[1] Also known as Lord Kelvin, who described the absolute temperature scale, from 1892.

DOI: 10.1201/9780429329524-3

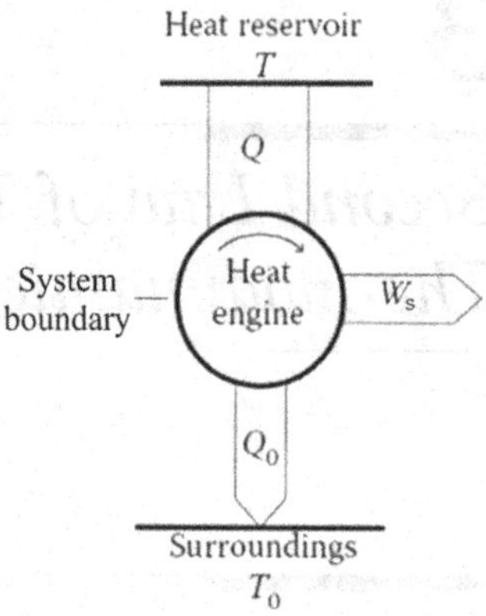

FIGURE 3.1
Basic scheme of a heat engine: Heat Q provided is partially converted to work W_s, another part Q_0 is discarded to the surroundings.

engine discussed in Chapter 6, the heat provided by the combustion of coal can only partially be converted into work. In heat engines, part of the energy provided in the form of heat always leaves the engine as "waste" heat at a low temperature.

This can be seen in Figure 3.1, which depicts the basic principle of a heat engine. Q indicates the heat typically provided by a combustion process at a high temperature T, whereas Q_0 indicates the heat lost at a low temperature T_0 to the environment. Therefore, only the difference may be obtained as useful mechanical (shaft) work W_s.

An energy balance shows

$$Q + W_s + Q_0 = 0 \quad \text{or} \quad |W_s| = Q - |Q_0|. \tag{3.1}$$

In Equation 3.1, W_s and Q_0 have negative values because they are leaving the system. In general, and if not otherwise specified, quantities added to the system have positive values while leaving quantities count negatively (see Questions 1.4.2 and 1.4.3).

The second form of Equation 3.1 has been written using absolute values for W_s and Q_0 because we are normally interested in these rather than in the negative values. As a result, Q represents the heat transferred to the system at high temperature, whereas $|W_s|$ and $|Q_0|$ stand for the shaft work obtained and the heat leaving the system at a low temperature. It follows that a thermal efficiency factor of the machine, defined as the ratio between the net work output, and the heat input may be computed as follows (see Chapter 6)

$$\eta_{th} \equiv \frac{|W_s|}{Q} = 1 - \frac{|Q_0|}{Q}. \tag{3.2}$$

3.2.1 How Did the Need for a New State Function Arise?

The big question in the early days of heat machines was how much heat had to be "wasted" as $|Q_0|$ because this would determine the amount of energy that could be retrieved as shaft work and also the efficiency of the machine: The lower this could be kept, the more efficient the process would be.

In a purely mathematical analysis of heat engines Sadi Carnot showed (Carnot 1824) that the fraction of wasted heat with respect to the heat provided must indeed reach a minimum and that the efficiency would reach a maximum value if the heat engine could be operated strictly reversibly. If there was a way to predict this "magical" heat minimum

$Q_{0\text{rev}}$, one could use the above equations to predict at least the theoretical maximum efficiency of heat engines. Unfortunately, this is not possible because heat is not a state function. Carnot has also shown that the efficiency must be exactly the same for any conceivable reversible heat engine operating between the same temperatures, and that they will be the more efficient the larger the temperature difference $(T - T_0)$.

Based on these results, the question arose whether a new state function could be found that would permit one to predict if not Q_0 then at least $Q_{0\text{rev}}$ in the case of reversible processes. Since the heat engine is a cyclic process (see Question 6.1), the amount of this state function in the engine cannot change over a complete cycle, because the cycle could only work repeatedly if the system returned to the exact same state after a complete cycle. The amount of this state function taken up would therefore have to be exactly equal to the amount released over a whole cycle.

A crucial element in the search for such a state function was probably the realization by Lord Kelvin around 1850 (see e.g. Levenspiel 1996), and in a preliminary form already by Carnot, that the amounts of heat reversibly exchanged with the environment are proportional to the absolute temperatures T and T_0:

$$\frac{|Q_{0\text{rev}}|}{Q_{\text{rev}}} = \frac{T_0}{T} \quad \text{or} \quad \frac{|Q_{0\text{rev}}|}{T_0} = \frac{Q_{\text{rev}}}{T}. \tag{3.3}$$

The last equation demonstrates that the same amount of Q_{rev}/T is released and taken up during a complete working cycle, so that the amount accumulated inside the engine stays constant. The function Q_{rev}/T can thus be considered as the desired state function. It can also be shown that the change of such a state function between two states of the system only depends on the two states and not on the path taken to bring the system from one state to the other, as is the case with any state function.

Clausius defined, around 1865, the new state function, which he named *entropy*, for the case of non-isothermal processes as

$$dS \equiv \frac{\delta Q_{\text{rev}}}{T}. \tag{3.4}$$

For isothermal heat transfer, this becomes

$$\Delta S \equiv \frac{Q_{\text{rev}}}{T}. \tag{3.5}$$

These definitions are not just applicable to heat engines but are absolutely general. In Equations 3.4 and 3.5 and in the following equations, δQ_{rev} and Q_{rev} do not therefore only signify the heat provided (reversibly) to a heat engine at high temperature, but any amount of heat reversibly exchanged between any closed system and its environment.

3.2.2 How Do We Formulate Entropy Balances and the Second Law of Thermodynamics?

The definitions of entropy (Equations 3.4 and 3.5) are obviously based on the second part of Equation 3.3. However, these expressions only contain reversible heat exchange, whereas an entropy balance must also accommodate irreversible processes. Let us thus analyze how

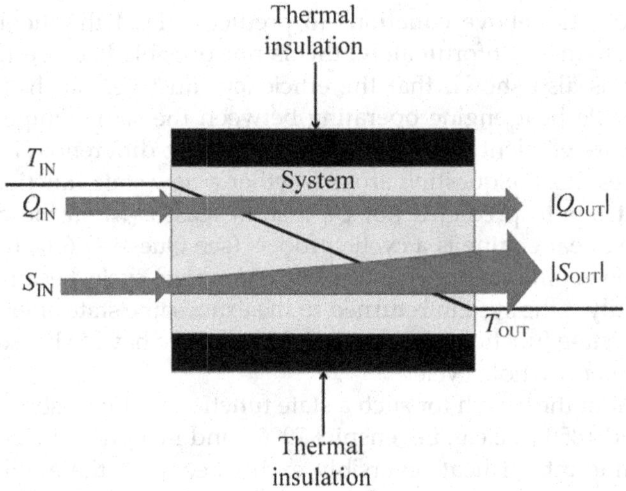

FIGURE 3.2
Analysis of heat transfer.

the second part of Equation 3.3 would have to be written for a very basic irreversible process such as heat conduction through a steady-state system, as shown in Figure 3.2.

It has been recognized for more than 200 years that heat always flows from a high temperature to a lower one and never the inverse. This is, therefore, an archetypical irreversible process. In Figure 3.2, T_{IN} must always be higher than T_{OUT}. If we again interpret Q/T as an exchange of entropy between the system and the environment, it is obvious that more entropy leaves the system than entered it. Since no other heat enters the system, some entropy must have been generated inside. The same argument may be applied to any slice of the system shown in Figure 3.2. The amount of heat flowing through the system is constant if the system is in a steady state (and if there are no heat sources nor sinks). But since the temperature decreases along the system, the ratio Q/T increases gradually from left to right. Because of the temperature gradient, the irreversible process of heat conduction clearly generates entropy within the system.

In Figure 3.2, we have considered the steady-state case where the entropy within the system is unchanged over time; we only have an increase of entropy from left to right. A general *entropy balance* for *a closed system* must account for entropy changes in the system (dS) owing to both the entropy exchanged with the environment and that generated within the system

$$dS = \frac{\delta Q}{T} + dS_{gen}, \tag{3.6}$$

where dS denotes the entropy accumulated in the system, $\delta Q/T$ the entropy added to the system from the outside in the form of heat, and dS_{gen} the entropy generated within the system owing to irreversibilities (see e.g. Sandler 1989).

In this form, the Second Law of Thermodynamics says: **The source term dS_{gen} can only be zero or positive,** i.e.

$$dS_{gen} \geq 0. \tag{3.7}$$

In the case of reversible processes, it is zero; and in real, irreversible processes it is positive. It should be noted that the Second Law does not prohibit the entropy from being reduced in a system, but it forbids entropy from disappearance, i.e. it cannot be destroyed. As a result, the entropy of a *closed, adiabatic* system can only remain constant (for reversible processes) or increase (for irreversible processes).

Considering the entropy balance in Equation 3.6, the definition of entropy in Equation 3.4 says that the effect of adding some entropy to a system on the properties of that system can be measured by adding an amount of heat reversibly to the system, i.e. by avoiding any additional entropy generation and by observing the effect on the system.

3.2.3 What Are the Immediate Consequences of the Second Law of Thermodynamics?

There are some immediate consequences of the Second Law of thermodynamics that we set out here explicitly before going any further.

3.2.3.1 What Are the Consequences of the Second Law for Heat Exchange in Any Process?

As we have seen, the First Law of Thermodynamics can predict how much energy a system will absorb or release in the form of heat and work during a defined change of state, but it cannot predict that for work or heat alone. By using the Second Law, one can at least say something about the heat exchange.

Imagine a *closed system* undergoing a defined change of state through an exchange of heat and work with the environment. Multiplying the entropy balance in the form of Equation 3.6 by the temperature T and solving it for δQ yields

$$\delta Q = TdS - TdS_{gen} \tag{3.8}$$

or, in integrated form,

$$Q = \int TdS - \int TdS_{gen}. \tag{3.9}$$

The change of state will also specify the value of the first integral on the right-hand side of Equation 3.9. The exchanged heat, Q, will be equal to this integral *if and only if* the process occurs reversibly, i.e. if $\int TdS_{gen} = 0$. In irreversible processes, $Q < \int TdS$ always holds because dS_{gen} can only be positive. This is independent of whether heat is absorbed ($Q > 0$) or heat is released ($Q < 0$).

This argument shows that the integral $\int TdS$ on the right-hand side of Equation 3.9 corresponds to the largest possible amount of energy that can be transferred to the system in the form of heat. Therefore, in irreversible endothermic ($Q > 0$) processes, less heat and more work will be transferred to the system than predicted by $\int TdS$. In irreversible exothermic ($Q < 0$) processes, more heat will be released than given by $\int TdS$.

3.2.3.2 What Are the Consequences of the Second Law for a Perpetual Motion Machine of the Second Kind?

A perpetual motion machine of the second kind is a heat engine that does nothing other than extract heat from an environment of uniform temperature and transform it integrally

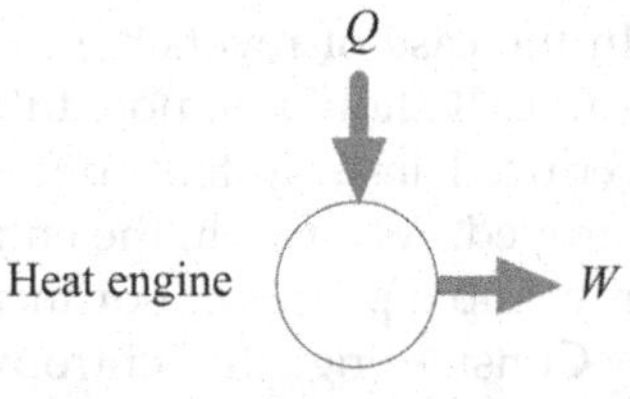

FIGURE 3.3
A perpetual motion machine of the second kind.

into useful work (see Figure 3.3). It would have a thermal efficiency of 100%. It would allow us, for example, to design engines driven solely by cooling their environment; so, for instance, ships could be powered solely by cooling down the water through which they navigate.

The entropy balance for such a machine (the round area is the system) would read

$$dS = \frac{\delta Q}{T} + dS_{gen}. \tag{3.10}$$

For a closed cycle, the final and initial values of S must be equal

$$\Delta S = \frac{Q}{T} + S_{gen} = 0. \tag{3.11}$$

Since both Q/T and S_{gen} are positive, Equation 3.11 can only be fulfilled for a heat engine that takes heat from a heat source, if there are **also** steps that reject heat to the surroundings; that is, where $\delta Q < 0$. This is even true if the engine operates reversibly, i.e. $dS_{gen} = 0$. It is thus an inevitable consequence of the Second Law that not all the heat provided may be turned into useful work.

In consequence, the additional entropy "generated" by the irreversibility must also be disposed of by rejecting heat to the surroundings with the same result that only a fraction of the heat provided is used to produce work.

3.2.4 How Can Heat Engines Convert at Least Part of the Heat Provided into Useful Work?

Heat engines avoid accumulating any excess entropy (see Question 3.2.3.2) by "wasting" a part of the heat provided at T at a low temperature T_0. The entropy balance thus reads

$$\Delta S = 0 = \frac{Q}{T} + \frac{Q_0}{T_0} + S_{gen} = 0. \tag{3.12}$$

The best results for heat engines are obviously reached if the heat, Q, serving as energy source is provided to the engine at high temperature T, in order to keep the entropy intake reasonably low, and when some of the heat, Q_0, is rejected at low temperature T_0, so that a large amount of entropy can be carried away with a relatively small amount of heat.

Theoretically, the best case would be a reversible operation and is known as a **Carnot process** (cf. Question 6.2). The analysis shows, however, that even for such an ideal process there is an upper limit for the thermal efficiency of a heat engine. This can be demonstrated using an entropy balance for a reversible case ($dS_{gen} = 0$)

$$\Delta S = 0 = \frac{Q}{T} + \frac{Q_0}{T_0} \quad \text{or} \quad \frac{Q}{T} = \frac{|Q_0|}{T_0}. \tag{3.13}$$

The minimum fraction of heat $|Q_0|/Q$ that must be "wasted" can be calculated from this as

$$\frac{|Q_0|}{Q} = \frac{T_0}{T}. \tag{3.14}$$

The efficiency of the Carnot machine may be calculated from this based on Equation 3.2 as

$$\eta_{\text{th,rev}} = 1 - \frac{T_0}{T}. \tag{3.15}$$

This is theoretically the highest possible efficiency of a heat engine. Not even in a reversible engine could all the heat be converted into useful work.

3.2.5 What Is the Entropy Balance for an Open System?

Following the rules laid out in Question 1.5 for the case of enthalpy (see Equations 1.69 and 1.81), the entropy balance for an open system would have to be formulated as

$$\mathrm{d}S = \frac{\delta Q}{T} + \sum_{\mathrm{B}} S_{\mathrm{B}} \mathrm{d}n_{\mathrm{B}} + \delta S_{\text{gen}} \tag{3.16}$$

or

$$\frac{\mathrm{d}S}{\mathrm{d}t} = \frac{\dot{Q}}{T} + \sum_{k} \sum_{\mathrm{B}} S_{\mathrm{B},k} \dot{n}_{\mathrm{B},k} + \dot{S}_{\text{gen}}, \tag{3.17}$$

where $S_{\mathrm{B},k}$ stands for the partial molar entropy of the species that are exchanged with the environment at the k-th mass exchange port, and $\dot{S}_{\text{gen}}$ and δS_{gen} for, respectively, the rate and the amount of entropy generation inside the system owing to irreversible processes.

3.3 What Is Gibbs Energy and Why Was It Introduced?

The First Law can predict how much energy in the form of heat and work must be added to or removed from a system in order to bring it from one specified state into another, but it cannot predict the same for work or heat separately. However, for many applications, one needs a way to estimate the amount of energy that can be exchanged between a system and its environment in the form of work alone. The Second Law can at least predict the heat exchange (Q_{rev}) for reversible processes, but it would be useful to be able to do something similar for the work in a reversible process.

3.3.1 Why Is There Another State Function Needed?

An idea for obtaining a prediction for W_{rev} would be to solve the entropy balance for Q_{rev} and substitute this into an energy balance. This would permit estimation of the minimum amount of energy that must be added in the form of work for a specified energy change of the system, or the maximum amount of energy that can be removed in the form of useful work. If the amount of entropy generated (S_{gen}) is known, this technique should also be able to predict the amount of work (W) exchanged.

Since neither heat nor work are state functions, the path of the process must be specified. The most useful option would be to constrain the analysis to *isobaric and isothermal* processes. The analysis is shown here for *closed* systems, but it can easily be extended to open systems.

Integrating the entropy balance (Equation 3.9) yields

$$Q = T\Delta S - T\Delta S_{gen}. \tag{3.18}$$

Substituting this into the energy balance (Equation 1.48) for p = const: $\Delta H = Q + W_o$ results in

$$\Delta U + p\Delta V - T\Delta S = W_o - T\Delta S_{gen}. \tag{3.19}$$

If a new state function called the *Gibbs energy* is defined as

$$G \equiv H - TS, \tag{3.20}$$

then $\Delta G = \Delta U + p\Delta V - T\Delta S$ for an isothermal system so that, based on Equation 3.19,

$$\Delta G = W_o - T\,\Delta S_{gen}. \tag{3.21}$$

The product $-T\,\Delta S_{gen}$ is the source term in the Gibbs energy balance. Since the amount of entropy generated can only be positive or zero, the *source term for Gibbs energy can only be zero or negative.* Gibbs energy can only be destroyed, or "dissipated", but never created.

3.3.2 What Is the Primary Use of Gibbs Energy?

The concept of Gibbs energy is utilized in countless different contexts, but the three following applications are perhaps amongst the most important ones. First of all, the Gibbs energy balance in Equation 3.21 may be solved for W_o and can be used for estimating the work that must be provided to effect a state change characterized by an increase of the Gibbs energy by ΔG, provided the system is closed, isothermal and isobaric and the entropy generation ΔS_{gen} is known. On the other hand, it will estimate the amount of useful energy that can be extracted from a system undergoing a decrease of Gibbs energy of $|\Delta G|$. As the entropy generation is rarely known, setting $T\,S_{gen}$ to zero in the equation will at least predict the minimum work needed in the first (endergonic) and the maximum amount of useful energy gained in the second (exergonic) case.

Second, Equation 3.21 also shows that if the system is not exposed to any external driving force, i.e. $W_o = 0$, a process can only occur if $\Delta G < 0$. The equation therefore predicts in which direction a spontaneous process will take place (see e.g. Questions 3.3.4

and 5.2.3). For such spontaneous processes, the negative Gibbs energy change can be considered as the driving force because it can be shown that the process will occur faster the more negative the Gibbs energy change is.

Third, no entropy can be generated in systems that are in equilibrium, and absolutely no driving force exists. The Gibbs energy change in Equation 3.21 may therefore be set to zero and used as an equilibrium condition (see e.g. Question 4.4).

3.3.3 What Is the Gibbs Energy Balance for an Open System?

Following the rules laid out in Question 1.5, a Gibbs energy balance for an open system would have to be formulated as

$$dG = \delta W_o + \sum_B \mu_B dn_B - T\delta S_{gen}, \tag{3.22}$$

where dn_B represents the amount of substance of B added or removed *but not the amount accumulated inside* the system, and μ_B stands for the partial molar Gibbs energy of the added chemical, which is also known as the *chemical potential*. An alternative form would be

$$\frac{dG}{dt} = \dot{W}_o + \sum_k \sum_B \mu_{B,k} \dot{n}_{B,k} - T\dot{S}_{gen}. \tag{3.23}$$

This balance is also valid for chemically reacting systems.

3.3.4 What Is the Chemical Potential?

For a quantitative exploitation of the Gibbs energy balances in Equations 3.22 and 3.23, we must link dG or dG/dt to measurable state variables by using total differentials, such as Equations 1.49 and 1.50. By using the analog of Equation 1.50 for dG, and designating temperature, pressure and the number of moles n_B of the mixture constituents as these measurable state variables, and by assuming a *closed*, isobaric and isothermal system ($dp = dT = 0$), Equation 3.22 would become

$$\sum_B \mu_B dn_B = dG = \delta W_o - T\delta S_{gen}, \tag{3.24}$$

where dn_B now stands for the change of the number of moles of B *within* the system. As noted after Equation 3.22, μ_B denotes the partial molar Gibbs energy, or *chemical potential of* B, and it is defined as

$$\mu_B \equiv \left(\frac{\partial G}{\partial n_B}\right)_{p,T,n_{i \neq B}}. \tag{3.25}$$

Equation 3.24 is a general Gibbs energy balance for isothermal, isobaric, closed systems.

For processes without external driving forces ($\delta W_o = 0$), dG can only be zero (for equilibrium systems) or negative (for irreversible processes). In this case, we can see from Equation 3.24 whether there is a tendency for a spontaneous change of the molar composition of the system at constant pressure and temperature. For those constituents with a positive μ_B there is a potential for a decrease of n_B because a negative dn_B will contribute to a negative dG and thus allow some entropy to be generated. Therefore, μ_B is called a **chemical potential**. In the case when μ_B is negative, a spontaneous change of composition would have to led to an increase of the number of moles of B.

The change of the number of moles of B could either occur because of an exchange of molecules of B with the environment or owing to a chemical reaction. The chemical potential therefore plays an absolutely central role in predicting chemical transformations and all kinds of diffusion effects.

Diffusion could, for instance, occur if a closed, adiabatic system consists of two phases α and β in contact with each other. If mixture constituents diffuse from one phase to the other, the Gibbs energy change of both phases can be estimated as the sum of the left-hand terms of Equation 3.24 for each phase. But because $dn_B^\alpha = -dn_B^\beta = dn_B$, the result for a chemically inert system without exchange of work ($\delta W_o = 0$) would be

$$\sum_B \left(\mu_B^\alpha - \mu_B^\beta\right) dn_B = -T\delta S_{gen} \le 0. \tag{3.26}$$

According to this equation, any chemical B whose potential is not the same in both phases opens the door for entropy generation and represents therefore a driving force for a flux of B from one phase to the other. If $\mu_B^\alpha > \mu_B^\beta$, dn_B will tend to adopt a negative value in order to reduce the Gibbs energy of the system, which leads to a flux of B from α to β. In the inverse case, dn_B would become positive, leading to a flux in the opposite direction. In both cases, the difference $\mu_B^\alpha - \mu_B^\beta$ will decrease until becoming zero, which represents an equilibrium state. Therefore, for phase α to be in equilibrium with phase β the chemical potentials of all B must be equal in both phases.

A similar analysis can be performed on a reactive mixture in a closed system at constant T and p. Any chemical reaction has a certain stoichiometry, which could for instance take the form

$$\nu_A A + \nu_B B \rightarrow \nu_C C + \nu_D D, \tag{3.27}$$

where A, B, C, D represent different chemical species. In general terms, we can write a chemical reaction as

$$\sum_B \nu_B B = 0, \tag{3.28}$$

where the stoichiometric coefficients, ν_B, would be counted negatively for the reactants and positively for the products. The Gibbs energy balance would still be represented by Equation 3.24 but the differential changes of the numbers of moles dn_B would now be linked to each other by molar balances of the form $dn_B = \nu_B\, d\xi$. For a system with only a single chemical reaction and without any external driving force, $\delta W_o = 0$, as was shown in Equation 1.36 when $r\, dV/dt$ is replaced by $d\xi$, the balance therefore reads

$$dG = \sum_B \nu_B \mu_B d\xi = -T\delta S_{gen} \le 0. \tag{3.29}$$

By analogy with the enthalpy of reaction (Equation 1.63), the sum $\sum_B \nu_B \mu_B$ is called the *Gibbs energy of reaction* and given the symbol $\Delta_r G$:

$$\Delta_r G = \sum \nu_B \mu_B. \tag{3.30}$$

Equations 3.29 and 3.30 show that $-\Delta_r G$ may be interpreted as the driving force for the extent of reaction (chemical flux) ξ. It thus indicates the direction in which the chemical reaction will occur. For the flux, ξ, to be positive, i.e. the reaction to proceed from the left to the right (reactants to products), $\Delta_r G$ must be negative. If $\Delta_r G$ is positive, ξ must be negative and the reaction will proceed backwards.

If the values of the chemical potentials are such that the Gibbs energy of reaction is zero ($\Delta_r G = \Sigma_B \nu_B \mu_B = 0$) the reaction can neither go forward nor backwards and the reaction is at equilibrium. These ideas will be developed further in Chapter 5.

It is important to point out here that the name *affinity*, with the symbol, A, is also used for the quantity we have designated as the Gibbs energy of reaction, although usually with a negative sign. We prefer the terminology of *energy of reaction* for consistency with other quantities in our text.

3.4 What Are Characteristic, or Fundamental, Equations?

The fundamental equation for the internal energy U may be obtained by reformulating its balance (see also Equation 1.68) for an *open non-reacting system at equilibrium* in molar terms

$$dU = \delta Q_{rev} - p dV + \delta W_o + \sum_B H_B dn_B. \tag{3.31}$$

In the following developments, we will assume that no work other than reversible volume work is exchanged with the environment. Solving the entropy balance (Equation 3.16) for δQ_{rev} and substituting the result into the above balance yields the following

$$dU = TdS - pdV + \sum_B \mu_B dn_B. \tag{3.32}$$

This is the *fundamental equation* for U. The variable μ_B is shorthand for the expression $H_B - TS_B$, which, according to the *definition of Gibbs energy* G (Equation 3.20), is identical with the partial molar Gibbs energy or the chemical potential μ_B.

Equation 3.32 may be directly compared with the following total differential of U as a function of the state variables S, V and n_B

$$dU = \left(\frac{\partial U}{\partial S}\right)_{V,n_B} dS + \left(\frac{\partial U}{\partial V}\right)_{S,n_B} dV + \sum_B \left(\frac{\partial U}{\partial n_B}\right)_{S,V,n_{i \neq B}} dn_B. \tag{3.33}$$

This comparison yields the following partial derivatives

$$\left(\frac{\partial U}{\partial S}\right)_{V,n_B} = T, \quad \left(\frac{\partial U}{\partial V}\right)_{S,n_B} = -p, \quad \left(\frac{\partial U}{\partial n_B}\right)_{S,V,n_{i \neq B}} = \mu_B. \tag{3.34}$$

If S and V are constants, the fundamental equation for U becomes

$$dU = \sum_{B} \mu_B dn_B. \tag{3.35}$$

Equation 3.32 could also be expressed in terms of dH by noting that $dH = dU + Vdp + pdV$ so that

$$dU = dH - pdV - Vdp. \tag{3.36}$$

Substituting this into the characteristic Equation 3.32 for U yields

$$dH = TdS + Vdp + \sum_{B} \mu_B dn_B, \tag{3.37}$$

which is the *characteristic equation for H.*

This equation may be directly compared with the total differential of H as a function of the state variables S, p and n_B. This yields the following partial derivatives

$$\left(\frac{\partial H}{\partial S}\right)_{p,n_B} = T, \quad \left(\frac{\partial H}{\partial p}\right)_{S,n_B} = V, \quad \left(\frac{\partial H}{\partial n_B}\right)_{S,p,n_{i\neq B}} = \mu_B. \tag{3.38}$$

If S and p are constants,

$$dH = \sum_{B} \mu_B dn_B. \tag{3.39}$$

Equation 3.37 could also be expressed in terms of dG by noting that $dG = dH - T\,dS - S\,dT$ so that $dH = dG + T\,dS + S\,dT$. Substituting this into the characteristic equation for H yields

$$dG = -SdT + Vdp + \sum_{B} \mu_B dn_B, \tag{3.40}$$

which is the *fundamental equation for G.*

This equation may be directly compared with the total differential of G as a function of the state variables T, p and n_B. This yields the following partial derivatives

$$\left(\frac{\partial G}{\partial T}\right)_{p,n_B} = -S, \quad \left(\frac{\partial G}{\partial p}\right)_{T,n_B} = V, \quad \left(\frac{\partial G}{\partial n_B}\right)_{T,p,n_{i\neq B}} = \mu_B. \tag{3.41}$$

If T and p are constants, the fundamental equation for G becomes

$$dG = \sum_{B} \mu_B dn_B. \tag{3.42}$$

These relationships indicate that associated with particular sets of thermodynamic variables a variety of different functions naturally arise. One that we have not yet introduced is the Helmholtz function, denoted by A. It is defined by the equation

$$A = U - TS. \tag{3.43}$$

The differential form of Equation 3.43, when combined with Equation 3.32, is the *fundamental equation for the Helmholtz function*

$$\mathrm{d}A = -S\mathrm{d}T - p\mathrm{d}V + \sum_{\mathrm{B}} \mu_{\mathrm{B}}\mathrm{d}n_{\mathrm{B}}. \tag{3.44}$$

For a closed system of *constant temperature, volume and composition* and in the absence of work other than volume work, using the methods outlined above we find

$$\mathrm{d}A < 0 \tag{3.45}$$

so that the Helmholtz function (or energy) can only decrease.

Euler's theorem (provided in Question 1.10.2) can be used to integrate Equations 3.32, 3.37 and 3.40 because T, p and μ are intensive quantities, while S, V, n, U, H and G are extensive quantities. The integrated form of Equation 3.32 is

$$U = TS - pV + \sum_{\mathrm{B}} \mu_{\mathrm{B}} n_{\mathrm{B}}, \tag{3.46}$$

while for Equation 3.37 it is

$$H = TS + \sum_{\mathrm{B}} n_{\mathrm{B}}\mu_{\mathrm{B}} \tag{3.47}$$

and for Equation 3.40 we obtain at constant p and T

$$G = \sum_{\mathrm{B}} n_{\mathrm{B}}\mu_{\mathrm{B}}. \tag{3.48}$$

Differentiation of Equation 3.48 gives

$$\mathrm{d}G = \sum_{\mathrm{B}} \mu_{\mathrm{B}}\mathrm{d}n_{\mathrm{B}} + \sum_{\mathrm{B}} n_{\mathrm{B}}\mathrm{d}\mu_{\mathrm{B}}. \tag{3.49}$$

Subtracting Equation 3.40 from Equation 3.49 yields

$$0 = S\mathrm{d}T - V\mathrm{d}p + \sum_{\mathrm{B}} n_{\mathrm{B}}d\mu_{\mathrm{B}}. \tag{3.50}$$

Equation 3.50 is the *Gibbs–Duhem* equation and has 0 as the characteristic value. When Equation 3.50 is divided by the total amount of substance n we obtain

$$0 = S_{\mathrm{m}}\mathrm{d}T - V_{\mathrm{m}}\mathrm{d}p + \sum_{\mathrm{B}} x_{\mathrm{B}}\mathrm{d}\mu_{\mathrm{B}}. \tag{3.51}$$

For a phase at constant temperature and pressure, Equation 3.51 can be written as

$$0 = \sum_{\mathrm{B}} x_{\mathrm{B}}d\mu_{\mathrm{B}}. \tag{3.52}$$

The Gibbs–Duhem equation is particularly useful for treating phase equilibrium: It provides the first step in determining whether some results are or are not thermodynamically consistent and a method for calculating the chemical potential differences or activity coefficients for a system from measurements (see Chapter 4).

The *absolute activity*, λ_B, of a substance B is another useful quantity that is defined in terms of the chemical potential by

$$\lambda_B = \exp\left(\frac{\mu_B}{RT}\right). \tag{3.53}$$

3.5 What Useful Thermodynamic Quantities Can I Calculate?

As a result of the formulations that flow from the characteristic functions of internal energy, we are able to do a number of things that prove useful in an engineering and experimental context. In this section, we illustrate some of these applications in the field of thermodynamics and the measurement of properties for both fluid systems and solids. We use the speed of sound as an example of a property of a material that can be measured with great accuracy and precision by modern means and relate it to the thermodynamic properties of fluids and solids.

3.5.1 How Do I Calculate Entropy, Gibbs Function and Enthalpy Changes?

The Second Law provides relationships that permit the determination of the dependence of the Gibbs function, entropy and enthalpy on important state variables such as pressure, volume and temperature for a nonreacting material of constant composition. For example, the first two partial derivatives of Equation 3.41 can be written for a **constant composition** as

$$\left(\frac{\partial G}{\partial T}\right)_p = -S, \tag{3.54}$$

$$\left(\frac{\partial G}{\partial p}\right)_T = V. \tag{3.55}$$

Equation 3.55 provides a means of determining differences in *Gibbs function* arising from a *pressure* change provided that the volume of the system is known as a function of pressure because

$$G(T_1, p_2) - G(T_1, p_1) = \int_{p_1}^{p_2} V\mathrm{d}p. \tag{3.56}$$

Variations of the *Gibbs energy* as a function of *temperature* at constant pressure may, in principle, be predicted in a similar fashion, using Equation 3.54 as follows

$$G(T_2, p_1) - G(T_1, p_1) = -\int_{T_1}^{T_2} S\mathrm{d}T. \tag{3.57}$$

However, because only differences in entropy can be measured and not absolute values, Equation 3.57 is of no practical value.

In order to predict the variation of *entropy* with *pressure* and *volume* we need some of the Maxwell relations.

Differentiation of Equation 3.54 with respect to p at constant T gives

$$\left\{\frac{\partial(\partial G/\partial T)_p}{\partial p}\right\}_T = -\left(\frac{\partial S}{\partial p}\right)_T, \tag{3.58}$$

and differentiation of Equation 3.55 with respect to T at constant p gives

$$\left\{\frac{\partial(\partial G/\partial p)_T}{\partial T}\right\}_p = \left(\frac{\partial V}{\partial T}\right)_p. \tag{3.59}$$

According to the rule of cross-differentiation from Equation 1.154, the left-hand sides of Equations 3.58 and 3.59 must be equal. Thus

$$\left(\frac{\partial S}{\partial p}\right)_T = -\left(\frac{\partial V}{\partial T}\right)_p, \tag{3.60}$$

which is called a *Maxwell equation*. Integration of Equation 3.60 gives

$$S(T_1, p_2) - S(T_1, p_1) = -\int_{p_1}^{p_2} \left(\frac{\partial V}{\partial T}\right)_p \mathrm{d}p. \tag{3.61}$$

Thus, measurements of V as a function of temperatures over a range around T_1 at pressures from p_1 to p_2 yield values of $(\partial V/\partial T)_p$ over the pressure range p_1 to p_2. The entropy difference of Equation 3.61 is then determined from the area beneath a plot of $(\partial V/\partial T)_p$ on the ordinate as a function of p on the abscissa.

Another Maxwell equation can be obtained by the same procedure starting with the fundamental equation for the Helmholtz energy (Equation 3.44) at constant composition to give

$$\left(\frac{\partial S}{\partial V}\right)_T = \left(\frac{\partial p}{\partial T}\right)_V. \tag{3.62}$$

Equation 3.62 can be applied to determine *entropy* changes with respect to *volume* at constant temperature.

For the prediction of *entropy* changes as a function of *temperature*, we need the *Gibbs–Helmholtz equation*. The definition

$$G = H - TS \tag{3.63}$$

can be recast with Equation 3.54 to be

$$H = G - T(\partial G/\partial T)_p \tag{3.64}$$

and is called the *Gibbs–Helmholtz equation.* It should be pointed out that $U = A - T(\partial A/\partial T)_V$ is also unfortunately referred to as the Gibbs–Helmholtz equation.

Differentiation of Equation 3.64, the Gibbs–Helmholtz equation, with respect to temperature at constant pressure leads to the result

$$(\partial H/\partial T)_p = -T(\partial^2 G/\partial T^2)_p. \tag{3.65}$$

Substituting C_p for the left-hand side and with Equation 3.54, Equation 3.65 becomes

$$(\partial S/\partial T)_p = (\partial H/\partial T)_p/T = C_p/T. \tag{3.66}$$

Equation 3.66 provides the basis for a calorimetric method for the measurement of entropy difference because

$$S(T_2, p) - S(T_1, p) = \int_{T_1}^{T_2} (C_p/T)\mathrm{d}T. \tag{3.67}$$

The variation of *entropy* with *pressure* for a phase of fixed composition is given by Equation 3.61 and combined with Equation 3.67 to obtain the dependence of entropy on both T and p from

$$S(T_2, p_2) - S(T_1, p_1) = \int_{T_1}^{T_2} (C_p/T)\mathrm{d}T - \int_{p_1}^{p_2} (\partial V/\partial T)_p \mathrm{d}p. \tag{3.68}$$

The variation of *enthalpy* with *pressure* may be predicted by differentiating the definition

$$H = G + TS \tag{3.69}$$

with respect to pressure at constant temperature. This gives

$$\left(\frac{\partial H}{\partial p}\right)_T = \left(\frac{\partial G}{\partial p}\right)_T + T\left(\frac{\partial S}{\partial p}\right)_T. \tag{3.70}$$

Substitution of Equation 3.55 and 3.60 into Equation 3.70 gives

$$\left(\frac{\partial H}{\partial p}\right)_T = V - T\left(\frac{\partial V}{\partial T}\right)_p, \tag{3.71}$$

and integration gives

$$H(T_1, p_2) - H(T_1, p_1) = \int_{p_1}^{p_2} \{V - T(\partial V/\partial T)_p\}\mathrm{d}p. \tag{3.72}$$

We have seen, for example, in Question 1.8.1 how the left-hand side of Equation 3.72 can be determined with a flow calorimeter, while the right-hand can be estimated from direct measurements of p, V and T. Thus, one is able to either perform measurements that confirm the thermodynamic consistency expressed in Equation 3.72 or determine one unknown quantity given a knowledge of others.

By substituting the ideal gas law into Equations 3.71 or 3.72, one finds that the enthalpy for an ideal gas does not depend on pressure, but only on temperature.

By performing an analogous analysis based on the definition and the fundamental equation for the Helmholtz energy (Equations 3.43 and 3.44), one finds an expression for the dependence of the internal energy on the volume analogous to Equation 3.72. By substituting the ideal gas law into it, one concludes that the internal energy U of an ideal gas does not depend on the system volume, but only on the temperature.

Expressions for the change in *chemical potential* with respect to pressure $\{\mu_B(T_1, p_2) - \mu_B(T_1, p_1)\}$ and the change in chemical potential with respect to temperature $\{\mu_B(T_2, p_1) - \mu_B(T_1, p_1)\}$ illustrate how *these differences* can be measured. From the definition of the chemical potential (Equation 3.25) and application of the rule of cross differentiation (Equation 1.154), we obtain

$$\begin{aligned}\left(\frac{\partial \mu_B}{\partial p}\right)_T &= \left\{\frac{\partial(\partial G/\partial n_B)_{T,p,n_A\neq n_B}}{\partial p}\right\}_T \\ &= \left\{\frac{\partial(\partial G/\partial p)_T}{\partial n_B}\right\}_{T,p,n_A\neq n_B} \\ &= \left\{\frac{\partial V}{\partial n_B}\right\}_{T,p,n_A\neq n_B} = V_B,\end{aligned} \tag{3.73}$$

where V_B is the partial molar volume of B. Integration of Equation 3.73 then gives

$$\mu_B(T_1, p_2) - \mu_B(T_1, p_1) = \int_{p_1}^{p_2} V_B \mathrm{d}p. \tag{3.74}$$

From Equation 3.53

$$\left(\frac{\partial \ln \lambda_B}{\partial p}\right)_T = \frac{V_B}{RT}. \tag{3.75}$$

Thus, the ratio of absolute activities is

$$\ln\left\{\frac{\lambda_B(T_1, p_2)}{\lambda_B(T_1, p_1)}\right\} = \int_{p_1}^{p_2} \left(\frac{V_B}{RT_1}\right) \mathrm{d}p. \tag{3.76}$$

The difference $\{\mu_B(T_2, p_1) - \mu_B(T_1, p_1)\}$ and ratio $\ln\{\lambda_B(T_2, p_1)/\lambda_B(T_1, p_1)\}$ are given by

$$\left(\frac{\partial \mu_B}{\partial T}\right)_p = -S_B. \tag{3.77}$$

or

$$\mu_{\rm B}(T_2, p_1) - \mu_{\rm B}(T_1, p_1) = -\int_{T_1}^{T_2} S_{\rm B}\,{\rm d}T \tag{3.78}$$

and

$$\left(\frac{\partial \ln \lambda_{\rm B}}{\partial T}\right)_p = -\frac{H_{\rm B}}{RT^2} \tag{3.79}$$

or

$$\ln\left\{\frac{\lambda_{\rm B}(T_2, p_1)}{\lambda_{\rm B}(T_1, p_1)}\right\} = -\int_{T_1}^{T_2}\left(\frac{H_{\rm B}}{RT^2}\right){\rm d}T \tag{3.80}$$

respectively.

The routes to obtain the difference $\{\mu_{\rm B}(T_2, p_1) - \mu_{\rm B}(T_1, p_1)\}$ and ratio $\ln\{\lambda_{\rm B}(T_2, p_1)/\lambda_{\rm B}(T_1, p_1)\}$ provided by Equations 3.78 and 3.80 are of no immediate use because neither $S_{\rm B}$ nor $H_{\rm B}$ can be measured; however, the equations themselves are useful, as we shall see in Chapter 4.

3.5.2 How Do I Calculate Expansivity and Compressibility?

The isobaric (constant pressure) expansivity or coefficient of thermal expansion α is defined by

$$\alpha = \frac{1}{V}\left(\frac{\partial V}{\partial T}\right)_p = \left(\frac{\partial \ln V}{\partial T}\right)_p. \tag{3.81}$$

α is usually positive but for water at temperatures between 273.15 K and 277.13 K, it is negative. The isothermal compressibility κ_T is defined by

$$\kappa_T = -\frac{1}{V}\left(\frac{\partial V}{\partial p}\right)_T = -\left(\frac{\partial \ln V}{\partial p}\right)_T, \tag{3.82}$$

and by the –1 rule (provided by Equation 1.153) is related to α

$$\frac{\alpha}{\kappa_T} = \left(\frac{\partial p}{\partial T}\right)_V. \tag{3.83}$$

The isentropic (constant entropy) compressibility κ_S is defined by

$$\kappa_S = -\frac{1}{V}\left(\frac{\partial V}{\partial p}\right)_S. \tag{3.84}$$

The difference between $(\partial V/\partial p)_T$ and $(\partial V/\partial p)_S$ can be written by use of the rule for changing a variable held constant (given by Equation 1.152) as

$$\left(\frac{\partial V}{\partial p}\right)_T - \left(\frac{\partial V}{\partial p}\right)_S = -\left(\frac{\partial V}{\partial T}\right)_p \left(\frac{\partial T}{\partial p}\right)_S. \tag{3.85}$$

Using the –1 rule (Equation 1.153) on $(\partial T/\partial p)_S$, Equation 3.85 becomes

$$\left(\frac{\partial V}{\partial p}\right)_T - \left(\frac{\partial V}{\partial p}\right)_S = \left(\frac{\partial V}{\partial T}\right)_p \left(\frac{\partial S}{\partial p}\right)_T \bigg/ \left(\frac{\partial S}{\partial T}\right)_p. \tag{3.86}$$

Substituting Equation 3.60 and 3.54 into Equation 3.86 gives

$$\left(\frac{\partial V}{\partial p}\right)_T - \left(\frac{\partial V}{\partial p}\right)_S = -T\left\{\left(\frac{\partial V}{\partial T}\right)_p\right\}^2 \bigg/ C_p, \tag{3.87}$$

which, with the definitions of Equations 3.81, 3.82 and 3.84, gives

$$\kappa_T - \kappa_S = \frac{T\alpha^2 V}{C_p}. \tag{3.88}$$

If we had independent means of measuring κ_S, κ_T, α, T, V and C_p of a phase, then Equation 3.88 could be used to test the measurements for thermodynamic consistency. If one parameter of Equation 3.88 cannot or has not been measured, then it can be calculated from measurements of the others from the same equation. Because α^2, T, V and C_p are all positive from Equation 3.88, κ_T is always greater then κ_S.

For a *closed* phase of fixed composition, the entropy balance of Equation 3.6 for a reversible process yields

$$\mathrm{d}S = \frac{\mathrm{d}Q}{T}, \tag{3.89}$$

and it follows from Equation 3.89 that if the process is adiabatic, so that $\delta Q = 0$, it must also be isentropic so that $\mathrm{d}S = 0$. This process can be realized for a *closed phase of fixed composition*. An expansion (or compression) of a known volume of fluid in a thermally insulated vessel can be achieved adiabatically by quickly changing the pressure and remeasuring the pressure and volume. The expansion can also be performed reversibly by changing the pressure slowly and, provided the thermal insulation is good, then the process is both adiabatic and reversible and so isentropic. This is easily achieved particularly with liquids to provide direct measurements of κ_S. For fluids, no one quantity in Equation 3.88 is more difficult to measure than the others, but for solids, κ_T is hard to determine and is therefore obtained from Equation 3.88 from measurements of κ_S, α, T, V and C_p. Measurements of the speed of sound are used to determine κ_S, which will be discussed in Question 3.5.3, and α is determined from the temperature dependence of the lattice constant by X-ray diffraction. We will return to expansion in Questions 3.5.6 and 3.5.7. Expansion and compression were considered in Questions 1.6.3 through 1.6.5.

3.5.3 What Can I Gain from Measuring the Speed of Sound in Fluids?

While the speed of sound in a phase is important in its own right in a number of applications, most of the interest in this quantity arises from its relationship with the thermodynamic properties of isotropic, Newtonian fluids and isotropic elastic solids. For fluid phases, as these usually support only a single longitudinal sound mode, the sound propagation speed u is given by (Herzfeld and Litovitz 1959)

$$u^2 = \left(\frac{\partial p}{\partial \rho}\right)_s = \frac{1}{\rho \kappa_S} = \frac{\gamma}{\rho \kappa_T}. \tag{3.90}$$

In Equation 3.90, all the symbols have been previously defined, including $\gamma = C_p/C_V$, which is now given in the form of C_p and C_V, the molar isobaric and isochoric heat capacities, respectively (see Question 1.8.3). Equation 3.90 is strictly valid only in the limits of vanishing amplitude and vanishing frequency (Herzfeld and Litovitz 1959; Morse and Ingard 1968; Goodwin and Trusler 2003). Although it is usually not possible to achieve both of these conditions directly in experiment, extrapolation of measurements made at finite amplitude and frequency are generally adequate. Equation 3.90 shows that the isentropic compressibility may be obtained from measurements of the speed of sound and the density, and that the isothermal compressibility may also be obtained if γ is known. Equation 3.90 forms the basis of almost all experimental determinations of the isentropic compressibility and is a convenient route to γ.

For independent variables of either (T, p) or (T, ρ_n), where ρ_n is the amount-of-substance density (which we distinguish from the mass density, ρ), Equation 3.90 can be recast for (T, p) as

$$u^2 = \frac{1}{M}\left[\left(\frac{\partial \rho_n}{\partial p}\right)_T - \frac{T}{\rho_n^2 C_p}\left(\frac{\partial \rho_n}{\partial T}\right)_p^2\right]^{-1}, \tag{3.91}$$

and for (T, ρ_n)

$$u^2 = \frac{1}{M}\left[\left(\frac{\partial p}{\partial \rho_n}\right)_T + \frac{T}{\rho_n^2 C_V}\left(\frac{\partial p}{\partial T}\right)_{\rho_n}^2\right]. \tag{3.92}$$

In Equations 3.91 and 3.92, M is the molar mass, C_p the isobaric molar heat capacity and C_V the isochoric molar heat capacity; to adhere strictly to the International Union of Pure and Applied Chemistry (IUPAC) (Quack et al. 2007), a subscript m should be included to indicate a molar quantity to give $C_{p,\mathrm{m}}$ and $C_{V,\mathrm{m}}$. We have not included them here to preserve simplicity and to avoid confusion.

In principle, these equations allow one to compute the speed of sound, u, from an equation of state in the form $\rho_n = \rho_n(T, p) = V_\mathrm{m}^{-1}$ or $p = p(T, \rho_n)$, although one requires a knowledge of the heat capacity in some reference state (see Question 5.2.1). An equation for C_p can be obtained by differentiation of Equation 3.71 with respect to T at constant p that gives

$$\left\{\frac{\partial(\partial H/\partial p)_T}{\partial T}\right\}_p = -T\left(\frac{\partial^2 V}{\partial T^2}\right)_p, \tag{3.93}$$

and using the rule for cross differentiation from Chapter 1, Equation 1.154, we obtain

$$\left(\frac{\partial C_p}{\partial p}\right)_T = -T\left(\frac{\partial^2 V}{\partial T^2}\right)_p, \tag{3.94}$$

which when integrated becomes

$$C_p = C_p^{\mathrm{o}}(T) - \int_{p^{\mathrm{o}}}^{p} T\left(\partial^2 \rho_n^{-1}/\partial T^2\right) dp, \tag{3.95}$$

where C_p^{o} is the molar isobaric specific heat capacity on a reference isobar $p = p^{\mathrm{o}}$ and $\rho_n = 1/V_{\mathrm{m}}$. The analogous expression for C_V is

$$C_V = C_V^{\mathrm{o}}(T) - \int_{\rho_n^{\mathrm{o}}}^{\rho_n} \left(T/\rho_n^2\right)\left(\partial^2 p/\partial T^2\right) d\rho_n, \tag{3.96}$$

where ρ_n^{o} is the amount-of-substance density on a reference isochore (line of constant amount of substance density).

3.5.4 What Can I Gain from Measuring the Speed of Sound in Solids?

The elastic properties of an isotropic solid may be specified by a pair of quantities such as the bulk modulus K and the shear modulus G. Other commonly used parameters are Young's modulus E, Poisson's ratio σ and the Lamé constants λ and μ. The shear modulus G is identical with the second Lamé constant μ and the other parameters are interrelated as follows (Landau and Lifshitz 1987)

$$\left.\begin{aligned} E &= 9GK/(3K + G) = 3(1 - 2\sigma)K \\ \sigma &= (3K - 2G)/(6K + 2G) = (E/2G) - 1 \\ \lambda &= K - 2G/3 \end{aligned}\right\}. \tag{3.97}$$

The symbols in Equation 3.97 are those commonly used in the field and have a different meaning than similar symbols in the list of nomenclature.

The elastic constants relate various types of stress and strain under isothermal conditions. In the case of pure shear stress, the resulting strain takes place without change of volume and so an isothermal and reversible shear process is also isentropic. Consequently, the shear modulus is the same for both static and dynamic processes in an elastic body. However, compressive stress gives rise to a change in volume so that an isothermal compression is not generally isentropic. The isothermal bulk modulus $K = 1/\kappa_T$ therefore differs from the isentropic bulk modulus, which is $K_S = 1/\kappa_S$, and the two are related by Equation 3.88 so that

$$1/K_S = 1/K - T\alpha^2/(\rho c_p), \tag{3.98}$$

where α is the coefficient of thermal expansivity, ρ the mass density and c_p the specific heat capacity. The isentropic analogue of Young's modulus, E_S, is given by

$$E_S = E/\{1 - ET\alpha^2/(9\rho c_p)\}. \tag{3.99}$$

Solids generally have both longitudinal or compressive sound modes, in which the direction of stress and strain is parallel to the direction of propagation, and two orthogonal shear or transverse wave modes in each of which the direction of shear stress is perpendicular to the direction of propagation. In an isotropic solid, the two shear modes are degenerate, each propagating with speed u_S given by

$$u_S^2 = G/\rho. \tag{3.100}$$

The speed u_L of longitudinal sound waves in a bulk specimen is given by

$$u_L^2 = (K_S + 4G/3)/\rho. \tag{3.101}$$

The actual phase speed, u, of a compression wave propagating along the axis of a solid bar generally depends upon the lateral dimension of the bar: u approaches u_L when the lateral extent (the dimension normal to the direction of propagation) of the bar is much greater than the wavelength of the sound. For bars of a smaller cross section, the phase speed, u, is generally smaller than u_L and, when the lateral dimensions are much smaller than the wavelength, it reaches a limit u_E given by

$$u_E^2 = E_S/\rho. \tag{3.102}$$

In an isotropic elastic solid $u_L > u_E > u_S$.

Acoustic, especially ultrasonic, methods are the most common means of determining the elastic constants of solids. High-frequency (that is 10 MHz) ultrasonic measurements typically provide directly values of u_S and u_L (Papadakis 1998). When combined with a measurement of the density, G and K_S may then be determined. The difference between κ_S and κ_T in a solid material is typically very small (<1%; see Ledbetter 1982) and so the isothermal bulk modulus K is easily obtained from K_S by means of a calculated correction according to Equation 3.98. Very approximate values of α and c_p will suffice for that purpose. Once G and K have been obtained, σ follows from Equation 3.97.

Low-frequency (i.e. 100 kHz) ultrasonic resonance experiments typically provide directly values of u_S and u_E from which G and E_S are obtained (Weston 1975), which then permits determination of the isothermal Young's modulus E.

3.5.5 Can I Evaluate the Isobaric Heat Capacity from the Isochoric Heat Capacity?

The difference $(C_p - C_V)$ is given by Equation 1.162 as

$$C_p - C_V = -T\left\{\left(\frac{\partial V}{\partial T}\right)_p\right\}^2 \Big/ \left(\frac{\partial V}{\partial p}\right)_T = T\alpha^2 V/\kappa_T, \tag{3.103}$$

where Equations 3.81 and 3.82 were used to obtain the second part of the equality.

Equation 3.103 is important because it interrelates independently measurable quantities which can be used to both experimentally verify the laws and permit the use of results for the most easily measured quantity to evaluate the least easily measured. For an ideal gas, $pV_{\mathrm{m}} = RT$ and substitution into Equation 3.103 gives $C_{p,\mathrm{m}}^{\mathrm{ig}} - C_{V,\mathrm{m}}^{\mathrm{ig}} = R$.

3.5.6 Why Use an Isentropic Expansion to Liquefy a Gas?

In Question 1.6.5, we discussed with specific quantities the work required for the adiabatic compression of a gas. Here we consider a fast, but not explosively fast (for which the pressure is neither uniform nor defined and the expansion irreversible) expansion or compression in a thermally insulated vessel. This process is isentropic and the temperature varies with pressure according to

$$\left(\frac{\partial T}{\partial p}\right)_S = -\frac{(\partial S/\partial p)_T}{(\partial S/\partial T)_p} = \frac{T(\partial V/\partial T)_p}{C_p} = \frac{T\alpha V}{C_p}. \tag{3.104}$$

For an ideal gas, integration of Equation 3.104 gives

$$T_2 = T_1(p_2/p_1)^{R/C_p}. \tag{3.105}$$

Question 1.6.5 provides an alternative derivation of Equation 3.105. For a perfect monatomic gas, for example, argon, for which the molar heat capacity is $C_p = 5R/2$, then starting at $T_1 = 300$ K, the expansion from $p_1 = 3.2$ MPa to $p_2 = 0.1$ MPa yields a final temperature $T_2 = 75$ K; this is a vivid explanation of why isentropic expansion is used to liquefy so-called permanent gases. The converse of this expansion, namely the isentropic compression of air leads to a temperature increase familiar to those who inflate their bicycle tires with a hand-pump.

3.5.7 Does Expansion of a Gas at Constant Energy Change Its Temperature?

The dependence of the temperature of a gas on its volume at constant energy (Joule 1845) $(\partial T/\partial V)_U$ can be obtained from the −1 rule (Equation 1.153) as

$$\left(\frac{\partial T}{\partial V}\right)_U = -\frac{(\partial U/\partial V)_T}{(\partial U/\partial T)_V}. \tag{3.106}$$

The numerator of Equation 3.106 can be obtained from Equation 3.32 as

$$(\partial U/\partial V)_T = T(\partial S/\partial V)_T - p. \tag{3.107}$$

Substitution of Equation 3.62 into Equation 3.107 and using $(\partial U/\partial T)_V = C_V$, Equation 3.106 becomes

$$\left(\frac{\partial T}{\partial V}\right)_U = \left\{p - T\left(\frac{\partial p}{\partial T}\right)_V\right\}\Big/ C_V. \tag{3.108}$$

Substitution of Equations 3.83 and 3.103 into Equation 3.108 gives

$$\left(\frac{\partial T}{\partial V}\right)_U = \frac{(p - T\alpha/\kappa_T)}{(C_p - T\alpha^2 V/\kappa_T)}. \tag{3.109}$$

An expression for the rate of change of temperature with respect to pressure at constant energy $(\partial T/\partial p)_U$ can also be obtained by first using the –1 rule Equation 1.153 to give

$$\left(\frac{\partial T}{\partial p}\right)_U = -\frac{(\partial U/\partial p)_T}{(\partial U/\partial T)_p}. \tag{3.110}$$

To find an expression for $(\partial U/\partial T)_p$ requires the use of the rule of change of variable held constant (Equation 1.152) to give

$$\left(\frac{\partial U}{\partial T}\right)_p = \left(\frac{\partial U}{\partial T}\right)_V + \left(\frac{\partial U}{\partial V}\right)_T \left(\frac{\partial V}{\partial T}\right)_p. \tag{3.111}$$

The first derivative on the right-hand side of Equation 3.111 is C_V; the second term requires Equation 3.107 for $(\partial U/\partial V)_T$ and Equation 3.81 to give

$$\left(\frac{\partial U}{\partial T}\right)_p = C_V + \left(\frac{T\alpha}{\kappa_T} - p\right)\alpha V = C_p - p\alpha V, \tag{3.112}$$

where Equation 3.103 has been used. The next requirement to solve Equation 3.110 is to find an expression for $(\partial U/\partial p)_T$. Combining Equations 3.20 and 3.43 gives

$$U = H - G + A, \tag{3.113}$$

which when differentiated with respect to p at constant T gives

$$\left(\frac{\partial U}{\partial p}\right)_T = \left(\frac{\partial H}{\partial p}\right)_T - \left(\frac{\partial G}{\partial p}\right)_T + \left(\frac{\partial A}{\partial p}\right)_T. \tag{3.114}$$

The first derivative on the right-hand side of Equation 3.114 is given by Equation 3.71 and the second derivative is given by Equation 3.55. However, an expression is required for $(\partial A/\partial p)_T$ that can be obtained by differentiation of

$$A = G - pV \tag{3.115}$$

with respect to pressure at constant temperature to give

$$\left(\frac{\partial A}{\partial p}\right)_T = \left(\frac{\partial G}{\partial p}\right)_T - V - p\left(\frac{\partial V}{\partial p}\right)_T = pV\kappa_T. \tag{3.116}$$

Equations 3.55 and 3.82 were also used to obtain the second equality of Equation 3.116. Substitution of Equations 3.55, 3.71 and 3.116 into Equation 3.114 using Equations 3.81 and 3.82 gives

$$\left(\frac{\partial U}{\partial p}\right)_T = V - T\left(\frac{\partial V}{\partial T}\right)_p - V + pV\kappa_T = V(p\kappa_T - \alpha T). \tag{3.117}$$

Finally, substitution of Equations 3.112 and 3.117 into Equation 3.110 gives

$$\left(\frac{\partial T}{\partial p}\right)_U = \frac{(T\alpha V - p\kappa_T V)}{(C_p - p\alpha V)}. \tag{3.118}$$

For a perfect gas, it is readily shown that $(\partial T/\partial V)_U = 0$ and also $(\partial U/\partial V)_T = 0$, $(\partial U/\partial p)_T = 0$, $(\partial H/\partial p)_T = 0$ and $(\partial H/\partial V)_T = 0$, so that changes of volume produce no temperature changes. Joule carried out measurements of the effect of an expansion at constant energy on air; he employed a thermometer of uncertainty ±0.01 K and observed no temperature change. However, a modern thermometer with uncertainty of <±0.001 K would reveal a temperature decrease of 0.003 K under the conditions he employed because for N_2(g) at a temperature of 293 K and pressure of 0.1 MPa $(\partial T/\partial p)_U \approx 0.003\ \text{K}\cdot\text{kPa}^{-1}$. The effects are evidently very small relative to the large heat capacity of any practical container and the method is not used to study the properties of gases. It should be remarked in the context of the discussions in Chapter 2 that the origin of the small temperature change in a real gas on expansion arises from the work done against the intermolecular potential at the expense of the kinetic energy.

3.5.8 What Is a Joule-Thomson Expansion?

In a Joule-Thomson isenthalpic expansion (Question 1.8.6) the temperature difference across a throttle or porous plug is measured as gas flows through the device from a higher pressure to a lower pressure. In the limit of very low pressures this experiment serves to determine the Joule-Thomson coefficient μ_{JT} defined by

$$\mu_{JT} = \left(\frac{\partial T}{\partial p}\right)_H = -\frac{\{V - T(\partial V/\partial T)_p\}}{C_p}, \tag{3.119}$$

which is zero for a perfect gas. As was indicated in Question 1.8.6, an alternative experiment is possible in which the gas exiting the porous plug is heated and restored to the inlet temperature by the application of electrical power. In that case, the experiment is isothermal and the change of enthalpy under those conditions in the limit of low pressure is defined as the isothermal Joule-Thomson coefficient ϕ_{JT} defined by

$$\phi_{JT} = \left(\frac{\partial H}{\partial p}\right)_T = V - T(\partial V/\partial T)_p. \tag{3.120}$$

For a perfect gas, ϕ_{JT} is zero.

3.6 How Are Thermal, Mechanical and Diffusive Equilibrium Ensured Between Two Phases?

In an isolated system comprising more than one phase, the condition of equilibrium means that nothing about the system will change with time. In addition, in each phase the properties of the system are uniform, so that they do not vary with position. Such systems are common in laboratory and engineering practice and it is therefore important to consider the conditions under which a multiphase system can be maintained in a stable equilibrium state. That is the purpose of this section where we consider thermal equilibrium, mechanical equilibrium and diffusive equilibrium separately.

3.6.1 How Are Thermal Equilibrium and Stability Ensured?

For an isolated system consisting of two phases α and β, of temperature T^α and T^β, separated by a *rigid impermeable diathermic* wall, we have

$$\mathrm{d}U^\alpha + \mathrm{d}U^\beta = 0, \tag{3.121}$$

and $\mathrm{d}V^\alpha = 0$, $\mathrm{d}V^\beta = 0$, $\mathrm{d}n^\alpha = 0$ and $\mathrm{d}n^\beta = 0$ for all substances. From Equation 3.34

$$\mathrm{d}S^\alpha = \mathrm{d}U^\alpha/T^\alpha \quad \text{and} \quad \mathrm{d}S^\beta = \mathrm{d}U^\beta/T^\beta, \tag{3.122}$$

and the sum of the entropy for both phases is

$$\mathrm{d}S = (\mathrm{d}U^\alpha/T^\alpha) + (\mathrm{d}U^\beta/T^\beta) = \mathrm{d}U^\alpha(T^\beta - T^\alpha)/(T^\beta T^\alpha). \tag{3.123}$$

If $T^\beta > T^\alpha$ and a process is underway (something is happening) then from the entropy balance of Equation 3.6 and from the Second Law, it follows for an isolated system undergoing an irreversible process that $\mathrm{d}S > 0$. Thus, the energy flows from the higher temperature phase to the lower temperature one. If $T^\alpha = T^\beta$ then nothing is happening and the system is in thermal equilibrium and $(\partial S/\partial t)_{U,V,\Sigma n} = 0$. This is an important statement because it can be shown as a consequence that the temperature is the thermodynamic temperature (see Question 1.7.2).

The condition that ensures the thermal stability of an isolated phase can be determined by considering a phase that is initially of energy $2U$ and volume $2V$. This phase is then divided in two with one part of energy $(U + \delta U)$ and volume V and the other part of energy $(U - \delta U)$ and volume V. The entropy change is given by

$$\delta S = S(U + \delta U, V) + S(U - \delta U, V) - S(2U, 2V). \tag{3.124}$$

Use of Taylor's theorem given by Equation 1.173 on Equation 3.34 gives

$$\delta S = (\partial^2 S/\partial U^2)_V(\delta U)^2 + \mathrm{O}(\delta U)^4, \tag{3.125}$$

where $\mathrm{O}(\delta U)^4$ means terms $(\delta U)^4$ and with higher powers of (δU). From Equation 3.34 we find $(\partial S/\partial U)_V = 1/T$ and by further substitution of Equation 3.34 we find Equation 3.125 becomes

$$(\partial^2 S/\partial U^2)_V = (\partial T^{-1}/\partial U)_V = -1/(TC_V). \tag{3.126}$$

Equations 3.126 and 3.125 mean that δS can only be positive so that the proposed change can only occur if C_V is negative. The condition of thermal stability of a phase is therefore

$$C_V > 0. \tag{3.127}$$

As a result, for example, the temperature of an isolated metal bar which is initially in thermal equilibrium will not spontaneously increase in temperature at one end while the temperature at the other decreases. Equally, if energy is added to a phase of constant volume and fixed composition, the temperature always increases.

3.6.2 How Are Mechanical Equilibrium and Stability Ensured?

For an *isolated* system at uniform temperature T consisting of two phases α and β separated by a *moveable impermeable diathermic* wall, where the pressures of the phases are p^{α} and p^{β}, we have

$$\mathrm{d}U^{\alpha} + \mathrm{d}U^{\beta} = 0, \tag{3.128}$$

$$\mathrm{d}V^{\alpha} + \mathrm{d}V^{\beta} = 0, \tag{3.129}$$

$\mathrm{d}n^{\alpha} = 0$ and $\mathrm{d}n^{\beta} = 0$ for all substances. From Equation 3.32 in a reversible process

$$T\mathrm{d}S^{\alpha} = \mathrm{d}U^{\alpha} + p^{\alpha}\mathrm{d}V^{\alpha} \text{ and } T\mathrm{d}S^{\beta} = \mathrm{d}U^{\beta} + p^{\beta}\mathrm{d}V^{\beta}, \tag{3.130}$$

and the sum of the entropy for both phases is

$$T\mathrm{d}S = (p^{\alpha} - p^{\beta})\mathrm{d}V^{\alpha}. \tag{3.131}$$

If $p^{\beta} > p^{\alpha}$ and a process is underway (something is happening), then from the entropy balance of Equation 3.6 and from the Second Law it follows that $\mathrm{d}S > 0$. According to Equation 3.131, $\mathrm{d}V^{\alpha}$ must be positive if $p^{\alpha} > p^{\beta}$ and negative in the reverse case. If $p^{\alpha} = p^{\beta}$, then nothing is happening and the system is in hydrostatic equilibrium. This is a purely mechanical result derived from the Second Law.

For an isolated system consisting of a phase, separated by a moveable impermeable diathermic wall, with each part initially at the same T and V, the condition that prevents a difference in the pressure of each part can be derived. To do so we assume that the partition moves so that one part has a volume of $(V + \delta V)$ and the other volume $(V - \delta V)$ both at the same temperature T. For variables of T and V, the Helmholtz function A should be used and, similarly to Equation 3.124, the change δA is given by

$$\delta A = A(V + \delta V, T) + A(V - \delta V, T) - A(T, 2V). \tag{3.132}$$

Use of Taylor's theorem given by Equation 1.173 on Equation 3.132 gives

$$\delta A = (\partial^2 A/\partial V^2)_V (\delta V)^2 + \mathrm{O}(\delta V)^4. \tag{3.133}$$

From the derivative, $(\partial A/\partial V)_T = -p$ so that

$$(\partial^2 A/\partial V^2)_V = -(\partial p/\partial V)_T. \tag{3.134}$$

If the partition were to move, Equations 3.133 and 3.134 show δA is negative and would require $(\partial p/\partial V)_T > 0$. It follows, the condition that ensures hydrostatic stability of an isolated phase in thermal equilibrium is

$$\left(\frac{\partial p}{\partial V}\right)_T < 0 \tag{3.135}$$

or

$$-\frac{1}{V}\left(\frac{\partial V}{\partial p}\right)_T = \kappa_T > 0 \tag{3.136}$$

so that when the pressure of a phase of fixed composition is increased at a constant temperature, the volume must always decreases.

3.6.3 How Are Diffusive Equilibrium and Stability Ensured?

When we consider an isolated system of uniform temperature T with two phases α and β separated by a *rigid diathermic wall permeable* to substance A of chemical potential μ_A^α and μ_A^β, the relation

$$dU^\alpha + dU^\beta = 0, \tag{3.137}$$

holds, and $dV^\alpha = 0$, $dV^\beta = 0$, $dn_A^\alpha + dn_A^\beta = 0$ and $dn_{B\neq A}^\alpha = 0$ and $dn_{A\neq B}^\beta = 0$. From Equation 3.32

$$TdS^\alpha = dU^\alpha + p^\alpha dV^\alpha - \mu_A^\alpha dn_A^\alpha \text{ and } TdS^\beta = dU^\beta + p^\beta dV^\beta - \mu_A^\beta dn_A^\beta, \tag{3.138}$$

so the entropy for the system is given by

$$TdS = \left(\mu_A^\beta - \mu_A^\alpha\right)dn_A^\alpha. \tag{3.139}$$

If $\mu_A^\beta > \mu_A^\alpha$ and a process is underway (something is happening), then from the entropy balance of Equation 3.6 and from the Second Law, it follows that $dS > 0$. According to Equation 3.139 dn_A^α must be positive if $\mu_A^\beta > \mu_A^\alpha$ and negative in the reverse case. So the substance A is flowing from the phase with higher chemical potential, μ_A^β, to the one with lower chemical potential, μ_A^α. If $\mu_A^\alpha = \mu_A^\beta$, then $dS = 0$ and nothing is happening, the system is said to be in *diffusive* equilibrium. When $\mu_B^\alpha = \mu_B^\beta$ for some, but not all, substances and $p^\alpha \neq p^\beta$, which is permitted since nothing was stated about pressure, the system is said to be in *osmotic* equilibrium.

What is the condition that prevents a substance being concentrated in one part of the system and depleted in another? To address this question, we propose a system that is initially at temperature T, pressure p and contains amount of substance $2n_A$ of A and $2n_B$ of B; the number of components can be infinite but for simplicity we have chosen two. As we have before, we will assume the phase can be split into two parts both with the same temperature T and pressure p with the following: (1) amount of substance of A of $(n_A + \delta n_A)$ and n_B of B; and (2) amount of substance of A of $(n_A - \delta n_A)$ and n_B of B. For variables n, T and p, the Gibbs function G should be used and the change arising from the proposed movement of substance A is given by

$$\delta G = G(T, p, n_A + \delta n_A, n_B) + G(T, p, n_A - \delta n_A, n_B) - G(T, p, 2n_A, 2n_B). \quad (3.140)$$

Use of Taylor's theorem given by Equation 1.173 on Equation 3.140 gives

$$\delta G = (\partial^2 G/\partial n_A^2)_{T,p,n_B}(\delta n_A)^2 + O(\delta n_A)^4. \quad (3.141)$$

From Equation 3.38 we have

$$(\partial^2 G/\partial n_A^2)_{T,p,n_B} = (\partial \mu_A/\partial n_A)_{T,p,n_B}. \quad (3.142)$$

The change proposed is possible and the amount of substance A can be greater in one part of a phase if δG is negative and that means only if

$$(\partial \mu_A/\partial n_A)_{T,p,n_B} < 0. \quad (3.143)$$

Thus, for an isolated phase at constant temperature and pressure *diffusional* stability (an absence of change) requires

$$(\partial \mu_A/\partial n_A)_{T,p,n_B} > 0, \quad (3.144)$$

and when substance A is added to a phase at constant temperature, pressure and fixed composition the chemical potential of A must increase.

3.7 What Are Standard Thermodynamic Quantities for Solutions?

In Chapter 1, in Question 1.9, we introduced both the concept of and need for a standard thermodynamic state. This standard state was defined by stating a particular state of aggregation, gas, (g), liquid (l) or solid (s) and for pure materials a standard pressure, $p^\ominus$, taken as 10^5 Pa (= 1 bar) and a standard temperature, variously defined but most often as $T = 298.15$ K or $T = 273.15$ K. For mixtures, the conventional means of stating the standard composition is through the molality. The standard molality is denoted by $m^\ominus$ and for standard thermodynamic quantities the recommended value, which is used almost exclusively, is 1 mol kg^{-1}.

As pointed out in Question 1.9, one very important use of the standard state is to enable the tabulation for a wide variety of chemical thermodynamic properties such as U, H, S at this specified standard state and to have a formal means of evaluating the same properties of the material under other thermodynamic conditions. Evidently, such tabulations greatly reduce the amount of material that must be specified for each chemical substance or mixture. Because it is not always possible to measure each of the requisite properties at the standard state, it is necessary to have a means of deriving the properties of materials in the standard state from the conditions under which they have been measured. It is the purpose of this question to provide just those relationships. This is the reason why the standard property appears on the left-hand side of many of the equations, while the right-hand side shows how to evaluate them from measured data. Indeed, we consider especially the standard thermodynamic functions for gas, liquid, or solid solutions because such mixtures often dominate practical applications. In the chapters of this book that are devoted to answering question of practical applications, such as Chapters 4 to 8, we shall have reason to make use of the same equations as are quoted here in order to evaluate the thermodynamic properties of mixtures and solutions in arbitrary thermodynamic states so that it is the latter which form the left-hand side of the relevant equations with the standard state values on the right. Chapter 9 lists some sources of the standard thermodynamic quantities for pure substances and mixtures.

3.7.1 What Are the Standard Thermodynamic Functions for a Gas Mixture?

The standard chemical potential for a gas B is related to the chemical potential at some other pressure and the same temperature using Equation 3.74 applied between the standard pressure and the total pressure of the system

$$\mu_B(g, T_1, p) - \mu_B(g, T_1, p^{\ominus}) = \int_{p^{\ominus}}^{p} V_B \mathrm{d}p. \tag{3.145}$$

So for a pure, perfect gas, the partial molar volume is equal to the molar volume of the pure substance, i.e. $V_B = (RT/p)$,

$$\mu_B(g, T_1, p^{\ominus}) = \mu_B(g, T_1, p) - RT \ln \frac{p}{p^{\ominus}}. \tag{3.146}$$

For a gas mixture containing the gas B and any number of other species, we show in Question 4.3.4 that the standard chemical potential of species B in the mixture where its mole fraction is x_B is

$$\mu_B^{\ominus}(g, T) = \mu_B(g, T, p, x) - RT \ln\left(\frac{x_B p}{p^{\ominus}}\right) - \int_0^p \left\{V_B(g, T, p, x) - \frac{RT}{p}\right\}\mathrm{d}p. \tag{3.147}$$

Equations 3.145 to 3.147, with the Roman character defining the state of the aggregation, with g for gas, l for liquid and s for solid is written according to the nomenclature established by IUPAC Green Book (Quack et al. 2007). This form of representation denotes, for example, the standard chemical potential as $\mu_B^{\ominus}(g, T)$, which indicates it is a function of the phase and temperature. However, as we recognize throughout this text, the language of the chemist is not familiar to all. Consequently, for the general audience we have adopted an approach that deviates from the formal IUPAC symbolism and indicates, when significant, the phase as a subscript after identifying the substance B, also a subscript. Thus, with these rules we now write, for example, $\mu_B^{\ominus}(g, T)$ of Equation 3.147 as $\mu_{B,g}^{\ominus}(T)$ and cast it in the form

$$\mu^{\ominus}_{\mathrm{B,g}}(T) = \mu_{\mathrm{B,g}}(T, p, x) - RT \ln\left(\frac{x_{\mathrm{B}} p}{p^{\ominus}}\right) - \int_0^p \left\{V_{\mathrm{B,g}}(T, p, x) - \frac{RT}{p}\right\} \mathrm{d}p. \tag{3.148}$$

In Equation 3.148, $\mu_{\mathrm{B,g}}(T, p, x)$ is the chemical potential and $V_{\mathrm{B,g}}(T, p, x)$ the partial molar volume of species B in a gas mixture of composition given by mole fractions x for which the mole fraction of B is x_{B} at a pressure p and temperature T. The pressure $p^{\ominus}$ is the standard pressure and is usually 0.1 MPa.[2]

In order to obtain expressions for the other standard thermodynamic functions for a gas mixture, we start from Equation 3.40:

$$\left(\frac{\partial G}{\partial T}\right)_{p,n_{\mathrm{B}}} = -S. \tag{3.149}$$

Differentiation of this with respect to n_{B} followed by reversal of the order of differentiation on the left leads to

$$\left(\frac{\partial \mu_{\mathrm{B}}}{\partial T}\right)_p = -S_{\mathrm{B}}, \tag{3.150}$$

when we recognize the definition of the chemical potential of species B in the mixture and the partial molar entropy, S_{B}. Use of Equation 3.150 following differentiation of Equation 3.148 with respect to temperature yields the following relationship between the standard partial molar entropy of species B and that at other conditions

$$S^{\ominus}_{\mathrm{B,g}}(T) = S_{\mathrm{B,g}}(T, p, x) + R \ln\left(\frac{x_{\mathrm{B}} p}{p^{\ominus}}\right) + \int_0^p \left[\left\{\frac{\partial V_{\mathrm{B,g}}(T, p, x)}{\partial T}\right\}_p - \frac{R}{p}\right] \mathrm{d}p. \tag{3.151}$$

Similarly, using Equation 3.64 for the enthalpy

$$H = G - T\left(\frac{\partial G}{\partial T}\right) \tag{3.152}$$

and following a similar procedure the relationship for the standard enthalpy is

$$H^{\ominus}_{\mathrm{B,g}}(T) = H_{\mathrm{B,g}}(T, p, x) - \int_0^p \left[V_{\mathrm{B}} - \left\{\frac{\partial V_{\mathrm{B,g}}(T, p, x)}{\partial T}\right\}_p\right] \mathrm{d}p. \tag{3.153}$$

Finally, using Equation 3.65 for the heat capacity at constant pressure and adopting the same procedure

[2] The value for $p^{\ominus}$ is 10^5 Pa and has been the IUPAC recommendation since 1982 and should be used to tabulate thermodynamic data. Prior to 1982 the standard pressure was usually taken to be $p^{\ominus}$ = 101 325 Pa (=1 bar or 1 atm), called the standard atmosphere. In any case, the value for $p^{\ominus}$ should be specified.

$$C_{p,\mathrm{B,g}}^{\ominus}(T) = C_{p,\mathrm{B,g}}(T, p, x) + \int_0^p T\left\{\frac{\partial^2 V_{\mathrm{B,g}}(T, p, x)}{\partial T^2}\right\}_p \mathrm{d}p. \tag{3.154}$$

For a perfect gas, the integrals in Equations 3.151, 3.153 and 3.154 vanish and for $p \approx p^{\ominus}$, $\Delta H_{\mathrm{m}} \approx \Delta H_{\mathrm{m}}^{\ominus}$ and $C_{p,\mathrm{m}} \approx C_{p,\mathrm{m}}^{\ominus}$ because the integrals in Equations 3.153 and 3.154 are a small fraction of the overall value.

3.7.2 What Is the Standard Chemical Potential for Liquid and Solid Solutions?

The standard chemical potential $\mu_{\mathrm{B,l}}^{\ominus}(T)$ for a pure liquid B is defined by

$$\mu_{\mathrm{B,l}}^{\ominus}(T) = \mu_{\mathrm{B,l}}^{*}(T, p^{\ominus}), \tag{3.155}$$

where $\mu_{\mathrm{B,l}}^{*}(T, p^{\ominus})$ is the chemical potential of pure B at the same temperature and at the standard pressure $p^{\ominus}$. Similarly, the standard chemical potential $\mu_{\mathrm{B,s}}^{\ominus}(T)$ for a pure solid B is

$$\mu_{\mathrm{B,s}}^{\ominus}(T) = \mu_{\mathrm{B,s}}^{*}(T, p^{\ominus}), \tag{3.156}$$

where the only change from Equation 3.155 is the state symbol. When $\mu_{\mathrm{B,l}}^{*}(T, p)$ is at a pressure $p \neq p^{\ominus}$, Equation 3.155 can be written as

$$\mu_{\mathrm{B,l}}^{\ominus}(T) = \mu_{\mathrm{B,l}}^{*}(T, p) + \int_p^{p^{\ominus}} V_{\mathrm{B,l}}^{*}(T, p)\mathrm{d}p, \tag{3.157}$$

where $V_{\mathrm{B,l}}^{*}(T, p)$ is the molar volume of the pure liquid B at temperature T and pressure p. Equation 3.157 can be cast in terms of the absolute activity as

$$\lambda_{\mathrm{B,l}}^{*}(T, p) = \lambda_{\mathrm{B,l}}^{\ominus}(T)\exp\left[\int_{p^{\ominus}}^{p} \{V_{\mathrm{B,l}}^{*}(T, p)/(RT)\}\mathrm{d}p\right]. \tag{3.158}$$

For $p \approx p^{\ominus} = 0.1$ MPa and $V_{\mathrm{B,l}}^{*} = 100\ \mathrm{cm^3\ mol^{-1}}$, the exponential term in Equation 3.158 is 1.004 ($100\ \mathrm{cm^3\ mol^{-1}}$ is typical for many organic liquids: Benzene has e.g. a molar volume of $89.24\ \mathrm{cm^3\ mol^{-1}}$). Therefore, approximate forms of Equations 3.157 and 3.158 that are useful for many purposes are

$$\mu_{\mathrm{B,l}}^{\ominus}(T) \approx \mu_{\mathrm{B,l}}^{*}(T, p) \tag{3.159}$$

and

$$\lambda_{\mathrm{B,l}}^{*}(T, p) \approx \lambda_{\mathrm{B,l}}^{\ominus}(T). \tag{3.160}$$

In solutions where small amounts of solute are present in large amounts of solvent, which are common in many (bio-)chemical processes, it is not always useful to use the standard state based upon the chemical potential of the pure substance:

$$\mu_{A,l}^{\ominus}(T) = \mu_{A,l}^{*}(T, p^{\ominus}). \tag{3.161}$$

Instead, it is preferable to use the chemical potential of the solute starting from the circumstance where it is infinitely diluted in the solvent. To approach this circumstance, we evaluate the chemical potential of the solute B by means of Equation 3.147 and extend the integral from the dilute gas phase to a saturated situation at the pressure p^{sat}, where the gas is in equilibrium with the dissolved phase, and then continue it in the dissolved phase up to the standard pressure, $p^{\ominus}$. For the solute B in solution, the result is

$$\mu_{B,sol}(T, p^{\ominus}, x_C) = \mu_{B,g}^{\ominus}(T) + RT \ln \frac{x_A p^{\text{sat}}}{p^{\ominus}} + \int_0^{p^{\text{sat}}} \left(V_{B,g} - \frac{RT}{p} \right) dp + \int_{p^{\text{sat}}}^{p^{\ominus}} V_{B,l} dp, \tag{3.162}$$

while for the pure solute we obtain

$$\mu_{B,sol}^{\ominus}(T) = \mu_{B,g}^{\ominus}(T) + RT \ln \frac{p^{\text{sat}}}{p^{\ominus}} + \int_0^{p^{\text{sat}}} \left(V_{B,g}^{*} - \frac{RT}{p} \right) dp + \int_{p^{\text{sat}}}^{p^{\ominus}} V_{B,l}^{*} dp. \tag{3.163}$$

In general, $V_{B,g}$ and $V_{B,g}^{*}$ as well as $V_{B,l}$ and $V_{B,l}^{*}$ are different because the former are concentration dependent. However, if each molecule of the solute B is surrounded by solvent molecules only (at infinite dilution), this difference vanishes and thus the difference between them yields

$$\mu_{B,sol}^{\ominus}(T) = \{\mu_{B,sol}(T, p^{\ominus}, x_C) - RT \ln x_B\}^{\infty}. \tag{3.164}$$

The situation called infinite dilution is indicated by the superscript ∞.

As explained in Chapter 1, it is usual for solutions to use molality (mol kg^{-1}) as the measure of concentration and an equation analgous to Equation 3.164, with a change in notation from mole fractions to molality m and the use of the standard molality $m^{\ominus} = 1$ mol kg^{-1} can be formulated as

$$\begin{aligned} \mu_{B,sol}^{\ominus}(T) &= \left\{ \mu_{B,sol}(T, p^{\ominus}, m_C) - RT \ln \frac{m_B}{m^{\ominus}} \right\}^{\infty} \\ &= \left\{ \mu_{B,sol}(T, p, m_C) - RT \ln \frac{m_B}{m^{\ominus}} \right\}^{\infty} + \int_p^{p^{\ominus}} V_{B,sol}^{*}(T, p) dp. \end{aligned} \tag{3.165}$$

In the two last equations, x_C and m_C represent, respectively, the mole fraction and the molality of all the components in the mixture.

3.8 What Is Exergy Good For?

While the *Carnot efficiency in* Equation 3.15 poses a fundamental upper limit for the thermal efficiency of a heat engine, practical machines of course exhibit lower efficiencies, and it is the art of the engineer to come as close to the ideal limit as possible with appropriate regard to economic constraints. The resulting actual efficiency is a product of two factors, namely the efficiency determined by the Second Law and, of necessity, a factor that represents the departures of the real machine from perfection. To balance better the actual technical achievement against what is possible in principle and to identify sources within a process where useful energy is destroyed or wasted, it has turned out to be beneficial to introduce a specific term for the useful energy. According to a suggestion of Rant (1956), this property is called *exergy E*. The term *exergy* does not introduce anything physically new that is different from the Second Law, yet it has proven to be a useful concept for technical purposes.

Thus, *exergy* is defined as that part of energy that – relative to a given reference state – can, without any restriction, be transformed into any other form of energy. Because the work that can be extracted from a process is the form of energy that is of primary interest, an alternative formulation of the definition is as follows: *exergy* is that part of energy that – relative to a given reference state – can, without any restriction, be transformed into useful work.

It is important to note that the definition relies on the specification of a reference state, and this is always taken to be the environment when we consider exergy. For example, a cold reservoir with temperature below ambient contains exergy or useful work. However, when a system is brought to the pressure and temperature of the environment, no potential exists for the extraction of useful work. Thus, the environmental state is a *dead state* and in this context its temperature is designated as T_0. An instructive and more detailed discussion of the term's immediate surroundings and environment is given by Çengel and Boles (2006).

To be complete, the other part of the energy, namely the one that in principle cannot be transformed into useful work, should also have a name: It is termed *anergy B*. Thus: *energy = exergy E + anergy B*.

The central question now is how to determine the exergy (and anergy) inherent in the different forms of energy. From the definition it is obvious that *mechanical (or electrical) energy* is pure exergy. Note that the limited conversion efficiency of an electrical motor that certainly is below unity does not contradict this statement, because ultimately this is an indication of the departure of the machine from perfection.

In the case of *heat*, the question is simple to answer, too, the maximum fraction of heat that can be transferred into other forms of energy and especially work is given by the Carnot efficiency

$$E_Q = \eta_{\mathrm{th,rev}} Q = \left(1 - \frac{T_0}{T}\right) Q. \tag{3.166}$$

Like heat itself, exergy may be expressed as a specific property $e = E/m$, in this case

$$e_q = \eta_{\mathrm{th,rev}} q = \left(1 - \frac{T_0}{T}\right) q. \tag{3.167}$$

The anergy is then

$$B_Q = \left(\frac{T_0}{T}\right) Q \quad \text{or} \quad b_q = \left(\frac{T_0}{T}\right) q. \tag{3.168}$$

Adopting the concept of exergy one may regard an ideal (i.e. reversible) heat engine as a machine that separates the heat provided to it into two parts: The useful part, exergy, is transferred into work, the remainder, anergy, is rejected to the surroundings as shown in Figure 3.4. In a real process, entropy is generated, which results in a loss of exergy (useful energy). We shall consider the connection of exergy loss and entropy generation later.

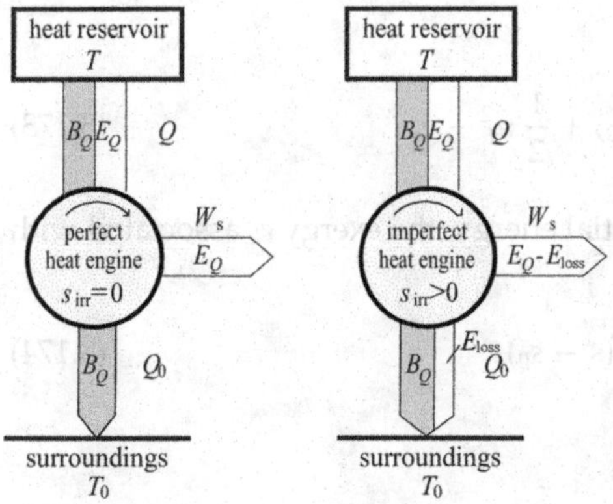

FIGURE 3.4
A heat engine may be regarded as a "separator" for exergy and anergy.

As another important example, we consider the exergy e_{FS} connected with a fluid flowing through a device representing an open, steady-state system that extracts useful work from the stream. The symbol e_{FS} designates the maximum amount of specific work that can be extracted when bringing the fluid stream from state 1 to the dead state, that is to equilibrium with the environmental state 0. Again, the maximum work can only be realized in an ideal, reversible process. We start with the First Law using specific quantities derived by applying Equation 1.77 or 1.78 to unit mass over a time dt

$$q + w_s = h_0 - h + \frac{1}{2}\left(u_{f0}^2 - u_f^2\right) + g\left(z_0 - z\right), \tag{3.169}$$

which states that heat q and (shaft) work w_s crossing the system boundaries result in a change of the total energy, namely of the sum of enthalpy, kinetic (velocity u_f) and potential energy (acceleration of free fall g and height z).

The exergy is, in magnitude, identical to the maximum work that can be extracted from the process:

$$e_{FS} = -w_s. \tag{3.170}$$

In a perfect process (Figure 3.4 left), all the exergy of the heat provided is turned into work that can be extracted from the engine, anergy as the remainder is rejected to the surroundings; in a real heat engine with irreversibilities part of the exergy is turned into anergy, it must be rejected as additional waste heat.

Considering that both kinetic and potential energy at the dead state are zero, we obtain

$$e_{FS} = h - h_0 + q + \frac{1}{2}u_f^2 + gz. \tag{3.171}$$

Because the process is to be performed in a reversible manner, the heat must be transferred at a vanishing temperature difference, that is, at the temperature T_0 of the surroundings. From the Second Law (Equation 3.16 reformulated per unit mass with dS = 0 and δS_{gen} = 0)

$$q = T_0\left(s_0 - s\right), \tag{3.172}$$

so that

$$e_{FS} = h - h_0 - T_0(s - s_0) + \frac{1}{2}u_f^2 + gz. \tag{3.173}$$

Neglecting the contributions of kinetic and potential energy, the exergy e_h associated with the enthalpy h may be expressed as

$$e_h = h - h_0 - T_0(s - s_0). \tag{3.174}$$

Consequently, the corresponding anergy b_h is

$$b_h = h_0 + T_0(s - s_0). \tag{3.175}$$

In a similar manner, the exergy e_u connected with the internal energy u may be obtained as (Çengel and Boles 2006)

$$e_u = u - u_0 - T_0(s - s_0) + p_0(v - v_0). \tag{3.176}$$

One of the motivations for the introduction of exergy was to identify whether energy is properly used within a process. This can be achieved by comparing the exergy provided to a process with the exergy available after the process has been performed. Ideally, the amount of exergy withdrawn from a process should equal the useful work extracted from this process.

As an example, let us consider the provision of a stream of hot water heated by an electrical resistance heater. Raising the temperature of the water stream from an ambient temperature of 298 K (or 25°C) to 333 K (or 60°C) increases the exergy connected with enthalpy from zero to an amount we denote by e_h, as given by Equation 3.174. With

$$(h - h_0) = c_p(T - T_0), \quad (s - s_0) = c_p \ln\left(\frac{T}{T_0}\right), \tag{3.177}$$

and $c_p \approx 4.2$ kJ kg^{-1} K^{-1}, we obtain

$$e_h = c_p(T - T_0) - T_0\, c_p \ln\left(\frac{T}{T_0}\right) = 147 \text{ kJ kg}^{-1} - 139 \text{ kJ kg}^{-1} = 8\text{kJ kg}^{-1}. \tag{3.178}$$

The exergy of 8 kJ kg^{-1} available after this process e_{out} must be compared with the exergy input e_{in}. In the case of an electric heater, this input is simply given by the work required to heat up the water (remember that electrical energy is made up of exergy only)

$$e_{in} = w_{el} = (h - h_0) = c_p(T - T_0) = 147 \text{ kJ kg}^{-1}. \tag{3.179}$$

How efficiently exergy is used may be judged by the exergetic (or Second Law) efficiency η_{ex}, defined as

$$\eta_{ex} = \frac{e_{out}}{e_{in}}. \tag{3.180}$$

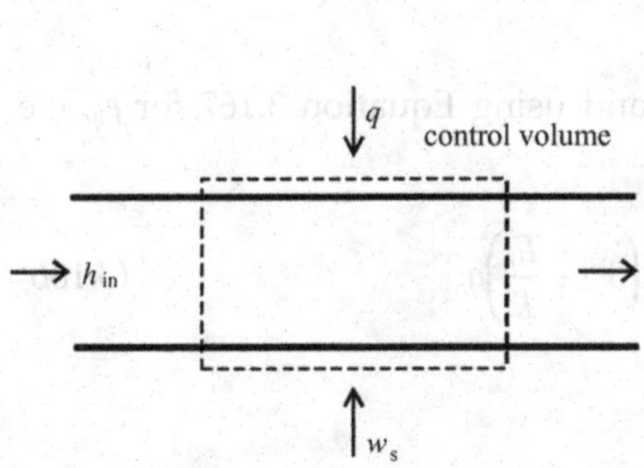

FIGURE 3.5
Schematic for determining the exergy loss for an open system: A steady flow associated with specific enthalpies h_{in} and h_{out} crosses the control volume; heat and shaft work transferred are summarized in the resulting quantities q and w_s, respectively.

In the present case, $\eta_{ex} = 8/147 = 5\%$. This poor result reflects the fact that energy of "high quality" (pure exergy) is (mis-) used to provide energy at a low temperature. For that purpose, waste heat from an engine or district heat from a power plant would suffice. The poor exergetic efficiency also implicitly takes into account the fact that electricity itself can be produced only with a certain thermal efficiency (ultimately limited by the Carnot factor) at an electrical power plant.

Because the concept of exergy is closely connected with the Second Law, it is obvious that it must be linked to another central term in connection with the Second Law, namely entropy. This fact can be seen, for example, from Equation 3.174, we can, however, obtain a more general relation between the two properties. In an irreversible process, entropy is generated, and at the same time, useful energy, that is, exergy is destroyed. For the derivation of such a relation we consider a steady-flow process (Figure 3.5; in a similar way, an identical relation may also be obtained for other systems).

The specific exergy loss, e_{loss}, may be simply obtained by setting up a control volume and balancing the exergy that enters into the system, e_{in}, against that leaving the system, e_{out},

$$e_{loss} = e_{in} - e_{out}. \tag{3.181}$$

In our example, a steady flow enters and leaves the control volume, associated with specific enthalpies h_{in} and h_{out}, respectively (kinetic and potential energies are neglected for simplicity because taking them into account would not alter the result). Heat entering and leaving the system is summarized into one specific quantity, q; the same holds for all forms of work, resulting in a term, w_s. Thus

$$e_{loss} = e_{in} - e_{out} = e_{h,in} - e_{h,out} + e_w + e_q. \tag{3.182}$$

From Equation 3.174

$$e_{h,in} - e_{h,out} = h_{in} - h_{out} - T_0(s_{in} - s_{out}), \tag{3.183}$$

and using the 1st Law, again ignoring kinetic and potential energy

$$q + w_s = h_{out} - h_{in}, \tag{3.184}$$

we obtain

$$e_{loss} = -q - w_s - T_0(s_{in} - s_{out}) + e_w + e_q. \tag{3.185}$$

Because the shaft work w_s is pure exergy, $w_s = e_w$, and using Equation 3.167 for e_q, we obtain

$$e_{loss} = -q - T_0(s_{in} - s_{out}) + \left(1 - \frac{T_0}{T}\right)q. \tag{3.186}$$

Performing an entropy balance

$$s_{out} - s_{in} = \frac{q}{T} + s_{gen}, \tag{3.187}$$

where q/T is the entropy connected with the net heat q and S_{gen} summarizes all sources of irreversibility (including that of heat transfer), we finally obtain

$$e_{loss} = -q - T_0\left(-\frac{q}{T} - s_{gen}\right) + \left(1 - \frac{T_0}{T}\right)q = T_0 s_{gen}. \tag{3.188}$$

This result, which is of general applicability, demonstrates that exergy loss is directly proportional to the entropy generated within a process.

Returning to our example of the electric resistance heater, this exergy loss shows up in the second term on the right-hand side of Equation 3.178 with a magnitude of 139 $kJ \cdot kg^{-1}$. Using the exergy loss, we may also write the exergetic efficiency of Equation 3.180 in a different form

$$\eta_{ex} = \frac{e_{out}}{e_{in}} = 1 - \frac{e_{loss}}{e_{in}}. \tag{3.189}$$

3.9 What Is the Minimum Work Required to Separate Air into Its Constituents?

To tackle the problem, it might seem straightforward to look for one or more processes that promise to separate air – or more generally a mixture of gases – and then seek to find the optimal conditions for each process under which they require the minimum amount of work. In general, it may be difficult to find any such process, and one can never be sure that the result obtained is actually the optimal choice; it may merely be the best from those selected. Thus, it is best to consider the problem from the other end: What is the amount of useful energy that is destroyed by the mixing of gases, or what is the exergy loss E_{loss} in such a process (See Question 3.8).

To simplify the analysis without losing the major thrust of the argument we consider air in the first instance as a mixture of only nitrogen and oxygen ($y_{N_2} = 0.79$, $y_{O_2} = 0.21$) and expand the problem to a more general case later. We further restrict the problem to treating dry air and neglect the varying humidity. When nitrogen and oxygen are mixed at standard conditions ($T = 298.15$ K, $p = 10^5$ Pa), these constituents may be treated as ideal gases. Thus, there is no enthalpy of mixing (nor a change in internal energy), and the mixing at constant pressure and temperature occurs in an adiabatic manner. If we

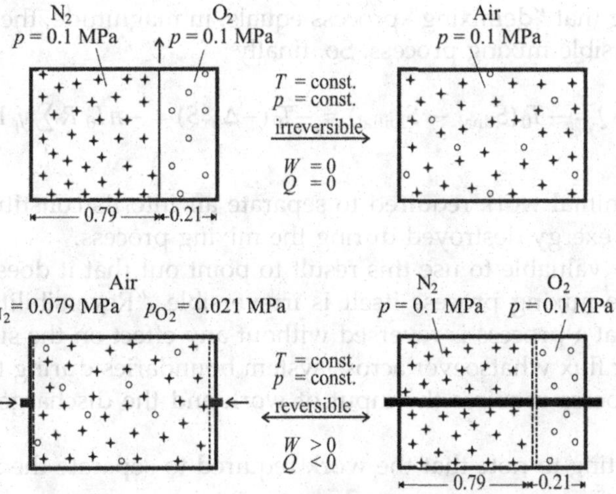

FIGURE 3.6
Mixing and separation of nitrogen and oxygen as the constituents of air at isothermal and isobaric conditions. Top: When removing the partition N_2 and O_2 mix irreversibly; there is no transfer of work or heat across the system boundaries. Bottom: In a hypothetical process the separation of air may be performed in a reversible manner: When work is applied to two semi-permeable pistons, the components are compressed from their respective partial pressures to system pressure under discharge of heat.

imagine that the two gases are held separately in a single rigid vessel and that we then remove the partition (as shown in Figure 3.6), the system undergoes a diffusion process toward a new equilibrium. This diffusion process is irreversible and accompanied by a rise of entropy (see Equation 2.57)

$$\Delta S_{gen} = -n\, R \sum_i y_i \ln y_i, \tag{3.190}$$

where n is the total amount of substance and y represents the mole fractions of species in the gas phase. The amount of useful energy destroyed by such a process, or in other words the exergy loss, may be generally described by $E_{loss} = T_0 \Delta S_{gen}$, where T_0 is the temperature of the surroundings to give

$$E_{loss} = -nT_0\, R \sum_i y_i \ln y_i. \tag{3.191}$$

Because of the irreversibility of the equilibration processes, there is no direct inverse of this process. However, it is obvious that the minimum work required to restore the initial state cannot be less in magnitude than the exergy loss in the mixing process: $W_{min} = E_{loss}$.

Now we can ask what such a separation process might look like. As the constituents are to be present at the original temperature the restoration process should obviously be performed in an isothermal way. From the First Law of Thermodynamics, $\delta W + \delta Q = \mathrm{d}U$, and $\mathrm{d}U = n\, C_{V,m} \mathrm{d}T = 0$ for a mixture of ideal gases, it follows that the amount of work applied to the system must be balanced by the same amount of heat rejected from the system. If we suppose that in an idealized circumstance the heat is rejected at constant temperature T_0 for both the system and the surroundings, then the heat transfer δQ is connected with the change of entropy $\mathrm{d}S$ via $\delta Q = T_0 \mathrm{d}S$. The total amount of entropy discharged during that "demixing" process equals, in magnitude, the entropy generated during the irreversible mixing process. So, finally

$$W_{min} = -Q = -T_0(S_{final} - S_{initial}) = -T_0(-\Delta_{irr}S) = -nT_0 R\sum_i y_i \ln y_i = E_{loss}. \quad (3.192)$$

Therefore, the minimal work required to separate air into its constituents has the same magnitude as the exergy destroyed during the mixing process.

It is particularly valuable to use this result to point out that it does not contradict the statement that the mixing process itself is irreversible. "Reversibility" (see Chapter 1) always implies that a process is reversed without any effect on the surroundings. While there is no energy flux whatsoever across system boundaries during the mixing process, the separation process requires the input of work and the discharge of heat across the system boundary.

It is also interesting to note that the work required to separate the mixture in the case that it is ideal is

$$W_{min} = -nT_0 R\sum_i y_i \ln y_i = -T_0 R\sum_i n_i \ln y_i = -T_0 R\sum_i n_i \ln(p_i/p), \quad (3.193)$$

which is equivalent to the total work required to compress the constituents from their partial pressures p_i to the system pressure p in an isothermal process (because $y_i = p_i/p$).

This analysis suggests that a hypothetical separation process might be as follows. The vessel containing the air (gas mixture) possesses two pistons, one on each side, as shown in Figure 3.6. The piston on the left consists of a semi-permeable membrane where only nitrogen molecules may pass through and oxygen molecules are withheld; for the piston on the right conditions are interchanged, so oxygen may pass and nitrogen is blocked. When we move the two pistons simultaneously in a way to achieve a final position where the left piston has traveled 79% of the total way (and accordingly the right one 21%), all the nitrogen is enclosed in the left compartment and all the oxygen in the right one. This process exactly corresponds to the compression of each component from its partial pressure in the mixture to a final pressure of 10^5 Pa, which then equals the system pressure.

Let us finally illustrate the process with a numerical example and then examine the effect of the real composition of air on the results. As an example, 1 m³ of air at standard conditions contains about 40 mol of an ideal gas mixture irrespective of composition. From Equation 3.193

$$W_{min} = -40.34\ \text{mol} \times 298.15\ \text{K} \times 8.314\ \text{J mol}^{-1}\text{K}^{-1}\sum_i y_i \ln y_i \approx -100.0\ \text{kJ}\sum_i y_i \ln y_i, \quad (3.194)$$

and we see that the actual work to separate the mixture depends only on the relative composition and not on the nature of the individual constituents. When we return to our simplest model of a binary mixture with $y_{N_2} = 0.79$, $y_{O_2} = 0.21$, we obtain

$$\begin{aligned} W_{min} &= -100.0\ \text{kJ}[0.79(-0.236) + 0.21(-1.56)] \\ &= 100.0\ \text{kJ}[0.186 + 0.328] = 51\ \text{kJ}. \end{aligned} \quad (3.195)$$

In a next approximation, we also consider argon as a constituent of air, now with a composition of $y_{N_2} = 0.781$, $y_{O_2} = 0.210$, $y_{Ar} = 0.009$, resulting in

$$\begin{aligned} W_{\min} &= -100.0\ \text{kJ}[0.781(-0.2472) + 0.210(-1.561) + 0.009(-4.711)] \\ &= 100.0\ \text{kJ}[0.1930 + 0.3277 + 0.0424] = 56.3\ \text{kJ}. \end{aligned} \tag{3.196}$$

This procedure may be expanded in a straightforward manner to include other components of air such as carbon dioxide, neon and so on. The main point about the numerical example, however, is to make clear that rather small or even spurious amounts of further components considerably increase the work required to separate the gas mixture. The reason behind this increase is the strong rise in the entropy of mixing with increasing dilution or – in other words – the comparatively large amount of work required to compress a volume containing a component at small partial pressure to system pressure. It should be obvious that the actual work for gas separation in a process is a multiple of the minimum work obtained from the idealized calculation.

Finally, the hypothetical separation process might – admittedly only theoretically – be inverted to obtain a mixing process producing work. The process is similar to the extraction of work from the isothermal expansion of a volume of a pure gas with a heat supply. In the case of the separation, the expansion is allowed through the movement of the two semipermeable pistons lowering the pressure from the system pressure p to the respective partial pressures $y_i p$ for each component.

References

Atkins P., 2007, *Four Laws That Drive the Universe*, Oxford University Press, Oxford.

Carnot N.L.S., 1824, *Reflections on the Motive Power of Heat*, English translation by. Thurston R. H. (1897), John Wiley, New York.

Çengel Y.A., and Boles M.A., 2006, *Thermodynamics—An Engineering Approach*, McGraw-Hill, Boston.

Clausius R., 1850a, "Über die bewegende Kraft der Wärme, Part I", *Ann. Phys. (Leipzig)* **79**:368–397. (also printed in 1851, "On the Moving Force of Heat, and the Laws regarding the Nature of Heat itself which are deducible therefrom. Part I," *Phil. Mag.* **2**:1–21).

Clausius R., 1850b, "Über die bewegende Kraft der Wärme, Part II", *Ann. Phys. (Leipzig)* **79**:500–524. (also printed in 1851, "On the Moving Force of Heat, and the Laws regarding the Nature of Heat itself which are deducible therefrom. Part II," *Phil. Mag.* **2**:102–119).

Goodwin A.R.H., and Trusler J.P.M., 2003, *Sound Speed*, Chapter 6, in *Experimental Thermodynamics, Volume VI, Measurement of the Thermodynamic Properties of Single Phases*, eds. Goodwin A.R.H., Marsh K.N., and Wakeham W.A., for IUPAC, Elsevier, Amsterdam.

Herzfeld K.F., and Litovitz T.A., 1959, *Pure and Applied Physics, Volume 7, Absorption and Dispersion of Ultrasonic Waves*, ed. Massey H.S.W., Academic Press, London.

Joule J.P., 1845, "LIV On the changes of temperature produced by the rarefaction and condensation of air", *Phil. Mag.* (series 3) **26**:369–383.

Landau L.D., and Lifshitz E.M., 1987, *Theory of Elasticity*, 2nd ed., Pergamon, Oxford.

Ledbetter H.M., 1982, "The temperature behavior of Young moduli of 40 engineering alloys", *Cryogenics* **22**:653–656.

Levenspiel O., 1996, *Understanding Engineering Thermodynamics*, Prentice Hall PTR, NJ, USA.

Margenau H., 1950, *The Nature of Physical Reality*, McGraw-Hill, New York, p. 215.

McGlashan M.L., 1979, *Chemical Thermodynamics*, Academic Press, London.

Morse P.M., and Ingard K.U., 1968, *Theoretical Acoustics*, McGraw-Hill, New York, p. 233.

Papadakis E.P., 1998, "Ultrasonic wave measurements of elastic moduli E, G, and MU for product development and design calculations", *J. Test. Eval.* **26**:240–246.

Quack M., Stohner J., Strauss H.L., Takami M., Thor A.J., Cohen E.R., Cvitas T., Frey J.G., Holström B., Kuchitsu K., Marquardt R., Mills I., and Pavese F., 2007, *Quantities, Units and Symbols in Physical Chemistry*, 3rd ed., RSC Publishing, Cambridge.

Rant Z., 1956, "Exergie, ein neues Wort für technische Arbeitsfähigkeit", *Forsch. Ingenieurwes.* **22**:36–37.

Rowlinson J.S., 2003, "The work of Thomas Andrews and James Thomson on the liquefaction of gases", *Notes Rec. R. Soc.* **57**:143–159.

Rowlinson J.S., 2005, "Which Kelvin?, Book review for degrees Kelvin: A tale of genius, invention, and tragedy", *Notes Rec. R. Soc.* **59**:339–341.

Sandler S.I., 1989, *Chemical and Engineering Thermodynamics*, 2nd ed., John Wiley and Sons, Inc., New York.

Westerhoff H.V., and van Dam K., 1987, *Thermodynamics and Control of Biological Free–Energy Transduction*, Chapter 2, Elsevier, Amsterdam.

Weston W.F., 1975, "Low-temperature elastic-constants of a superconducting coil composite", *J. Appl. Phys.* **46**:4458–4465.

4

Multi-Phase Systems

4.1 What Are the Characteristic Properties of Multi-Phase Systems?

In the previous chapters, we have been concerned with the thermodynamic relationships for a homogeneous phase. This chapter extends our questions, examples and discussion to a heterogeneous system; that is one containing more than one phase. This chapter introduces the thermodynamic concepts required for the treatments of the equilibrium of any system with independent variables of temperature, pressure and amount of substance of the components within it; this includes a pure substance and multicomponent mixtures. When these are combined with the rules of thermodynamics provided in Chapters 1 and 3, we have methods to determine changes of entropy, energy and enthalpy with temperature, pressure and composition in each phase and also methods to determine changes in Gibbs function and chemical potential (also absolute activity) with respect to pressure and composition (Guggenheim 1985; McGlashan 1979). In principle, these are sufficient to determine the equilibrium between phases of a pure substance or mixtures; however, other methods will need to be introduced to expedite such calculations and that is the purpose of this introductory section.

4.1.1 What Are the Important Equations for Systems With Many Phases?

In both laboratory experiments to determine thermodynamic properties and in chemical or biochemical process plant, systems are generally considered under conditions where temperature and pressure are the independent variables. In such cases, the Gibbs function is the appropriate thermodynamic quantity to examine. Equation 3.40 for a single phase of a system containing various species B, can be repeated for each of many phases and the sum then becomes

$$\sum_{\alpha} \mathrm{d}G^{\alpha} = -\sum_{\alpha} S^{\alpha}\mathrm{d}T + \sum_{\alpha} V^{\alpha}\mathrm{d}p + \sum_{\alpha}\sum_{\mathrm{B}} \mu_{\mathrm{B}}\mathrm{d}n_{\mathrm{B}}^{\alpha}, \qquad (4.1)$$

where the Σ means the sum over all phases included in the system. In Equation 4.1, we have purposely omitted the superscript α on the uniform intensive properties of T, p and μ_{B} because, for now, we will only consider systems that are in thermal, hydrostatic and diffusive equilibrium (see Question 3.6).

Removing the summation over all phases in Equation 4.1 and, for simplicity, replacing it with a superscript Σ for the system, Equation 4.1 can be cast as follows:

$$\mathrm{d}G^{\Sigma} = -S^{\Sigma}\mathrm{d}T + V^{\Sigma}\mathrm{d}p + \sum_{\mathrm{B}} \mu_{\mathrm{B}}\mathrm{d}n_{\mathrm{B}}^{\Sigma}. \qquad (4.2)$$

DOI: 10.1201/9780429329524-4

Equation 4.2.2 does not include one situation that arises when two phases are separated by a partition permeable to some substances but not others in the system. In this case, $p^\alpha \neq p^\beta$ and this special case is called osmotic equilibrium for which the absolute difference $|p^\alpha - p^\beta|$ is the osmotic pressure that will be discussed further in Question 4.4.3.

When the system is of fixed chemical composition, $dn_B^\Sigma = 0$ and, in that case, Equation 4.2.2 becomes

$$dG^\Sigma = -S^\Sigma dT + V^\Sigma dp. \tag{4.3}$$

For a *closed* system with a chemical reaction, Equation 4.3.3 can be modified by addition of the term $\Delta_r G d\xi = -(\sum_B \nu_B \mu_B)d\xi$ where, as discussed in Chapter 3, ξ is the extent of reaction and $\Delta_r G$ is the Gibbs energy of reaction, identical to A, the affinity for a chemical reaction.

4.1.2 What Is the Phase Rule?

The Gibbs–Duhem equation for a single phase α in thermal, hydrostatic and diffusive equilibrium is, according to Equation 3.51, given by

$$0 = S_m^\alpha dT - V_m^\alpha dp + \sum_B x_B^\alpha d\mu_B^\alpha. \tag{4.4}$$

The number of independent, intensive variables in Equation 4.4 is $(C + 1)$, where C is the number of components in the phase and that equals the number of terms in the summation. If there are P phases present in the system then there are P Equations 4.4 that provide $(P - 1)$ restrictions. Thus, the number of independent intensive variables or *degrees of freedom* of the system are

$$F = (C + 1) - (P - 1) = C + 2 - P. \tag{4.5}$$

If there are N_R chemical reactions occurring in the system and, if these are all at equilibrium, then there is one additional equation $\sum_B \nu_B \mu_B = 0$ for each and so the F must be reduced by N_R so that Equation 4.5 becomes

$$F = C + 2 - P - N_R. \tag{4.6}$$

Equation 4.6 is the *phase rule*. We are now armed with sufficient information to start the discussion of the phase equilibrium of a pure substance. Before doing so, we draw some conclusions from Equation 4.4 that at constant temperature and pressure becomes

$$\sum_B x_B^\alpha d\mu_B^\alpha = 0, \tag{4.7}$$

and for a binary mixture $(1 - x)$A + xB, Equation 4.7 is

$$(1 - x)d\mu_A + x d\mu_B = 0, \tag{4.8}$$

or

$$\mu_A(T,p,x^\beta) - \mu_A(T,p,x^\alpha) = \int_{\mu_B(T,p,x^\alpha)}^{\mu_B(T,p,x^\beta)} \frac{x}{1-x} d\mu_B, \tag{4.9}$$

where x^α and x^β refer to the mole fraction of species B in the phases α and β, respectively. Equation 4.9 provides a route to the determination of $\mu_A(T,p,x^\beta) - \mu_A(T,p,x^\alpha)$ from measurements of the difference $\mu_B(T,p,x) - \mu_B^*(T,p)$, at mole fractions, x, that include x^α and x^β. Here, the superscript asterisk denotes a pure substance and we should note that the measurements are to be performed with x^α constant.

In the case of a multicomponent mixture, only (C – 1) of the differences in chemical potential are independent; the necessary work is slightly reduced.

4.2 What Is Phase Equilibrium of a Pure Substance?

For two coexisting phases α and β of a pure, nonreacting substance in equilibrium $C = 1$, $P = 2$ and $R = 0$ so that according to Equation 4.6, $F = 1$; the equilibrium of three phases of a pure substance results in $F = 0$ and the system has no independent intensive variables. The equilibrium temperature is then called the triple-point temperature while the equilibrium pressure is the triple-point pressure and both are fixed. A $p(T)$ projection for the phase equilibrium of a pure substance is shown schematically in Figure 4.1 (Goodwin and Ambrose 2005).

The curves AB, BD and BC meet at the triple point B; for a solid with more than one solid phase, there is more than one triple point. The curve AB depicting the s = g equilibrium tends to zero pressure at low temperatures while the curve BD (representing the s = l equilibrium) continues upward indefinitely and has a large and positive slope for most substances; water is an exception and for this the slope is large and negative (see next question). If vapor pressure is plotted as a function of temperature, as it is represented schematically in Figure 4.1, the curves for the solid (AB) and liquid (BC) intersect at the triple point (point B in Figure 4.1) with a discontinuity of slope and the latter terminates at a higher temperature at the critical point C where the properties of vapor and liquid become identical, and at this temperature the vapor pressure is known as the critical pressure; at $T = T_c$ $(\partial p/\partial V_m)_T = 0$ and $(\partial^2 p/\partial V_m^2)_T = 0$. The critical temperature is the highest temperature at which two fluid phases of liquid and gas for a pure substance can coexist. Supercooled liquid, which is metastable, has a higher vapor pressure than that of the stable solid.

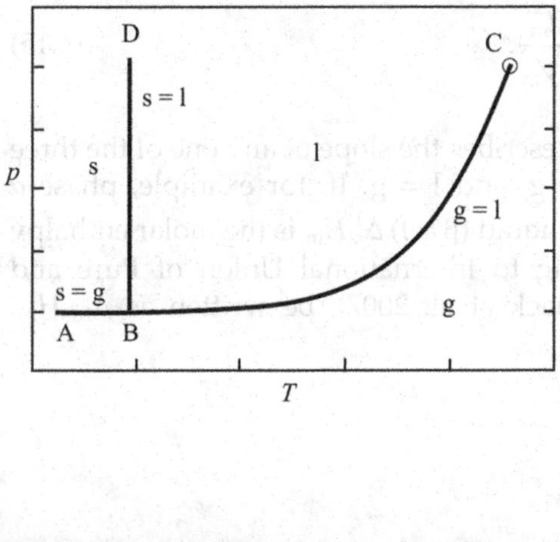

FIGURE 4.1
Pressure p of a pure substance as function of temperature T. The figure shows the solid (s), liquid (l) and gaseous (g) phases. The lines are defined as follows: A to B sublimation line where solid is in equilibrium with vapor (s = g); B to C liquid in equilibrium with vapor (l = g); B to D the melting line where solid is in equilibrium with liquid (s = l); and, C, the critical point. B is the triple point where solid, liquid and vapor coexist (s = l = g).

4.2.1 What Is Clapeyron's Equation?

For a pure substance, Equation 3.20 is

$$G_m = H_m - TS_m = \mu, \tag{4.10}$$

where the definition from Chapter 3, $G_m = \mu$ of Equation 3.25, has been used. We remind the reader that the subscript m denotes a molar quantity. For the equilibrium of two phases α and β, $\mu^\alpha = \mu^\beta$ so that $G_m^\alpha = G_m^\beta$ and thus the differences of the molar enthalpy and entropy between the two phases, denoted by Δ_α^β, are related as

$$\Delta_\alpha^\beta H_m = T\Delta_\alpha^\beta S_m. \tag{4.11}$$

The $\Delta_\alpha^\beta H_m$ of Equation 4.11 can be obtained experimentally from Equations 1.128 and 3.72 while $\Delta_\alpha^\beta S_m$ can be obtained from Equation 3.68.

For two phases (solid, liquid, or gas) α and β of a pure substance in equilibrium, Equation 4.6 gives $F = 1$ so there is a relationship between, for example, the temperature T and pressure p. If the temperature T is chosen as the independent variable, then pressure is dependent and, in this text, will be denoted by p^{sat} for the case of liquid and gas equilibrium (written as l = g). For two phases α and β, which could be solid and liquid (s = l), solid and gas (s = g) or liquid and gas (l = g), of a pure substance in equilibrium the Gibbs–Duhem Equation 4.4 for each of the phases is

$$0 = S_m^\alpha dT - V_m^\alpha dp + d\mu \tag{4.12}$$

and

$$0 = S_m^\beta dT - V_m^\beta dp + d\mu. \tag{4.13}$$

The chemical potential μ of the substance B and therefore the partial molar Gibbs function, must be equal in both phases so that Equations 4.12 and 4.13 can be written as

$$\frac{dp^{sat}}{dT} = \frac{S_m^\beta - S_m^\alpha}{V_m^\beta - V_m^\alpha} = \frac{\Delta_\alpha^\beta S_m}{\Delta_\alpha^\beta V_m}, \tag{4.14}$$

or, in view, of Equation 4.11 as

$$\frac{dp^{sat}}{dT} = \frac{\Delta_\alpha^\beta H_m}{T\Delta_\alpha^\beta V_m}. \tag{4.15}$$

Equation 4.15 is called *Clapeyron*'s equation and describes the slope of any one of the three saturation lines shown in Figure 4.1: s = l, s = g, and l = g. If, for example, phase α represents the solid phase (α = s) and phase β the liquid (β = l) $\Delta_s^l H_m$ is the molar enthalpy of fusion, which for chemists should, according to International Union of Pure and Applied Chemistry (IUPAC) nomenclature (Quack et al. 2007), be written as $\Delta_{fus}H_m$. Similarly, $\Delta_s^l V_m$ should be written as $\Delta_{fus}V_m$.

4.2.2 How Do we Represent the Vapor Pressure of a Substance?

If one of the phases is a dilute gas, so that it can be considered ideal with $pV_m = RT$, and the molar volume of the gas phase $V(g) \gg V(l)$ or $V(g) \gg V(s)$ so that $V(l)$ or $V(s)$ can be neglected then Equation 4.15 becomes

$$\frac{dp^{sat}}{dT} \approx \frac{p^{sat}\Delta_\alpha^\beta H_m}{RT^2}. \tag{4.16}$$

Further, simplification can be obtained by assuming $\Delta_\alpha^\beta H_m$ is independent of temperature over a range of temperatures T_1 to T_2, then Equation 4.16 can be written as

$$\ln\left\{\frac{p^{sat}(T_2)}{p^{sat}(T_1)}\right\} \approx \frac{\Delta_\alpha^\beta H_m (T_2 - T_1)}{RT_1T_2}. \tag{4.17}$$

When T_1 is fixed by selecting p_1, for example, $p_1 = p^\ominus = 0.1$ MPa,[1] then Equation 4.17 reduces to

$$\ln\left\{\frac{p^{sat}(T_2)}{p^\ominus}\right\} \approx \frac{\Delta_\alpha^\beta H_m}{RT_1} - \frac{\Delta_\alpha^\beta H_m}{RT_2} = a - \frac{b}{T_2}. \tag{4.18}$$

Over a range of temperature close to the normal boiling temperature, the observed vapor pressure may often be fitted to an equation of the form suggested by Equation 4.18 that is

$$\ln\left(\frac{p^{sat}}{p^\ominus}\right) = a + \frac{b}{T}, \tag{4.19}$$

where a and b are substance dependent parameters for each phase. The Antoine equation given by

$$\ln\left(\frac{p^{sat}}{p^\ominus}\right) = e + \frac{f}{(g + T)}, \tag{4.20}$$

where e, f and g are also substance dependent parameters for each phase, provides a better representation over a slightly wider temperature range about the normal boiling temperature.

To represent measurements of the vapor pressure within experimental error from the triple point temperature to the critical temperature requires a more complicated equation. One representation that has been extremely successful is the so-called Wagner equation

$$\ln\left(\frac{p}{p_c}\right) = (n_1\tau + n_2\tau^{1.5} + n_3\tau^c + n_4\tau^d)\left(\frac{T_c}{T}\right), \tag{4.21}$$

[1] The value for $p^\ominus$ is 10^5 Pa and has been the IUPAC recommendation since 1982 and should be used to tabulate thermodynamic data. Prior to 1982 the standard pressure was usually taken to be $p^\ominus$ = 101 325 Pa (=1 bar or 1 atm), called the standard atmosphere). In any case, the value for $p^\ominus$ should be specified.

where T_c and p_c are the critical temperature and pressure, respectively, $\tau = (1 - T/T_c)$, the n_i, with $i = 1, 2, 3$, and 4, are parameters for each substance that are adjusted to the available measurements and, typically, $c = 2.5$ and $d = 5$. Equation 4.21 reduces to Equation 4.19 when truncated after the first term $n_1\tau$. The vapor pressure of a liquid is, in practice, affected by the curvature of the surface from which evaporation takes place, and the vapor pressure of microscopic droplets is higher than the normal value; this affects the formation of clouds and rain.

Engineers prefer the use of specific quantities rather than molar quantities so that then Equation 4.15 can be written as

$$\frac{dp^{sat}}{dT} = \frac{\Delta_\alpha^\beta h}{T\Delta_\alpha^\beta v}. \tag{4.22}$$

It is also common practice in engineering to use the specific gas constant $R_s = R/M$ so that the same approximations used to derive Equation 4.16 result in

$$\frac{dp^{sat}}{dT} = \frac{p^{sat}\Delta_\alpha^\beta h}{R_s T^2}. \tag{4.23}$$

The vapor pressures of different substances vary widely. At $T = 298.15$ K, for example, the vapor pressures of many involatile substances are very low indeed ($<10^{-5}$ Pa) and are exceedingly difficult to measure, whereas that of a volatile substance such as carbon dioxide is about 6 MPa. The temperature at which the vapor pressure of a substance is 0.101325 MPa is defined as its normal boiling temperature T^b; normal boiling temperatures range from 4.2 K for helium up to, for example, 6,000 K for tantalum.

For l = g, the vapor pressure at a temperature T close to the normal boiling temperature, T^b, can be obtained from a modified form of Equation 4.18 as

$$p^{sat}(T)/\text{MPa} \approx 0.1\exp\left\{\frac{10(T - T^b)}{T}\right\}, \tag{4.24}$$

where it is assumed $\Delta_l^g H_m/(RT^b) = 10$. Equation 4.24 is called Trouton's rule and with the assumption that the critical pressure of all substances is the same is a result of the principle of corresponding states discussed in Chapter 2.

From a plot of temperature of saturation, T^{sat}, as a function of amount of substance density $\rho_n = (1/V_m)$ for a l = g phase boundary it is found empirically that $\{\rho_{n,l} + \rho_{n,g}\}/2$ lies on a straight line and this is referred to as the law of the rectilinear diameter that is obeyed for all pure substances. For mixtures, the reduced orthobaric densities $\rho_g/\rho_c = V_{m,c}/V_{m,g}$ and $\rho_l/\rho_c = V_{m,c}/V_{m,l}$ plotted against reduced temperature T/T_c follow the Cailleter and Mathias's law of the rectilinear diameter as

$$\frac{1}{2}\frac{\{\rho_l + \rho_g\}}{\rho_c} = 1 + 0.797\left(1 - \frac{T}{T_c}\right), \tag{4.25}$$

a further consequence of the principle of corresponding states for a two-phase fluid mixture.

The critical pressures of most substances do not exceed 5 MPa, although a few are much higher than this, for example, water with $p_c = 22.05$ MPa. Critical properties of only a few hundred elements and compounds have been measured because for many, particularly the involatile elements, the temperatures are too high to be experimentally accessible, and most compounds decompose before the critical temperature is reached. So far, consideration has been restricted to substances that vaporize without decomposition. If the substance vaporizing decomposes irreversibly, as do many inorganic compounds at high temperatures, there are different chemical species in the liquid and vapor phases.

The classical equations of state, for example, the van der Waals equation described later, can at least qualitatively provide estimates of the $p(\rho)$ isotherms at temperatures close to critical. The van der Waals equation is an analytic equation; that is, one for which the expansion as a Taylor series about the critical point converges about that point. Analytic equations are unable to predict the behavior observed at the critical point and indeed they cannot even fit observations near the critical point. For example, all analytic equations of state predict a finite value of C_V at the critical temperature, and that is inconsistent with direct measurement that shows divergence to infinity at T_c. Such observations require advanced theories of critical are beyond the scope of this text and the interested reader is referred to, for example, Behnejard et al. (2010).

Close to the critical point there is a characteristic opalescence (cloudiness) in the fluid; this arises because the correlation length or mean distance over which the molecules' order increases by several orders of magnitudes from about 1 nm through the wavelengths of visible light (between about 380 to 780 nm) and these correlated fluctuations scatter the light.

4.2.3 What Does Clapeyron's Equation Have to Do with Ice Skating?

Now, how is Clapeyron's equation (Equation 4.22) related or at least, allegedly, related to ice skating? It seems obvious that if a film of liquid water forms at the solid surface between a skate and ice then it can explain the lubrication process that makes it so simple to slip on ice. However, the origin of this liquid film is not that obvious and historically there have been debates on the underlying mechanism (Rosenberg 2005; Dash et al. 2006). One potential explanation for the observation is the decrease of the melting temperature of water with pressure; indeed, this has become a myth; "pressure melting became the standard textbook explanation, and it has been propagated through generations of students" (Dash et al. 2006). We can quickly demonstrate, however, that the resulting effect is minimal. An estimate of the decrease of melting temperature with pressure may be directly obtained from Clapeyron's equation.

To that end, we apply Equation 4.22 to the equilibrium between liquid and solid water, and obtain

$$\frac{\mathrm{d}p^{\mathrm{sat}}}{\mathrm{d}T} = \frac{\Delta_{\mathrm{s}}^{\mathrm{l}}h}{T\Delta_{\mathrm{s}}^{\mathrm{l}}v}, \tag{4.26}$$

where $\Delta_{\mathrm{s}}^{\mathrm{l}}h$ is the specific enthalpy of the transition from solid to liquid (melting) and $\Delta_{\mathrm{s}}^{\mathrm{l}}\,v$ is the difference in the specific volumes v^{l} and v^{s} of the liquid and solid phases, respectively. Inserting into Equation 4.26 the respective values for these properties from Feistel and Wagner (2006) at a temperature 273.15 K and ambient pressure of 0.101325 MPa, we obtain

$$\frac{\mathrm{d}p^{\mathrm{sat}}}{\mathrm{d}T} = \frac{333 \times 10^{3}\,\mathrm{J\,kg^{-1}}}{-90.7 \times 10^{-6}\,\mathrm{m^{3}\,kg^{-1}}273\,\mathrm{K}} \approx -1.34 \times 10^{7}\mathrm{Pa\,K^{-1}} \approx -13.4\,\mathrm{MPa\,K^{-1}}. \tag{4.27}$$

Because water exhibits the peculiarity that the specific volume of the solid phase is larger than that of the liquid phase, the melting curve has a negative slope, meaning that the melting point is shifted toward lower temperatures with increasing pressure. For the sake of completeness, it should be mentioned that the estimate provided in Equation 4.27 refers only to a specific temperature and one form of ice (ice exhibits a variety of phases) of importance here is the common hexagonal structure. From Equation 4.27 we deduce that a considerable pressure of 13.4 MPa is needed to lower the melting temperature by only 1 K. It is certainly difficult to give a precise estimate for the pressure exerted by an ice skater as the contact area might be a topic of debate. However, if we consider a skater of mass 120 kg (that would be considered rather heavy), skating on two blades each with dimensions of about 40 cm × 1.5 mm, we obtain an area of $6{\cdot}10^{-4}$ m^2 and a force of about 600 N (on each blade), resulting in a pressure of 1 MPa. Even when skating just on one leg or when using special blades, it becomes obvious that the resulting effect on the melting point will be well below 1 K. If pressure melting was the decisive effect for making ice skating possible no one could skate at temperatures at a few degrees below 0°C. Yet we know – perhaps from our own painful experience – that one may also slip on ice with normal shoes that have a much larger area than ice skates.

In consequence, we may answer the question from the headline by stating that Clapeyron's equation has little to do with ice skating. There are other mechanisms mainly responsible for a liquid film on the ice surface, namely frictional melting, see, for example, Colbeck (1995), and most importantly "premelting," indicating that a liquid-like layer is formed on the ice surface well below the normal bulk melting temperature. The thickness and structure of this layer have been topics of intense research, including the effects of impurities and confinement, for reviews the reader may consult Dash et al. (2006), Wettlaufer and Grae Worster (2006) and Slater and Michaelides (2019).

4.3 How Do I Calculate the Chemical Potential?

Formally, as we have seen in Chapter 3, the chemical potential is related to the partial molar volume as

$$\left(\frac{\partial \mu_{\mathrm{B,g}}(T,p,y_C)}{\partial p}\right)_{T,y_C} = V_{\mathrm{B}}. \tag{4.28}$$

In this equation, which is explicitly formulated for the gas phase g, and all subsequent equations we use y for mole fractions in the gas phase and x for mole fractions in the liquid phase; the symbols y_C and x_C represent the complete set of molefractions for the mixture containing C components.

In this section, we examine how to calculate the chemical potential for a range of materials and introduce several new quantities related to it. We shall show that the chemical potential and quantities derived from it have a vital part in the description and evaluation of the phase behavior of multicomponent systems. In particular, we shall touch upon the engineering application of these quantities and how an equation of state, which provides a relationship between the pressure of a fluid, its volume, composition and temperature, can deliver a thermodynamically consistent approach to that complex engineering task. The detailed implementation of these and other methods for treating phase behavior are beyond the scope of this text and the reader may consult specialized texts, e.g. see Walas (1985) or Tosun (2012).

4.3.1 What Are the Chemical Potential and Fugacity of a Pure Gas?

Here we consider a pure gas B and for simplicity use the notation V_m for its molar volume so that the chemical potential itself is obtained as the integral

$$\mu_{B,g}(T, p) = \int_0^p V_m \mathrm{d}p. \tag{4.29}$$

This integral diverges at its lower limit and the evaluation of the absolute value of the potential for even a gas is not straightforward. It is, however, possible to evaluate the differences of the chemical potential from its standard value $\mu_{B,g}^{\ominus}(T)$.

An ideal gas is in its standard state when its pressure is the standard pressure, $p^{\ominus}$. For this case, the chemical potential for some other pressure is evaluated as

$$\mu_{B,g}(T, p) - \mu_{B,g}^{\ominus}(T) = \int_{p^{\ominus}}^{p} \frac{RT}{p} \mathrm{d}p = RT \ln \frac{p}{p^{\ominus}}. \tag{4.30}$$

If the gas is **not** an ideal-gas then the standard state is a hypothetical one. This state is perhaps best conceived as one in which the real gas is at the standard pressure, $p^{\ominus}$, but where all intermolecular interactions have been turned off so that the gas behaves ideally and obeys the perfect gas equation.

We then introduce a new quantity, the *fugacity* $\tilde{p}$, which is defined by an equation of the same form as Equation 4.30, so that

$$\mu_{B,g}(T, p) = \mu_{B,g}^{\ominus}(T) + RT \ln \frac{\tilde{p}_{B,g}}{p^{\ominus}}. \tag{4.31}$$

The fugacity has the dimension of pressure. It is the pressure that the hypothetical ideal gas would need to have in order for its chemical potential at the given temperature to be the same as the chemical potential of the real gas.

To evaluate the chemical potential, we must evaluate the fugacity through the pressure-volume characteristics of the gas. The difference of chemical potential between p and p' is

$$\mu_{B,g}(T, p) - \mu_{B,g}(T, p') = RT \ln \frac{\tilde{p}_{B,g}}{\tilde{p}'_{B,g}}. \tag{4.32}$$

Now we also have that

$$\mu_{B,g}(T, p) - \mu_{B,g}(T, p') = \int_{p'}^{p} V_m \mathrm{d}p, \tag{4.33}$$

from which it follows that

$$\ln\frac{\tilde{p}_{B,g}}{\tilde{p}'_{B,g}} = \int_{p'}^{p}\frac{V_m}{RT}dp. \tag{4.34}$$

We now use the identity

$$\ln\frac{p}{p'} = \int_{p'}^{p}\frac{dp}{p}, \tag{4.35}$$

and subtract it from both sides of Equation 4.34 to yield

$$\ln\frac{\tilde{p}_{B,g}p'}{\tilde{p}'_{B,g}p} = \int_{p'}^{p}\left(\frac{V_m}{RT} - \frac{1}{p}\right)dp'. \tag{4.36}$$

Now we take the limit of both sides of this equation as p' approaches zero. In this limit, the gas approaches ideal gas behavior, $\tilde{p}'_{B,g}$ approaches p' and the ratio $\tilde{p}_{B,g}p'/\tilde{p}'_{B,g}p$ approaches $\tilde{p}_{B,g}/p$ so that

$$\ln\frac{\tilde{p}_{B,g}}{p} = \int_{0}^{p}\left(\frac{V_m}{RT} - \frac{1}{p}\right)dp. \tag{4.37}$$

The chemical potential of a pure gas at a pressure p is therefore, from Equation 4.31

$$\begin{aligned}\mu_{B,g}(T,p) &= \mu^{\ominus}_{B,g}(T) + RT\ln\frac{\tilde{p}_{B,g}}{p^{\ominus}} \\ &= \mu^{\ominus}_{B,g}(T) + RT\ln\frac{p}{p^{\ominus}} + \int_0^p\left(V_m - \frac{RT}{p}\right)dp.\end{aligned} \tag{4.38}$$

It is also useful to express this result as integral over the volume. Beginning from Equation 4.34, we can change the integration variable to volume using integration by parts so that

$$\ln\frac{\tilde{p}_{B,g}}{\tilde{p}'_{B,g}} = \frac{1}{RT}[pV]^{p,V_m}_{p',V'_m} - \frac{1}{RT}\int_{V'_m}^{V_m}p\,dV, \tag{4.39}$$

where V_m and V'_m are the respective molar volumes of the gas. Using the identity

$$\ln\frac{V_m}{V'_m} = \int_{V'_m}^{V_m}\frac{dV}{V_m} \tag{4.40}$$

and addition to both sides of Equation (4.39) yields

$$\ln\frac{\tilde{p}_{B,g}V_m}{\tilde{p}'_{B,g}V'_m} = \frac{1}{RT}[pV]^{pV_m}_{p'V'_m} - \frac{1}{RT}\int_{V'_m}^{V_m}\left(p - \frac{RT}{V_m}\right)dV. \tag{4.41}$$

Rearranging Equation 4.41 to make $\ln \tilde{p}_{B,g}$ the subject allows it to be substituted into Equation 4.31 for the chemical potential. Then, taking the limit that V'_m approaches infinity, the gas behaves according to the perfect gas equation of state and $\tilde{p}'_{B,g}$ approaches p' so that $p'V'_m = RT$ and we can finally deduce that

$$\mu_{B,g}(T, V_m) = \mu^{\ominus}_{B,g}(T) + pV_m - RT - RT\ln\left(\frac{p^{\ominus}V_m}{RT}\right) + \int_{\infty}^{V_m}\left(p - \frac{RT}{V_m}\right)dV_m. \quad (4.42)$$

4.3.2 What Is the Chemical Potential for a van der Waals Gas?

Using this form of Equation 4.42, we can readily evaluate the chemical potential for a pure gas that obeys the van der Waals equation of state

$$p = \frac{RT}{V_m - b} - \frac{a}{V_m^2}. \quad (4.43)$$

In Equation 4.43, the parameters b and a are given by

$$b = \frac{V_{m,c}}{3}. \quad (4.44)$$

and

$$a = \frac{9RT_cV_{m,c}}{8}. \quad (4.45)$$

Substituting Equation 4.43 in Equation 4.42, we obtain after integration

$$\mu_{B,g}(T, V_m) = \mu^{\ominus}_{B,g}(T) + \frac{RTV_m}{V_m - b} - \frac{2a}{V_m} - RT - RT\ln\left(\frac{p^{\ominus}V_m}{RT}\right) + RT\ln\left(\frac{V_m}{V_m - b}\right). \quad (4.46)$$

4.3.3 What Are the Conditions for Equilibria of the Gas and Liquid for the van der Waals Equation?

Equilibria of the gas with the liquid are obtained by solving the simultaneous equations

$$\mu_{B,g}(T, V_{m,g}) = \mu_{B,l}(T, V_{m,l}) \quad (4.47)$$

and

$$p(T, V_{m,g}) = p(T, V_{m,l}), \quad (4.48)$$

with, from Equation 3.135,

$$\left(\frac{\partial p}{\partial V_{m,g}}\right)_T < 0 \quad (4.49)$$

and

$$\left(\frac{\partial p}{\partial V_{\mathrm{m,l}}}\right)_T < 0. \tag{4.50}$$

In this chapter and in Equations 4.46 through 4.50, we have used the nomenclature introduced in Question 1.9. The van der Waals Equation 4.46 when substituted into Equation 4.47 gives

$$\frac{(V_{\mathrm{m,g}} - V_{\mathrm{m,l}})b}{(V_{\mathrm{m,g}} - b)(V_{\mathrm{m,l}} - b)} + \ln\left(\frac{V_{\mathrm{m,g}} - b}{V_{\mathrm{m,l}} - b}\right) = \frac{2a(V_{\mathrm{m,g}} - V_{\mathrm{m,l}})}{RTV_{\mathrm{m,g}}V_{\mathrm{m,l}}}, \tag{4.51}$$

while use of Equation 4.43 in Equation 4.48 provides

$$\frac{V_{\mathrm{m,g}}V_{\mathrm{m,l}}}{(V_{\mathrm{m,g}} - b)(V_{\mathrm{m,l}} - b)} = \frac{2a(V_{\mathrm{m,g}} + V_{\mathrm{m,l}})}{RTV_{\mathrm{m,g}}V_{\mathrm{m,l}}}. \tag{4.52}$$

Use of Equation 4.43 in Equations 4.49 and 4.50 shows that for mechanical equilibrium of a van der Waals fluid, the following conditions on temperature and volume must be met

$$\frac{RT}{(V_{\mathrm{m,g}} - b)^2} > \frac{2a}{(V_{\mathrm{m,g}})^3} \tag{4.53}$$

and

$$\frac{RT}{(V_{\mathrm{m,l}} - b)^2} > \frac{2a}{(V_{\mathrm{m,l}})^3}. \tag{4.54}$$

For mixtures, to use the van der Waals equation, parameters a and b are required for each substance in a phase.

4.3.4 How Do Chemical Potentials Depend on Concentration for a Gas Mixture?

The standard state of substance B in a gas mixture is the same as the standard state of the pure gas described in Question 4.3.1. It is the hypothetical state in which pure gaseous B has the same temperature as the mixture, is at the standard pressure $p^{\ominus}$ and behaves as an ideal gas.

Consider that we have pure A, an ideal gas, in a rigid box at a pressure p. We then slide a rigid membrane into the box so as to divide the box into two compartments. The membrane is permeable to molecules of A, so that molecules of A pass freely through its pores. The pressure on both sides remains equal to p. Then, without changing the volume of either compartment, we add a second substance B to one side of the membrane to form an ideal gas mixture. The membrane is impermeable to molecules B so that they stay on one side of the membrane and cause a pressure increase there. Since the mixture is an ideal gas, the molecules of A and B do not interact and the addition of gas B causes no change in the amounts of A on either side of the membrane. Thus, the pressure of A in the pure phase and the partial pressure of A in the mixture are both equal to p. Thus, the partial pressure of gas A in an ideal gas mixture in equilibrium with pure ideal A is equal to the pressure of the pure gas.

Because the system is in equilibrium, gas A must have the same chemical potential on both sides of the membrane (essentially in the two phases). Since the chemical potential of the pure ideal gas is given by Equation 4.30, and we assume p_A, the partial pressure of A in the mixture is equal to p in the pure gas, then the chemical potential of A in the mixture is given by

$$\mu_{A,g}(T, p, y_A) = \mu^{\ominus}_{A,g}(T) + RT\ln\frac{p_A}{p^{\ominus}}. \tag{4.55}$$

This can easily be generalized to a mixture containing C components and we have

$$\mu_{A,g}(T, p, y_C) = \mu^{\ominus}_{A,g}(T) + RT\ln\frac{p_A}{p^{\ominus}} = \mu^{\ominus}_{A,g}(T) + RT\ln\frac{y_A p}{p^{\ominus}}, \tag{4.56}$$

where the second part arises from the fact that the partial pressure $p_A = y_A p$ for an ideal gas mixture. We see that although a constituent of an ideal gas mixture with a partial pressure that is $p^{\ominus}$ is not in its standard state, it has the same chemical potential as in its standard state.

Following the argument used for a pure, real gas, the fugacity, $\tilde{p}_{B,g}$, of species B in a *real gas mixture* is defined by

$$\mu_{B,g}(T, p, y_C) = \mu^{\ominus}_{B,g}(T) + RT\ln\frac{\tilde{p}_{B,g}}{p^{\ominus}} \tag{4.57}$$

or

$$\tilde{p}_B(T, p, y_C) \stackrel{\text{def}}{=} p^{\ominus}\exp\left[\frac{\mu_B(T, p, y_C) - \mu^{\ominus}_{B,g}(T)}{RT}\right]. \tag{4.58}$$

To relate the fugacity to the (p, V) behavior of the substance, we follow the same steps as set out in Question 4.3.1 for a pure gas, and we obtain

$$ln\frac{\tilde{p}_{B,g}}{p_B} = \int_0^p\left(\frac{V_B}{RT} - \frac{1}{p}\right)dp \tag{4.59}$$

as the analog of Equation 4.37.

We can also write down the chemical potential of the species B in the gas phase using Equations 4.57 and 4.59 as

$$\mu_{B,g}(T, p, y_C) = \mu^{\ominus}_{B,g}(T) + RT\ln\frac{p_B}{p^{\ominus}} + \int_0^p\left(V_B - \frac{RT}{p}\right)dp \tag{4.60}$$

or

$$\mu_{\mathrm{B,g}}(T,p,y_{\mathrm{C}}) = \mu_{\mathrm{B,g}}^{\ominus}(T) + RT\ln\frac{y_{\mathrm{B}}p}{p^{\ominus}} + \int_0^p\left(V_{\mathrm{B}} - \frac{RT}{p}\right)\mathrm{d}p. \tag{4.61}$$

4.3.5 What Is the Fugacity Coefficient?

The *fugacity coefficient*, $\phi_{\mathrm{B,g}}(T,p,y_{\mathrm{C}})$, is defined by

$$\phi_{\mathrm{B,g}}(T,p,y_{\mathrm{C}}) \stackrel{\mathrm{def}}{=} \frac{\tilde{p}_{\mathrm{B,g}}(T,p,y_{\mathrm{C}})}{y_{\mathrm{B}}p}. \tag{4.62}$$

The fugacity coefficient is preferred to the fugacity because the fugacity coefficient varies less than the fugacity with respect to changes in pressure and composition and reflects deviations from the ideal gas state more transparently. Equation 4.59 can be cast as

$$\ln\phi_{\mathrm{B}} = \ln\left(\frac{\tilde{p}_{\mathrm{B,g}}}{y_{\mathrm{B}}p}\right) = \int_0^p\left\{\frac{V_{\mathrm{B,g}}(T,p,y_{\mathrm{C}})}{RT} - \frac{1}{p}\right\}\mathrm{d}p. \tag{4.63}$$

For a pure substance $y_{\mathrm{B}} = 1$ and Equation 4.63 becomes

$$\ln\phi_{\mathrm{B}} = \ln\left(\frac{\tilde{p}^{*}_{\mathrm{B,g}}}{p}\right) = \int_0^p\left\{\frac{V_{\mathrm{m,g}}(T,p)}{RT} - \frac{1}{p}\right\}\mathrm{d}p. \tag{4.64}$$

For an ideal gas, for which $pV_{\mathrm{m}} = RT$, Equation 4.64 reduces to $\phi_{\mathrm{B}} = 1$ and

$$\tilde{p}_{\mathrm{B}}^{\mathrm{pg}} = p. \tag{4.65}$$

For a *binary* gas mixture at sufficiently low pressure, the properties can be represented by the virial equation (Chapter 2) truncated after the second virial coefficient so that

$$Z = \frac{pV_{\mathrm{g}}(T,p,y)}{RT} = 1 + B_{\mathrm{mix}}(T,y)\frac{p}{RT}, \tag{4.66}$$

in which the second virial coefficient of a binary mixture is given by

$$\begin{aligned} B_{\mathrm{mix}}(T,y) &= y_{\mathrm{A}}^2 B_{\mathrm{A}} + 2y_{\mathrm{A}}y_{\mathrm{B}}B_{\mathrm{AB}} + y_{\mathrm{B}}^2 B_{\mathrm{B}} \\ &= y_{\mathrm{A}}B_{\mathrm{A}} + y_{\mathrm{B}}B_{\mathrm{B}} + y_{\mathrm{A}}y_{\mathrm{B}}\delta_{\mathrm{AB}}, \end{aligned} \tag{4.67}$$

and

$$\delta_{\mathrm{AB}} = 2B_{\mathrm{AB}} - (B_{\mathrm{A}} + B_{\mathrm{B}}). \tag{4.68}$$

It is often a good approximation to assume

$$B_{AB} = \frac{1}{2}(B_A + B_B), \tag{4.69}$$

and then $\delta_{AB} = 0$. It follows from Equation 4.66 that the partial molar volume of species B is

$$V_{B,g}(T, p, y_B) = \left(\frac{\partial V_g(T, p, y_B)}{\partial y_B}\right)_{T,p,y_C \neq y_B} = \left(\frac{\partial B_{mix}(T, y_B)}{\partial y_B}\right)_{T,y_C \neq y_B}. \tag{4.70}$$

Equation 4.63 can then be written, using Equations 4.67 and 4.70, as

$$\ln \phi_B = \left(\frac{B_B p}{RT}\right) \tag{4.71}$$

and

$$\tilde{p}_{B,g}(T, p, y_B) = (y_B p)\exp\left(\frac{B_B p}{RT}\right), \tag{4.72}$$

which is often used to correct equilibrium constants to different thermodynamic conditions.

4.4 Equilibrium Between Phases

The equilibrium between phases is a very important circumstance and is the basis of many processes in both the natural world as well as practice in the chemical industry.

4.4.1 What Is the Condition of Equilibrium Between Two Phases of a Mixture of Substances?

For equilibrium between two coexisting phases, there must be equality between the chemical potentials of each of the components. The mole fractions x^α and x^β of two coexisting phases α and β of a mixture for which the independent variables are temperature T and pressure p are therefore determined by solution of the simultaneous equations where the notation x^α represents the mole fraction of each component in the phase,

$$\mu_{A,\alpha}(T, p, x^\alpha) = \mu_{A,\beta}(T, p, x^\beta) \tag{4.73}$$

and

$$\mu_{B,\alpha}(T, p, x^\alpha) = \mu_{B,\beta}(T, p, x^\beta). \tag{4.74}$$

The diffusional stability conditions (Question 3.6.3) are

$$\left(\frac{\partial \mu_{\mathrm{A}}}{\partial x}\right)_{T,p}^{\alpha} < 0, \quad \left(\frac{\partial \mu_{\mathrm{B}}}{\partial x}\right)_{T,p}^{\alpha} > 0, \quad \text{and} \quad \left(\frac{\partial \mu_{\mathrm{A}}}{\partial x}\right)_{T,p}^{\beta} < 0\ , \quad \left(\frac{\partial \mu_{\mathrm{B}}}{\partial x}\right)_{T,p}^{\beta} > 0. \tag{4.75}$$

The mole fractions x^{α} and x^{β} of two coexisting phases α and β of a *binary* mixture with independent variables of temperature T and molar volume V_{m} are determined by simultaneous solution of the equations

$$\mu_{\mathrm{A},\alpha}(T, V_{\mathrm{m}}^{\alpha}, x^{\alpha}) = \mu_{\mathrm{A},\beta}(T, V_{\mathrm{m}}^{\beta}, x^{\beta}), \tag{4.76}$$

$$\mu_{\mathrm{B},\alpha}(T, V_{\mathrm{m}}^{\alpha}, x^{\alpha}) = \mu_{\mathrm{B},\beta}(T, V_{\mathrm{m}}^{\beta}, x^{\beta}) \tag{4.77}$$

and

$$p(T, V_{\mathrm{m}}^{\alpha}, x^{\alpha}) = p(T, V_{\mathrm{m}}^{\beta}, x^{\beta}). \tag{4.78}$$

The diffusional stability can be written using results of Question 1.10.1, for example, as

$$\left(\frac{\partial \mu_{\mathrm{A}}}{\partial x}\right)_{T,p}^{\alpha} = \left(\frac{\partial \mu_{\mathrm{A}}}{\partial x}\right)_{T,V_{\mathrm{m}}}^{\alpha} - \left(\frac{\partial \mu_{\mathrm{A}}}{\partial V_{\mathrm{m}}}\right)_{T,x}^{\alpha} \frac{\left(\frac{\partial p}{\partial x}\right)_{T,V_{\mathrm{m}}}^{\alpha}}{\left(\frac{\partial p}{\partial V_{\mathrm{m}}}\right)_{T,x}^{\alpha}} < 0, \tag{4.79}$$

and others analogous to Equation 4.75.

4.4.2 How Is Gas-Liquid Equilibrium in a Binary Mixture Described?

For a *binary* mixture of species A and B at constant temperature with both gas and liquid phases, equilibrium means that $\mu_{\mathrm{A,g}} = \mu_{\mathrm{A,l}} = \mu_{\mathrm{A}}$, $\mu_{\mathrm{B,g}} = \mu_{\mathrm{B,l}} = \mu_{\mathrm{B}}$ and $p = p^{\mathrm{sat}}$. The Gibbs–Duhem Equation 4.4 becomes for the liquid phase

$$(1 - x_{\mathrm{B}})(\mathrm{d}\mu_{\mathrm{A}} - V_{\mathrm{A,l}}\mathrm{d}p^{\mathrm{sat}}) + x_{\mathrm{B}}(\mathrm{d}\mu_{\mathrm{B}} - V_{\mathrm{B,l}}\mathrm{d}p^{\mathrm{sat}}) = 0, \tag{4.80}$$

which may be solved for x_{B}, the mole fraction of B, to give

$$x_{\mathrm{B}} = \frac{\mathrm{d}\mu_{\mathrm{A}} - V_{\mathrm{A,l}}\mathrm{d}p^{\mathrm{sat}}}{\mathrm{d}\mu_{\mathrm{A}} - V_{\mathrm{A,l}}\mathrm{d}p^{\mathrm{sat}} - \mathrm{d}\mu_{\mathrm{B}} + V_{\mathrm{B,l}}\mathrm{d}p^{\mathrm{sat}}}. \tag{4.81}$$

4.4.3 How Can I Determine Chemical Potential Differences?

For a gas mixture, the chemical potential difference, $\mu_{\mathrm{B,\,g}}(T, p, y_{C}^{\beta}) - \mu_{\mathrm{B,\,g}}(T, p, y_{C}^{\alpha})$, for two phases can be obtained from Equation 4.61 and is given by

$$\begin{aligned}\mu_{\mathrm{B,g}}\left(T, p, y_{C}^{\beta}\right) - \mu_{\mathrm{B,g}}\left(T, p, y_{C}^{\alpha}\right) = {} & \int_0^p \left\{V_{\mathrm{B,g}}\left(T, p, y_{C}^{\beta}\right) - V_{\mathrm{B,g}}\left(T, p, y_{C}^{\alpha}\right)\right\}\mathrm{d}p \\ & + RT \ln\left(\frac{y_{\mathrm{B}}^{\beta}}{y_{\mathrm{B}}^{\alpha}}\right)\end{aligned}, \tag{4.82}$$

where $V_{B,g}$ is the partial molar volume of gaseous substance B. Measurements of partial molar volume in each phase can be used to determine $\mu_{B,g}(T, p, y_C^{\beta}) - \mu_{B,g}(T, p, y_C^{\alpha})$. The partial molar volumes in a single phase of a mixture can be determined using very precise measurements of density (Goodwin et al. 2003; Weir and de Loos 2005). We note that the measurement of the partial molar volume also yields

$$\left(\frac{\partial \ln \lambda_B}{\partial p}\right)_T = \frac{V_B}{RT}, \tag{4.83}$$

and in turn this allows the use of the relationship

$$\left(\frac{\partial \ln \lambda_B}{\partial T}\right)_p = -\frac{H_B}{RT^2}, \tag{4.84}$$

as discussed in Question 4.6.4.

Consider two liquid mixtures denoted as phases α and β containing the same C components separated by membrane permeable to solely to substance B. At equilibrium, the chemical potentials of substance B in phases α and β are equal and given by

$$\mu_{B,l}(T, p^{\beta}, x_C^{\beta}) = \mu_{B,l}(T, p^{\alpha}, x_C^{\alpha}). \tag{4.85}$$

If both phases are held at a pressure, p, the difference $\mu_{B,l}(T, p, x_C^{\beta}) - \mu_{B,l}(T, p, x_C^{\alpha})$ is given by Equation 4.61 as

$$\mu_{B,l}(T, p, x_C^{\beta}) - \mu_{B,l}(T, p, x_C^{\alpha}) = \int_{p^{\beta}}^{p} V_{B,l}(T, p, x_C^{\beta})dp + \int_{p^{\alpha}}^{p} V_{B,l}(T, p, x_C^{\alpha})dp. \tag{4.86}$$

If phase α is pure B, denoted by superscript $*$, Equation 4.85 can be written as

$$\mu_B(T, p + \Pi, x_C^{\beta}) = \mu_B^*(T, p), \tag{4.87}$$

where Π is the pressure that arises to ensure diffusive equilibrium (no movement of species B) and is called the osmotic pressure of the mixture for substance B. In this case, the chemical potential difference is given by Equation 4.86 as

$$\mu_B(T, p + \Pi, x_C^{\alpha}) - \mu_B^*(T, p) = -\int_p^{p+\Pi} V_B(T, p, x_C^{\beta})dp. \tag{4.88}$$

Thus, measurements of the osmotic pressure allow evaluation of the chemical potential difference. The difficulty in carrying this through in practice arises from the selection of a membrane permeable solely to substance B.

4.4.4 How Do Chemical Potentials Depend on Concentration for a Liquid Mixture?

Equation 4.61 for the chemical potential of a species B is quite general and not confined to the gas phase. However, to extend it to the liquid phase it is necessary to extend the integration over pressure from the dilute gas phase to the saturated situation, where the liquid mixture is in equilibrium with its vapor at p^{sat}, then continues in the liquid phase up to the requisite pressure of the system, p. The common standard state is still the chemical potential of pure B at the standard pressure. Thus, we have

$$\begin{aligned}\mu_{\text{B,l}}(T,p,x_{\text{C}}) &= \mu^{\ominus}_{\text{B,g}}(T) + RT\ln\left(\frac{y_{\text{B}}p^{\text{sat}}}{p^{\ominus}}\right) + \int_0^{p^{\text{sat}}}\left\{V_{\text{B,g}}(T,p,y_{\text{C}}) - \frac{RT}{p}\right\}\text{d}p \\ &+ \int_{p^{\text{sat}}}^{p} V_{\text{B,l}}(T,p,x_{\text{C}})\text{d}p.\end{aligned} \tag{4.89}$$

Here, $V_{\text{B,g}}$ and $V_{\text{B,l}}$ are the partial molar volumes of B in gas and liquid phases and x_{C} denotes the set of mole fractions of all species in the liquid mixture. Evidently, to use an equation such as this to evaluate the chemical potential or fugacity coefficients in a liquid mixture it is necessary to have an equation that relates the partial molar volume to the pressure and temperature of the system over the whole range of pressures, from 0 to p. This is an *equation of state* and models used for such calculations are known as *EoS models.* Their development requires extensive volumetric data from experiment. It should be noted that EoS models are often pressure explicit, i.e. the pressure is expressed as a function of volume and temperature. For that reason, Equation 4.89 is often used in a form where the integration variable is changed to V (Prausnitz et al. 1998).

The treatment of the chemical potential of solids is analogous to that of liquids.

4.4.5 Can Fugacity Be Used to Calculate (Liquid + Vapor) Phase Equilibrium?

The chemical potential of a liquid mixture is given by Equation 4.89 and, at equilibrium between a liquid mixture and its vapor,

$$\mu_{\text{B,g}}(T,p,y_c) = \mu_{\text{B,1}}(T,p,x_c) \tag{4.90}$$

or

$$\ln\{\lambda_{\text{B,g}}(T,p,y_{\text{C}})\} = \ln\{\lambda_{\text{B,l}}(T,p,x_{\text{C}})\} \tag{4.91}$$

holds, so that for all B

$$\tilde{p}_{\text{B,g}}(T,p^{\text{sat}},y_{\text{C}}) = \tilde{p}_{\text{B,l}}(T,p^{\text{sat}},x_{\text{C}}). \tag{4.92}$$

Equation 4.89 can be recast using the ratio of the absolute activities, each defined by Equation 3.53, (which is discussed in Question 4.6.2; see Equations 4.145 and 4.146), that are much more convenient to use than the chemical potential in this case, as

$$\frac{p^{\ominus}\lambda_{\text{B,l}}(T,p,x_{\text{C}})}{\lambda^{\ominus}_{\text{B,g}}(T)} = (y_{\text{B}}p^{\text{sat}})\exp\left[\int_0^{p^{\text{sat}}}\left\{\frac{V_{\text{B,g}}(T,p,y_{\text{C}})}{RT} - \frac{1}{p}\right\}\text{d}p + \frac{1}{RT}\int_{p^{\text{sat}}}^{p} V_{\text{B,l}}(T,p,x_{\text{C}})\text{d}p\right]. \tag{4.93}$$

Substitution of Equation 4.63 into the previous equation gives

$$\frac{\tilde{p}_{\mathrm{B,l}}(T, p^{\mathrm{sat}}, x_{\mathrm{C}})}{x_{\mathrm{B}} p^{\mathrm{sat}}} = \frac{\tilde{p}_{\mathrm{B,g}}(T, p^{\mathrm{sat}}, y_{\mathrm{C}})}{y_{\mathrm{B}} p^{\mathrm{sat}}} \exp\left\{ \frac{1}{RT} \int_{p^{\mathrm{sat}}}^{p} V_{\mathrm{B,l}}(T, p, x_{\mathrm{C}}) \mathrm{d}p \right\}, \tag{4.94}$$

and with use of the definition of a fugacity coefficient of substance B in a mixture given by Equation 4.62 we can write

$$\tilde{p}_{\mathrm{B,l}}(T, p^{\mathrm{sat}}, x_{\mathrm{C}}) = \phi_{\mathrm{B,g}}(T, p^{\mathrm{sat}}, y_{\mathrm{C}}) x_{\mathrm{B}} p^{\mathrm{sat}} \exp\left[\frac{1}{RT} \int_{p^{\mathrm{sat}}}^{p} V_{\mathrm{B,l}}(T, p, x_{\mathrm{C}}) \mathrm{d}p \right]. \tag{4.95}$$

The fugacity coefficient of the liquid is defined as

$$\phi_{\mathrm{B,l}}(T, p, x_{\mathrm{C}}) \stackrel{\mathrm{def}}{=} \frac{\tilde{p}_{\mathrm{B,l}}(T, p, x_{\mathrm{C}})}{x_{\mathrm{B}} p^{\mathrm{sat}}}. \tag{4.96}$$

Equations precisely analogous to Equations 4.89 and 4.93 apply to equilibrium between a vapor and a solid or between a liquid and a solid. The problem of calculating phase equilibria has been changed from estimating chemical potentials to the determination of the saturation pressure, liquid volumes and the fugacity coefficients of the system.

The fugacity coefficient of Equation 4.96 requires an equation of state to evaluate the partial molar volume. However, equations of state usually have temperature and volume as the independent variables and not pressure, so that it is usually more convenient to work with a different form of result for the fugacity coefficient. This can be obtained by repeating the analysis of Question 4.3.1 from the equivalent of Equation 4.39 for a mixture where we converted an integral over pressure to one over volume. After some lengthy algebra, one finds the fugacity-coefficient of a substance in a mixture at constant temperature and composition is given by

$$RT \ln \phi_{\mathrm{B}}(x) = \int_{V}^{\infty} \left[\left(\frac{\partial p}{\partial n_{\mathrm{B}}} \right)_{T,V,n_{\mathrm{C}} \neq n_{\mathrm{B}}} - \frac{RT}{V} \right] \mathrm{d}V - RT \ln Z, \tag{4.97}$$

in which, $Z = pV/nRT$ is the compression factor of the mixture; both it and the partial derivative in Equation 4.97 can be obtained from the equation of state.

For a pure substance, Equation 4.97 becomes

$$RT \ln \phi_{\mathrm{B}}^{*} = RT \ln \frac{f}{p} = \int_{V}^{\infty} \left[\frac{p}{n_{\mathrm{B}}} - \frac{RT}{V} \right] \mathrm{d}V - RT \ln Z + RT(Z - 1). \tag{4.98}$$

For an extensive treatment of the fugacity concept, the reader should refer to Van Ness and Abbott (1982), Tester and Modell (1996), Prausnitz et al. (1998) and Smith et al. (2018).

4.4.6 What Is the Poynting Factor?

The exponential term in Equation 4.95 is called the Poynting factor and is widely used by engineers; it is denoted by F_{B} and is given by

$$F_{\mathrm{B}}(T,x)=\exp\left[\int_{p^{\mathrm{sat}}}^{p}\frac{V_{\mathrm{B,l}}(T,p,x_{\mathrm{C}})}{RT}\mathrm{d}p\right]. \tag{4.99}$$

The liquid molar volume $V_{\mathrm{B,l}}(T,p,x)$ in Equation 4.99 is usually a weak function of pressure at temperatures below the critical temperature and thus

$$\int_{p^{\mathrm{sat}}}^{p}V_{\mathrm{A,l}}(T,p,x_{\mathrm{C}})\mathrm{d}p\approx V_{\mathrm{A,l}}(T,x_{\mathrm{C}})(p-p^{\mathrm{sat}}) \tag{4.100}$$

so that

$$F_{\mathrm{B}}=\exp\left\{\frac{V_{\mathrm{A,l}}(T,x_{\mathrm{C}})(p-p^{\mathrm{sat}})}{RT}\right\}, \tag{4.101}$$

which is a form accurate enough for many purposes.

4.4.7 What Is the Gibbs-Duhem Equation for Fugacity Coefficients?

At constant temperature and pressure, it follows from Equation 4.4 and the definition of fugacity coefficients that the Gibbs–Duhem equation for fugacity coefficients can be obtained as

$$\sum_{\mathrm{B}}x_{\mathrm{B}}\mathrm{d}[\ln\{\phi_{\mathrm{B}}(T,p,x_{\mathrm{C}})\}]=0. \tag{4.102}$$

An equation of state provides a thermodynamically consistent route to the evaluation of the fugacity of components in both vapor and liquid phases. It thus offers a very convenient basis for phase-equilibrium calculations. The most well-known application of such methods in chemical engineering lies in the field of high-pressure vapor + liquid equilibria (VLE) where the equation-of-state approach is the method of choice for the vast majority of systems (Walas, 1985; de Nevers, 2012; Narayanan, 2013).

4.5 What Are Ideal Liquid Mixtures?

We first define an ideal mixture by the requirement that for each component in the mixture the absolute activity $\lambda_{\mathrm{A}}(T,p,x_{\mathrm{C}})$ of the component in the mixture is equal to the mole fraction of the component multiplied by the absolute activity of the pure species at the same temperature and pressure $\lambda_{\mathrm{A}}^{*}(T,p)$, that is, for all species

$$\lambda_{\mathrm{A}}^{\mathrm{id}}(T,p,x_{\mathrm{C}})=x_{\mathrm{A}}\lambda_{\mathrm{A}}^{*}(T,p). \tag{4.103}$$

In what follows we confine ourselves to an *ideal binary* mixture because the generalization to multicomponent systems is straightforward but understanding can be obscured by the algebraic complexity for multicomponent mixtures.

For the *ideal binary* mixture in the liquid phase, therefore

$$\lambda_A^{id}(T, p, x_A) = x_A \lambda_A^*(T, p) \tag{4.104}$$

and

$$\lambda_B^{id}(T, p, x_B) = x_B \lambda_B^*(T, p). \tag{4.105}$$

The molar Gibbs energy for the liquid mixture is, from Equation 3.48,

$$G_{m,l}(T, p, x_A) = x_A \mu_{A,l}(T, p, x_A) + x_B \mu_{B,l}(T, p, x_B). \tag{4.106}$$

The molar Gibbs energy of mixing is therefore obtained from Equation 1.19 as

$$\Delta_{mix} G_m^{id}(T, p, x) = x_A \mu_A^{id}(T, p, x_A) + x_B \mu_B^{id}(T, p, x_B) - x_A \mu_A^*(T, p) - x_B \mu_B^*(T, p). \tag{4.107}$$

Now, given that

$$\mu_{A,l}^{id}(T, p, x) = \mu_{A,l}^{\ominus}(T) + RT \ln \lambda_{A,l}^{id}(T, p, x_A), \tag{4.108}$$

$$\mu_{A,l}^*(T, p) = \mu_{A,l}^{\ominus}(T) + RT \ln \lambda_{A,l}^*(T, p), \tag{4.109}$$

where $\mu_{A,l}^{\ominus}(T)$ is the standard chemical potential for pure liquid A. Equivalent expressions for species B can easily be written. It follows immediately from the subtraction of Equation 4.109 from Equation 4.108 with the use of Equation 4.104 that

$$\mu_{A,l}^{id}(T, p, x) = \mu_{A,l}^*(T, p) + RT \ln x_A. \tag{4.110}$$

This concept is often expressed in terms of fugacities. The term $RT \ln x_A$ of Equation 4.110 can be linked to a ratio of fugacities based on the definition of fugacity given in Equation 4.58 as

$$\mu_{A,l}^{id}(T, p, x) - \mu_{A,l}^*(T, p) = RT \ln \frac{\tilde{p}_{A,l}^{id}(T, p, x)}{\tilde{p}_{A,l}^*(T, p)} = RT \ln x_A, \tag{4.111}$$

which leads to

$$\tilde{p}_{A,l}^{id}(T, p, x) = x_A \tilde{p}_{A,l}^*(T, p). \tag{4.112}$$

This equation is known as the *Lewis fugacity rule.*

Substitution of Equations 4.103 into 4.108 and 4.109, and using these equations and the corresponding equations for species B, leads to the result that

$$\Delta_{mix} G_m^{id} = RT \{x_A \ln x_A + x_B \ln x_B\}. \tag{4.113}$$

It also follows from Equations 3.41, 3.77 and 3.79 after some algebra that

$$\Delta_{\text{mix}} S_{\text{m}}^{\text{id}} = -R\{x_{\text{A}} \ln x_{\text{A}} + x_{\text{B}} \ln x_{\text{B}}\}, \tag{4.114}$$

$$\Delta_{\text{mix}} H_{\text{m}}^{\text{id}} = 0 \tag{4.115}$$

and

$$\Delta_{\text{mix}} V_{\text{m}}^{\text{id}} = 0 \tag{4.116}$$

for an ideal liquid mixture.

4.5.1 What Are Excess Functions?

In practice, no real mixture is ideal, but mixtures formed from similar chemical substances do behave in large measure as ideal. When mixtures are not ideal, it is usual to discuss their behavior in terms of the *excess molar functions* X_{m}^{E} defined by

$$X_{\text{m}}^{\text{E}} = \Delta_{\text{mix}} X_{\text{m}} - \Delta_{\text{mix}} X_{\text{m}}^{\text{id}}, \tag{4.117}$$

so that the excess property is expressed with respect to the ideal mixture denoted by id and defined by Equations 4.113 to 4.116.

Thus, for a *binary* mixture, the excess molar Gibbs energy is

$$\begin{aligned} G_{\text{m}}^{\text{E}}(T, p, x_{\text{A}}) &= \Delta_{\text{mix}} G_{\text{m}}(T, p, x_{\text{A}}) - RT\{x_{\text{A}} \ln x_{\text{A}} + x_{\text{B}} \ln x_{\text{B}}\} \\ &= x_{\text{A}}\mu_{\text{A}}(T, p, x_{\text{B}}) + x_{\text{B}}\mu_{\text{B}}(T, p, x_{\text{A}}) - [x_{\text{A}}\mu_{\text{A}}^{*}(T, p) + x_{\text{B}}\mu_{\text{B}}^{*}(T, p)] \\ &\quad - RT\{x_{\text{A}} \ln x_{\text{A}} + x_{\text{B}} \ln x_{\text{B}}\}, \end{aligned} \tag{4.118}$$

using the definitions of $G_{\text{m}}(T, p, x)$ from Equation 4.106 and $\Delta_{\text{mix}} G_{\text{m}}(T, p, x)$ from Equation 4.107. This can be written as

$$G_{\text{m}}^{\text{E}} = x_{\text{A}}\mu_{\text{A}}^{\text{E}} + x_{\text{B}}\mu_{\text{B}}^{\text{E}}, \tag{4.119}$$

where

$$\mu_{\text{A}}^{\text{E}}(T, p, x_{\text{A}}) = \mu_{\text{A}}(T, p, x_{\text{A}}) - \mu_{\text{A}}^{*}(T, p) - RT \ln x_{\text{A}} \tag{4.120}$$

and

$$\mu_{\text{B}}^{\text{E}}(T, p, x_{\text{B}}) = \mu_{\text{B}}(T, p, x_{\text{B}}) - \mu_{\text{B}}^{*}(T, p) - RT \ln x_{\text{B}}. \tag{4.121}$$

The results for the other thermodynamic quantities are

$$S_{\text{m}}^{\text{E}} = \Delta_{\text{mix}} S_{\text{m}} + R\{x_{\text{A}} \ln x_{\text{A}} + x_{\text{B}} \ln x_{\text{B}}\}, \tag{4.122}$$

$$H_{\text{m}}^{\text{E}} = \Delta_{\text{mix}} H_{\text{m}} \tag{4.123}$$

and

$$V_{\mathrm{m}}^{\mathrm{E}} = \Delta_{\mathrm{mix}} V_{\mathrm{m}}. \tag{4.124}$$

4.6 What Uses Are Made of Activity and Activity Coefficients?

Many mixtures of interest in the chemical industry exhibit strong nonideality, for example, acetone + water, and these have traditionally been described by activity coefficient models (or as we will see equivalently the excess molar Gibbs function) for the liquid phase and an equation of state for the vapor phase. It is to the introduction of the activity coefficient that we now turn.

4.6.1 What Are Activity Coefficients?

The chemical potential of substance A in a binary liquid mixture can be obtained from its value at saturation from the equation

$$\mu_{\mathrm{A,l}}(T, p, x_{\mathrm{A}}) = \mu_{\mathrm{A,l}}(T, p^{\mathrm{sat}}, x_{\mathrm{A}}) + \int_{p^{\mathrm{sat}}}^{p} V_{\mathrm{A,l}}(T, p, x_{\mathrm{A}})\mathrm{d}p, \tag{4.125}$$

which follows from its definition in Equations 3.25 and 3.73.

At equilibrium (saturation), when the pressure in the system is the saturation vapor pressure of the mixture p^{sat}, the liquid and gas phase chemical potentials of substance A are equal so that

$$\mu_{\mathrm{A,l}}(T, p^{\mathrm{sat}}, x_{\mathrm{A}}) = \mu_{\mathrm{A,g}}(T, p^{\mathrm{sat}}, y_{\mathrm{A}}). \tag{4.126}$$

In view of Equation 4.61, Equation 4.126 can be written as

$$\mu_{\mathrm{A,l}}(T, p^{\mathrm{sat}}, x_{\mathrm{A}}) = \mu_{\mathrm{A,g}}^{\ominus}(T) + RT\ln\left\{\frac{y_{\mathrm{A}}p^{\mathrm{sat}}}{p^{\ominus}}\right\} + \int_{0}^{p^{\mathrm{sat}}}\left\{V_{\mathrm{A,g}}(T, p, x_{\mathrm{A}}) - \frac{RT}{p}\right\}\mathrm{d}p. \tag{4.127}$$

Equation 4.125 is then

$$\mu_{\mathrm{A,l}}(T, p, x_{\mathrm{A}}) = \mu_{\mathrm{A,g}}^{\ominus}(T) + RT\ln\left\{\frac{y_{\mathrm{A}}p^{\mathrm{sat}}}{p^{\ominus}}\right\} + \int_{0}^{p^{\mathrm{sat}}}\left\{V_{\mathrm{A,g}}(T, p, y_{\mathrm{A}}) - \frac{RT}{p}\right\}\mathrm{d}p + \int_{p^{\mathrm{sat}}}^{p} V_{\mathrm{A,l}}(T, p, x_{\mathrm{A}})\mathrm{d}p. \tag{4.128}$$

For substance B, the equivalent equation is

$$\mu_{\mathrm{B,l}}(T, p, x_{\mathrm{B}}) = \mu_{\mathrm{B,g}}^{\ominus}(T) + RT\ln\left\{\frac{y_{\mathrm{B}}p^{\mathrm{sat}}}{p^{\ominus}}\right\} + \int_{0}^{p^{\mathrm{sat}}}\left\{V_{\mathrm{B,g}}(T, p, y_{\mathrm{B}}) - \frac{RT}{p}\right\}\mathrm{d}p + \int_{p^{\mathrm{sat}}}^{p} V_{\mathrm{B,l}}(T, p, x_{\mathrm{B}})\mathrm{d}p. \tag{4.129}$$

The chemical potential of pure A is obtained by use of Equation 4.130:

$$\mu_{A,l}^*(T,p) = \mu_{A,g}^\ominus(T) + RT\ln\left\{\frac{p_A^{sat}}{p^\ominus}\right\} + \int_0^{p_A^{sat}}\left\{V_{A,g}^*(T,p) - \frac{RT}{p}\right\}dp + \int_{p^{sat}}^{p} V_{A,l}^*(T,p)dp + \int_{p_A^{sat}}^{p^{sat}} V_{A,l}^*(T,p)dp, \tag{4.130}$$

where, as usual, the asterisk indicates the pure material and now p_A^{sat} indicates the saturation vapor pressure of pure A at the temperature T. There is an exactly equivalent expression for pure B.

For a binary mixture, the part of the excess chemical potential in the liquid phase designated in Equation 4.120 as μ_A^E is

$$\begin{aligned}\mu_{A,l}^E(T,p,x_A) = {} & \mu_{A,g}^\ominus(T) + RT\ln\left\{\frac{y_A p^{sat}}{p^\ominus}\right\} + \int_0^{p^{sat}}\left\{V_{A,g}(T,p,y_A) - \frac{RT}{p}\right\}dp \\ & + \int_{p^{sat}}^{p} V_{A,l}(T,p,x_A)dp - \mu_{A,g}^\ominus(T) - RT\ln\left\{\frac{p_A^{sat}}{p^\ominus}\right\} \\ & - \int_0^{p_A^{sat}}\left\{V_{A,g}^*(T,p) - \frac{RT}{p}\right\}dp \\ & - \int_{p_A^{sat}}^{p} V_{A,l}^*(T,p)dp - \int_{p_A^{sat}}^{p^{sat}} V_{A,l}^*(T,p)dp - RT\ln x_A,\end{aligned} \tag{4.131}$$

so that

$$\begin{aligned}\mu_{A,l}^E(T,p,x_A) = {} & RT\ln\left\{\frac{y_A p^{sat}}{x_A p_A^{sat}}\right\} + \int_0^{p^{sat}}\left\{V_{A,g}(T,p,y_A) - \frac{RT}{p}\right\}dp + \int_{p^{sat}}^{p} V_{A,l}(T,p,x_A)dp \\ & - \int_0^{p_A^{sat}}\left\{V_{A,g}^*(T,p) - \frac{RT}{p}\right\}dp - \int_{p^{sat}}^{p} V_{A,l}^*(T,p)dp - \int_{p_A^{sat}}^{p^{sat}} V_{A,l}^*(T,p)dp,\end{aligned} \tag{4.132}$$

with an exactly equivalent expression for $\mu_{B,l}^E(T,p,x)$. The G_m^E can be obtained from Equation 4.119.

The *activity coefficients* $\gamma_{A,l}$ and $\gamma_{B,l}$ for liquids (or solids) are defined by

$$RT\ln\gamma_{A,l} \overset{\text{def}}{=} \mu_A^E = RT\ln\left\{\frac{\lambda_A(T,p,x)}{x_A\lambda_A^*(T,p)}\right\} \tag{4.133}$$

and

$$RT\ln\gamma_{B,l} \overset{\text{def}}{=} \mu_B^E = RT\ln\left\{\frac{\lambda_B(T,p,x)}{x_B\lambda_B^*(T,p)}\right\}, \tag{4.134}$$

where μ_A^E is given by Equation 4.132, and μ_B^E by an equivalent expression exchanging B for A everywhere.

From Equation 4.119, we then have

$$G_{\mathrm{m}}^{\mathrm{E}}/RT = x_{\mathrm{A}} \ln \gamma_{\mathrm{A},1} + x_{\mathrm{B}} \ln \gamma_{\mathrm{B},1}. \tag{4.135}$$

The left-hand side is a molar quantity for the mixture, and the right-hand side is therefore a sum of the product of a mole fraction and a partial molar quantity. Therefore, the logarithm of the activity coefficient is a partial molar quantity and

$$\ln \gamma_{\mathrm{A},1} = \left(\frac{\partial G_{\mathrm{m}}^{\mathrm{E}}/RT}{\partial x_{\mathrm{A}}} \right)_{T,p,x_{\mathrm{B}}} = \frac{\mu_{\mathrm{A},1}^{E}}{RT} \tag{4.136}$$

with an equivalent expression for the other activity coefficient

$$\ln \gamma_{\mathrm{B},1} = \left(\frac{\partial G_{\mathrm{m}}^{\mathrm{E}}/RT}{\partial x_{\mathrm{B}}} \right)_{T,p,x_{\mathrm{A}}} = \frac{\mu_{\mathrm{B},1}^{\mathrm{E}}}{RT}. \tag{4.137}$$

In order to work with activity coefficients in either phase equilibrium or the treatment of chemical reactions it is necessary to have a model for the dependence of $\mu_{\mathrm{B},1}^{\mathrm{E}}(T, p, x)$ upon composition; such models are called *excess Gibbs energy models or activity coefficient models.* Examples of such models can be found, for example in Prausnitz and Tavares (2004).

In general, for real liquids

$$\mu_{\mathrm{B},1}(T, p, x) - \mu_{\mathrm{B},1}^{*}(T, p) = RT \ln \frac{\tilde{p}_{\mathrm{B},1}}{\tilde{p}_{\mathrm{B},1}^{*}}, \tag{4.138}$$

using fugacities and, by introducing the fugacity coefficient, this becomes

$$\mu_{\mathrm{B},1}(T, p, x) - \mu_{\mathrm{B},1}^{*}(T, p) = RT \ln \frac{x_{\mathrm{B}} p \phi_{\mathrm{B},1}}{p \phi_{\mathrm{B},1}^{*}}. \tag{4.139}$$

Then, from the definition of the activity coefficient in Equation 4.134 it follows that

$$\gamma_{\mathrm{B},1}(T, p, x) = \frac{\phi_{\mathrm{B},1}(T, p, x)}{\phi_{\mathrm{B},1}^{*}(T, p)}. \tag{4.140}$$

It is obvious from this equation that in an ideal mixture, where the activity coefficient is unity, the fugacity coefficient of B in the mixture is equal to that of pure liquid B. The latter need not itself be unity, which means that pure B need not be ideal in the sense of the ideal gas law even if the activity coefficient is unity.

It is sometimes easier to use as the standard state that for the liquid rather than for the gas as we have in the development to this point. From Equation 3.157, we can write the standard chemical potential of species B for example as

$$\mu_{\mathrm{B},1}^{\ominus}(T) = \mu_{\mathrm{B},1}^{*}(T, p) + \int_{p}^{p^{\ominus}} V_{\mathrm{B},1}^{*}(T, p)\mathrm{d}p, \tag{4.141}$$

or in terms of the absolute activity from Equation 3.158

$$\lambda_{\mathrm{B},1}^{\ominus}(T) = \lambda_{\mathrm{B},1}^{*}(T, p)\exp\left\{\frac{1}{RT}\int_{p}^{p^{\ominus}} V_{\mathrm{B},1}^{*}(T, p)\mathrm{d}p\right\}. \tag{4.142}$$

Equation 4.134 can be written for the chemical potential as

$$RT \ln x_{\mathrm{B}}\gamma_{\mathrm{B},1}(T, p, x_{\mathrm{B}}) = \mu_{\mathrm{B},1}(T, p, x_{\mathrm{B}}) - \mu_{\mathrm{B},1}^{\ominus}(T) + \int_{p}^{p^{\ominus}} V_{\mathrm{B},1}^{*}(T, p)\mathrm{d}p \tag{4.143}$$

or for the absolute activity

$$RT \ln \gamma_{\mathrm{B},1}(T, p, x_{\mathrm{B}}) = RT \ln\left(\frac{\lambda_{\mathrm{B},1}(T, p, x_{\mathrm{B}})}{x_{\mathrm{B}}\left[\lambda_{\mathrm{B},1}^{\ominus}(T)\exp\left\{\frac{1}{RT}\int_{p^{\ominus}}^{p} V_{\mathrm{B},1}^{*}(T, p)\mathrm{d}p\right\}\right]}\right). \tag{4.144}$$

4.6.2 What Is the Relative Activity?

The chemical potential difference, $\mu_{\mathrm{B}} - \mu_{\mathrm{B}}^{\ominus}$, or the corresponding ratio of absolute activities, $\lambda_{\mathrm{B}}/\lambda_{\mathrm{B}}^{\ominus}$, occurs frequently and it is called the *relative activity* a defined by

$$RT \ln a \overset{\mathrm{def}}{=} \mu_{\mathrm{B}} - \mu_{\mathrm{B}}^{\ominus} \tag{4.145}$$

or

$$a = \frac{\lambda_{\mathrm{B}}}{\lambda_{\mathrm{B}}^{\ominus}}. \tag{4.146}$$

In Equation 4.145, $\mu^{\ominus}$ is the standard chemical potential (see Question 1.9 and Question 3.7.2), while in Equation 4.146, $\lambda_{\mathrm{B}}^{\ominus}$ is the standard absolute activity; both evidently depend on the choice of the standard state. For example, the reference state of infinite dilution, which is related to molarity, is often used in (bio-)chemistry for reactions in liquid solvents such as water. In this case, the relative activity approaches the concentration for the reference state of infinite dilution.

4.6.3 What Is Raoult's Law?

When the pressure is close to $p^{\ominus}(p \to p^{\ominus})$, the integral makes a negligible contribution to each of Equations 4.141 through 4.144 and is often taken to be zero so that, for example, Equation 4.144 becomes

$$RT \ln \gamma_{B,1} \approx RT \ln \left\{ \frac{\lambda_{B,1}(T, p, x_C)}{x_B \lambda_{B,1}^{\ominus}(T)} \right\}, \tag{4.147}$$

and, with Equation 4.146, Equation 4.147 becomes

$$RT \ln \left\{ \frac{\lambda_{B,1}(T, p, x_C)}{x_B \lambda_{B,1}^{\ominus}(T)} \right\} = RT \ln \left(\frac{a_B}{x_B} \right) = RT \ln \gamma_{B,1}. \tag{4.148}$$

From Equation 4.148, the definition

$$a_B = \gamma_{B,1} x_B \tag{4.149}$$

emerges, which is equivalent to

$$\frac{\lambda_{B,1}(T, p, x_C)}{\lambda_{B,1}^{*}(T, p)} = \frac{\tilde{p}_{B,1}(T, p, x_C)}{\tilde{p}_{B,1}^{*}(T, p)} = \gamma_{B,1} x_B. \tag{4.150}$$

If the partial molar liquid volume is much less than that of the gas, as is often the case, $V_{B,1}(T, p, x_B) \ll V_{B,g}(T, p, x_B)$ so that $V_{B,1}(T, p, x_B)$ can be neglected. If we further assume that the gas phase is ideal so that $V_{m,g} = RT/p$, then Equation 4.132 reduces to

$$\mu_A^E = RT \ln \left\{ \frac{y_A p^{sat}}{x_A p_A^{sat}} \right\}, \tag{4.151}$$

and for species B we have

$$\mu_B^E = RT \ln \left(\frac{y_B p^{sat}}{x_B p_B^{sat}} \right) \tag{4.152}$$

For an ideal liquid mixture we know that $\mu_A^E = 0$ and $\mu_B^E = 0$; thus, Equations 4.151 and 4.152 become

$$y_A p^{sat} = x_A p_A^{sat} \tag{4.153}$$

and

$$y_B p^{sat} = x_B p_B^{sat}. \tag{4.154}$$

The left-hand side of Equations 4.153 and 4.154 is the partial pressure of each component in the gas phase as given by Dalton's law for the ideal gas. This result is more general and the right-hand side of the two equations yield the partial pressure in the gas phase even if that phase is not ideal. It is in those more general circumstances that Equations 4.153 and 4.154 are known as Raoult's law (Narayanan 2013).

At slightly higher pressures when perhaps the vapor phase cannot be treated as an ideal gas, the (p, V_m, T) can be represented adequately by a virial expansion up to the second virial coefficient – see Equation 4.66. This allows the integral in the gas phase of Equation 4.127 to be evaluated (employing Equation 4.70) as

$$\int_0^{p^{sat}} \left\{ V_{A,g}(T, p, y_A) - \frac{RT}{p} \right\} dp = B_A p^{sat} + y_B \delta_{AB} p^{sat}, \tag{4.155}$$

where δ_{AB} is defined by Equation 4.68. Provided p^{sat} is not very different from the pressure p at which we require the chemical potential, we can assume liquid incompressibility, so that the integral of Equation 4.129 will be

$$\int_{p^{sat}}^{p} V_{A,l}(T, p, x_A) dp = V_{A,l}(T, x_A)(p - p^{sat}). \tag{4.156}$$

Thus, Equation 4.132 for μ_A^E can be approximated by

$$\begin{aligned} \mu_A^E &= +RT \ln\left\{ \frac{y_A p^{sat}}{x_A p_A^{sat}} \right\} + (B_A - V_{A,l}^*(T))(p^{sat} - p_A^{sat}) \\ &\quad + y_A \delta_{AB} p^{sat} + \{V_{A,l}(T, x_A) - V_{A,l}^*(T)\}(p - p^{sat}) \end{aligned}, \tag{4.157}$$

and equivalently

$$\begin{aligned} \mu_B^E &= RT \ln\left\{ \frac{y_B p^{sat}}{x_B p_B^{sat}} \right\} + (B_B - V_{B,l}^*(T))(p^{sat} - p_B^{sat}) \\ &\quad + y_B \delta_{AB} p^{sat} + \left\{V_{B,l}(T, x_B) - V_{B,l}^*(T)\right\}(p - p^{sat}) \end{aligned}, \tag{4.158}$$

and from Equation 4.118 the molar excess Gibbs function is given by

$$G_m^E(T, p, x_A) = x_A \mu_A^E + x_B \mu_B^E, \tag{4.159}$$

where for a binary mixture of course $x_A + x_B = 1$.

The use of models using activity coefficients for the liquid phase and an equation of state for the vapor phase is quite common. This is because very many systems encountered in practice involve associating substances, most especially, but not confined to, water and aqueous solutions. For the liquid state of such systems there are no general equations of state models that are available. The activity coefficient models therefore provide the only practical way to deal with the liquid phase of such systems. However, the use of two different schemes in gaseous and liquid phases has a number of drawbacks that include (1) it is not possible to define standard states for supercritical components; (2) critical phenomena cannot be predicted because a different model is used for the liquid and vapor phases; (3) the model parameters are highly temperature-dependent; and (4) it is not possible to predict values for the density, enthalpy and entropy from the same model. It should also be emphasized that no matter what model is used to describe deviations from ideality, it remains necessary to satisfy the Gibbs–Duhem equation under all circumstances.

4.6.4 How Do You Evaluate Colligative Properties?

From Question 3.5.1, Equations 3.73, 3.75, and 3.79 we have

$$\left(\frac{\partial \mu_B}{\partial p}\right)_T = V_B, \tag{4.160}$$

$$\left(\frac{\partial \ln \lambda_B}{\partial p}\right)_T = \frac{V_B}{RT}, \tag{4.161}$$

and

$$\left(\frac{\partial \ln \lambda_B}{\partial T}\right)_T = \frac{H_B}{RT^2}, \tag{4.162}$$

and they can be used to examine what are called *colligative properties*, namely the depression of the freezing point of a solvent by the addition of a solute and the elevation of the boiling point of a liquid by a solute.

We shall consider first the depression of the freezing point, that is of the phase transition temperature between solid and liquid. We consider two binary systems. In the first, which consists of the pure solvent, A, there are two phases in equilibrium at the temperature T_A^*, which is the phase transition temperature of the pure solvent. We denote the solid phase by β* and the liquid phase by α*. In the second system, the liquid phase consists of the same solvent with the addition of a solute, B, and we denote this phase by α. It is in equilibrium with a solid phase denoted by β at temperature T_A, which is the phase transition temperature for this system. We assume that the solute, B, does not dissolve in the solid phase β so that in fact the two solid phases are identical except for their temperature.

The equality of the chemical potential of the solvent between the two phases in each system leads to the following two equalities of the absolute activities,

$$\lambda^*_{A,\alpha^*}(T^*_A, p) = \lambda^*_{A,\beta^*}(T^*_A, p), \tag{4.163}$$

and

$$\lambda_{A,\alpha}(T_A, p, x_B) = \lambda^*_{A,\beta}(T_A, p), \tag{4.164}$$

where x_B denotes the mole fraction of the solute in the phase α. In view of these results, we can write an expression for the ratio of the absolute activities for the solvent in the solution and in its pure liquid state at the temperature T_A as

$$\frac{\lambda^*_{A,\alpha^*}(T_A, p)}{\lambda_{A,\alpha}(T_A, p, x_B)} = \frac{\lambda^*_{A,\beta^*}(T^*_A, p)}{\lambda^*_{A,\beta}(T_A, p)}\,\frac{\lambda^*_{A,\alpha^*}(T_A, p)}{\lambda^*_{A,\alpha^*}(T^*_A, p)} \tag{4.165}$$

or

$$\ln\left\{\frac{\lambda^*_{A,\alpha^*}(T_A, p)}{\lambda_{A,\alpha}(T_A, p, x_B)}\right\} = \ln\left\{\frac{\lambda^*_{A,\beta^*}(T^*_A, p)}{\lambda^*_{A,\beta}(T_A, p)}\right\} - \ln\left\{\frac{\lambda^*_{A,\alpha^*}(T^*_A, p)}{\lambda^*_{A,\alpha^*}(T_A, p)}\right\}. \tag{4.166}$$

We now recognize that phases β and β*are identical, so that

$$\ln\left\{\frac{\lambda^*_{A,\alpha^*}(T_A, p)}{\lambda_{A,\alpha}(T_A, p, x_B)}\right\} = \int_{T_A}^{T^*_A}\left\{\left(\frac{\partial \ln\lambda_{A,\beta^*}(T, p)}{\partial T}\right)_p - \left(\frac{\partial \ln\lambda_{A,\alpha^*}(T, p)}{\partial T}\right)_p\right\}. \tag{4.167}$$

In the last equation, we prepare to use Equation 4.162 so that we can write

$$\ln\left\{\frac{\lambda^*_{A,\alpha^*}(T_A, p)}{\lambda_{A,\alpha}(T_A, p, x_B)}\right\} = \int_{T_A}^{T^*_A}\left\{\frac{(H^{\alpha *}_A - H^{\beta *}_A)}{RT^2}\right\}\mathrm{d}T = \int_{T_A}^{T^*_A}\left\{\frac{\Delta^\alpha_\beta H^*_A}{RT^2}\right\}\mathrm{d}T. \tag{4.168}$$

The molar enthalpy difference, $\Delta^\alpha_\beta H^*_A$, is the molar enthalpy of melting of the solvent A, $\Delta^\alpha_\beta H^*_A$. The ratio $\lambda^*_{A,\alpha^*}(T_A, p)/\lambda_{A,\alpha}(T_A, p, x_B)$ must always be greater than unity, as a consequence of the Inequality 3.143, $(\partial\mu_A/\partial n_A)_{T,p,n_C \neq n_A} > 0$. Thus, the transition temperature from liquid to solid in the case of the solution must be less than the transition temperature for the pure solvent, i.e. any solute depresses the freezing temperature, $T_A < T^*_A$.

In the case when α is a liquid and β a gas, the same argument can be followed through with appropriate changes for an involatile solvent which is therefore not present in the gas phase. In that case, $\Delta^g_l H^*_A$ is the molar enthalpy of evaporation and it follows from Equation 4.168 that $T_A > T^*_A$, so that the boiling temperature of a solvent is always increased by the addition of an involatile solute.

It is remarkable that the temperature difference in either of these phase transition temperatures that is generated by a solute depends only upon the mole fraction of the solute but not its nature as Equation 4.168 demonstrates.

4.6.5 What Is a Test of Thermodynamic Consistency?

For a binary mixture, the Gibbs–Duhem equation at constant p and T is Equation 4.80, and, when written using the absolute activity, it becomes

$$0 = x_A \mathrm{d}\ln\lambda_A + x_B \mathrm{d}\ln\lambda_B, \tag{4.169}$$

or, in terms of the activity coefficients we have through Equations 4.133 and 4.134, it becomes

$$0 = x_A\left(\frac{\partial \ln \gamma_A}{\partial x_A}\right)_{T,p,x_B} + x_B\left(\frac{\partial \ln \gamma_B}{\partial x_A}\right)_{T,p,x_B}. \tag{4.170}$$

For a multicomponent mixture, the generalization is

$$\sum_B x_B\left(\frac{\partial \ln \gamma_B}{\partial x_B}\right)_{T,p,x_C \neq x_B} = 0. \tag{4.171}$$

From Equations 4.161 and 4.162, for a binary mixture, we also obtain

$$\left(\frac{\partial \ln \gamma_{\mathrm{B}}}{\partial p}\right)_T = \frac{(V_{\mathrm{B}} - V_{\mathrm{B}}^*)}{RT}, \tag{4.172}$$

and

$$\left(\frac{\partial \ln \gamma_{\mathrm{B}}}{\partial T}\right)_p = -\frac{(H_{\mathrm{B}} - H_{\mathrm{B}}^*)}{RT^2}. \tag{4.173}$$

Integration of Equation 4.170 gives

$$\int_0^1 \ln(\gamma_{\mathrm{B}}/\gamma_{\mathrm{A}})\,\mathrm{d}x_{\mathrm{A}} = 0. \tag{4.174}$$

It is entirely possible to measure the activity coefficients of the two species in a binary mixture at a particular temperature and pressure separately. Equation 4.174 is therefore often used (or at least should be used) to test the validity of those independent measurements because, if $\ln(\gamma_{\mathrm{A}}/\gamma_{\mathrm{B}})$ is plotted against x_{A}, the net area under the curve should be zero. It should be noted that Equation 4.174 is a necessary but not sufficient condition for thermodynamic consistency.

Values of the activity coefficient are usually determined from measurements of x, y and p^{sat} at each temperature, for binary liquid mixtures. Because the phase rule yields $F = 2$ for a binary mixture, only two quantities are required but the measurements must be tested for thermodynamic consistency and this can be done through measurements of the third quantity.

4.6.6 How Do I Use Activity Coefficients Combined with Fugacity to Model Phase Equilibrium?

For a system at constant and uniform temperature and pressure, and of constant amount of substance, we have seen (Question 4.4.2) that the equilibrium conditions in terms of the chemical potential result in

$$\mu_{\mathrm{B},\alpha} = \mu_{\mathrm{B},\beta} = \cdots = \mu_{\mathrm{B},\pi}, \tag{4.175}$$

where (A, B, …. C) label the components and (α, β, …., π), the phases. Generalization of Equation 4.92 then permits phase equilibrium to be defined in terms of the fugacity $\tilde{p}$ for each substance B of the mixture of C components {A, B, …} by

$$\tilde{p}_{\mathrm{B},\alpha} = \tilde{p}_{\mathrm{B},\beta} = \cdots = \tilde{p}_{\mathrm{B},\pi}. \tag{4.176}$$

For vapor + liquid equilibrium, Equation 4.176 becomes:

$$\tilde{p}_{\mathrm{i,g}}(T, p, y) = \tilde{p}_{\mathrm{i,l}}(T, p, x), \quad \mathrm{i} = \mathrm{A},\ \mathrm{B},\ \cdots, C. \tag{4.177}$$

Using Equation 4.62 in the left-hand side of the equation above and Equation 4.95 in the right-hand side, along with Equation 4.99, the equilibrium condition of Equation 4.177 becomes, for each substance B of the mixture, of C components

$$y_B\, \phi_{B,g}(T, p, y_B)p = x_B\, \gamma_{B,l}(T, p, x_B)\phi_{B,l}\left(T, p_B^{sat}, x_B\right) p_B^{sat}(T)\, F_B(T, x_B), \qquad (4.178)$$

where F_B is the Poynting factor for species B.

The equilibrium of the mixture of C components requires C equations of the type of Equation 4.178, one for each component. This formalism is known as the gamma-phi approach for calculating vapor-liquid equilibria. The fugacity coefficient $\phi_{B,g}(T, p, y_B)$ that accounts for the nonideality of the vapor phase of each component can be evaluated from an equation of state as can the fugacity coefficient $\phi_{B,l}(T, p, x_B)$ in the liquid phase, while the activity coefficient $\gamma_{B,l}(T, p, x_B)$ used to describe the nonideal behavior of the liquid phase can be determined from an excess Gibbs function model. Given both sets of models, Equation 4.178 allows one, at a particular temperature and pressure, to evaluate the mole fraction of a species in one phase given its value in another. For further details, the reader should refer to Van Ness and Abbott (1982), Walas (1985), Tester and Modell (1996), Prausnitz et al. (1998), Poling et al. (2001), de Nevers (2012), Narayanan (2013) and Smith et al. (2018).

4.6.7 How Do We Obtain Activity Coefficients?

The experimental methods used to acquire values of the activity coefficient have been alluded to in Question 4.6.5. Other methods rely on the use of Equation 4.178. For the majority of cases, it is necessary to have experimental values of the activity coefficients for substance B in a binary mixture. A typical experimental determination of the activity coefficient therefore requires measurements of the total pressure p, as well as the mole fractions, y_B and x_B, in the vapor and liquid phase, respectively, for a binary mixture at vapor-liquid equilibrium at a particular temperature.

Measurements as a function of mole fraction for the liquid phase are used to determine the parameters in a suitable activity-coefficient model. As an example, Figure 4.2 shows the measured $(p, x, y)_T$ at T = 318.15 K for (nitromethane + tetrachloromethane) while, in Figure 4.3, the corresponding activity coefficients of both components are shown also as a function of liquid composition. The mixture (nitromethane + tetrachloromethane) is not ideal and, as expected, the activity coefficients for both substances are greater than unity.

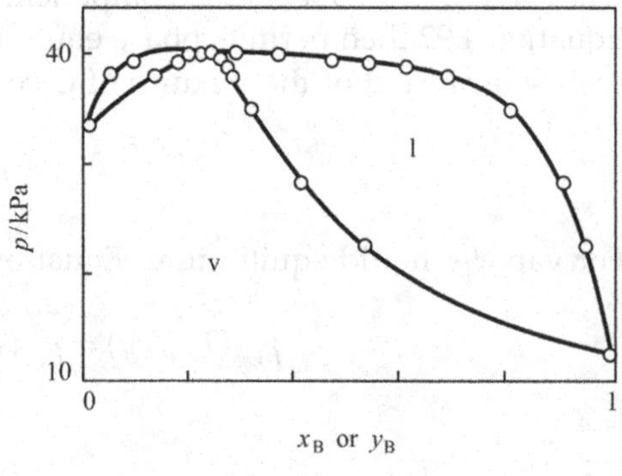

FIGURE 4.2
$(p, x)_T$ section for {tetrachloromethane(A) + nitromethane(B)} at T = 318.15 K. Symbols denote experimental values. Curves represent values calculated using Wilson's equation (Wilson 1964).

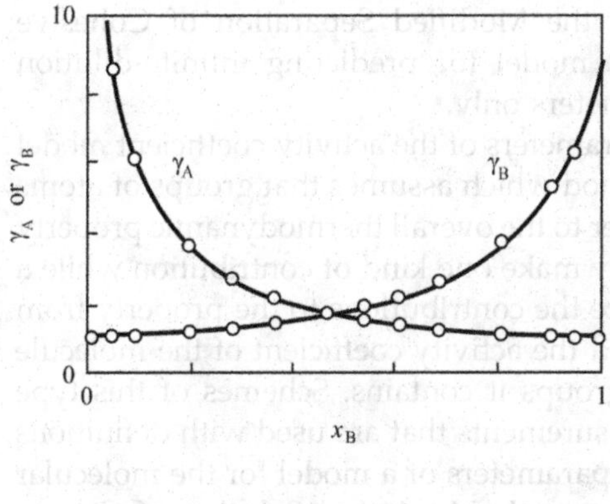

FIGURE 4.3
Activity coefficients γ_A and γ_B for {tetrachloromethane(A) + nitromethane(B)} at T = 318.15 K, as a function of mole fraction x_B. O, experimental values; —, values obtained from Wilson's equation (Wilson 1964).

The Poynting factor given by Equation 4.101 is set equal to unity, which is a reasonable assumption provided the pressure does not differ significantly from the vapor pressure of the pure components.

4.6.8 Activity Coefficient Models

The first model of this type was reported by Margules (1895) and represented the logarithm of the activity coefficient by a power series in composition for each component. van Laar (1910 and 1913) proposed a model based on van der Waals's equation of state with two adjustable parameters; predictive capabilities of that scheme have been found to be limited.

Typically, the model requires the measurement of (vapor + liquid) equilibria at a given temperature for all possible binary mixtures formed from the components of the fluid. The parameters of the activity coefficient model are then fitted to experimental data for binary mixtures. The resulting model can be applied to predict the activity coefficients of a multicomponent mixture over a range of temperature and pressure. For binary mixtures, the model is used to extrapolate the measured values with respect to temperature and pressure. For multicomponent mixtures, the model also exploits extrapolation of the composition. Examples of this approach are Wilson (1964), T-K-Wilson (Tsuboka and Katayama 1975), the Non-Random Two-Liquid model (NRTL) of Renon (1968 and 1969) and UNIQUAC (Abrams and Prausnitz 1975). Certainly, the most reliable procedure for the determination of parameters in any activity-coefficient model involves a fit to experimental data over a range of liquid compositions. The solution of the model for the parameters which best represent the data is a matter for nonlinear regression analysis. However, whatever the solution eventually found, it must still conform to the Gibbs–Duhem Equation 4.174. A description of activity coefficient models has been given by Assael et al. (1996), or Kontogeorgis and Folas (2010).

The requirement to measure the (vapor + liquid) equilibria for all binary mixtures can be rather onerous and it will be no surprise to learn that engineers have created other approximate routes that either reduce or eliminate recourse to specific measurements.

In the absence of sufficient measurements, the model parameters are often estimated from Equations 4.157 to 4.159. In this case, the activity coefficients of each of the components A and B in a binary mixture in the limit as their mole fractions approach unity (often called infinite-dilution) are used, because access to the parameters of an empirical model is simplified. For example, the Wilson method may be implemented from the two infinite-dilution activity coefficients for a binary pair. Other models of this type have been proposed by Pierotti et al. (1959) and Helpinstill and van Winkle (1968) for polar mixtures. Thomas and Eckert (1984) proposed the Modified Separation of Cohesive Energy Density (given the acronym MOSCED) model for predicting infinite-dilution activity coefficients from pure component parameters only.

In the absence of specific measurements, the parameters of the activity-coefficient model can be estimated using a group-contribution method which assumes that groups of atoms within a molecule contribute in an additive manner to the overall thermodynamic property for the entire molecule. Thus, a methyl group may make one kind of contribution while a hydroxyl group makes another contribution. Once the contributions to the property from each group of the molecule have been determined the activity coefficient of the molecule can be obtained from the contributions of the groups it contains. Schemes of this type ultimately rely on (vapor + liquid) equilibria measurements that are used with definitions of the groups within molecules to determine the parameters of a model for the molecular group by regression. Examples of this approach are the Analytical Solution of Groups (ASOG) (Wilson and Deal 1962; Wilson 1964; Kojima and Toshigi 1979) and the Universal Functional Group Activity Coefficients (UNIFAC) (Fredenslund et al. 1975; 1977) models; the UNIFAC method is widely used (Kontogeorgis and Folas 2010).

4.6.9 How Can I Estimate the Equilibrium Mole Fractions of a Component in a Phase?

To complete the description of phase equilibrium, a means of determining the distribution of the substance B between the liquid and gas phases is required. This can be done by analogy with the methods used for chemical equilibrium discussed in Chapter 5, in terms of the standard equilibrium constant. Of course, in the (vapor + liquid) equilibrium the components of the mixture are unchanged by the vaporization and condensation, so that the equilibrium constant describes the distribution of the components between the various phases. For that reason, the equilibrium constant under these circumstances is often known as the partition coefficient and its use will be illustrated in Question 7.9.

When the (vapor + liquid) equilibrium can be represented by fugacity coefficients, the distribution of species B is determined for each species from the ratio of the fugacity coefficients for the liquid, $\phi_{\mathrm{B,l}}(T, p, x_{\mathrm{C}})$, to that of the gas, $\phi_{\mathrm{B,g}}(T, p, y_{\mathrm{C}})$ (given by Equations 4.96 and 4.62 where the fugacity of the liquid $\tilde{p}_{\mathrm{B,l}}(T, p, x_{\mathrm{C}})$ and gas $\tilde{p}_{\mathrm{B,g}}(T, p, y_{\mathrm{C}})$ are given by Equations 4.94 and 4.58) by

$$K_p = \prod_{\mathrm{B}} \frac{\phi_{\mathrm{B,l}}(T, p, x_{\mathrm{C}})}{\phi_{\mathrm{B,g}}(T, p, y_{\mathrm{C}})} = \prod_{\mathrm{B}} \frac{\tilde{p}_{\mathrm{B,l}}(T, p, x_{\mathrm{C}}) y_{\mathrm{B}}}{\tilde{p}_{\mathrm{B,g}}(T, p, y_{\mathrm{C}}) x_{\mathrm{B}}} = \prod_{\mathrm{B}} \frac{y_{\mathrm{B}}}{x_{\mathrm{B}}}, \tag{4.179}$$

because at equilibrium, $\tilde{p}_{\mathrm{B,l}}(T, p, x_{\mathrm{C}}) = \tilde{p}_{\mathrm{B,g}}(T, p, y_{\mathrm{C}})$. In Equation 4.179, y_{B} and x_{B} are the mole fractions in the gas and liquid phases, respectively, of substance B.

For (vapor + liquid) equilibrium that requires the use of activity coefficients, Equation 4.178 can be used so that for each substance B

$$K_{\mathrm{B}} = \frac{y_{\mathrm{B}}}{x_{\mathrm{B}}} = \frac{\gamma_{\mathrm{B,l}}(T, p, x_{\mathrm{B}}) \tilde{p}_{\mathrm{B,l}}(T, p, x_{\mathrm{B}}) p_{\mathrm{B}}^{\mathrm{sat}} F_{\mathrm{B}}}{\phi_{\mathrm{B,g}}(T, p^{\mathrm{sat}}, y_{\mathrm{B}}) p}, \tag{4.180}$$

and thus again we find

$$K_p = \prod_{\mathrm{B}} \frac{\gamma_{\mathrm{B},1}(T, p, x_{\mathrm{B}})\tilde{p}_{\mathrm{B},1}(T, p, x_{\mathrm{B}})p_{\mathrm{B}}^{\mathrm{sat}}F_{\mathrm{B}}}{\phi_{\mathrm{B,g}}(T, p^{\mathrm{sat}}, y_{\mathrm{B}})p} = \prod_{\mathrm{B}} \frac{y_{\mathrm{B}}}{x_{\mathrm{B}}}. \tag{4.181}$$

4.7 How do I Calculate Vapor + Liquid Equilibrium?

The coexisting phases of liquid and gas of a pure component are of considerable importance in both chemistry and engineering applications, so we devote here some space to particular aspects of the behavior of these two -phase systems. For the initial examples in Question 4.7.1, water and air are used because of their considerable importance in practical applications. However, the issues raised in Question 4.7.1 have relevance to every system.

The reader interested specifically in the computation of phase boundaries for nonpolar and polar fluid mixtures should consult Questions 7.5.4 and 7.5.5, respectively, as well as Question 7.5.6.

4.7.1 Is There a Difference Between a Gas and a Vapor?

When water is boiled one observes water above the liquid in a form that is commonly referred to as "vapor" or "steam". Thermodynamically, this nomenclature is incorrect. Steam refers to gaseous water that is a clear colorless substance invisible to the human eye. The observer actually sees a mist of water droplets formed from condensed steam and they are thus liquid water. Before continuing to address the question posed by this section heading, we digress to consider evaporation.

Figure 4.4 illustrates the concept of the vaporization of a liquid of fixed amount of substance and initial mass m at constant pressure achieved by a piston and added force given by a mass and local acceleration of free fall. The corresponding points on a $p(v_c)$ section are shown in Figure 4.5, where v_c denotes the specific volume at the critical point. When energy is provided to the liquid it expands from points 1 to 2 as shown in Figure 4.4 and Figure 4.5.

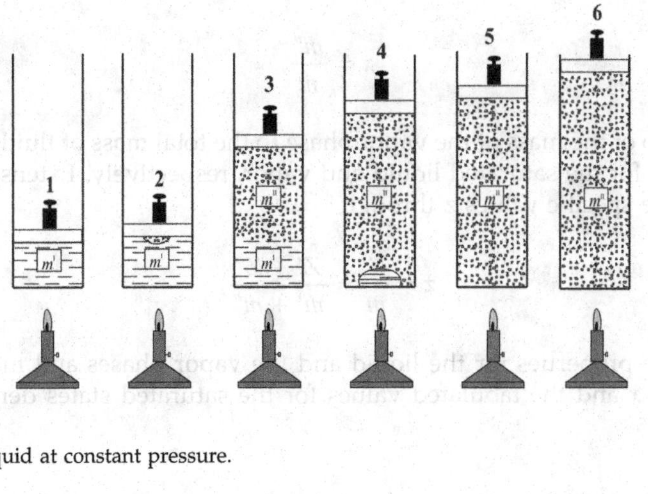

FIGURE 4.4
Vaporization of a liquid at constant pressure.

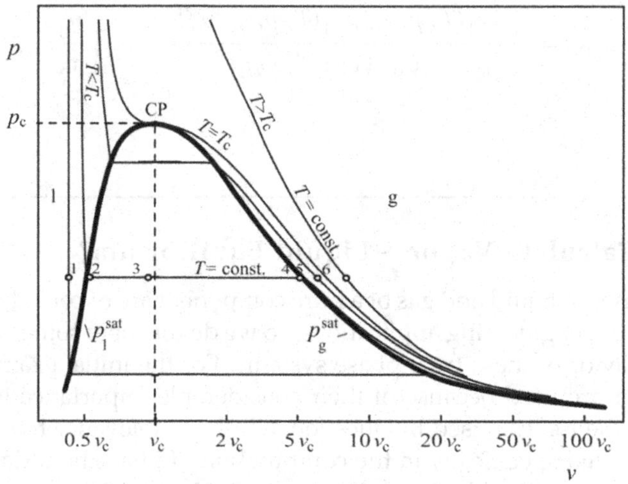

FIGURE 4.5
$p(v_c)$ section for an isobaric vaporization process where v_c is the specific critical volume. The saturated liquid (bubble curve) and saturated vapor (dew curve) are shown along with items 1 through 6 of Figure 4.4.

At step 3 of Figure 4.4 and Figure 4.5, the vessel contains a mixture of saturated liquid of mass m' and a mass of saturated vapor designated m''. During evaporation the volume occupied by the fluid increases because the vapor phase requires a much larger volume than the liquid phase. The mass m' decreases while m'' increases, as illustrated in step 4 of Figure 4.4 and Figure 4.5. This process continues until $m' = 0$ and all the liquid has evaporated (just after point 4 of Figures 4.4 and 4.5). Addition of energy to a purely gaseous phase results in an increase in temperature of the phase and also the volume occupied as illustrated in steps 5 and 6 of Figures 4.4 and 4.5. The temperature of steps 2 through 4 of Figures 4.4 and 4.5 is constant and equal for both gas and liquid owing to the absorption of heat $\Delta_l^g h$, which is the specific enthalpy of evaporation.

In points 2 to 4 of Figures 4.4 and 4.5 the temperature and pressure are insufficient to unambiguously determine the state of the system as it is possible for the states to be in either one-phase region. To specify the state of the two-phase system requires introduction of the *quality* x given by

$$x = \frac{m''}{m}, \tag{4.182}$$

which is the ratio of the mass of the vapor phase to the total mass of fluid and has a value between 0 and 1 for the saturated liquid and vapor, respectively. Extensive properties Z are related to the specific values z through

$$z = \frac{Z}{m} = \frac{Z' + Z''}{m' + m''}, \tag{4.183}$$

that combine the properties for the liquid and the vapor phases and may be expressed with the quality x and the tabulated values for the saturated states denoted by ′and ″ using

$$z = (1 - x)z' + x\,z'' = z' + (z'' - z'). \tag{4.184}$$

This relation is routinely used for the specific volume v, specific internal energy u, specific enthalpy h and specific entropy s. We will now return to address the question posed regarding the difference between vapor and gas.

In common understanding, the term *vapor* implies that it has emerged from the evaporation of a liquid. But one can also vaporize liquid nitrogen and would hardly speak about air containing nitrogen vapor. We can get closer to an answer if we reverse the vaporization process and compress to liquefy a vapor. Compression of gases is usually performed isothermally, as discussed in Chapter 1. If we start at point 5 in the $p(v)$ diagram of Figure 4.5 and compress the vapor isothermally, the system reaches the saturation line and the vapor begins to condense. If the compression commenced at point 6 of Figure 4.5, the isotherm will follow the line to infinite pressure without crossing the saturation line and forming liquid. From Figure 4.5, the resulting difference between the compression starting at point 5 or point 6 arises because the starting temperature 5 is below the critical temperature $T = T_c$ while point 6 is above the critical temperature. The word *vapor* may be defined as a gas at a temperature below its critical temperature, and steam is therefore simply water vapor.

4.7.2 What Is Humidity and How Does Air Conditioning Work?

We conclude from the previous section that moist air is a mixture of air and water given by a gaseous phase (air and water vapor) and a condensed phase liquid. The condensed phase consists essentially of pure water in either liquid or solid form; at a pressure of p = 0.1 MPa, if the system temperature $T > T(H_2O, s + l + g)$ = 273.16 K liquid water is the phase while if $T < T\,(H_2O, s + l + g)$, the condensed phase is ice.

For many technical applications, and especially for air conditioning, the gaseous phase may be approximated by a mixture of two components that behave as ideal gases, these are dry air, which here will be given the subscript a, and will be treated as a pure component, and water vapor, given the subscript v, which because of the low partial pressure relative to atmospheric pressure of about 0.1 MPa for air can also be considered an ideal gas. In the ideal mixture the total pressure p of the gas phase is simply the sum of the partial pressures of the two constituents given by $p = p_a + p_v$. Condensation of water occurs when the water content of the moist air increases to saturation that is when the partial pressure of water vapor (hypothetically) exceeds the maximum permissible value $p_{v,max}$ that is equal to the vapor pressure of pure water at the specified temperature $p^{sat}_{H_2O}(T)$. The reasoning behind this statement is that each component in an ideal gas mixture behaves as if it existed alone. As the vapor pressure of water, which may be taken from steam tables (see Chapter 9) or calculated from Equation 4.20 (the Antoine equation of Question 4.2), depends on temperature and the temperature affects the capacity of air to maintain water vapor before it condenses as illustrated in Figure 4.1 and Figure 4.6.

Moist air, as Figure 4.6 shows, is characterized by a partial pressure p_v of water vapor in air. Isothermal addition of water is shown in Figure 4.6 by a vertical line connecting p_v to $p^{sat}_v(T)$, while isobaric cooling of moist air is shown in Figure 4.6 by a horizontal line connecting p_v to $p^{sat}_v(T_d)$ at the dewpoint temperature, T_d. Water condenses when the saturation line is reached. Dehumidification of moist air is achieved by cooling to a temperature below the dew point temperature.

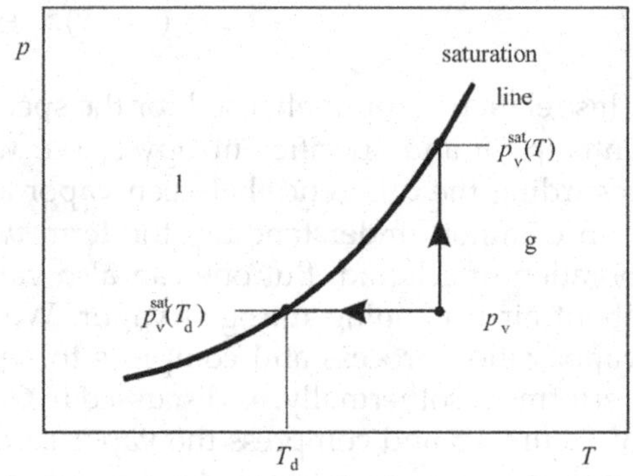

FIGURE 4.6
Schematic of the $p(T)$ section for the evaporation of water.

Condensation of water vapor occurs in everyday life when the temperature of the system is lowered below the saturation temperature corresponding to the partial pressure of the water in atmospheric air. For example, condensation happens when a person wearing spectacles enters a heated room from the external environment in winter. Because the lenses of the spectacles are cold the chilled air near the surface cannot hold the same amount of water as the air in the heated room, and small water droplets start to form on the lenses. The same phenomenon may occur at the inner surface of the windows of a house in winter, or when the windscreen in your car fogs up from the water vapor content of your warm breath. You will often find dew on the lawn after a cool night in summer or on the outer surface of a container holding a chilled drink. It is also possible then to see that a similar process happens in our initial example with water boiling in a kettle: hot steam at a temperature of about 100°C exits the kettle, and the air in the room at a temperature below 100°C is locally supersaturated, so that small water droplets form, which are observed as fog or mist.

For completeness, two important variables characterizing moist air are introduced. The first is the relative humidity Φ (or sometimes φ), a property you can read from a device called a hygrometer and which is expressed as the ratio of the actual partial pressure of water vapor in air $p_{w,a}$ at a particular temperature to its saturation value, $p_w^{sat}(T)$, at the same temperature T defined by

$$\Phi = \frac{p_{w,a}(T)}{p_w^{sat}(T)}. \tag{4.185}$$

The quantity Φ in Equation 4.185 varies between 0 and 1. Cooling moist air increases Φ up to unity when liquid water (or ice) forms.

Another quantity, which relates the mass of vapor m_v to the mass of dry air m_a, has been given several names, including the absolute or specific humidity, the humidity ratio or the moisture content, usually with symbol ω or X (sometimes – and very unfortunately – also x, which may be easily confused with the quality defined by Equation 4.182), thus

$$\omega = \frac{m_v}{m_a}. \tag{4.186}$$

Equation 4.186 is used with the mass of dry air because in air conditioning the mass of dry air often remains constant while the total mass of humid air varies. In some cases, the moisture content is extended to include all water, now designated by a subscript w, and typically given the symbol X defined by

$$X = \frac{m_w}{m_a}, \tag{4.187}$$

where the moisture content is also given for supersaturated air or pure water (where $X \to \infty$). When moist air is heated or cooled, absolute humidity is not altered, but in contrast, relative humidity is.

The values of the absolute and relative humidity can be interrelated. Assuming moist air is an ideal gas this is given by

$$\omega = \frac{m_v}{m_a} = \frac{p_v V M_v/(RT)}{p_a V M_a/(RT)} = \frac{p_v M_v}{p_a M_a} = \frac{18.02\, p_v}{28.96\, p_a} = 0.622 \frac{p_v}{p_a}, \tag{4.188}$$

where m_v is the mass of water vapor, m_a is the mass of air, p_v is the pressure of the water vapor, p_a is the air pressure, R is the gas constant, V is the volume occupied and M_v and M_a are the molar mass of water vapor and air, respectively. Combining Equation 4.188 with Equation 4.185, we obtain

$$\omega = 0.622 \frac{p_v}{p_a} = 0.622 \frac{p_v}{p - p_v} = 0.622 \frac{\Phi\, p_w^{sat}}{p - \Phi\, p_w^{sat}} = 0.622 \frac{p_w^{sat}}{p/\Phi - p_w^{sat}}. \tag{4.189}$$

Cooling moist air below the dew point temperature is of vital importance in the air conditioning process where, as well as the maintenance of a specific temperature, control of humidity is required. In engineering terms, it is a relatively easy task to add water but the reverse process is more challenging. Dehumidification is required to remove the moisture generated by human beings in a room, and can also be employed to defog the windscreen in your car on a cold winter day using the A/C rather than the heater. While it is of course possible to avoid moisture, for example, in the packaging of electronic equipment, by adding some hygroscopic material, this approach is not practicable for a continuous process, because the material would have to be removed and dried in some batch-process for reuse.

As a consequence, in A/C applications, moist air is drawn out of a room into a machine, cooled below its dew point, the condensate removed, and the air with lower moisture is reheated to the desired temperature before being ejected back into the room. To reduce energy consumption the heating is, or should be, achieved using a heat exchanger between the two air streams.

4.7.3 Which Equations of State Are Used in Engineering VLE Calculations?

Equations of state are used in engineering to predict thermodynamic properties, in particular the phase behavior of pure substances and mixtures. However, since there is neither an exact statistical-mechanical solution relating the properties of dense fluids to their intermolecular potentials, nor detailed information available on intermolecular potential functions, all equations of state are, at least partially, empirical in nature. The equations of state in common use within both industry and academia can be classified as follows: (1) cubic equations such as that of van der Waals that are described by Economou

(2010); (2) those based on the virial equation discussed by Trusler (2010) and in Chapter 2 of this volume; (3) equations based on general results obtained from statistical mechanics and computer simulations mentioned, including the many forms of Statistical Associating Fluid Theory known by the acronym SAFT as described briefly in Question 2.9.2 and by McCabe and Galindo (2010); and (4) those obtained by selecting, based on statistical means, terms that best represent the available measurements obtained from a broad range of experiments as outlined by Lemmon and Span (2010). Forms corresponding to item 3 are particularly advantageous when one of the phases includes water.

The development of an equation of state typically commences with the representation of the thermodynamic properties of pure fluids and the functions are then extended to provide estimates of the properties of mixtures by the introduction of mixing and combining rules.

Mixing rules are used to obtain numerical estimates for the parameters in an equation of state for a specified mixture from the same parameters when the same equation of state is used to represent the properties of the pure substance. However, in the description of a mixture, parameters appear that result from the interactions between unlike species, for example, the second virial coefficient B_{AB} used in Equation 4.67. These parameters are often obtained using combining rules. By using mixing and combining rules, measurements are only required for the pure substances and for the relatively small number of mixtures that it has been possible to study experimentally. When these mixing and combining rules are used with $p(V_m, T)$ equations of state they provide the link between the microscopic and the macroscopic properties. The uncertainty of the predictions that result from the use of an equation of state with its mixing and combining rules can be evaluated using tests against experimental data and additional adjustable parameters are added when there is sufficient experimental data. Therefore, the development of an equation of state for mixtures is largely reduced to the establishment of the mixing and combining rules to describe the thermodynamic properties, especially the phase boundaries.

The plethora of both equations of state and of mixing and combining rules means there is a multitude of options available and that some adopted are purely empirical. Consequently, the task of providing a comprehensive list of all equations of state, mixing and combining rules is rather daunting. The basis for the inclusion of those selected herein was their frequent appearance in the archival literature, which does not necessarily imply that the rules are optimal or even correct. The reader requiring a rather more extensive review of equations of state should consult Goodwin and Sandler (2010) and the work of Kontogeorgis and Folas (2010) for mixing and combining rules.

The methods most frequently used to predict the properties of mixtures for over 100 years have inevitably undergone only minor additions and corrections to, it is claimed, improve the representation of experimental data for specific categories of substances. It is, however, possible that completely different alternatives to these traditional approaches are required, particularly for a method to be both predictive and applicable over a wide range of fluids and conditions (Heideman and Fredenslund 1989). Such methods might arise from future research and methods based on statistical mechanics and quantum-mechanical calculations (Leonhard et al., 2007; Singh et al., 2007; Dalarsson et al., 2011; Snurr et al., 2015), are ultimately sought rather than empiricism.

For the purpose of elucidating calculations in the remainder of this section we will consider the cubic equation of state of the form of Equation 4.43 with Equations 4.44 and; however, we wish to emphasize that our analysis is much more general in reality. We employ the van der Waals one-fluid theory for mixtures. This assumes that the properties of a mixture can be represented by a hypothetical pure fluid. Thus, the thermodynamic behavior of a mixture of constant composition is assumed to be identical to that of a one-component fluid; this assumption cannot hold true near the critical point where the thermodynamic behavior of a mixture at constant thermodynamic potential is most definitely not identical with that of a one-component fluid.

In this section, we use the symbol x_i to represent the mole fraction of species i, no matter what its phase. The van der Waals one-fluid theory gives the following for the mixing rules for the van der Waals equation of state

$$a(x) = \sum_{i=0}^{C} \sum_{j=0}^{C} x_i x_j a_{ij}, \tag{4.190}$$

and

$$b(x) = \sum_{i=0}^{C} \sum_{j=0}^{C} x_i x_j b_{ij}. \tag{4.191}$$

The equations are quadratic in mole fraction x for the parameters a and b of Equations 4.44 and 4.45 of substances i and j. Equation 4.191 is often approximated by

$$b(x) = \sum_{i=0}^{C} x_i b_i. \tag{4.192}$$

Before the introduction of combining rules, we digress to return to intermolecular potentials and, in particular, the Lennard-Jones model of the intermolecular potential (Lennard-Jones 1931), which accounts for the repulsive and attractive forces. For the interaction of spherical substances A and B in a mixture of (A + B), $\phi_{AB}(r)$ is given by

$$\phi_{AB}(r) = 4\varepsilon_{AB}\left\{\left(\frac{\sigma_{AB}}{r_{AB}}\right)^{12} - \left(\frac{\sigma_{AB}}{r_{AB}}\right)^{6}\right\} \tag{4.193}$$

and is frequently used in computer simulation. For a ternary mixture of spherical molecules, it is assumed that ϕ (r_{AB}, r_{BC}, r_{CA}) is given by the sum of three pair-interaction energies {ϕ (r_{AB}) + ϕ(r_{BC}) + ϕ (r_{CA})} of which the first term in the summation is given by Equation 4.193. The parameter ε_{AB} of Equation 4.193 defines the depth of the potential well and σ_{AB} is the separation distance at the potential minimum. Combining rules at the molecular level are required to determine ε_{AB} and σ_{AB} from the pure-component values, and it is the discussion of these that we now turn to because they provide background information for this and other sections of this chapter.

The parameter σ_{AB} for unlike interactions between molecules A and B is most often determined from the rule proposed by Lorentz (1881), which is based on the collision of hard spheres; the result is that σ_{AB} is given by the arithmetic mean of the pure-component values with

$$\sigma_{AB} = \frac{\sigma_A + \sigma_B}{2}. \tag{4.194}$$

The parameter ε_{AB} is obtained from the expression of Berthelot (1898) for the geometric mean of the pure-component parameters of:

$$\varepsilon_{AB} = (\varepsilon_A \varepsilon_B)^{1/2}. \tag{4.195}$$

Equation 4.195 arises from consideration of the London (1937) theory of dispersion (Hirschfelder et al. 1954; Rowlinson and Swinton 1982; Henderson and Leonard 1971; Maitland et al. 1981).

Equations 4.194 and 4.195 are collectively known as the Lorentz-Berthelot combining rules and have been widely used (Maitland et al. 1981); they are known to fail particularly in the case of highly nonideal mixtures (Delhommelle and Millié 2001; Ungerer et al. 2004; Al-Matar and Rockstraw 2006, Haslam et al. 2008; Goodwin and Sandler 2010).

Because the core volume b of Equation 4.43 is proportional to σ^3 of Equation 4.194 and a is proportional to the depth of the potential well given by Equation 4.195, Equations 4.194 and 4.195 can be written as

$$b_{AB} = \frac{(b_A^{1/3} + b_A^{1/3})^3}{8} \tag{4.196}$$

and

$$a_{AB} = (a_A a_B)^{1/2}, \tag{4.197}$$

respectively. Equations 4.196 and 4.197 provide the means to estimate both a_{AB} and b_{AB}. Of course, in reality molecules are not hard spheres so that Equation 4.196 requires empirical correction by the addition of a parameter β_{AB}, particularly if the model is used to estimate phase boundaries. Equation 4.197 is also modified by a parameter k_{AB} for the same reason. These semi-empirical arguments and modifications lead to the forms of Equations 4.196 and 4.197 that are routinely used in engineering calculations,

$$b_{AB} = (1 - \beta_{AB})\frac{(b_A^{1/3} + b_A^{1/3})^3}{8} \tag{4.198}$$

and

$$a_{AB} = (1 - k_{AB})(a_A a_B)^{1/2}. \tag{4.199}$$

The parameters β_{AB} of Equation 4.198 and k_{AB} of Equation 4.199 are frequently called binary interaction parameters. Equation 4.198 is often written as

$$b_{AB} = 0.5(1 - \beta_{AB})(b_A + b_B), \tag{4.200}$$

because, in this form, the combined equation of state, mixing and combining rules provide estimates of the properties of the mixture that differ less from the experimental measurements than when Equation 4.198 is used. The importance of the binary interaction parameter k_{AB} of Equation 4.199 in the estimation of phase equilibria can be illustrated for the system $xCO_2 + (1 - x)C_2H_6$ for which the $p(x)_T$ section has been estimated with $k_{12} = 0$ and $k_{12} = 0.124$, as shown in Figure 4.7, where the data are compared with the measured values. The system $xCO_2 + (1 - x)C_2H_6$ exhibits azeotropic behavior, which will be discussed in Question 4.10.4. As a rule, increasing the molecular complexity increases the sensitivity of the calculation to the interaction parameter. Hence, in complicated

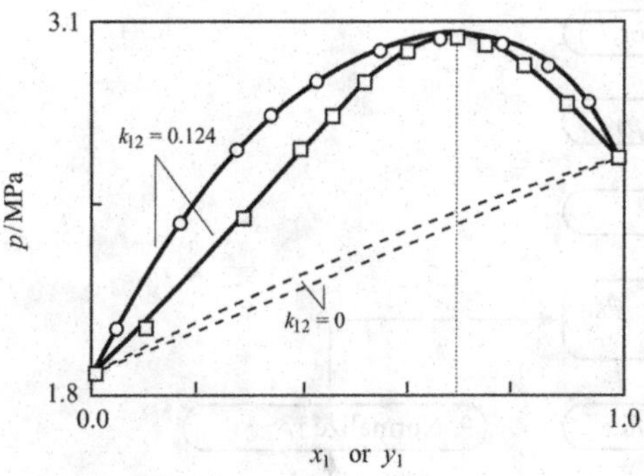

FIGURE 4.7
$p(x)_T$ section for the vapor + liquid equilibrium of $\{CO_2(1) + C_2H_6(2)\}$ as a function of mole fraction x of the liquid and y of the gas phases. O: liquid phase measured bubble pressure (Fredenslund and Mollerup 1974); □: measured dew pressure; ——, dew pressures (Fredenslund and Mollerup 1974) estimated from the Peng-Robinson equation of state with $k_{12} = 0.124$;------, dew pressure estimated from the Peng-Robinson equation of state with $k_{12} = 0$; vertical ········, indicates the azeotropic mixture at $x = 0.7$.

mixtures, the availability of the binary interaction parameters for a particular equation of state might be the overwhelming criterion for choosing a particular functional form for the equation of state.

The cubic equations of state of Peng-Robinson (Peng and Robinson 1976) and Redlich-Kwong-Soave (Soave 1972) are the most commonly used in these calculations. However, other equations that might be categorized as virial equations or as hard sphere approximations with up to 53 adjustable parameters, such as the modified Benedict-Webb-Rubin equation, as originally proposed by Strobridge (1962), have also been employed.

4.7.4 What Is a Bubble-Point or Dew-Point Calculation and Why Is It Important?

A specific example of a dew temperature was provided in Question 4.7.1 for gaseous water and in air. This concept will be generalized herein to vapor + liquid equilibrium (VLE) and extended to also include the bubble pressure. We recall that the dew point is the point of a thermodynamic surface at which liquid first forms and, by analogy, the bubble point is the point at which vapor first forms in a system.

The basic "engine" of most phase equilibrium calculations is an algorithm to calculate the dew or bubble pressure for a mixture of specified composition and temperature. The kind of calculation to be made (dew or bubble) may be specified by giving the *vapor fraction* β, which is defined as the amount of substance in the vapor phase divided by the total amount of substance. It follows that this quantity is unity at a dew point and zero at a bubble point. The phase rule (defined in Question 4.1.2) then requires specification of either the temperature or the pressure in addition to the composition of the bulk phase. It is then our task to calculate the remaining variables; these are either p (for specified T) or T (for specified p) and the composition of the coexisting phase at the dew or bubble point. This problem should have either one solution, when two phases are possible under the specified conditions, or no solution when they are not. Whether or not this is the case with a specific thermodynamic model remains to be proven because the model may or may not accord with reality.

The calculation commences with Equation 4.175 or more often for engineers with Equation 4.177, with one equation for each of the n substances in the mixture to give n simultaneous equations to be solved to determine equilibrium. It is also possible to establish the simultaneous equations using the equality of the product of the fugacity coefficients and mole fractions of the gas and liquid phases given by Equations 4.62 and 4.96. At a specified temperature and pressure, the fraction of vapor for a component B is given by one element of the continued product of Equation 4.179. At equilibrium, Equation 4.179 can also be cast as the ratio of the activity coefficient of a liquid to that of the gas.

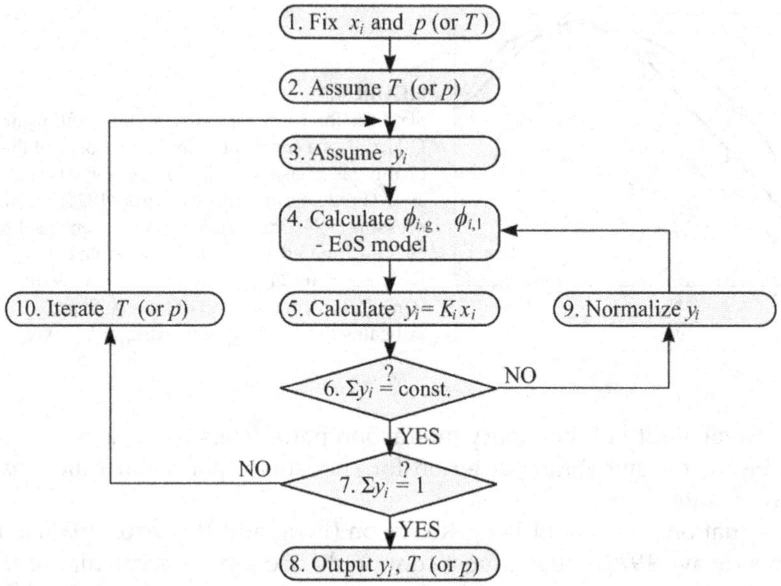

FIGURE 4.8
Bubble-point algorithm using an equation of state for both phases.

We can now proceed to describe a basic algorithm for determining the bubble point of a fluid mixture of n components, on the basis of Equations 4.62 and 4.96 and Equations 4.63 and 4.94 employing an equation of state for both phases.

There are many ways in which one might set about solving the phase-equilibrium problem, but the strategy outlined in Figure 4.8 is a simple and reliable approach to the problem and involves the following steps:

1. The liquid composition x_i (i = 1, 2, n) and either the pressure p or the temperature T must be specified.
2. An initial value is assumed for the unknown bubble-point temperature or pressure; often, Raoult's law in Equation 4.153 is employed for this purpose.
3. Initial values for vapor composition y_i (i = 1, 2, n) are assumed. Unless the system is known to exhibit nearly ideal behavior, one often sets $y_i = x_i$. The sum $s = y_i$ should be initialized to unity at this stage.
4. Next, the fugacity coefficients $\phi_{i,g}(T, p, y_n)$ and $\phi_{i,1}(T, p, x_n)$ of Equations 4.62 and 4.96, respectively, of each component i in the vapor and liquid phases are calculated at the assumed temperature, pressure and phase compositions. To do so requires the equation of state for the molar volume of each component in each phase as provided by Equations 4.63 and 4.94, respectively.
5. New approximations to the vapor mole fractions are estimated from one product of Equation 4.179 using $y_i = x_i K_i$ with $K_i = \phi_{i,1}(T, p, x_n)/\phi_{i,g}(T, p, y_n)$.
6. The new sum $s = \Sigma y_i$ is calculated. If this is equal to that for the previous iteration, then proceed to step 7; otherwise, go to step 9.

7. Once a constant value of s is obtained subject to the presently assumed estimate of the unknown bubble-point temperature or pressure, test to see if $s = 1$. If this condition is satisfied, then proceed to step 8; otherwise, go to step 10.
8. A solution has been found satisfies the thermodynamic requirements for thermal, hydrostatic and phase equilibrium.
9. Normalized values of the vapor-phase mole fractions are calculated, $y'_i = y_i/s$, and used in another iteration starting at step 4.
10. A new estimate of the unknown bubble-point temperature or pressure must be made. If $s > 1$, then the assumed temperature (pressure) is too high (low) while if $s < 1$, then the reverse applies. The simplest method for updating the unknown T or p is by means of a bisection algorithm; this requires that upper and lower limits of the unknown be established at the start of the procedure.

The interaction parameters, the k_{ij}'s in the equation of state mixing rules, are usually obtained by regression to measurements of dew and bubble pressures for the binary subsystems.

The determination of the dew-point temperature or pressure and the composition of the coexisting liquid is almost identical to that for the bubble-point problem. In this case, the vapor composition is specified, and iterations are performed over the liquid mole fractions and the unknown temperature or pressure. The algorithm shown in Figure 4.8 may be used after obvious changes. It might be interesting to note that, since a bubble-point routine returns the composition of the coexisting vapor, it may be used as it stands to generate points on the dew-point surface (although not at predetermined vapor compositions).

An equivalent algorithm can be employed in the case of an activity-coefficient model for the liquid phase and an equation of state for the vapor (Assael et al 1996). If there is no equation of state for the liquid phase, this alternative route is the only one available. In such a case, Equation 4.178 for the ith component is used for $\gamma_{i,1}(T, p, x_n)$, which is determined from an activity coefficient model, while $\phi_{i,g}(T, p, y_n)$ is obtained from Equation 4.62, $\tilde{p}_{i,1}(T, p, x_n)$ is found from Equation 4.95 and F_i from Equation 4.99 using an equation of state.

4.7.5 What Is a Flash Calculation?

The modeling of flash processes is probably the single most important application of chemical engineering thermodynamics. A flash process is one in which a fluid stream of known overall composition and flow rate passes through a throttle, turbine or compressor and into a vessel (flash drum) where liquid and vapor phase are separated before each passes through the appropriate outlet. Such a process may be operated under many different sets of conditions, including the following: (1) constant temperature and pressure (isothermal flash); (2) constant enthalpy and pressure (isenthalpic flash); and (3) constant entropy and pressure (isentropic flash). The thermodynamic modeling of these processes requires in each case determination of the vapor fraction and the vaporization equilibrium ratio for the components in the system. It is also important in general to determine the heat absorbed or liberated in the flash process (the heat duty), although this is zero by definition in an isenthalpic or isentropic flash. In performing VLE calculations, we may choose to employ an equation of state for both phases or, where necessary, an activity-coefficient model for the liquid and an equation of state for the vapor.

4.7.5.1 What Is an Isothermal Flash?

The isothermal flash (constant temperature and pressures), illustrated schematically in Figure 4.9, is one of the most common features encountered in chemical engineering practice. The feed, at temperature T_F and pressure p_F, passes through a throttle and enters the flash vessel, where liquid and vapor phases may separate. The operating pressure p of the unit is controlled in some way and heat is supplied or removed at rate $\dot{Q}$ though a heat exchanger so as to maintain isothermal conditions at temperature T. The molar flow rate $\dot{n}$ of the feed to the unit is specified, together with the overall composition (mole fractions z_i) and the temperature and pressure at which the unit operates. The objectives of the calculation are to determine the compositions (y_i and x_i) and the molar flow rates ($\dot{n}_v$ and $\dot{n}_l$) of the vapor and liquid streams leaving the unit.

From the known composition of the mixture, a material balance is used for each of the n components to distribute the substance between the phases:

$$\dot{n}\, z_i = \dot{n}_l\, x_i + \dot{n}_V\, y_i \tag{4.201}$$

with

$$y_i = K_i\, x_i. \tag{4.202}$$

Combining Equation 4.201 with Equation 4.202 and eliminating the flow rates in favor of the vapor fraction $\beta = \dot{n}_V / \dot{n}$, the so-called flash condition may be written as

$$f(\beta) = \sum_{i=1}^{n} x_i - 1 = \sum_{i=1}^{n} \frac{z_i}{1 + \beta(K_i - 1)} - 1 = 0. \tag{4.203}$$

Equation 4.203 may be solved for β with a Newton-Raphson algorithm (Lee and Wang, 2003) that gives for successive iterations

$$\beta_{k+1} = \beta_k + \left[\sum_{i=1}^{n}\left(\frac{z_i}{1 + \beta_k(K_i - 1)}\right) - 1\right]\left[\sum_{i=1}^{n}\left(\frac{(K_i - 1)z_i}{[1 + \beta_k(K_i - 1)]^2}\right)\right]^{-1}. \tag{4.204}$$

Typically, commencing with $\beta_1 = 1$, the convergence is rapid. The phase compositions are then given by

$$x_i = \frac{z_i}{1 + \beta(K_i - 1)} \quad \text{and} \quad y_i = K_i x_i. \tag{4.205}$$

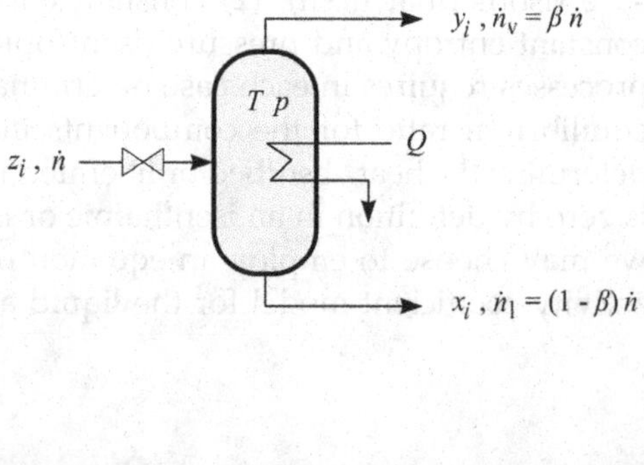

FIGURE 4.9
Isothermal flash unit.

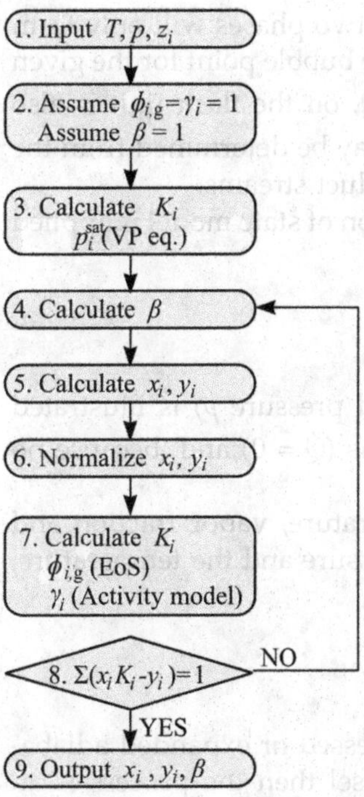

FIGURE 4.10
Isothermal flash algorithm using an activity-coefficient model for the liquid phase.

Of course, both β and the K_i's are unknown and the latter are therefore evaluated during each cycle of Equation 4.204. An algorithm for solving this flash problem is shown in Figure 4.10 for the case in which an activity coefficient model is applied for the liquid phase.

The isothermal flash algorithm involves the following steps:

1. The temperature, pressure and overall mixture composition are specified.
2. Initial values of unity are assumed for the vapor-phase fugacity coefficient and liquid-phase activity coefficient of each component. β is initialized with the value unity.
3. A first approximation to the K_i is calculated for each component from Equation 4.179 using $y_i = x_i K_i$ with $K_i = \phi_{i,1}(T, p, x_n)/\phi_{i,g}(T, p, y_n)$ with p_i^{sat} determined from a suitable representation of the vapor pressure and $\phi_{i,1}(T, p, x_n)$ calculated from an equation of state.
4. A new value of β is determined from a single iteration of Equation 4.204.
5. The compositions of each phases are determined from Equations 4.205.
6. The mole fractions are normalized so that $\sum x_i = \sum y_i = 1$.
7. New vaporization equilibrium ratios are calculated from Equation 4.180 with $\gamma_{i,1}(T, p, x_n)$ determined from the activity coefficient model. $\phi_{i,g}(T, p, y_n)$ is obtained from Equation 4.62, $\tilde{p}_{B,1}(T, p, x_B)$ from Equation 4.95 and F_B from Equation 4.99 via the equation of state.
8. We now test to see if the new vapor composition differs from that of the previous iteration. If it does, begin a new iteration at step 4; otherwise, go to step 9.
9. A solution to the problem has been found.

One rather obvious point that should not be forgotten is that two phases will only form when the specified pressure lies between the dew point and the bubble point for the given temperature and feed composition. Usually the heat duty, $\dot{Q}$, on the flash unit is also required. $\dot{Q}$ (which is positive for heat supplied to the unit) may be determined from the molar flow rates and the molar enthalpy of the feed and product streams.

The method is a good deal simpler to implement if an equation of state model is applied consistently to both phases during the entire calculation.

4.7.5.2 What Is an Isenthalpic Flash?

In Figure 4.11, an isenthalpic flash (constant enthalpy H and pressure p) is illustrated schematically. The unit is operated under adiabatic conditions ($\dot{Q} = 0$) and, because no work is done on the fluid, the process is isenthalpic.

The objective of the flash calculation is to find the temperature, vapor fraction and product compositions for the case in which the operating pressure and the temperature, pressure and composition of the feed are specified.

4.7.5.3 What Is an Isentropic Flash?

If, instead of expanding through a throttle, the feed is compressed or expanded adiabatically and reversibly before entering the adiabatic flash vessel then the process is an isentropic flash (constant entropy S and pressure p). An isentropic flash unit is illustrated schematically in Figure 4.12.

The objective of the flash calculation is to find the temperature, vapor fraction and product compositions for the case in which the operating pressure and the temperature, pressure and composition of the feed are specified. Both an isenthalpic and an isentropic flash can be solved with methods analogous to Figure 4.10 and details are given in Assael et al. (1996).

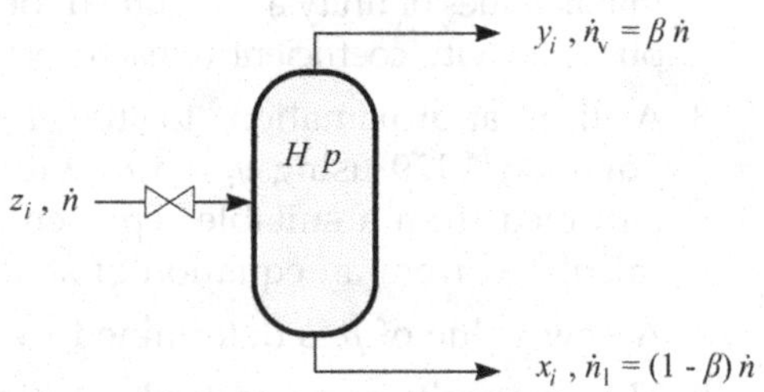

FIGURE 4.11
Isenthalpic flash unit.

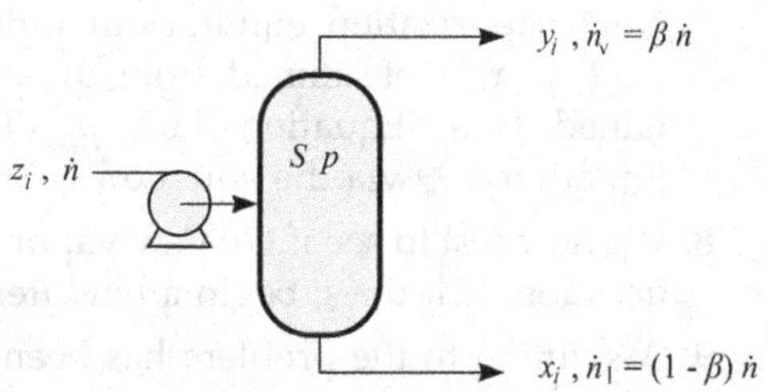

FIGURE 4.12
Isentropic flash.

4.8 How Does the Temperature of the Liquid Change When I Dilute Whiskey with Water?

This example is intended to illustrate in a simple manner the nature of the calculations that the preceding material makes possible.

The concepts required to describe the properties of a mixture of two liquids have been introduced in Question 4.5 and these include ideal mixtures and the definition of the excess properties given by Equations 4.118 through 4.124. For an ideal mixture, the molar volume of mixing and the molar enthalpy of mixing are zero, as given by Equations 4.116 and 4.115, respectively. The excess molar enthalpy and excess molar volume are given in Equations 4.123 and 4.124. Normally, $\Delta_{\text{mix}} H_{\text{m}}$ and $\Delta_{\text{mix}} V_{\text{m}}$ are nonzero.

Atkins, de Paula and Keeler (2018) have a description of a "corrupt barman" and their argument will be used here as an example. The barman mixes, at a temperature of 298.15 K, a volume of 100 cm^3 of substance he intends to sell as pseudowhiskey. This barman uses 40 cm^3 of ethanol and 60 cm^3 of water to make the drink. The negative volume of mixing, $\Delta_{\text{mix}} V_{\text{m}}$ for this system results in a volume of 96 cm^3 of ethanol + water. We note here parenthetically that, based on the densities of the two pure substances at a temperature of 298.15 K and a pressure of 0.1 MPa, an amount of substance of ethanol $n(C_2H_5OH) = 0.71$ mol and an amount of substance of water of $n(H_2O) = 3.45$ mol would actually be required to provide 100 cm^3 of (ethanol + water).

If a further volume of 50 cm^3 of water is then added to the original mixture, the volume will change again and, if the dilution is prepared adiabatically, so will the temperature of the resulting mixture. Adiabatic conditions can be approximated adequately for our purposes by rapid mixing or by the use of a polystyrene-foam cup.

The temperature change can be determined from the First Law of Thermodynamics for a closed system in the absence of external work from stirring and energy transfer from the surroundings because the internal energy U of the system remains unaltered. However, for mixtures, it is more common to consider the enthalpy $H = U + pV$; in the case of liquids, practically no difference arises. Dilution of the pseudowhiskey results in a volume of mixing $\Delta_{\text{mix}} V_{\text{m}}$ that is less then 1 cm^3; the corresponding change of enthalpy at atmospheric pressure of about 0.1 MPa (10^5 Pa) is $p\Delta V = 10^5$ Pa x 10^{-6} m^3 = 0.1 J and this is negligible compared to the other energies involved in the mixing.

Because the enthalpies of mixing (equivalent to the excess enthalpies) are defined and measured for constant temperature (and pressure) and because we expect a change in temperature during our mixing process, we notionally split up the process into two steps:

1. First, we perform the dilution step at constant temperature of 298.15 K by rejecting exactly an amount of heat Q_{D} to render the temperature unaltered. (We anticipate the heat is released during the dilution step but note that the sign of Q_{D} does not affect the following calculations.)
2. Then we use exactly this heat to increase the temperature of the resulting mixture to a final temperature.

First of all, we calculate the mole fractions of ethanol in the respective mixtures before the dilution (initial, i) and after dilution (final, f) and obtain $x_{\text{i}} = 0.17$ and $x_{\text{f}} = 0.10$ (with an amount of substance $n_{\text{a}} = 2.77$ mol of water added). Because the formal IUPAC nomenclature adopted by chemists (Quack et al. 2007) is cumbersome for this example, as shown in the first edition (Assael et al. 2011), we use here a simplified notation. In particular, we use n_E for the amount of substance of ethanol, $H_{m,E}$ for its molar enthalpy with n_W and $H_{m,W}$ for the corresponding quantities for water so that the energy balance for the first step reads

$$\begin{aligned} &n_E H_{m,E} + n_W H_{m,W} + n_a H_{m,W} + (n_E + n_W + n_a) H_m^E(x_f, 298.15\text{K}) - \\ &[n_E H_{m,E} + n_W H_{m,W} + (n_E + n_W) H_m^E(x_i, 298.15\ \text{K}) + n_a H_{m,W}] = Q_D, \end{aligned} \tag{4.206}$$

or

$$(n_E + n_W + n_a) H_m^E(x_f,\ 298.15\text{K}) - (n_E + n_W) H_m^E(x_i, 298.15\text{K}) = Q_D. \tag{4.207}$$

Figure 4.13 shows the variation of the excess molar enthalpy for ethanol + water as a function of composition as determined experimentally. By careful interpolation in the data that support Figure 4.13, we can obtain H_m^E ($x_f = 0.10$, $T_i = 298$ K) = −711 J mol^{-1} and $H_m^E(x_f = 0.17, T_i = 298\ \text{K}) = -784\ \text{J mol}^{-1}$. Thus

$$\begin{aligned} Q_D &= (0.71 + 3.45 + 2.77)(-711\ \text{J}) - (0.71 + 3.45)(-784\ \text{J}) \\ &= -4.93\ \text{kJ} + 3.26\ \text{kJ} = -1.67\ \text{kJ}, \end{aligned} \tag{4.208}$$

which is negative; this means that heat must be discarded to hold the temperature constant.

In the second step, we add this heat to increase the temperature of the mixture to the final temperature T_f obtained from

$$H_m(x_f,\ T_f) - H_m(x_f,\ 298.15\ \text{K}) = (n_E + n_W + n_a)\, C_{p,m}(T_f - 298.15\ \text{K}) = -Q_D, \tag{4.209}$$

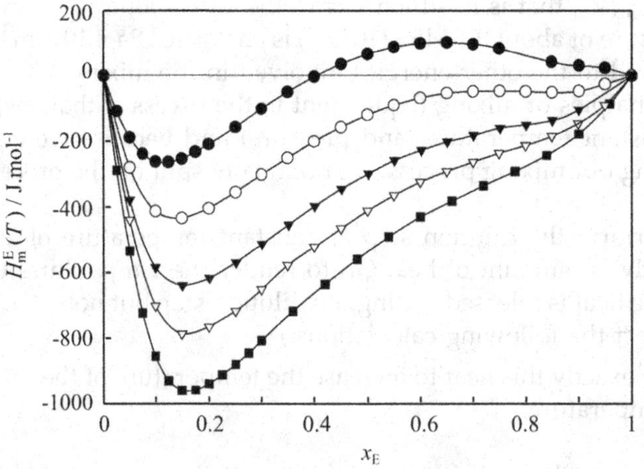

FIGURE 4.13
Molar excess enthalpy $H_m^E(T)$ for (ethanol + water) as a function of mole fraction of ethanol x_E and temperature T. •, T = 338.15 K; ○, T = 338.15 K; ▼, T = 338.15 K; ∇, T = 338.15 K; and ■, T = 338.15 K. Data from Friese et al. (1998, 1999).

where $C_{p,\mathrm{m}}$ is the molar heat capacity at constant pressure for the mixture. In this step, however, we must proceed with some caution. From Figure 4.13, we recognize that H_m^E is a function of both temperature and composition and, from the definition of the heat capacity at constant pressure, we have

$$C_{p,\mathrm{m}} = \left(\frac{\partial H_\mathrm{m}}{\partial T}\right)_{p,x}. \tag{4.210}$$

We can now split up the heat capacity into two parts

$$\begin{aligned} C_{p,\mathrm{m}} &= \left[\frac{\partial\{(n_\mathrm{E}H_\mathrm{m,E} + n_\mathrm{W}H_\mathrm{m,W} + n_\mathrm{a}H_\mathrm{m,W})/n + H_\mathrm{m}^\mathrm{E}\}}{\partial T}\right]_{p,x} \\ &= \left[\frac{\partial\{(n_\mathrm{E}H_\mathrm{m,E} + n_\mathrm{W}H_\mathrm{m,W} + n_\mathrm{a}H_\mathrm{m,W})/n\}}{\partial T}\right]_{p,x} + \left(\frac{\partial H_\mathrm{m}^\mathrm{E}}{\partial T}\right)_{p,x} \\ &= C_{p,\mathrm{m}}^\mathrm{id} + C_{p,\mathrm{m}}^\mathrm{E}, \end{aligned} \tag{4.211}$$

where $n = n_\mathrm{E} + n_\mathrm{W} + n_\mathrm{a}$ is the total amount of substance. Here, $C_{p,\mathrm{m}}^\mathrm{id}$ is the ideal part of the heat capacity, which can be easily obtained by summing up the heat capacities of the individual components weighted with their respective mole fractions, and $C_{p,\mathrm{m}}^\mathrm{E}$ is the excess part of the heat capacity, which takes the nonideality of the solution into account. From the tabulated heat capacities of the pure substances, we obtain $C_{p,\mathrm{m}}^\mathrm{id} = 79.1\ \mathrm{J\ mol^{-1}\cdot K^{-1}}$, which is assumed to be constant over the small temperature range of interest. On the other hand, $C_{p,\mathrm{m}}^\mathrm{E}$ has to be estimated from Figure 4.13. If we consider the excess enthalpies for $x_\mathrm{f} = 0.10$ and temperatures of 285.65 K, 298.15 K and 308.15 K we see that H_m^E is almost linear with temperature in this range and from a fit through these three points we obtain for the gradient $C_{p,\mathrm{m}}^\mathrm{E} = 12.2\ \mathrm{J\ mol^{-1}\ K^{-1}}$.

From Equation 4.209, it then follows that

$$(T_\mathrm{f}/\mathrm{K} - 298.15\ \mathrm{K}) = \frac{1.67\ \mathrm{kJ}}{6.93\ \mathrm{mol}(79.1 + 12.2)\mathrm{J\ mol^{-1}\ K^{-1}}} = 2.6\ \mathrm{K}. \tag{4.212}$$

In this case of dilution, we therefore observe a moderate, yet easily measurable temperature increase of the mixture of 2.6 K.

The question may arise what would happen if we added ethanol to the pseudowhiskey. In general, when we start with a mixture of pure substances (water and ethanol), a negative excess enthalpy means that we have to discharge heat to keep the solution at constant temperature and the temperature would rise if the system was adiabatic. At a temperature of $T = 338.15$ K and depending on the final composition the opposite effect is also possible.

When we dilute our original mixture with water at $T = 298.15$ K, the corresponding point for the final concentration on the connecting line (between the original state and that for pure water) is above the curve for H_m^E ($T_\mathrm{i} = 298.15$ K), thus the temperature will rise for adiabatic mixing as shown in Figure 4.14. On the other hand, one recognizes that when connecting this initial point with the point of pure ethanol there are portions of the connecting line (roughly in a range $0.27 < x_\mathrm{E} < 0.75$), that are below the H_m^E curve, which

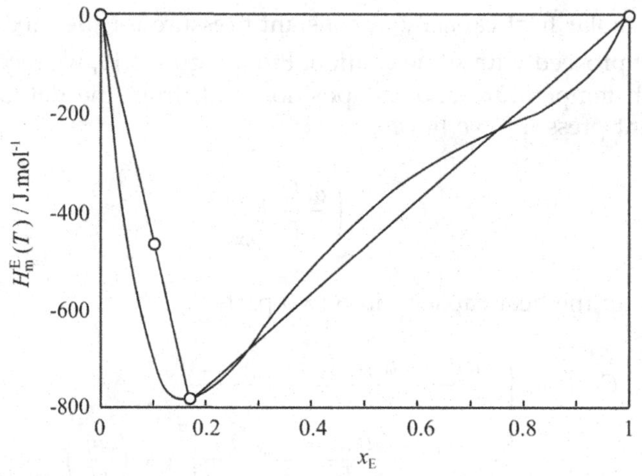

FIGURE 4.14
Molar excess enthalpy $H_m^E(T)$ for (ethanol + water) as a function of mole fraction of ethanol x_E at $T = 298.15$ K. Illustration of the effect of adding water or ethanol, respectively, to a mixture with an initial concentration $x_E = 0.17$ at an initial temperature $T_i = 298.15$ K of all components. The temperature of the mixture will rise (fall) if the point on the connecting line for the final concentration is above (below) the curve H_m^E (298.15 K).

means the mixture would cool down during mixing. We note, however, that neither the cooling effect nor the high alcohol content of the drink would encourage consumption.

4.9 What Are the Characteristics of Liquid + Liquid and Solid + Liquid Equilibria?

We now return to the discussion of liquids and in particular some issues regarding liquid + liquid equilibrium and solid + liquid equilibrium. These are certainly important industrially and to our way of life.

4.9.1 What Are Conformal Mixtures?

In many circumstances, it is assumed that the pair interaction energy, ϕ_{AB}, between the molecules of substances A and B of a mixture is solely a function of the intermolecular separation r and given by

$$\phi_{AB}(r) = \varepsilon_{AB}\ \Phi(r/r^*_{AB}), \tag{4.213}$$

where ε_{AB} is the well depth at the equilibrium r and r^*_{AB} is a characteristic separation. The function Φ also represents the like interaction between molecules of species A and B. Strictly, the dependence of ϕ_{AB} solely on r means the theory is limited to mixtures of spherical molecules while the requirement on the function Φ means that the bulk properties of the systems conform to the principle of corresponding states described in Chapter 2. This means that the macroscopic properties of substances that conform to Equation 4.213 behave identically when appropriately scaled by energy, ε, or volume, r^{*3}, parameters. Systems and, in particular, mixtures that satisfy Equation 4.213 or its consequences are known as *conformal mixtures*.

There are many routes that can be followed from Equation 4.213 that depend on the method used to describe the mixture. The most common is the one-fluid theory as applied to the van der Waals equation, as discussed earlier (Question 4.3.1). In the one fluid theory, the liquid mixture is assumed to be represented by a hypothetical pure fluid that itself also conforms to Equation 4.213. To complete the theory, we then require an equation of state and a selection of mixing and combining rules. In addition to the van der Waals equation, an equation introduced by Carnahan and Starling (1972) is commonly employed for liquids. It has a set of mixing rules and combining rules analogous to, but not identical with, those for the van der Waals one-fluid model. This set of rules together with the equation of state allows prediction of the properties of liquid mixtures such as G_m^E, H_m^E and V_m^E from the critical properties of the pure substances A and B.

4.9.2 What Are Simple Mixtures?

For nonelectrolytes, a simple mixture of two substances is defined by the fact that its excess molar Gibbs function that can be written as

$$G_m^E = x_A x_B L w(T, p), \tag{4.214}$$

where L is the Avagadro constant, x_A and x_B are the mole fractions of the two species and w is an interaction parameter with the dimensions of energy. This parameter depends on temperature and pressure and is obtained from a molecular theory or a molecular model, which is known as Porter's model. For the total amount of substance present, we can write:

$$G_m^E = \frac{n_B n_A L w(T, p)}{(n_A + n_B)^2}. \tag{4.215}$$

From Equations 4.133 we have

$$\left(\frac{\partial G_m^E}{\partial n_A}\right)_{T,p,n_B} = RT \ln \gamma_A = L w(T, p) n_B \left[\frac{(n_A + n_B) - n_A}{(n_A + n_B)^2}\right] = x_B^2 L w(T, p). \tag{4.216}$$

The activity coefficients of a simple mixture are given by

$$\ln \gamma_A = \frac{x_B^2 w(T, p)}{k_B T} \tag{4.217}$$

and

$$\ln \gamma_B = \frac{x_A^2 w(T, p)}{k_B T}. \tag{4.218}$$

One method that has been used for such systems uses Hildebrand's theory of solubility (Barton, 1991). It makes use of a parameter δ_A of a substance A, which can be estimated from the properties of pure A using

$$\delta_A = \left\{ \frac{\Delta_l^g H_A^* - RT}{V_A^*} \right\}^{1/2}, \tag{4.219}$$

and it can then be used to estimate w from the properties of pure substances. For a binary mixture, the expression is

$$w(T, p) = \frac{V_A^* V_B^* (\delta_A - \delta_B)^2}{L\{x_A V_A^* + x_B V_B^*\}}. \tag{4.220}$$

4.9.3 What Are Partially Miscible Liquid Mixtures?

Liquid mixtures can separate into two liquid phases. The two phases appear at a temperature below what is called an upper critical solution temperature (UCST) or above a lower critical solution temperature (LCST). Mixtures with a LCST can also have a USCT at higher temperature and exhibit what is called closed loop miscibility. It is possible to have a LCST at a temperature greater than the UCST so that at temperature between the liquids are miscible. An example of such a circumstance is provided by the system water and 2-butoxy-ethanol, $CH_2(OH)CH_2OC_4H_9$. Figure 4.15 shows examples of UCST and LCST.

For a simple mixture, as defined by Equation 4.214, the chemical potentials of substances A and B of a binary liquid mixture, using Equation 4.121, are given by

$$\mu_{A,1}(T, p, x_A) = \mu_{A,1}^*(T, p) + RT \ln x_A + x_B^2 L\, w(T, p) \tag{4.221}$$

and

$$\mu_{B,1}(T, p, x_B) = \mu_{B,1}^*(T, p) + RT \ln x_B + x_A^2 L\, w(T, p). \tag{4.222}$$

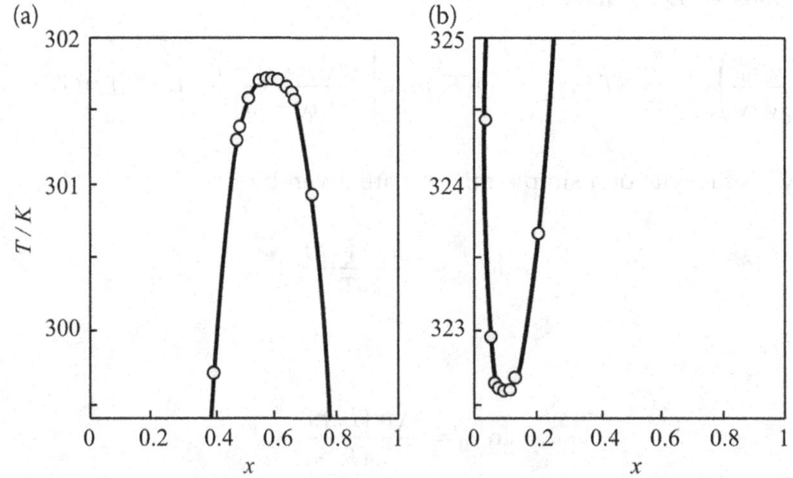

FIGURE 4.15
$p(x)$ section illustrating the partial miscibility. (a): $(1-x)C_6H_{12} + xCH_2I_2$ showing a UCST. o: experimental values;—: estimated. (b): $(1-x)H_2O + x\, CH_3(C_2H_5)_2N$ illustrating the LCST. o: experimental values;—, estimated.

The condition for coexisting phases α and β is given by

$$\ln\left\{\frac{x_A^\alpha}{x_A^\beta}\right\} = \frac{(x_B^\beta - x_B^\alpha)(x_B^\beta + x_B^\alpha)w}{k_B T} \tag{4.223}$$

and

$$\ln\left(\frac{x_B^\alpha}{x_B^\beta}\right) = \frac{(x_B^\beta - x_B^\alpha)(x_B^\beta + x_B^\alpha - 2)w}{k_B T}. \tag{4.224}$$

Equations 4.223 and 4.224 do not represent well the measured properties of real mixtures, but they nevertheless provide estimates that are qualitatively correct.

4.9.4 When Do Critical Points in Liquid Mixtures Occur?

The criteria for a critical point of a mixture are the same for both (vapor + liquid) equilibria and solutions. Thus, the critical temperature T_c is always defined by

$$\left(\frac{\partial^2 G_m}{\partial x^2}\right)_{T,p} = 0, \tag{4.225}$$

$$\left(\frac{\partial^3 G_m}{\partial x^3}\right)_{T,p} = 0 \tag{4.226}$$

and

$$\left(\frac{\partial^4 G_m}{\partial x^4}\right)_{T,p} > 0. \tag{4.227}$$

For the simple mixture defined by Equation 4.214

$$\begin{aligned} G_{m,l}(T, p, x_A) &= x_A G_A^*(T, p) + (1 - x_A)G_B^*(T, p) \\ &\quad + RT\,[x_A \ln x_A + (1 - x_A)\ln(1 - x_A)] + x_A(1 - x_A)Lw(T, p). \end{aligned} \tag{4.228}$$

Thus

$$\left(\frac{\partial G_{m,l}}{\partial x_A}\right)_{T,p} = G_A^* - G_B^* + RT\ln\left[\frac{x_A}{1 - x_A}\right] + (1 - 2x_A)Lw \tag{4.229}$$

and

$$\left(\frac{\partial^2 G_{m,l}}{\partial x_A^2}\right)_{T,p} = \frac{RT}{x_A(1 - x_A)} - 2Lw. \tag{4.230}$$

Thus, $\left(\frac{\partial^2 G_{m,l}}{\partial x_A^2}\right)_{T,p} > 0$ and the single liquid phase is the equilibrium state if

$$T > \frac{2x_A x_B w}{k_B}. \tag{4.231}$$

On the other hand, when $\left(\frac{\partial^2 G_{m,l}}{\partial x_A^2}\right)_{T,p} < 0$ phase separation occurs if

$$T < \frac{2x_A x_B w}{k_B}, \tag{4.232}$$

the limit of stability occurs when

$$T = \frac{2x_A x_B w}{k_B}, \tag{4.233}$$

and the highest temperature at which phase separation occurs for any composition is when $x_A = x_B = 0.5$ and this is the upper critical solution temperature T_{uc} given by

$$T_{uc} = \frac{w}{2k_B}. \tag{4.234}$$

So T_{uc} only exists if $w > 0$. This is a useful approximation for estimating the conditions under which a UCST will occur.

For associating liquid and fluid mixtures, that is, those mixtures that form compounds through, for example, hydrogen bonding, the reader should refer to the methods of SAFT (see Chapter 2) and the so-called Cubic Plus Association equation of state (Economou 2010).

4.10 What Particular Features Do Phase Equilibria Have?

Excess molar functions, for example, as given in the simplified forms of Equations 4.157 to 4.159, are useful for mixtures of liquids of similar volatility at pressures that are about $p^\ominus = 0.101325$ MPa ≈ 0.1 MPa and do not exceed $2p^\ominus$. For a mixture of liquids of similar volatility at higher pressure p, however, a virial expansion is inadequate. Unfortunately, in just those cases, insufficient information is usually available to determine an equation of state for the coexisting gas phase. This means it is difficult to use Equation 4.127 and these circumstances severely limit the temperature range over which G_m^E can be determined. Furthermore, the activity coefficient includes the absolute activity of each pure substance and that requires it to be a liquid at the relevant temperature and pressure. Thus, for mixtures of substances of very different volatility, it is possible that one component at the relevant temperature and pressure may be either a gas or a solid. Evidently in such cases the approach of activity coefficients is rather difficult to apply. For example, at $T = 300$ K for $(1 - x)C_6H_6 + xN_2$ nitrogen is a gas, and it would be necessary to use an equation of state method for it. In the case of $(1 - x)C_6H_6 + xC_{14}H_{10}$ (anthracene), the modified-UNIFAC method can be employed.

4.10.1 What Is a Simple Phase Diagram?

Phase diagrams for mixtures are at least three dimensional (p, T, x). These are usually shown as two-dimensional projections of $p(T)$. In this case, the pressure as a function of temperature, $p(T)_x$ at constant composition (these are called isopleths) would reveal the dew and bubble pressure that meet on the critical line, $p(x)_T$ that are isotherms and $T(x)_p$ isobars. For a $p(x)$ diagram, it is possible to show several x's and the critical line is then the locus of the maxima of $p(x)$ isothermal sections. For a particular temperature, the lines joining the mole fractions of the coexisting fluid phases are called tie lines and, in very simple mixtures, define two curves one for the gas the other the liquid. Here we will restrict comments to those dealing with fluid phases and exclude the formation of solids.

For a binary mixture, the vapor + liquid phase equilibria is simple when the critical points of the two pure substances are joined by a continuous curve and there is neither azeotropy (Question 4.10.4) nor three fluid phases.

4.10.2 What Is Retrograde Condensation (or Evaporation)?

Retrograde condensation is a phenomenon that occasioned great surprise when it was first observed as recorded by McGlashan (1979) because it seems to contradict the intuitive notion that an increase of pressure leads to a reduction in volume in a system. The phenomenon can be illustrated by reference to Figure 4.16, which shows the high-pressure region of the phase diagram for mixtures of argon and krypton [$(1 - x)$Ar + xKr] for a temperature T = 177.38 K. The curve labeled "g" in the diagram represents the dew curve for the mixture outside of which a gaseous mixture exists while inside it a two-

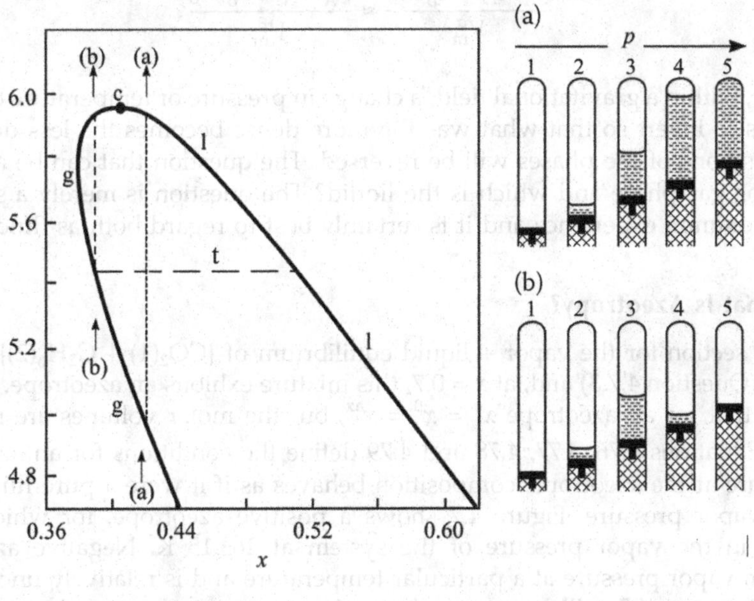

FIGURE 4.16
LEFT: The $p(x)_T$ section at T = 177.38 K for $(1 - x)$Ar + xKr illustrates retrograde condensation. c denotes the critical point; l is the bubble curve at $x \geq x_c$; and g labels the dew curve. RIGHT: Illustrates the relative volumes of liquid and gas obtained for changing pressure with a piston for the paths along the dotted lines (a) for the upper set and along the dashed lines (b) for the lower set.

phase system of gas and liquid exists. The curve labeled "l" marks the boundary between a liquid mixture and the two-phase region. At a given pressure, the line labeled "t" is a tie line that connects the gas and liquid phase compositions that are in equilibrium at that pressure and the specified temperature. The point designated (c) is the critical point for the mixture at this particular temperature. If a gas mixture with a composition of $x = 0.42$ designated by (a) is compressed from 4.8 MPa by means of a piston, as illustrated in schematic diagrams (a) to the right, then it will follow the dotted path shown in the figure. We begin at (a)1 where we are below the dew curve and the system is entirely in the gas phase. As the pressure is increased, we cross the dew curve, g, when condensation from vapor to liquid begins as illustrated at (a)2, as we progress to higher pressures, more condensation occurs, as indicated in (a)3 and 4. Eventually at (a)5 we cross the liquid line, l, and all the gas disappears. On the other hand, if we begin the same process with a composition $x = 0.39$, then we follow the second dashed line and the schematic diagrams (b) starting from (b)1 where the system is entirely gaseous. Again, we intersect the vapor line, g, at (b)2 and condensation begins so some liquid forms. As the pressure increases, more liquid forms as in (b)3; however, as the pressure further increases, the liquid volume decreases as in (b)4 until the vapor line is reached again where all the liquid disappears and there is again only a single vapor phase. This is the phenomenon of retrograde condensation.

4.10.3 What Is the Barotropic Effect?

This occurs when a mixture of two substances is at a temperature and pressure such that the molar mass of the pure substances and the molar volumes of the coexisting phases give nearly equal mass densities for the phases. This means that

$$\frac{x_A^\alpha M_A + x_B^\alpha M_B}{V_m^\alpha} \approx \frac{x_A^\beta M_A + x_B^\beta M_B}{V_m^\beta}. \tag{4.235}$$

In this case, within a gravitational field, a change in pressure or temperature can cause the two phases to invert so that what was the more dense becomes the less dense and the vertical positions of the phases will be reversed. The question that can be asked then is which is the gas phase and which is the liquid? The question is merely a semantic one based on common experience and it is certainly best to regard both as fluid phases.

4.10.4 What Is Azeotropy?

The $(p, x)_T$ section for the vapor + liquid equilibrium of $\{CO_2(1) + C_2H_6(2)\}$ is shown in Figure 4.7 (Question 4.7.3) and, at $x = 0.7$, this mixture exhibits an azeotrope. We see from Figure 4.7 that for an azeotrope $x^\alpha = x^\beta = x^{az}$, but the molar volumes are not identical, $V_m^\alpha \neq V_m^\beta$. Equations 4.76, 4.77, 4.78 and 4.79 define the conditions for an azeotrope. The fluid mixture at the azeotropic composition behaves as if it were a pure fluid and it has a unique vapor pressure. Figure 4.7 shows a positive azeotrope, for which there is a maximum in the vapor pressure of the system at 263.15 K. Negative azeotropy has a minimum vapor pressure at a particular temperature and is relatively uncommon. The diagram of Figure 4.7 will be repeated at other temperatures and thus an azeotropic line (the line joining azeotropic points) can persist to the critical line. However, this is not always the case and, when it is not, the maximum of the curve occurs at a mole fraction that attains $x_B = 1$ at a temperature below the critical temperature of pure substance B.

4.11 What Are Solutions?

Chapter 1 defines a solution as a mixture for which it is convenient to distinguish between the solvent and the solutes. The amount of substance of solvent is often much greater than that of the solutes and this is called a dilute solution. It is usual in solutions to use molality m_B of a solute B rather than mole fraction x_B in a solvent A of molar mass M_A where the molality is given by

$$x_B = \frac{M_A m_B}{1 + M_A \sum_B m_B} \tag{4.236}$$

and

$$m_B = \frac{x_B}{M_A (1 - \sum_B x_B)}. \tag{4.237}$$

4.11.1 What Is the Osmotic Coefficient of a Solvent?

The osmotic coefficient ϕ_m of a solvent, A, is defined by

$$\phi_m = \frac{\mu_A^*(T, p) - \mu_A(T, p, x_C)}{RTM_A \sum_{B \neq A} m_B}, \tag{4.238}$$

in terms of chemical potential and the molality, m. The chemical potential of the solvent in the solution is related to the fugacity in a gas phase with which it is in equilibrium by

$$\mu_{A,l}(T, p, x_C) = \mu_{A,g}^{ref} + RT \ln\left(\frac{\tilde{p}_{A,g}(T, p, x_C)}{p}\right), \tag{4.239}$$

while for the pure solvent

$$\mu_{A,l}^*(T, p) = \mu_{A,g}^{ref} + RT \ln\left(\frac{\tilde{p}_{A,g}^*(T, p)}{p}\right). \tag{4.240}$$

Subtraction of these two equations leads to

$$\mu_{A,l}^*(T, p) - \mu_{A,l}(T, p, x_C) = RT \ln\left(\frac{\tilde{p}_{A,g}^*(T, p)}{\tilde{p}_{A,g}(T, p, x_C)}\right), \tag{4.241}$$

which shows how ϕ can be obtained from measurements of the fugacity.

Now for a binary solution we have

$$\phi_m = \frac{\mu_A^*(T,p) - \mu_A(T,p,x_A)}{RTM_A m_B}. \tag{4.242}$$

Solving for μ_A and differentiating at constant T and p we obtain

$$d\mu_A = -RTM_A(\phi_m dm_B + m_B d\phi_m). \tag{4.243}$$

Using

$$\mu_{B,1}(T,p) = \mu_{B,1}^{\ominus} + RT\ln\left(\gamma_{m,B}(T,p)\frac{m_B}{m^{\ominus}}\right), \tag{4.244}$$

we obtain

$$d\mu_B = RT\left(d\ln\gamma_{m,B} + \frac{dm_B}{m_B}\right). \tag{4.245}$$

We substitute these expressions for $d\mu_A$ and $d\mu_A$ into the Gibbs–Duhem equation in the form $n_A d\mu_A + n_B d\mu_B = 0$ and use the fact that $n_A M_A = n_B/m_B$ and then rearrange to obtain

$$d\{(1-\phi_m)m_B\} + m_B d\ln\gamma_{m,B} = 0. \tag{4.246}$$

Integration over m_B then results in

$$\ln\gamma_{m,B}(m_B) = (\phi_m - 1) + \int_0^{m_B}\frac{(\phi_m - 1)}{m_B}dm_B. \tag{4.247}$$

Thus, once ϕ_m has been measured as a a function of molality from zero up to the molaity of interest, this equation can be used to evaluate the activity coefficient $\gamma_{m,B}$ at that same molality.

4.11.2 What Are Henry's Law and Infinite Dilution?

For a solution using molality as a basis, we write the chemical potential of a solute as

$$\mu_{B,1}(T,p,m_C) = \mu_{B,1}^{\text{ref}}(T,p) + RTln\left(\gamma_{m,B}\frac{m_B}{m^{\ominus}}\right), \tag{4.248}$$

where the subscript m on the activity denotes the molality basis. For a solution in equilibrium with a vapor phase, we can then write

$$\mu_B(T,p,m_{B,1}) = \mu_{B,1}^{\text{ref}}(T,p) + RT\ln\left(\gamma_{m,B}\frac{m_B}{m^{\ominus}}\right) = \mu_{B,g}^{\text{ref}}(T,p) + RT\ln\frac{\tilde{p}_{B,g}}{p} \tag{4.249}$$

or

$$\gamma_{m,B} m_B = m^{\ominus} \frac{\tilde{p}_B}{p} \exp\left(\frac{\mu_{B,l}^{ref}(T,p) - \mu_{B,g}^{ref}(T,p)}{RT} \right). \tag{4.250}$$

Now we can evaluate the coefficient of $\tilde{p}_B$ on the right-hand side by considering the limit as m_B approaches zero when the activity coefficient approaches unity, and the ratio $\tilde{p}_B/m_B$ approaches the Henry's law constant, H_m.

Thus

$$\gamma_{m,B} = \tilde{p}_{B,g}/H_m m_B, \tag{4.251}$$

which reveals how the activity coefficient of the solute behaves as a function of molality on the assumption that the infinitely dilute behavior extends to non-zero molalities. It is obeyed over only a short range of molalities for most systems.

Of course, in the limit of infinite dilution, the activity coefficient is unity, and we recover exactly Henry's law:

$$\tilde{p}_{B,g} = H_m m_B. \tag{4.252}$$

4.11.3 What Is the Minimum Work Required for the Desalination of Seawater?

In recent decades, the steadily growing world population and the increasing scarcity of natural freshwater reserves have led to a continuous demand for seawater desalination plants. In 2018, the world's production of desalinated water was more than 95 million m^3/day (Jones et al. 2019). There is therefore very considerable interest in the energy that is required to achieve this desalination and it is a topic where thermodynamics can contribute useful insight. Here we can only give an indication of how that can be achieved but it is hoped the section will illustrate the power of thermodynamics in the field of energy utilization.

Natural seawater exhibits a wide range of salinities, i.e. mass fractions of dissolved salts, ranging from below 8 g/kg (Baltic Sea) to above 40 g/kg (Red Sea) while the global mean value is about 35 g/kg. Depending on the location, the composition also varies, but the dominant component is always sodium chloride (NaCl). To enable a comparison of the thermophysical properties of seawater, relevant, e.g. for the design of desalination plants, a reference seawater composition was defined (Millero et al. 2008). This definition is derived from the composition of "standard seawater", which traditionally was obtained from North Atlantic surface water, and contains the following approximate mass fractions of major constituents: Cl^- 1.9%, Na^+ 1.1%, SO_4^{2-} 0.3%, Mg^{2+} 0.1%.

At first sight, this question is similar to Question 3.9, dealing with the minimum amount of work for the separation of air into its constituents, yet there is a major difference, namely that here not a true separation of seawater into its constituents (various salts and pure water) is required but, practically, the only desired product is pure water, while brine, containing a higher salt concentration, is rejected to the sea. We could again, as in Question 3.9, calculate an exergy loss in order to determine the minimum work to obtain the isolated constituents. The exergy loss is, of course, correlated with the generation of entropy but, in this instance, the calculation is more complicated because we not only have to consider the entropy of mixing, but also the entropy of solution. This approach to the evaluation was considered by Spiegler and El-Sayad (2001).

Here we take an approach different to that in Question 3.9 and look at two of the basic processes for desalination that are also used in practical desalination plants though, of course, in a rather more sophisticated manner than we set out here. We base our calculations on an average salt concentration of 35 g/kg and assume NaCl to be the only constituent. The actual composition of seawater enters twofold into the calculation, once through the molar mass of the constituents and secondly through the number of ions produced in solution, which is (approximately) $i = 2$ for NaCl. Because sodium chloride is the dominant component, this is a rather good approximation, and the calculation may directly be expanded by using the exact composition.

The first example considers the process of reverse osmosis, RO (cf. Question 4.4.3), which uses a partially permeable membrane and where a certain pressure must be applied to overcome the osmotic pressure Π.

According to van't Hoff, the osmotic pressure may be calculated by (Atkins et al., 2018)

$$\Pi = icRT \approx 3.0 \text{ MPa}, \tag{4.253}$$

where c is the molar concentration ($\approx$ 0.61 kmol/m^3) of NaCl and a temperature T of 298.15 K is assumed. The work required to obtain 1 m^3 of pure water, assuming the idealized case when Π does not increase during the process, i.e. that there is a continuous supply of seawater, then amounts to

$$W = \Pi V = 3.0 \text{ MPa } 1 \text{ m}^3 = 3.0 \text{ MJ} \approx 0.8 \text{ kWh}. \tag{4.254}$$

When taking real osmotic pressures for seawater, exhibiting a more complicated composition, a somewhat lower value of 2.5 MJ results (Spiegler and El-Sayad 2001).

Another group of desalination techniques relies on distillation, where a conceptually straightforward approach is that of mechanical vapor compression, MVC (see Figure 4.17). As compared to pure water, salt water has a lower vapor pressure by an amount $|\Delta p^{\text{sat}}|$ (cf. Question 4.6.4), which is related to the vapor pressure of water at the same temperature, p^{sat}, as

$$\Delta p^{\text{sat}} = -i\, x_{\text{NaCl}} p^{\text{sat}}, \tag{4.255}$$

where x_{NaCl} is the mole fraction of sodium chloride. The amounts of substance for 1 kg of salt water are

$$n_{\text{NaCl}} = 35.00 \text{ g}/(58.44 \text{ g/mol}) = 0.599 \text{ mol}$$

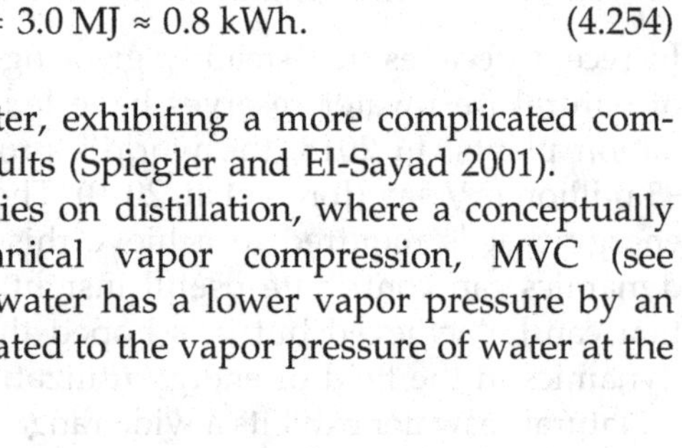

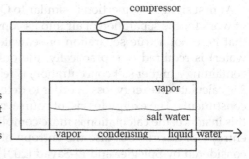

FIGURE 4.17
Principle of mechanical vapor compression: water vapor is compressed to a higher pressure (and temperature) just enough for evaporating salt water; because of the associated heat transfer the vapor condenses and leaves as desalted water.

and

$$n_{H2O} = 965.0\,\text{g}/(18.02\,\text{g/mol}) = 53.6\,\text{mol},$$

resulting in $x_{NaCl} = 0.011$ and

$$\Delta p^{sat} = -2 \times 0.011\, p^{sat} = -0.022 p^{sat}.$$

Now, for simplicity, we consider the limiting case when this reduced pressure is exactly compensated by a mechanical compressor, so the vapor within the pipe condenses when evaporating salt water. The shaft work required for isothermal compression can be calculated as (cf. Equation 1.94)

$$(W_s)_{rev}^{isoth.} = n \int_{p^{sat} - |\Delta p^{sat}|}^{p^{sat}} V_m\, dp = nRT \ln\left(\frac{p^{sat}}{p^{sat} - |\Delta p^{sat}|}\right) = -nRT \ln\left(1 - \frac{|\Delta p^{sat}|}{p^{sat}}\right)$$

$$\approx nRT \frac{|\Delta p^{sat}|}{p^{sat}}. \qquad (4.256)$$

When entering the respective numbers from above, a work of $W_s = 3.0$ MJ for the provision of 1 m^3 of pure water, containing $n = 54.9$ kmol, results, in accordance with the result above. Thus, irrespective of the actual underlying processes, the theoretical minimal amounts of energies for seawater desalination are identical. While desalination plants based on both principles are in operation worldwide, the last decade has seen a strong increase in the market share of RO plants, which has been made possible through major improvements in membranes and "energy recovery" to partially reuse the energy required for the provision of high pressure.

As well as the two types of processes we have considered explicitly, there are two competing types of plants in operation with a market share that is significant and they mainly use thermal energy. They are based on multi-stage flash (MSF) or multi-effect distillation (MED) principles. Both use steam at a certain temperature level (~120°C for MSF, ~80°C for MED), which is usually provided in the form of bleed steam from conventional power plants. An interesting feature of these desalination plants is that "renewable energy sources", such as concentrated solar radiation, may be used as driving sources of operation. Both of these types of seawater desalination plants use various consecutive stages at different temperature and pressure levels to significantly reduce the high energy demand to evaporate water. In MSF plants there are a number of flash stages, part of the seawater entering a chamber at lower pressure is flash-evaporated and the resulting steam is used to preheat the incoming seawater. In MED plants, seawater is trickled over numerous vapor-heated horizontal tubes, and the vapor produced serves as heating steam for the next effect.

Having looked at the various types of plant and the minimum energy required for the desired separation, it is interesting to compare that figure with the energy actually required to achieve the desalination in practice. Table 4.1 gives the results of a survey of rough numbers of the energy demands of actual plants on the market for the various principles together with their market shares. The electric and thermal energies provided are mean values from literature sources (Stillwell and Webber 2016; Alkaisi et al. 2017), which each also give a quite wide range of values. Thermal desalination plants

TABLE 4.1

Overview of energy requirements and market shares of desalination technologies. Electric and thermal energies are mean values from literature (Stillwell and Webber, 2016; Alkaisi et al., 2017). The total electric equivalent is approximated using the exergy of thermal energy. Market shares of produced water are from Jones et al. (2019). RO: reverse osmosis, MVC: mechanical vapor compression, MSF: multi-stage flash, MED: multiple-effect distillation

Technology	Specific energy requirement / (MJ/m³)			Market share
	electric	thermal	total "electric equivalent"	(2018)
RO	11–23	–	11–23	69%
MVC	25–49	–	25–49	N/A
MSF	12–20	170–410	42–91	13%
MED	6–9	170–410	25–56	7%

normally use bleed steam from power plants, which is then not available for the generation of electrical energy. This reduction in the production of electric energy would be the most appropriate number to compare the thermal energy expended to an electrical equivalent. However, this amount very much depends on the specific configuration; in any event, the calculation is not straightforward, see, e.g. Altmann et al. (2019). To construct Table 4.1, a simplified approach has been taken: for both MSF and MED the exergies of saturated steam at the respective temperature levels have been calculated according to Equation 3.174 but with an assumed factor of 0.8 for the exergetic efficiency, Equation 3.180. While all numbers for energies in Table 4.1 only have indicative character, it becomes obvious that all realized processes, and especially those based on thermal principles, need a large multiple of the theoretical minimum energy because of inevitable losses occurring in practice. For example, there is scaling and fouling in these plants, for RO these losses are also related to necessarily imperfect membranes, for thermal processes to the necessity of maintaining a certain temperature difference for driving heat transfer and to thermal losses.

4.12 What Happens at Interfaces Between Phases?

We have so far considered the properties of individual phases, gas, liquid or solid, but there is also considerable interest and importance for the boundary between the phases, where one phase stops and another phase begins: the interface. Here we consider the gas-liquid interface mainly with a brief digression to answer one question about the liquid solid interface. First we note that liquids always seek to minimize the total energy of a free system of liquid molecules and this is is less if many of them are in the bulk of the liquid than on the surface. This means that in the absence of any other forces, for example in the microgravity environment on the space station, droplets of liquid are spherical because this shape has the smallest surface area for a given volume. On earth, where gravitational forces are stronger, the spherical shape is distorted, but the shape is always affected by the aim to reach the minimum surface area.

In thermodynamics, surface effects are described using chemical potentials. Changing the area of a surface that is part of a thermodynamic system and encloses it involves doing work. It is therefore another contribution to the work done on a system dW (Question 1.3.6). The work involved in creating a surface dA is proportional to the area of the surface and

$$dW^{surf} = \gamma dA, \tag{4.257}$$

where γ is the coefficient of proportionality known as *the interfacial tension* in general or, in the particular gas of the vapor-liquid interface, *the surface tension*.

The work of forming a surface is additional to all other kinds of work and so it must be regarded as a contribution to the Gibbs function (or if expressed per mole, the chemical potential) for a system. For a pure component therefore, the change of Gibbs energy (Equation 4.3.3) must be written

$$dG^{\Sigma} = -S^{\Sigma}dT + V^{\Sigma}dp + \gamma dA. \tag{4.258}$$

Here we see that for constant temperature and pressure a smaller Gibbs energy is obtained if the area of an interface is decreased. The condition for a spontaneous change, $dG < 0$ means that $d\sigma < 0$ so that surfaces have a natural tendency to contract, as indicated above.

4.12.1 What Happens with Bubbles and Drops?

For our purposes here, bubbles are volumes of vapor (often air and vapor) that are encased in a thin film of liquid or cavities full of vapor in a liquid. These two cases are distinguished by the fact that the first kind of bubbles have two interfaces, whereas the cavities have just one. This factor of two between the two cases is important. Liquid drops on the other hand are ideally spheres of a liquid in equilibrium with its vapor.

Bubbles are at equilibrium because the drive to reduce the surface area of the interface by contraction of the bubble is balanced by the increasing pressure inside the closed system that this causes. We first write a force balance for a spherical cavity containing phase α separated by an interface from a phase β

$$4\pi r^2 p^{\alpha} = 4\pi r^2 p^{\beta} + 8\pi\gamma r, \tag{4.259}$$

where r is the radius of the cavity and p^{α} and p^{β} the pressures in the respective phases. The surface tension produces an energy $4\pi r^2\gamma$ for a spherical bubble and to find the force we evaluate the work required to stretch this surface through dr, which is just $d(\gamma A) = 8\pi\gamma r dr$ so that the force is $8\pi\gamma r$. From Equation 4.259, we then find that

$$p^{\alpha} - p^{\beta} = 2\gamma/r, \tag{4.260}$$

which is known as the Laplace equation.

When the bubble has vapor or gas as phase α, and phase β is just a thin layer of liquid with again phase α beyond it (a child's bubble-blowing is an example), then there are two vapor-liquid surfaces of essentially the same radius and the pressure difference across the film is $4\gamma/r$.

4.12.2 What Is Capillary Rise?

Consider the situation in Figure 4.18 where phase α is a liquid and has a tendency to stick to the walls of a glass capillary tube of radius r as a result of the attractive forces between the solid wall and the liquid molecules. The system is surrounded by phase β, which is a gas. The forces between the molecules of the solid wall and those of the liquid act to draw the liquid up the solid wall, while the attractive forces between the molecules of the liquid retain the surface intact. As a result, a thin film of liquid climbs the wall of the tube and, as it does so, it has the effect of curving the surface of the phase α inside the tube (we assume the surface to be a hemisphere of radius r). This means that the pressure just under the curving meniscus in phase α, p_α, is less than the pressure in phase β by approximately $2\gamma/r$. Thus, there is an extra force acting externally that causes the liquid phase α to rise in the capillary tube until the weight of the supported column of phase α is equal to the force acting on the column from the interfacial tension so that

$$2\gamma/r = (\rho^\alpha - \rho^\beta)\pi r^2 gh/\pi r^2 = (\rho^\alpha - \rho^\beta)gh. \quad (4.261)$$

This simple configuration and result offer a remarkably accurate method of measuring the surface tension of liquids.

The result can be generalized to any two fluid phases and to cases where phase α, rather than being tangential to the glass wall as we have discussed so far, makes an angle θ with the solid surface. In this more general case, we have to allow for the fact that only the vertical component of the force due to surface tension moves the liquid in the tube and we find

$$2\gamma \cos\theta/r = (\rho^\alpha - \rho^\beta)gh. \quad (4.262)$$

In the unusual, but observable case, when $\theta > \pi/2$, the surface of the fluid in the tube is convex and $h < 0$, so that the surface in the capillary will be below the surface of the bulk phase; one such case is when phase α is mercury.

4.12.3 Why Does Rain Bead on a Freshly Polished Car?

The application of special coatings on a surface can be used to alter the chemical characteristics and change the contact angle θ, between a liquid and a solid surface. An example

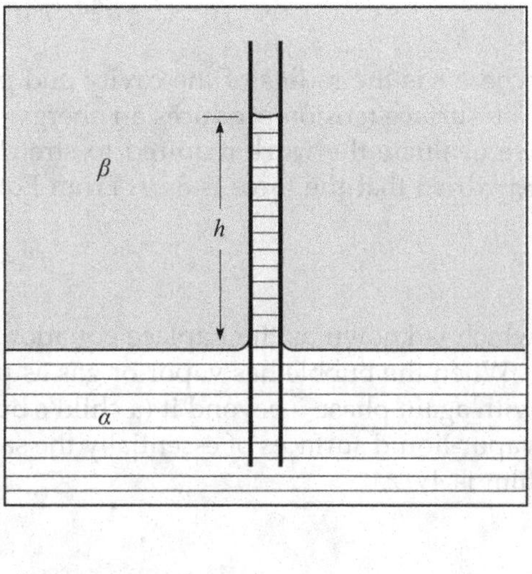

FIGURE 4.18
Capillary-rise method to measure interfacial tension.

of this is the effect of polish on the painted surface of a car. When there is no polish, the water of a rain drop, with a contact angle of zero, will cause the drop to spread over the entire surface in a thin layer as it seeks to minimize its surface energy. On the other hand, a surface coating of polish modifies the surface chemistry and changes the contact angle to near $\pi/2$. The minimum energy configuration is then the near-spherical drop discussed above as modified by gravity and it constitutes a bead of water on the solid surface.

References

Abrams D.S., and Prausnitz J.M., 1975, "Statistical thermodynamics of liquid mixtures: A new expression for the excess Gibbs energy of partly or completely miscible systems", *AIChE J.* **21**:116–128.

Al-Matar A., and Rockstraw A., 2006, "Assessment of the effect of mixing rules on predicting the second virial coefficient and a further evidence of the Inadequacy of the Lorentz-Berthelot rules", *Dirasat Eng. Sci.* **33**:27–36.

Alkaisi A., Mossad R., and Sharifian-Barforoush A., 2017, "A review of the water desalination systems integrated with renewable energy", *Energy Procedia* **110**:268–274.

Altmann T., Robert J., Bouma A., Swaminathan, J., and Lienhard V.J.H., 2019, "Primary energy and exergy of desalination technologies in a power-water cogeneration scheme", *Appl. Energy* **252**: 113319.

Assael M.J., Trusler J.P.M., and Tsolakis Th.F., 1996, Thermophysical Properties of Fluids. *An Introduction to Their Prediction* 1st Edition, Imperial College Press, London.

Assael M.J., Wakeham W.A., Goodwin A.R.H., Will S., and Stamatoudis M., 2011, *Commonly Asked Questions in Thermodynamics*, CRC Press., Boca Raton, U.S.A.

Atkins, P. W., de Paula, J., and Keeler, J., 2018, *Atkins' Physical Chemistry*, 11th ed., Oxford University Press, Oxford.

Barton A.F.M., 1991, *CRC Handbook of Solubility Parameters and Other Cohesion Parameters*, 2nd ed., CRC Press, Florida.

Behnejard H., Sengers J.V., and Anisimov M.A., 2010, *Thermodynamic Behavior of Fluids Near Critical Points, Ch. 10, in Experimental Thermodynamics Volume VII: Applied Thermodynamics of Fluids*, eds. Goodwin A.R.H., Sengers J.V., and Peters C.J., for IUPAC, Royal Society of Chemistry, Cambridge, UK.

Berthelot D., 1898, "Sur le Mélange des Gaz", *C. R. Acad. Sci.* (Paris) **126**:1703.

Carnahan N.F., and Starling K.E., 1972, "Intermolecular repulsions and the equation of state for fluids", *AIChE J.* **18**:1184–1189.

Colbeck S.C., 1995, "Pressure melting and ice skating", *Am. J. Phys.* **63**:888–890.

Dalarsson N., Dalarsson M., and Golubovic L., 2011, *Introductory Statistical Thermodynamics*, Elsevier, USA.

Dash J.G., Rempel A.W., and Wettlaufer J.S., 2006, "The physics of premelted ice and its geophysical consequences", *Rev. Mod. Phys.* **78**:695–741.

de Nevers N., 2012, *Physical and Chemical Equilibrium for Chemical Engineers*, 2nd ed., Wiley, New Jersey.

Delhommelle J., and Millié P., 2001, "Inadequacy of the Lorentz-Berthelot combining rules for accurate predictions of equilibrium properties by molecular simulation", *Mol. Phys.* **99**:619–625.

Economou I.G., 2010, *Cubic and Generalized van der Waals Equations of State, Chapter 4, Experimental Thermodynamics Volume VIII: Applied Thermodynamics of Fluids*, eds. Goodwin A.R.H., Sengers J.V., and Peters C.J., for IUPAC, Royal Society of Chemistry, Cambridge, UK.

Feistel R., and Wagner W., 2006, "A new equation of state for H_2O ice Ih", *J. Phys. Chem. Ref. Data* **35**:1021–1047.

Fredenslund A., Gmehling J., and Rasmussen P., 1977, *Vapor-Liquid Equilibria Using UNIFAC*, Elsevier, Amsterdam.

Fredenslund A., Jones R.L., and Prausnitz J.M., 1975, "Group-contribution estimation of activity coefficients in nonideal liquid mixtures", *AIChE J.* **21**:1086–1099.

Fredenslund A., and Mollerup J., 1974, "Measurement and prediction of equilibrium ratios for the C2H6+CO2 system", *J. Chem. Soc. Faraday Trans. I* **70**:1653–1660.

Friese T., Ulbig P., Schulz S., and Wagner K., 1998, "Effect of NaCl on the excess enthalpies of binary liquid systems," *Thermochim. Acta* **310**:87–94.

Friese T., Ulbig P., Schulz S., and Wagner K., 1999, "Effect of NaCl or KCl on the excess enthalpies of alkanol plus water mixtures at various temperatures and salt concentrations", *J. Chem. Eng. Data* **44**:701–714.

Goodwin A.R.H., and Ambrose D., 2005, *Vapor Pressure, in Encyclopedia of Physics, Volume* 2, 3rd ed., eds. Lerner R.G., and Trigg G.L., Wiley-VCH, Berlin, pp. 2846–2848.

Goodwin A.R.H., Marsh K.N., and Wakeham W.A., Eds., 2003, *Experimental Thermodynamics, Volume VI, Measurement of the Thermodynamic Properties of Single Phases*, for IUPAC, Elsevier, Amsterdam.

Goodwin A.R.H., and Sandler S.I., 2010, *Mixing and Combining Rules, Ch. 5, in Experimental Thermodynamics Volume VIII: Applied Thermodynamics of Fluids*, eds. Goodwin A.R.H., Sengers J.V., and Peters C.J., for IUPAC, Royal Society of Chemistry, Cambridge, UK.

Guggenheim E.A., 1985, *Thermodynamics: An Advanced Treatment for Chemists and Physicists*, 7th ed., North-Holland Publishing Company, Amsterdam.

Haslam A.J., Galindo A., and Jackson G., 2008, "Prediction of binary intermolecular potential parameters for use in modelling fluid mixtures", *Fluid Phase Equilib.* **266**:105–128.

Heideman R., and Fredenslund Aa., 1989, "Vapor-liquid equilibria in complex mixtures", *Chem. Eng. Res. Des.* **67**:145–158.

Helpinstill J.G., and van Winkle M., 1968, "Prediction of infinite dilution activity coefficients for polar-polar binary systems", *Ind. Eng. Chem. Proc. Des. Dev.* **7**:213–220.

Henderson D., and Leonard P.J., 1971, *Liquid Mixtures, in Physical Chemistry -an Advanced Treatise, Volume 8B, The Liquid State*, eds. Eyring H., Hederson D., and Jost W., Academic Press, New York.

Hirschfelder J.O., Curtis C.F., and Bird R.B., 1954, *Molecular Theory of Gases and Liquids*, Wiley, New York.

Jones E., Qadir M., van Vliet M.T., Smakhtin V., and Kang, S.M., 2019, "The state of desalination and brine production: A global outlook", *Sci. Total Environ.* **657**:1343–1356.

Kojima K., and Toshigi K., 1979, *Prediction of Vapor-LiquidEquilibria by the ASOGMethod, Physical Sciences Data 3*, Elsevier Publishing Company, Tokyo.

Kontogeorgis G., and Folas G., 2010, *Thermodynamic Models for Industrial Applications: From Classical and Advanced Mixing Rules to Association Theories*, Wiley, Chichester.

Lee E.T., and Wang J.W., 2003, *Statistical Methods for Survival Data Analysis*, 3nd ed., Wiley, New Jersey.

Lemmon E.W., and Span R., 2010, *Multi-parameter Equations of State for Pure Fluids and Mixtures, Chapter 12, Experimental Thermodynamics Volume VIII: Applied Thermodynamics of Fluids*, eds. Goodwin A.R.H., Sengers J.V., and Peters C.J., for IUPAC, Royal Society of Chemistry, Cambridge, UK.

Lennard-Jones J.E., 1931, "Cohesion", *Proc. Phys. Soc.* **43**:461–482.

Leonhard K., Nguyen V.N., and Lucas K., 2007, "Making equation of state models predictive—Part 2: An improved PCP-SAFT equation of state," *Fluid Phase Equilib.* **258**:41–50.

London F., 1937, "The general theory of molecular forces", *Trans. Faraday Soc.* **33**:8–26.

Lorentz H.A., 1881, "Über die Anwendung des Satzes vom Virial in der kinetischen Theorie der Gase", *Ann. Phys.* **12**:127–136.

Maitland G.C., Rigby M., Smith E.B., and Wakeham W.A., 1981, *Intermolecular Forces: Their Origin and Determination*, Clarendon Press, Oxford.

Margules S., 1895, "On the composition of the saturated vapours of mixtures", *Math. Naturw. Akad. Wiss. (Vienna)* **104**:1243–1249.

McCabe C., and Galindo A., 2010, *SAFT Associating Fluids and Fluid Mixtures, Ch. 8, Experimental Thermodynamics Volume VIII: Applied Thermodynamics of Fluids*, eds. Goodwin A.R.H., Sengers J.V., and Peters C.J., for IUPAC, Royal Society of Chemistry, Cambridge, UK.

McGlashan M.J., 1979, *Chemical Thermodynamics* **vol 2**, Academic Press, London.

Millero F.J., Feistel R., Wright D.G., and McDougall T.J., 2008, "The composition of Standard Seawater and the definition of the reference-composition salinity scale", *Deep Sea Res. Pt. I* **55**:50–72.

Narayanan K.V., 2013, *A Textbook of Chemical Engineering Thermodynamics*, 2nd ed., PHI Learning Private Limited, Delhi.

Peng D.Y., and Robinson D.B., 1976, "A new two-constant equation of state", *Ind. Eng. Chem.* Fundam. **15**:59–64.

Pierotti G.J., Deal C.H., and Derr E.L., 1959, "Activity coefficients and molecular structure," *Ind. Eng. Chem.* **51**:95–102.

Poling B.E., Prausnitz J.M., and O'Connell J.P., 2001, *The Properties of Gases and Liquids*, 5th ed., McGraw-Hill, New York.

Prausnitz J.M., Lichtenthaler R.N., and Gomes de Azevedo E., 1998, *Molecular Thermodynamics of Fluid-Phase Equilibria*, 3rd ed., Prentice Hall, Englewood Cliffs.

Prausnitz, J.M., Tavares, F.W. 2004 "Thermodynamics of fluid-phase equilibria for standard chemical engineering operations", *AIChE J.* **50**: 739–761.

Quack M., Stohner J., Strauss H.L., Takami M., Thor A.J., Cohen E.R., Cvitas T., Frey J.G., Holmstrom B., Kuchitzu K., Marquardt R., Mills I., and Pavese F., 2007, *Quantities, Units and Symbols in Physical Chemistry*, 3rd ed., RSC Publishing, Cambridge.

Renon H., and Prausnitz J.M., 1968, "Local compositions in thermodynamic excess functions for liquid mixtures", *AIChE J.* **14**:135–144.

Renon H., and Prausnitz J.M., 1969, "Estimation of parameters for NRTL equation for excess Gibbs energies of strongly nonideal liquid mixtures," *Ind. Eng. Chem. Process Des. Dev.* **8**:413–419.

Rosenberg R., 2005, "Why is ice slippery?," *Phys. Today* **58**:50–55.

Rowlinson J.S., and Swinton F.L., 1982, *Liquids and Liquid Mixtures*, 3nd ed., eds. Perlmutter P., Buckingham A.D., and Danishefksy S., Butterworth Publishers, London.

Singh M., Leonhard K., and Lucas K., 2007, "Making equation of state models predictive: Part 1: Quantum chemical computation of molecular properties", *Fluid Phase Equilib.* **258**:16–28.

Slater B., and Michaelides, A., 2019, "Surface premelting of water ice," *Nat. Rev. Chem* **3**:172–188.

Smith J.M., van Ness H.C., Abbott M.M., and Swihart M.T., 2018, *Introduction to Chemical Engineering Thermodynamics*, 8th ed., McGraw-Hill, New York.

Snurr R.Q., Adjiman C.l., and Kofke D.A., Eds., 2015, *Foundations of Molecular Modeling and Simulation. Select Papers from FOMMS*, Springer, London.

Spiegler K.S., and El-Sayad Y.M., 2001, "The energetics of desalination processes", *Desalination* **134**:109–128.

Stillwell A.S. and Webber M.E., 2016, "Predicting the specific energy consumption of reverse osmosis desalination", *Water* **8**:601.

Soave G., 1972, "Equilibrium constants from a modified Redlich-Kwong equation of state", *Chem. Eng. Sci.* **27**:1197–1203.

Strobridge T.R., 1962, "The thermodynamic properties of nitrogen from 64 to 300 K between 0.1 and 200 Atmospheres", NBS Technical Note No. 129.

Tester J.W., and Modell M., 1996, *Thermodynamics and Its Applications*, 2nd ed., Prentice Hall, New York.

Thomas E.R., and Eckert C.A., 1984, "Prediction of limiting activity coefficients by a modified separation of cohesive energy density model and UNIFAC", *Ind. Eng. Chem. Process Des. Dev.* **23**:194–209.

Tosun, I., 2012, *The Thermodynamics of Phase and Reaction Equilibria*, 1st ed., Elsevier, New York.

Trusler J.P.M., 2010, *The Virial Equation of State, Chapter 3, Experimental Thermodynamics Volume VIII: Applied Thermodynamics of Fluids*, eds. Goodwin A.R.H., Sengers J.Y., and Peters C.J., for IUPAC, Royal Society of Chemistry, Cambridge, UK.

Tsuboka T., and Katayama T., 1975, "Modified Wilson equation for vapor-liquid and liquid-liquid equilibria", *J. Chem. Eng. Japan* **8**:181–187.

Ungerer P., Wender A., Demoulin G., Bourasseau E., and Mougin P., 2004, "Application of Gibbs ensemble and NPT Monte Carlo simulation to the development of improved processes for H2S-rich gases", *Mol. Simul.* **30**:631–648.

van Laar J.J., 1910, "Über Dampfspannungen von binären Gemischen" *Z. Physik Chem.* **72**:723–751.

van Laar J.J., 1913, "Zur Theorie der Dampfspannungen von binären Gemischen", *Z. Physik Chem.* **83**:599–609.

Van Ness H.C., and Abbott M.M., 1982, *Classical Thermodynamics of Nonelectrolyte Solutions*, McGraw-Hill, New York.

Walas, S.M., 1985, *Phase Equilibria in Chemical Engineering*, Butterworth Publishers, Boston.

Wettlaufer J.S., and Grae Worster M., 2006, "Premelting dynamics", *Ann. Rev. FluidMech.* **38**:427–452.

Weir R.D., and de Loos T.W., (Eds) 2005, *Experimental Thermodynamics, Volume VII, Measurement of the Thermodynamic Properties of Multiple Phases*, for IUPAC, Elsevier, Amsterdam.

Wilson G.M., 1964, "Vapor-liquid equilibrium. 11. A New expression for the excess free energy of mixing", *J. Am. Chem. Soc.* **86**:127–130.

Wilson G.M., and Deal C.H., 1962, "Activity coefficients and molecular structure. Activity coefficients in changing environments-Solutions of Groups", *Ind. Eng. Chem. Fundam.* **1**:20–23.

5

Reactions and Electrolytes

5.1 Introduction

In this chapter, we consider a number of aspects of chemically reacting systems at both equilibrium and as they approach equilibrium. We consider the thermodynamic variables as they relate to chemical reactions including the enthalpy of reaction, the Gibbs energy of reaction and the criteria for equilibrium. Given that we cannot measure the absolute values of any thermodynamic energy, it is evident that we shall need to consider means which enable us to work with energy differences and thus with a common state of reference for such differences. One of those is the standard state that we have already introduced in Chapter 1 and employed in Chapters 3 and 4 but here we also describe the standard states appropriate for partially soluble solutes. We shall also consider the role of the equilibrium constant for reacting gas, condensed phase and multiphase systems and how it is affected by temperature and pressure. We then give some attention to electrochemistry and electrolyte solutions (Robinson and Stokes 2002), although in this case we simply provide enough material to enable an understanding of aspects of energy conversion and storage discussed in Chapter 7 and the peculiarities of the thermodynamic treatment of biological systems discussed in Chapter 8. A brief discussion is given of the rates at which reacting systems not in equilibrium proceed (their kinetics rather than thermodynamics – for an account of irreversible thermodynamics see Kjelstrup and Bedeaux 2010), with some examples of how to calculate the variation of substance concentration with time. Finally, we examine the thermodynamics of gas adsorption on solid surfaces and explore what this can tell us about solid-catalyzed reactions.

5.2 What Are Chemical Reaction Variables and How Do We Use Them?

On either the laboratory or the industrial chemical engineering scales, many processes involve chemical reactions as well as flow, work and heat transfer. It is therefore important for us to consider thermodynamic principles and practice as they apply to systems in which there are chemical reactions. In a reaction vessel, a number of chemical components (reactants) are mixed together and the reaction conditions in terms of pressure, temperature, catalyst and auxiliary compound concentrations are adjusted in order to favor the chemical transformation. A chemical reaction or chemical reactions then take place that produce different chemical species (products).

In general, there is an incomplete conversion of reactants to products. After some time, a point is reached where there is no change with respect to time of the amount of substance of

DOI: 10.1201/9780429329524-5

either the reactants or the products. A fixed amount of each substance from the reactants exists simultaneously with a fixed amount of substance of the products. This state is called *chemical equilibrium*. Thermodynamics is paramount in predicting the conversion at equilibrium, and from this to predict the product yield that can be expected. But thermodynamics also permits engineers to manipulate the driving forces for the chemical reactions such that the reaction is forced to go on.

In general, energy is released or absorbed in a chemical reaction and a knowledge of that energy is vital to an understanding of the chemistry of reactions and to their utilization in a chemical engineering context. A typical example is the need to control the temperature of large-scale chemical reactors in view of this heat generation or consumption. Thermodynamics permits precise predictions of the heat flux that has either to be removed and/or introduced and represents therefore an indispensable basis for reactor design and operation. Failure to control the temperature adequately can entail a sudden explosion-like release of heat and/or an instantaneous catastrophic pressure increase, leading to disaster.

For a *closed* system (see Question 1.3.1) the amount of substance of a species B n_B can only change if the extent of one or more chemical reactions changes. Chemical reactions may be formalized as (see Equations 3.27 and 3.28)

$$0 = \sum_{B} \nu_B B, \tag{5.1}$$

where ν_B is the stoichiometric coefficient of chemical compound B. In closed systems, molar balances have only an accumulation and a source term (see Equations 1.31 and 1.36)

$$dn_B = \sum_j \nu_{B,j} V r_j dt = \sum_j \nu_{B,j}\, d\xi_j, \tag{5.2}$$

where V is the volume of the closed system, r_j is the rate and ξ_j the extent of advancement of the jth chemical reaction (in mol) and ν_B is the stoichiometric coefficient of B in the jth reaction. The sources of B are of course the chemical reactions. For reactants, the source is negative because their stoichiometric coefficients are negative by definition and this is called a sink.

Chemical reaction variables describe the effect of the chemical reaction on state functions of the system such as internal energy, Gibbs energy, enthalpy, entropy and so on, assuming that the other independent state variables stay constant. We are going to develop this concept for *closed systems* with *only one chemical reaction*.

5.2.1 What Is the Heat of Reaction and How Do We Calculate It?

The enthalpy, or heat, of reaction, has briefly been introduced in Question 1.5.4. It was shown there that the heat of reaction predicts the heat generation of an isothermal and isobaric system per one mole of reaction advancement. The concept of a reaction variable is based on the total differential for the state function concerned. For enthalpy, this would be Equation 1.50 (Question 1.5.2)

$$dH = \left(\frac{\partial H}{\partial T}\right)_{p,n_B} dT + \left(\frac{\partial H}{\partial p}\right)_{T,n_B} dp + \sum_B H_B dn_B, \tag{5.3}$$

where H_B stands for the partial molar enthalpy of substance B in the reaction mixture (see Question 1.3.13).

If there is only one reaction, the variations of the number of moles of the reacting species dn_B are interlinked by the stoichiometry of this reaction and can thus be all tied to the variation of the extent of reaction, ξ, as expressed in the second part of Equation 5.2

$$dH = \left(\frac{\partial H}{\partial T}\right)_{p,n_B} dT + \left(\frac{\partial H}{\partial p}\right)_{T,n_B} dp + \sum_B H_B \nu_B d\xi. \tag{5.4}$$

The sum $\sum_B H_B \nu_B d\xi$ may be interpreted as the partial differential of H with respect to the reaction advancement and is called *enthalpy of reaction* or heat of reaction

$$\sum_B \nu_B H_B = \left(\frac{\partial H}{\partial \xi}\right)_{T,p} \equiv \Delta_r H. \tag{5.5}$$

Equation 5.5 defines the enthalpy of reaction but it also shows by its left-hand side how it can be calculated. However, the question still arises as to how the partial molar enthalpy H_B can be estimated or measured. As is usually the case with energetic quantities, absolute values of enthalpy do not exist and cannot be measured, so that a **reference state common to all reaction partners** has to be defined. The energy of this reference state is arbitrarily defined as zero. It has been shown in Question 1.9 that the situation is similar when the height of mountains is considered, where the sea level serves as the common reference.

In the case of partial molar enthalpies and chemical reactions, the enthalpy of the constituent elements of the compound B in their natural state is often used as such a reference. The partial molar enthalpy of B must thus be evaluated as the energy difference between a mole of B in the reaction mixture and the energy of its constituent elements. This difference is called *the energy or heat of formation* $\Delta_f H_B$ (see Figure 5.1). For many chemical compounds it is negative, because the compounds are often more stable than the elements. In consequence, the partial molar enthalpies are replaced by the enthalpies of formation in Equation 5.5, giving

$$\Delta_r H = \sum_B \nu_B \Delta_f H_B. \tag{5.6}$$

The enthalpies of formation would have to be measured in a calorimetric experiment. This could, in principle, be done according to the procedure explained in Question 1.8.2.

Unfortunately, the reactions corresponding to the formation of compounds cannot be easily carried out in calorimeters. It is generally much easier to burn the compounds to carbon dioxide, water and the combustion products of the other elements, such as sulfur dioxide. Therefore, the totally combusted state of the compounds is used as an alternative reference state. The energy difference between the compound and the products of combustion in terms of CO_2, H_2O, SO_2, etc. is called *the energy or heat of combustion* $\Delta_c H_B$ and can conveniently be determined in a combustion calorimeter such as the one described for measuring the food energy in a chocolate bar in Question 1.8.4. Heats of combustion are usually negative because the totally oxidized state is usually even more stable than the chemical compound (see Figure 5.1).

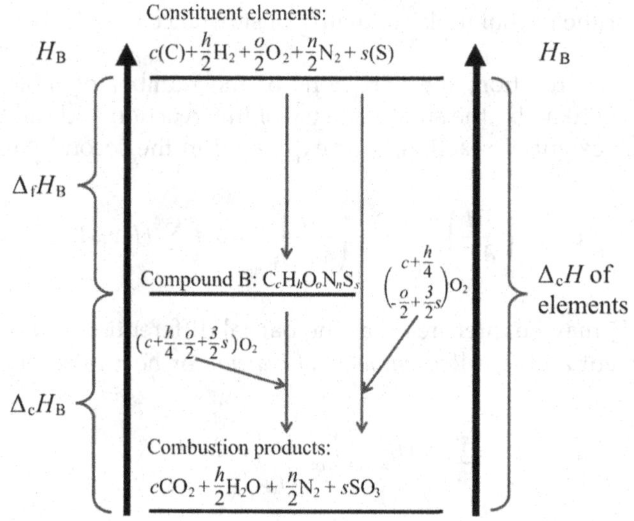

FIGURE 5.1
The relation between the enthalpy of formation and the enthalpy of combustion of an imaginary compound B of the chemical formula $C_cH_hO_oN_nS_s$. The horizontal lines indicate the energy levels of the compound and of the two reference states. The enthalpy of formation is usually determined as $\Delta_f H_B = \Delta_c H_{elements} - \Delta_c H_B$.

The enthalpy of reaction could, in principle, be computed directly based on heats of combustion as

$$\Delta_r H = -\sum_B \nu_B \Delta_c H_B. \tag{5.7}$$

The negative sign in Equation 5.7 comes from the fact that heats of combustion do not represent the difference of the energy of the compound minus the energy of the reference, but the other way round. These considerations are represented graphically in Figure 5.1.

Usually, the heats of combustion are, however, recalculated and tabulated as heats of formation. This can be done easily by applying The *Law of Hess,* which says that enthalpies of reaction may be computed directly as shown in Equation 5.6, or by decomposing the reaction into several partial reactions yielding finally the same products. The enthalpy of reaction is then obtained simply as the sum of the enthalpies of the partial reactions.

Thus, the formation of a compound may be decomposed into a complete oxidation of the elements yielding the combustion products of the compound, followed by the reduction (reverse oxidation) of these to yield the desired compound. The enthalpy of formation of the corresponding compound can thus be computed by the sum of the enthalpies of formation of the combustion products of the compound of interest plus the negative value of the heat of combustion (see Figure 5.1). The first step will consume some oxygen, which will be released again in the reduction step. The oxygen plays no role in the calculation because the same amount is consumed in the original combustion of the elements as is released in the reverse process. Besides, oxygen is an element and has a zero energy by virtue of our current definition.

Partial molar enthalpies, i.e. heats of formation or combustion of reaction partners, depend on their exact physical-chemical conditions, such as aggregation state (gaseous, liquid or solid), temperature, pressure and composition of the reaction mixture. It was explained in Questions 1.9 and 3.7 that in order to tabulate their values one needs to standardize these conditions by introducing *standard states.* Enthalpies of formation and combustion at standard conditions are called *standard enthalpy of formation* $\Delta_f H_B^{\ominus}$ and *standard enthalpy of combustion* $\Delta_c H_B^{\ominus}$, of B, respectively. From these, the standard enthalpy of reaction $\Delta_r H^{\ominus}$ can be computed

$$\Delta_r H^{\ominus} = \sum_B \nu_B \Delta_f H_B^{\ominus} = -\sum_B \nu_B \Delta_c H_B^{\ominus}. \tag{5.8}$$

The calculations for correcting the results to non-standard states are explained for the Gibbs reaction energy $\Delta_r G$ in Question 5.3.1. The standard molar enthalpies of formation at 298.15 K are available for most common substances (Thermodynamic Research Centre 1942–2007 (NSRDS-NIST-75) and 1955–2007 (NSRDS-NIST-74)).

5.2.2 What Is the Gibbs Energy of Reaction and How Do We Calculate It?

By analogy with the heat of reaction, the Gibbs energy of reaction is based on a total differential of the relevant state function, which is G in this case. Thus, by analogy with Equation 5.4, it can be written

$$dG = \left(\frac{\partial G}{\partial T}\right)_{p,n_B} dT + \left(\frac{\partial G}{\partial p}\right)_{T,n_B} dp + \sum_B \mu_B dn_B. \tag{5.9}$$

Assuming again only one chemical reaction, the changes of the amount of the reacting species dn_B may be linked to the changes of the extent of the reaction $d\xi$ by expressing them through molar balances, Equation 5.2, much in the same way as for the enthalpy in Equation 5.4:

$$dG = \left(\frac{\partial G}{\partial T}\right)_{p,n_B} dT + \left(\frac{\partial G}{\partial p}\right)_{T,n_B} dp + \sum_B \nu_B \mu_B d\xi. \tag{5.10}$$

The sum $\Sigma_B \nu_B \mu_B$ is again interpreted as the partial molar differential of G with respect to the extent of reaction and is called the *Gibbs energy of reaction*

$$\Delta_r G \equiv \sum_B \nu_B \mu_B = \left(\frac{\partial G}{\partial \xi}\right)_{T,p}. \tag{5.11}$$

The first part of the equation defines the Gibbs energy of reaction $\Delta_r G$ and also indicates how to compute it. By using the same reference states as for the heats of reaction, the $\Delta_r G$ may be calculated either using the Gibbs energies of formation $\Delta_f G_B$ or the Gibbs energies of combustion $\Delta_c G_B$

$$\Delta_r G = \sum_B \nu_B \Delta_f G_B, \tag{5.12}$$

$$\Delta_r G = -\sum_B \nu_B \Delta_c G_B. \tag{5.13}$$

Neither the Gibbs energy of formation nor of combustion can be measured in calorimeters. Instead, they are usually determined by measuring the compositions, and hence the equilibrium constants, of reaction mixtures at equilibrium, where $\Delta_r G$ is zero (see Question 5.3.1). In performing such experiments, the standardization of the reaction and the suitable choice of standard states play a considerably more important role than for reaction enthalpies, because the Gibbs energies and chemical potentials usually depend much more sensitively on the composition of the reaction mixture than the partial molar enthalpies. The results of such measurements are therefore always tabulated in terms of standard Gibbs energies of formation $\Delta_f G_B^\ominus$ or standard Gibbs energies of combustion $\Delta_c G_B^\ominus$, from which the standard Gibbs energies of reaction may be determined as

$$\Delta_r G^\ominus \equiv \sum_B \nu_B \mu_B^\ominus = \sum_B \nu_B \Delta_f G_B^\ominus = -\sum_B \nu_B \Delta_c G_B^\ominus. \tag{5.14}$$

The choice of these standard states depends on convenience and can be different for each compound participating in the reaction. As these standard states do not serve as reference states, their values disappear in the calculation of actual partial molar Gibbs energy of formation and of combustion. Indeed, the actual molar Gibbs energy of formation and combustion will be computed as their standard values, representing the difference between the chemical potential in the chosen standard state minus the one in the reference states, plus a correction term for the difference between the actual and the standard conditions. The calculations needed for correcting the results to non-standard conditions are explained in Question 5.3.1.

5.2.3 What Does the Gibbs Energy of Reaction Tell Us About How Chemical Reactions Proceed?

The total differential of G for a closed system (Equation 5.10) combined with the definition of the Gibbs reaction energy (Equation 5.11) is given as

$$dG = \left(\frac{\partial G}{\partial T}\right)_{p,n_B} dT + \left(\frac{\partial G}{\partial p}\right)_{T,n_B} dp + \Delta_r G d\xi. \tag{5.15}$$

If this is substituted into a Gibbs energy balance for an *isothermal, isobaric and closed* system (Equation 3.23) one obtains

$$\Delta_r G d\xi = dG = \delta W_o - T\delta S_{gen}. \tag{5.16}$$

In cases *without external driving forces* ($\delta W_o = 0$), this equation enables a judgment to be made as to whether a reaction can occur spontaneously if properly catalyzed and in which direction it will proceed. Since δS_{gen} must be positive for anything to happen spontaneously, the Gibbs reaction energy must be negative in order to allow $d\xi$ to be positive and for the reaction to proceed in a forward direction, i.e. from left to right. If $\Delta_r G$ is positive, $d\xi$ would have to be negative in order to allow entropy to be generated and the reaction would tend to proceed backwards.

In the first case where $\Delta_r G < 0$, the concentrations and the chemical potentials of the reactants will decrease and those of the products will increase (see Question 5.3.1) as the reaction proceeds. Since the reactants have negative stoichiometric coefficients, the Gibbs reaction energy $\Delta_r G$ will increase and tend towards zero. In the second case with a positive $\Delta_r G$, the inverse considerations apply and $\Delta_r G$ will decrease, thereby also tending to zero. In both cases, $\Delta_r G$ will finally reduce to zero and entropy production is no longer possible according to Equation 5.16 ($\delta W_o = 0$). In the absence of a driving force for spontaneous processes ξ will also come to a halt and the reaction will remain in this equilibrium position.

Equation 5.16 is often used to predict the equilibrium composition of a reaction mixture by setting δS_{gen}, and often also δW_o, to zero. The calculation of an equilibrium composition necessitates however the knowledge of how the chemical potential depends on the composition of the mixture (see Question 5.3.1).

Equation 5.16 also shows that a chemical reaction can be forced to proceed even if the Gibbs energy of reaction is positive if one can introduce an external driving force ($\delta W_o > 0$). This is typically the case in electrochemical systems, where the potential difference applied between the two electrodes acts as such a driving force (see Question 7.2.4, water electrolysis). Photosynthesis is another example (see Question 8.3.3).

5.2.4 What Are Other Thermodynamic Variables of Reaction?

Besides the enthalpy and the Gibbs energy of reaction, other variables of reaction need often to be considered, such as those for S, V, U, A and C_p. By analogy with the enthalpy and Gibbs energy of reaction, these are defined as

$$\begin{aligned}
\Delta_r S &\equiv \sum_B \nu_B S_B = \left(\frac{\partial S}{\partial \xi}\right)_{p,T} \\
\Delta_r \varsigma &\equiv \sum_B \nu_B V_B = \left(\frac{\partial V}{\partial \xi}\right)_{p,T} \\
\Delta_r U &\equiv \sum_B \nu_B U_B = \left(\frac{\partial U}{\partial \xi}\right)_{p,T} \\
\Delta_r A &\equiv \sum_B \nu_B A_B = \left(\frac{\partial A}{\partial \xi}\right)_{p,T} \\
\Delta_r C_p &\equiv \sum_B \nu_B C_{p,B} = \left(\frac{\partial C_p}{\partial \xi}\right)_{p,T} .
\end{aligned} \tag{5.17}$$

For the reasons described in Questions 5.2.1 and 5.2.2, variables of reaction are tabulated for standardized conditions. Therefore, the functions defined by Equations 5.17 can be found in the literature in the form of standard variables of reaction, defined as follows:

$$\Delta_r S^\ominus \equiv \sum_B \nu_B S_B^\ominus$$
$$\Delta_r V^\ominus \equiv \sum_B \nu_B V_B^\ominus$$
$$\Delta_r U^\ominus \equiv \sum_B \nu_B U_B^\ominus \qquad (5.18)$$
$$\Delta_r A^\ominus \equiv \sum_B \nu_B A_B^\ominus$$
$$\Delta_r C_p^\ominus \equiv \sum_B \nu_B C_{p,B}^\ominus.$$

The standard reaction entropy, volume, and heat capacity can be computed directly as indicated in Equation 5.18 because the partial molar quantities of these functions for the compounds B can be measured experimentally. The other partial molar quantities in Equation 5.18 are not known in an absolute way and must be replaced by their standard variables of formation or combustion. The procedures needed for correcting the results of such calculations to non-standard conditions are explained for the Gibbs reaction energy $\Delta_r G$ in Question 5.3.1. An analogous procedure may be applied to the changes in S, V, U, A and C_p.

5.3 How Can We Predict the Equilibrium Composition of Reaction Mixtures?

A closed system in which a single chemical reaction can occur will reach equilibrium when the chemical potentials of the reactants and the products of the reaction have changed to the point where the Gibbs energy of reaction $\Delta_r G$ becomes zero (see Equation 5.16 in Question 5.2.3). In order to use this criterion to compute the equilibrium composition of the reaction mixture, one must know how the chemical potentials of the chemical compounds depend on their concentrations. This is described in detail in Chapter 4; in this question, we use the equations derived there to characterize the equilibrium state of a reacting mixture.

5.3.1 How Can We Use the Equilibrium Criterion to Calculate the Equilibrium Constant?

As explained in Question 5.2.3, the equilibrium criterion for a chemical reaction follows from Equation 5.16 by setting δS_{gen} and δW_o equal to zero

$$\Delta_r G \, d\xi = 0, \qquad (5.19)$$

which is only possible if the reaction Gibbs energy $\Delta_r G$ is zero, so that the equation yields zero no matter what the value of $d\xi$ is.

The Gibbs energy of reaction $\Delta_r G$ can be calculated according to its definition Equation 5.11

$$\Delta_r G = \sum_B \nu_B \mu_B. \qquad (5.20)$$

In order to estimate the composition of the reaction mixture that renders $\Delta_r G$ in Equation 5.11 equal to zero, one has to substitute for the chemical potentials the functions showing their dependence on the substance concentrations developed in Chapter 4. As an example, we will assume that the same standard state, pure gaseous B at pressure $p^\ominus$, may be used for all reaction partners. Then the relation in Equation 4.61 may be used

$$\mu_{B,g}(T,p,y_C) = \mu^\ominus_{B,g}(T) + RT\ln\left(\frac{y_B p}{p^\ominus}\right) + \int_0^p \left(V_B - \frac{RT}{p}\right)dp. \tag{5.21}$$

The integral term at the end of this expression represents RT times the logarithm of the fugacity coefficient ϕ_B as given in Equation 4.63. In order to abbreviate the equation this term can therefore be combined with the second term to yield

$$\mu_{B,g}(T,p,y_C) = \mu^\ominus_{B,g}(T) + RT\ln\left(\frac{y_B p\phi_B}{p^\ominus}\right). \tag{5.22}$$

The equilibrium criterion thus becomes

$$\sum_B \nu_B\mu_B = 0 = \sum_B \nu_B\mu_B^\ominus + \sum_B \nu_B RT\ln\left(\frac{y_B p\phi_B}{p^\ominus}\right). \tag{5.23}$$

With the definition of Equation 5.11, this can be written in a more compact form

$$\frac{\Delta_r G^\ominus}{RT} + \ln\prod_B\left(\frac{y_B^{eq} p\phi_B}{p^\ominus}\right)^{\nu_B} = 0. \tag{5.24}$$

A *standard equilibrium constant*, $K^\ominus$, is usually defined as follows (see also Equation 2.68):

$$K^\ominus \equiv \prod_B (\lambda_B^\ominus)^{-\nu_B} = \exp\left(-\frac{\Delta_r G^\ominus}{RT}\right) = \prod_B\left(\frac{y_B^{eq} p\phi_B}{p^\ominus}\right)^{\nu_B}. \tag{5.25}$$

Here, $\lambda_B^\ominus$ is the standard absolute activity as defined in Equations 3.53 and 5.27 and $y_B{}^{eq}$ is the equilibrium mole fraction of component B. The constant defined in the above equation, $\prod_B(\lambda_B^\ominus)^{-\nu_B}$, is also called the *equilibrium reaction coefficient*. Similar expressions for reaction mixtures not at equilibrium are known as reaction coefficients or reaction quotients, Q_r. (see also Equation 7.39). The constant $K^\ominus$ is only a function of temperature and not of pressure or composition.

Equation 5.25 can of course also be written as

$$\Delta_r G^\ominus = -RT\ln K^\ominus. \tag{5.26}$$

Equation 5.26, known as *the van't Hoff isotherm*, shows that the *standard Gibbs energy of reaction* determines the standard equilibrium constant for the reaction at a specific constant temperature. If $\Delta_r G^\ominus$ is negative, $K^\ominus$ will be greater than unity, meaning that according to Equation 5.25 the y_B of the products, which have positive stoichiometric coefficients ν_B, will be high and those of the reactants small. If $\Delta_r G^\ominus$ is positive, the reverse is true. Hence, the equilibrium yield of a reaction is high when the standard Gibbs energy of reaction has a large negative value and low when it has a positive value.

Equation 5.25 permits the calculation of the equilibrium constant from the standard Gibbs energy of reaction and links the constant to the distribution of the mole fractions of the reacting substances in the gas phase. Expressions for reaction mixtures in which other standard states must be used for some reaction participants in condensed phases are developed in Question 5.4. Once the equilibrium constant $K^{\ominus}$ is known, the extent of reaction at the equilibrium can be determined by expressing all mole fractions in terms of extent of reaction advancement ξ and solving the resulting function for it (see Question 5.3.5).

5.3.2 How Can We Use Measured Standard Equilibrium Constants for Calculating Other Reaction Variables?

The partial molar quantities that appear in the definitions of the various standard thermodynamic reaction functions (Equations 5.18) may be linked directly to the chemical potentials by using the *standard absolute activity* $\lambda_B^{\ominus}$ defined as follows (see Equation 3.53 for a general definition of absolute activity)

$$\mu_B^{\ominus} = RT \ln \lambda_B^{\ominus}. \tag{5.27}$$

Once the relevant standard chemical potentials, $\mu_B^{\ominus}$, or standard absolute activities, $\lambda_B^{\ominus}$, are known, all other standard thermodynamic functions are obtained by using thermodynamic relations developed in Chapter 3. Applying Equation 3.77 to the standard chemical potential yields the standard molar entropy

$$S_B^{\ominus} = -\frac{d\mu_B^{\ominus}}{dT} = -R \ln \lambda_B^{\ominus} - RT\frac{d \ln \lambda_B^{\ominus}}{dT}. \tag{5.28}$$

The standard molar enthalpy follows from its relationship with the standard chemical potential through the Gibbs–Helmholtz equation (Equation 3.64)

$$H_B^{\ominus} = \mu_B^{\ominus} - T\frac{d\mu_B^{\ominus}}{dT} = -RT^2\frac{d \ln \lambda_B^{\ominus}}{dT}. \tag{5.29}$$

The standard molar heat capacity at constant pressure can be calculated as follows

$$C_{p,B}^{\ominus} = \frac{dH_B}{dT} = -T\frac{d^2\mu_B^{\ominus}}{dT^2} = -2RT\frac{d\ln\lambda_B^{\ominus}}{dT} - RT^2\frac{d^2 \ln \lambda_B^{\ominus}}{dT^2}, \tag{5.30}$$

while the chemical potential itself is linked to the standard absolute activity, as shown by Equation 5.27.

By substituting the right-hand side of the second equality of Equations 5.27 – 5.30 into the definitions of the various standard reaction variables (Equations 5.18), one obtains relations between the latter and the standard equilibrium constant, which from Equation 5.24 is seen to be

$$K^{\ominus}(T) = \exp\left(-\frac{\sum_{B}\nu_{B}\mu_{B}^{\ominus}(T)}{RT}\right) = \prod_{B}(\lambda_{B}^{\ominus}(T))^{-\nu_{B}}. \tag{5.31}$$

These relations are:
for the standard molar reaction entropy

$$\Delta_{r}S^{\ominus} = \sum_{B}\nu_{B}S_{B}^{\ominus} = -R\ln K^{\ominus} - RT\frac{d\ln K^{\ominus}}{dT}, \tag{5.32}$$

for the standard molar reaction enthalpy (heat of reaction)

$$\Delta_{r}H^{\ominus} = \sum_{B}\nu_{B}H_{B}^{\ominus} = RT^{2}\frac{d\ln K^{\ominus}}{dT}, \tag{5.33}$$

for the standard molar reaction Gibbs energy

$$\Delta_{r}G^{\ominus} = \sum_{B}\nu_{B}\mu_{B}^{\ominus} = \sum_{B}\nu_{B}G_{B}^{\ominus} = -RT\ln K^{\ominus}, \tag{5.34}$$

which is the van't Hoff isotherm of Equation 5.25.

Equation 5.33 is usually written as

$$\frac{d\ln K^{\ominus}}{dT} = \frac{\Delta_{r}H^{\ominus}}{RT^{2}}, \tag{5.35}$$

and called the *van't Hoff isochore,* since it applies at constant pressure. It is important because if a value of $K^{\ominus}$ for a reaction is available at one temperature T_1 the value at another temperature T_2 can be determined from an integration of this equation, as shown in Question 5.3.3. It can also be used to determine $\Delta_{r}H^{\ominus}$ from measurements of $K^{\ominus}$ as a function of temperature.

5.3.3 How Do We Calculate the Change of Equilibrium Constant with Temperature?

Equation 5.35 (the van't Hoff isochore derived in Question 5.3.2) provides the temperature dependence of the standard equilibrium constant and is the starting point for evaluating $K^{\ominus}(T)$ at any temperature T_2 from its known value at a particular temperature T_1. We can integrate Equation 5.35 to give

$$\ln\{K^{\ominus}(T_2)\} = \ln\{K^{\ominus}(T_1)\} + \int_{T_1}^{T_2}\left\{\frac{\Delta_{r}H_{m}^{\ominus}(T)}{RT}\right\}dT. \tag{5.36}$$

To evaluate the integral in Equation 5.36, we need to know the temperature dependence of $\Delta_{r}H^{\ominus}(T)$. This standard enthalpy of reaction, usually evaluated using Equation 5.8, is related to the standard enthalpies of the individual reaction components, $H_{B}^{\ominus}$, by Equation 5.5. Since from Equation 1.128, defining C_p, we can relate $H_{B}^{\ominus}$ at any temperature T to its value at a reference temperature T_3 by

$$H_{\mathrm{B}}^{\ominus}(T) = H_{\mathrm{B}}^{\ominus}(T_3) + \int_{T_3}^{T} C_{p,\mathrm{B}}^{\ominus}(T)\mathrm{d}T, \tag{5.37}$$

we can insert $H_{\mathrm{B}}^{\ominus}(T)$ for each component into Equation 5.5 to give

$$\Delta_{\mathrm{r}} H^{\ominus}(T) = \Delta_{\mathrm{r}} H^{\ominus}(T_3) + \int_{T_3}^{T} \sum_{\mathrm{B}} \nu_{\mathrm{B}} C_{p,\mathrm{B}}^{\ominus}(T)\mathrm{d}T. \tag{5.38}$$

This equation, which relates the standard enthalpy of reaction $\Delta_{\mathrm{r}}H^{\ominus}(T)$ at any temperature T to its value at a temperature T_3 (often 298 K) at which its value is known (via the heat capacities of the reaction components $C_{p,\mathrm{B}}^{\ominus}(T)$, which are themselves temperature dependent), is known as *Kirchoff's law*. It can be seen from Equation 5.18 that the function inside the integral on the right-hand side of Equation 5.38 is in fact the standard reaction heat capacity $\Delta_{\mathrm{r}}C_p^{\ominus}(T)$, so that an alternative form of Kirchoff's law is

$$\Delta_{\mathrm{r}} H^{\ominus}(T) = \Delta_{\mathrm{r}} H^{\ominus}(T_3) + \int_{T_3}^{T} \Delta_{\mathrm{r}}C_p^{\ominus}(T)\mathrm{d}T. \tag{5.39}$$

Substitution of either Equations 5.38 or 5.39 into Equation 5.33 then gives an equation that enables us to determine $K^{\ominus}(T_2)$ from $K^{\ominus}(T_1)$, provided that we know the temperature dependence of $\Delta_{\mathrm{r}}H^{\ominus}(T)$ through the heat capacities and their own temperature dependence

$$\ln\{K^{\ominus}(T_2)\} = \ln\{K^{\ominus}(T_1)\} + \frac{\Delta_{\mathrm{r}}H^{\ominus}(T_3)(T_2 - T_1)}{RT_1T_2} + \int_{T_1}^{T_2}\left\{\int_{T_3}^{T}\sum_{\mathrm{B}}\nu_{\mathrm{B}}C_{p,\mathrm{B}}^{\ominus}(T)\mathrm{d}T\right\}\left(\frac{1}{RT^2}\right)\mathrm{d}T. \tag{5.40}$$

The temperature dependence of substance heat capacities is commonly represented as a polynomial function, whose coefficients are available from a range of thermodynamic databases and modeling packages (see Question 9.3). So the integrals in Equation 5.40 may usually be evaluated in full, although at the cost of some tedious arithmetic.

In order to evaluate Equation 5.40 analytically and more simply, various assumptions concerning the temperature dependence of $\Delta_{\mathrm{r}}H^{\ominus}$ (called the Ulich approximations) are required. The first Ulich approximation is that $\Delta_{\mathrm{r}}H^{\ominus}$ is assumed to be temperature independent in the considered temperature range. In the second Ulich approximation, $\Delta_{\mathrm{r}}H^{\ominus}$ is a function of temperature but $C_p^{\ominus}$ is assumed temperature independent and in the third approximation, the overall temperature range is divided into different temperature intervals, which are considered separately and either the first or the second approximation applied to each interval. For example, the second approximation applied to Equation 5.40, using Equation 5.18 for $\Delta_{\mathrm{r}}C_p^{\ominus}(T)$ and taking the same reference temperature T_0 for $K^{\ominus}(T)$ and $\Delta_{\mathrm{r}}H^{\ominus}(T)$ (so that $T_1 = T_3 = T_0$), yields

$$\ln(K^{\ominus}(T)) = \ln(K^{\ominus}(T_0)) - \frac{\Delta_{\mathrm{r}}H^{\ominus}(T_0)}{R}\left(\frac{1}{T} - \frac{1}{T_0}\right) + \frac{\Delta_{\mathrm{r}}C_p^{\ominus}(T_0)}{R}\ln\left(\frac{T}{T_0}\right). \tag{5.41}$$

The analytical solution for the first Ulich approximation simply follows from Equation 5.41 when $\Delta_{\mathrm{r}}C_p^{\ominus}(T_0)$ is set equal to zero.

Using Equation 5.35, we can see that for an exothermic reaction, where $\Delta_r H^\ominus$ is negative, $K^\ominus$ decreases with temperature and hence the equilibrium shifts to the left in favor of the reactants. Conversely, for endothermic reactions, $K^\ominus$ increases with temperature and the equilibrium yield of products increases. This observation is one manifestation of *Le Chatelier's principle*, which states that if a change is imposed on any system at equilibrium, the position of the equilibrium will shift in a direction which tends to counteract that change. Hence where the forward reaction produces heat, an increase in temperature will result in suppression of this reaction with a lower yield of products, whereas lower temperatures will have the opposite effect. See also Question 5.3.4 for the effect of pressure changes.

5.3.4 What Is the Equilibrium Constant for a Reacting Gas Mixture?

To derive the equilibrium constant for reacting gas mixtures, Equation 5.25 can be used

$$K^\ominus = \exp\left(\frac{-\Delta_r G^\ominus}{RT}\right) = \prod_B \left(\frac{y_B^{eq} p \phi_B}{p^\ominus}\right)^{\nu_B}. \tag{5.42}$$

If the fugacity coefficient is replaced by the solution of Equation 4.63 for the fugacity coefficient ϕ_B, the following result is obtained:

$$K^\ominus = \prod_B \left(\frac{y_B^{eq} p}{p^\ominus}\right)^{\nu_B} \exp\left(\sum_B \left\{\nu_B \int_0^p \left(\frac{V_B(T, p, y_B^{eq})}{RT} - \frac{1}{p}\right) dp\right\}\right). \tag{5.43}$$

For a mixture of ideal gases or for real gas mixtures at low pressure ($p \to 0$), the integral in Equation 5.43 becomes zero and the equation simplifies to

$$K^\ominus \approx \prod_B \left(\frac{y_B^{eq} p}{p^\ominus}\right)^{\nu_B}. \tag{5.44}$$

A simplified, commonly applied equation results when $p^\ominus$ is taken into the equilibrium constant to form the pressure independent K_P, the partial pressure equilibrium constant

$$K_p \equiv K^\ominus (p^\ominus)^{\Sigma_B \nu_B} = \prod_B (y_B p)^{\nu_B} = \prod_B p_B^{\nu_B}, \tag{5.45}$$

where p_B is the partial pressure of the species B and K_P has dimensions of $(\text{pressure})^{\Sigma_B \nu_B}$.

We can use Equation 5.45 to examine the effect of a pressure change on the position of the equilibrium of a gas phase reaction by re-writing it as

$$K_y = \prod_B y_B^{\nu_B} = K_p \, p^{-\Sigma_B \nu_B}, \tag{5.46}$$

where K_y is the equilibrium constant written in terms of mole fractions y_B. Since K_p does not depend on pressure, if the change in the number of moles on reaction, $\Sigma_B \nu_B$, is positive, then an increase in pressure will decrease K_y and hence the mole fraction of products in the gas mixture at equilibrium. Conversely, if the number of moles decreases as the reaction proceeds from left to right, so that $\Sigma_B \nu_B$ is negative, an increase in pressure will increase both K_y and the equilibrium mole fractions of the products. This is another example of Le Chatelier's principle (see Question 5.3.3), this time applied to pressure changes. Increasing the pressure will cause the equilibrium to shift in the direction that tends to nullify this change by reducing the gas pressure through reducing the number of moles of gas present in the mixture.

Equation 5.46 can be further developed by writing at constant temperature

$$\frac{\mathrm{d}\ln K_y}{\mathrm{d}p} = \frac{\mathrm{d}\ln K_p}{\mathrm{d}p} - \sum_B \nu_B \frac{\mathrm{d}\ln p}{\mathrm{d}p}. \tag{5.47}$$

Since for an ideal gas mixture $p\Delta_r V^\ominus = \Sigma_B \nu_B RT$, K_P is independent of pressure and $(\mathrm{d}\ln p/\mathrm{d}p)$ is $1/p$, then

$$\frac{\mathrm{d}\ln K_y}{\mathrm{d}p} = \frac{-\Sigma_B \nu_B}{p} = \frac{\Delta_r V^\ominus}{RT}, \tag{5.48}$$

where $\Delta_r V^\ominus$ is the volume change for one mole of reaction with all reactants and products in their standard states. A similar equation can be applied to equilibria in solution.

In situations where the integral in Equation 5.43 cannot be neglected, the pressure dependency can be described by the truncated virial expression (see Question 2.8)

$$V_B = \frac{RT}{p} + B_B, \tag{5.49}$$

where B_B is the second virial coefficient of the species B. With this approximation, the integral in Equation 5.43 can be solved analytically, resulting in

$$K^\ominus(T) \approx \prod_B \left(\frac{y_B^{eq} p}{p^\ominus}\right)^{\nu_B} \exp\left(\frac{\Sigma_B V_B B_B p}{RT}\right). \tag{5.50}$$

If no approximation for the dependence of $V_B(T, p, y_C^{eq})$ on pressure is possible, the definition of fugacity coefficient (Equation 4.63)

$$\ln \phi_B = \int_0^p \left(\frac{V_B}{RT} - \frac{1}{p}\right)\mathrm{d}p, \tag{5.51}$$

can be inserted in Equation 5.43, resulting in

$$K^\ominus(T) = \prod_B \left(\frac{y_B^{eq} p}{p^\ominus}\right)^{\nu_B} \exp\left[\sum_B V_B \ln \phi_B\right]. \tag{5.52}$$

5.3.5 What Is the Difference Between G, $G^\ominus$, $\Delta_r G$ and $\Delta_r G^\ominus$ for Reacting Mixtures?

Unfortunately, it often happens that experimentalists determine (and also users employ) $\Delta_r G^\ominus$ rather than $\Delta_r G$. Also, the difference between changes in G and $\Delta_r G$ is often not clear. This can lead to considerable misinterpretations about the thermodynamic feasibility of chemical and biochemical reactions. The question which then arises is whether the differences between G, $\Delta_r G$, $G^\ominus$ and $\Delta_r G^\ominus$ are significant? The best way to answer this question is by means of an example, specifically ammonia synthesis in the Haber-Bosch process, a gas phase reaction carried out at high temperatures and pressures:

$$N_2(g) + 3H_2(g) \xrightarrow{T = 500K, p = 0.5MPa} 2NH_3(g). \tag{5.53}$$

According to Equations 1.31 and 5.1, for an extent of reaction ξ the amount of substance B, $n_B(\xi)$, is given by

$$n_B(\xi) = n_B(\xi = 0) + \nu_B \xi, \tag{5.54}$$

where ν_B is the stoichiometric number. The Gibbs energy for the reacting system, G, can be calculated using Euler's theorem (see Question 1.10.2) as follows, assuming the fugacity coefficients are unity

$$G = \sum_B n_B \mu_B = \sum_B n_B(\xi)\left(\mu_B^\ominus + RT\left(\frac{p}{p^\ominus}\right) + RT\ln(y_B)\right) = G^\ominus + G^p + G^{mix}. \tag{5.55}$$

The course that G takes as the reaction proceeds is shown in Figure 5.2 (solid line). Here, it has been assumed that the pressure remains constant during the reaction. Constant pressure can be achieved, for example, by a moving piston in a cylinder. The plot of G against ξ passes through a minimum that corresponds to the equilibrium state, $\xi = \xi^{eq}$. According to Equation 5.55, the Gibbs energy function, G, consists of a concentration-independent term, $G^\ominus$, a pressure-dependent term, G^p, and a concentration-dependent term, G^{mix}. The concentration independent term $G^\ominus$ is the standard Gibbs energy and can be written as

$$G^\ominus = \sum_B n_B \mu_B^\ominus = \sum_B n_B(\xi = 0)\mu_B^\ominus + \xi \Delta_r G_m^\ominus. \tag{5.56}$$

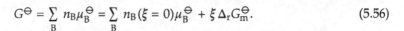

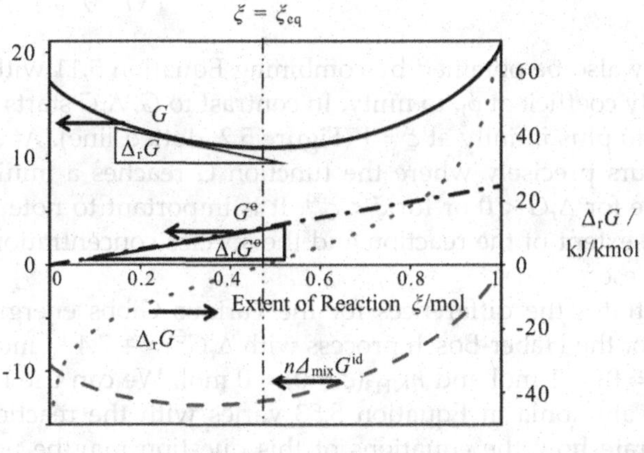

FIGURE 5.2
Differences between the various forms of Gibbs energy using the Haber-Bosch process as an example. (G, $G^\ominus$, $\Delta_r G$, $\Delta_r G^\ominus$, $n\Delta_{mix}G^{id}$) G: solid line (Equation 5.55); $G^\ominus$: dash-dotted line, Equation 5.55; $\Delta_r G$: short dashed line (Equation 5.60, the slope of G vs ξ); $\Delta_r G^\ominus$: slope of $G^\ominus$ vs ξ; $n\Delta_{mix}G^{id}$: long-dashed line (Equation 5.59). Arrows indicate the y-axis relevant to that quantity. The symbols and conditions are explained in the text.

$G^{\ominus}$ is shown as the dash-dotted line in Figure 5.2. Thus, the plot of $G^{\ominus}$ against ξ is a straight line with the slope of $\Delta_r G^{\ominus}$. For this reaction this line starts at a value of zero for $\xi = 0$ because both pure hydrogen and nitrogen have a standard chemical potential of zero. The line also ends (for $\xi = 1$) at a value of $\Delta_r G^{\ominus}$, which amounts to twice the standard chemical potential of ammonia at the temperature of 500 K.

Using Equations 5.54 and 5.55, the pressure-dependent term can be written as

$$G^{p} = \sum_{B} n_B(\xi = 0)RT \ln\left(\frac{p}{p^{\ominus}}\right) + \xi\ RT \ln \prod_B \left(\frac{p}{p^{\ominus}}\right)^{\nu_B}. \quad (5.57)$$

The concentration-dependent part of G is a result of the effect of mixing on the Gibbs energy

$$G^{mix} = \sum_{B} n_B(\xi)RT \ln(y_B). \quad (5.58)$$

G^{mix} is shown in Figure 5.2 by the long stroke line. If G^{mix} is divided by the mole number of the system, $n = \sum_B n_B(\xi)$, the well-known expression (Equation 4.113 in Question 4.5)

$$\Delta_{mix}G^{id} = \frac{G^{mix}}{n} = \sum_{B} y_B RT \ln(y_B), \quad (5.59)$$

for the ideal Gibbs energy of mixing results, which confirms that this term is caused by mixing effects.

The Gibbs energy of reaction, $\Delta_r G$, is calculated by differentiating G (Equation 5.55) with respect to ξ according to Equation 5.11 in Question 5.2.2. Here, to simplify the differentiation, it is assumed that the pressure remains constant during the reaction. The result is

$$\Delta_r G = \Delta_r G^{\ominus} + RT \ln \prod_B y_B^{\nu_B} + RT \ln\left(\frac{p}{p^{\ominus}}\right) \sum_{B} \nu_B. \quad (5.60)$$

Equation 5.60 may also be obtained by combining Equation 5.11 with Equation 5.22 and setting the fugacity coefficient ϕ_B to unity. In contrast to G, $\Delta_r G$ starts from minus infinity at $\xi = 0$ and runs to plus infinity at $\xi = 1$ (Figure 5.2, dotted line). At equilibrium ($\xi = \xi_{eq}$) $\Delta_r G = 0$; this occurs precisely where the function G reaches a minimum. The forward reaction is feasible for $\Delta_r G < 0$ or for $\xi < \xi^{eq}$. It is important to note that $\Delta_r G$ is strongly dependent on the extent of the reaction and the species concentrations, whereas $\Delta_r G^{\ominus}$ is independent of these.

Figure 5.2 illustrates the differences for the various Gibbs energies (i.e. G, $G^{\ominus}$, $\Delta_r G$, $\Delta_r G^{\ominus}$, $n\Delta_{mix}G^{id}$) for the Haber-Bosch process with $\Delta_r G^{\ominus} = +7.4$ kJ mol^{-1} at 500 K, $n_{H2}(\xi = 0) = 3$ mol, $n_{N2}(\xi = 0) = 1$ mol and $n_{NH3}(\xi = 0) = 0$ mol. We can use Figure 5.2 to explore how the yield of ammonia in Equation 5.53 varies with the reaction temperature and pressure to illustrate how the equations of this question may be used in practice. The stoichiometric coefficients are

$$\nu_{N_2} = -1 \quad \nu_{H_2} = -3 \quad \nu_{NH_3} = 2 \quad \text{and} \quad \sum_{B} \nu_B = -2, \quad (5.61)$$

and the number of moles of each species at equilibrium when $\xi = \xi^{eq}$ are, using Equation 5.53: $n_{N_2} = 1 - \xi^{eq}$ mol, $n_{H_2} = 3 - 3\xi^{eq}$ mol and $n_{NH_3} = 2\xi^{eq}$ mol, with the total number of moles, $n_{tot} = 4 - 2\xi^{eq}$. The corresponding mole fractions, y_B, are therefore:

$$y_{N_2} = \frac{(1-\xi^{eq})}{4-2\xi^{eq}} \quad y_{H_2} = \frac{(3-3\xi^{eq})}{4-2\xi^{eq}} \quad y_{NH_3} = \frac{2\xi^{eq}}{4-2\xi^{eq}}. \tag{5.62}$$

At equilibrium, Equation 5.60 becomes

$$\Delta_r G = 0 = \Delta_r G^{\ominus} + RT \ln \prod_B \left(y_B^{eq}\right)^{\nu_B} + RT \ln\left(\frac{p}{p^{\ominus}}\right) \sum_B \nu_B, \tag{5.63}$$

$$\Delta_r G^{\ominus} = -RT \ln \prod_B \left(y_B^{eq}\right)^{\nu_B} - RT \ln\left(\frac{p}{p^{\ominus}}\right) \sum_B \nu_B, \tag{5.64}$$

which is a combination of Equations 5.34, 5.44 and 5.45, with

$$\Delta_r G^{\ominus} = -RT \ln K^{\ominus} \quad \text{where} \quad K^{\ominus} = K_y \left(\frac{p}{p^{\ominus}}\right)^{\sum_B \nu_B}, \tag{5.65}$$

K_y being the mole fraction equilibrium constant given by the quotient in the first term of Equation 5.64 (see also Equation 5.46). Using the mole fractions from Equation 5.62 and with $\Sigma_B \nu_B = -2$, we can write for the reaction in Equation 5.53:

$$\begin{aligned} K^{\ominus} &= \frac{(2\xi^{eq})^2}{(4-2\xi^{eq})^2} \frac{(4-2\xi^{eq})}{(1-\xi^{eq})} \frac{(4-2\xi^{eq})^3}{(3-3\xi^{eq})^3} \frac{(p^{\ominus})^2}{p^2} \\ &= \frac{(2\xi^{eq})^2}{(1-\xi^{eq})} \frac{(4-2\xi^{eq})}{(3-3\xi^{eq})^3} \frac{(p^{\ominus})^2}{p^2} = \exp\left(-\frac{\Delta_r G^{\ominus}}{RT}\right). \end{aligned} \tag{5.66}$$

Since for the conditions for Figure 5.2, $\Delta_r G^{\ominus} = 7.4$ kJ/mol, $p = 0.5$ MPa, $p^{\ominus} = 0.1$ MPa and $T = 500$ *K*, we can solve Equation 5.66 for ξ^{eq}. This give $\xi^{eq} = 0.48$, in agreement with the value indicated in Figure 5.2 where $\Delta_r G$ and $dG/d\xi$ both become zero, corresponding to a mole fraction of ammonia in the reactor, $y_{NH_3} = 2\xi^{eq}/(4 - 2\xi^{eq})$, of 0.32. The corresponding equilibrium constants are $K_y = 4.215$ and $K^{\ominus} = 0.169$.

Under these conditions, $\Delta_r G^{\ominus}$ is positive and $K^{\ominus} < 1$ so the reaction does not proceed spontaneously and the yield of ammonia is comparatively low. In fact under standard conditions at 298.15 K, $\Delta_r G^{\ominus}$ is −32.82 kJ/mol and $K^{\ominus}$ is 5.62×10^5 (i.e. ≫1), so thermodynamics tells us that the reaction should proceed spontaneously with essentially complete conversion to ammonia. So why is the Haber process not conducted under these very mild conditions? The answer is that the activation energy, ΔG^* (see Question 5.7.1), for the reaction to proceed is extremely high (~400 kJ/mol) because both nitrogen and oxygen require a lot of energy to dissociate. This means that at 298.15 K, although thermodynamics indicates that the reaction should proceed spontaneously, the thermal energy available is insufficient to overcome the energy barrier ΔG^* and consequently the speed with which the reaction proceeds is negligible; the reaction is said to be kinetically inhibited. Increasing the temperature means that a higher proportion of molecules will dissociate and overcome the activation barrier. However, since the reaction is exothermic ($\Delta_r H^\ominus = -32.82$ kJ/mol), the equilibrium constant decreases with temperature and so the thermodynamic yield of ammonia will decrease the higher the temperature is increased in order to enhance the rate of the reaction. $\Delta_r G^\ominus$ remains negative until 464 K but temperatures of over 1,000 K would be required to give viable rates, at which the mole fraction of ammonia would be negligible. The great discovery that Fritz Haber made in 1909 was that by carrying out the reaction in the presence of a metal catalyst (originally expensive osmium but later inexpensive industrial iron), the activation energy ΔG^* can be reduced significantly. The catalyst accelerates the reaction rate so that we can obtain ammonia under conditions (250–400°C, 500–650 K) where the equilibrium conversion is large enough to be useful. By 1913, with co-inventor Carl Bosch, this was developed into the first commercial ammonia plant in Oppau, Germany, and the so-called Haber-Bosch process (Leigh 2004; Flavell-While 2010) has proved to be one of the most significant inventions of the twentieth century, with fertilizers made from ammonia estimated to be responsible for providing food for half the world's population (Smil 2001).

However, elevated temperatures using metal/oxide catalysts are not sufficient to produce economic yields of ammonia on their own. In Question 5.3.4 we saw that for reactions where the number of moles of substance decreases from reactants to products, so that $\Sigma_B \nu_B$ is negative as it is in this case, an increase in pressure will increase both K_y and the equilibrium mole fractions of the products. In Equation 5.66, $K^\ominus(T)$ is determined just by the temperature and as p^2 in the denominator is increased, ξ^{eq} (and K_y) adjust upwards to compensate. Hence, if we increase the pressure from 0.5 MPa as in Figure 5.2 to 20 MPa, ξ^{eq} increases to 0.91 (with $K_y = 6{,}750$) and the ammonia mole fraction y_{NH_3} to 0.83, giving the much higher conversions required for a viable commercial process.

The effect of increasing the temperature further can be calculated using the van't Hoff isochore in the form given in Equation 5.37. Using Kirchoff's law (Equation 5.38), we find that $\Delta_r H^\ominus$ at 573 K is −103 kJ/mol and if we assume that this remains constant over the range 500–650 K (the first Ulich approximation of Question 5.3.3), then solving Equation 5.37 knowing $K^\ominus$ (T_0 = 500 K) gives $K^\ominus$ (T = 650 K) = 4.96×10^{-4}, a dramatic decrease of almost three orders of magnitude. At a reactor pressure of 0.5 MPa, ξ^{eq} is now only 0.08 with $y_{NH_3} = 0.04$. Raising the pressure to 20 MPa increases the ammonia yield to $y_{NH_3} = 0.45$ ($\xi^{eq} = 0.62$), but using catalysts that are effective for lower temperatures is clearly preferable.

5.4 How Are Equilibrium Constants Defined for Reacting Condensed Phase (Solid or Liquid) Mixtures?

As we have seen in Question 5.3, the equilibrium constant K of a (bio-)chemical reaction is the value of its reaction quotient (see Equation 7.39 and the text following Equation 5.25) at equilibrium – that is, in a state in which a system has no measurable tendency to further change and the driving forces approaches zero ($\Delta_r G = 0$, see Question 5.3.1). So far in Chapter 5, we have mainly used the pure state of a compound B at the standard pressure $p^\ominus$ and at the system temperature as the standard state and corrected the chemical potentials for non-ideality, usually using *fugacity coefficients* (see for example Equation 5.22 for chemical potential and Equations 5.50 or 5.52 for the consequent equilibrium constants). However, for liquid phases it is often more convenient to use the pure liquid compound B as the standard state and correct for non-idealities with an *activity coefficient*.

The activity coefficient has been defined in Equation 4.133 and may be seen as a correction factor that corrects Equation 4.103, which defines ideal systems, for real behavior:

$$\gamma_B \equiv \frac{\lambda_B}{x_B \lambda_B^*}. \tag{5.67}$$

Here, λ_B is the absolute activity of B in the mixture and λ_B^* stands for the absolute activity of pure B. The activity coefficient γ_B may also be interpreted as a correction factor of the Lewis fugacity rule (Equation 4.112 $\tilde{p}_B = x_B \tilde{p}_B^*$)

$$\gamma_B \equiv \frac{\tilde{p}_B}{x_B \tilde{p}_B^*}. \tag{5.68}$$

For example, if ethanol is one of the reactants or products in aqueous solution, Figure 5.3 illustrates how its fugacity varies with mole fraction and how its activity coefficient, evaluated using Equation 5.68, varies with mole fraction, approaching unity for the pure solvent. This behavior, is typical of many liquid mixtures, following Raoult's law as component mole fraction approaches unity and Henry's law as it approaches zero (see Questions 4.6.3 and 4.11.2).

Using Equation 5.68 for the fugacity $\tilde{p}_B$, Equation 4.138 leads to the following relationship of the chemical potential of B to its mole fraction

$$\mu_{B,1} - \mu_{B,1}^* = RT \ln \frac{\gamma_B x_B \tilde{p}_{B,1}^*}{\tilde{p}_{B,1}^*} = RT \ln \gamma_B x_B. \tag{5.69}$$

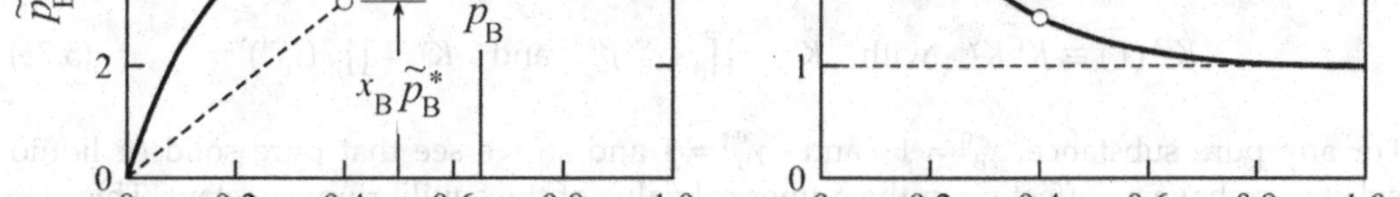

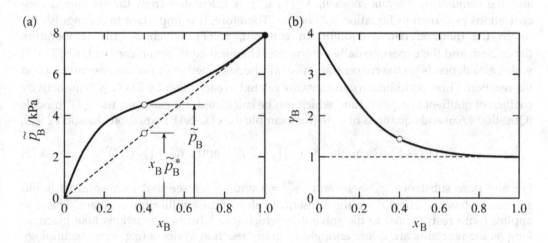

FIGURE 5.3
(a) Ethanol fugacity $\tilde{p}_B$ as a function of its mole fraction x_B in a binary mixture with water at 298 K and 0.1 MPa. The dashed line shows the (ideal solution) Raoult's law behavior and the filled circle at $x_B = 1$ is the pure liquid standard state. (b) Ethanol activity coefficients γ_B as a function of mole fraction x_B calculated using Equation 5.68 as illustrated in Figure 5.3(a).

The product $\gamma_B x_B$ is known as the activity of compound B, a_B, and its standard chemical potential $\mu_B^\ominus$ here is μ_B^*, that of pure liquid B at its saturation fugacity. We use the generic symbol $\mu_B^\ominus$ for the standard chemical potential of substance B and in this and subsequent questions show how we define it for specific conditions.

Solving this equation for μ_B, and introducing it into the definition of the Gibbs reaction energy (the driving force, Equation 5.11) and setting this driving force to zero yields the following equilibrium criterion for reacting liquid mixtures

$$\Delta_r G = 0 = \sum_B \nu_B \mu_B^{eq} = \sum_B \nu_B \left(\mu_B^\ominus + RT \ln a_B^{eq}\right) = \Delta_r G^\ominus + RT \ln \prod_B (a_B^{eq})^{\nu_B}, \quad (5.70)$$

where a_B^{eq} is called relative activity at equilibrium and is defined as

$$a_B^{eq} = \frac{\lambda_B^{eq}}{\lambda_B^\ominus} = \gamma_B^{eq}\, x_B^{eq}. \quad (5.71)$$

γ_B^{eq}, x_B^{eq} are the activity coefficient (discussed in Question 4.6.1) and the equilibrium mole fraction of the component B, respectively. The product $\prod_B (a_B^{eq})^{\nu_B}$ in Equation 5.70 is the thermodynamic equilibrium constant of the condensed phase reaction and is related to the standard Gibbs energy of reaction according to

$$K^\ominus(T) \equiv \prod_B (a_B^{eq})^{\nu_B} = \prod_B (\gamma_B^{eq} x_B^{eq})^{\nu_B} = \exp\left(-\frac{\Delta_r G^\ominus}{RT}\right) = \exp\left(-\frac{\sum_B \nu_B \mu_B^\ominus}{RT}\right). \quad (5.72)$$

Equation 5.72 results from Equations 5.70 and 5.71 by rearrangement and is also equivalent to Equation 5.26.

In practice, the equilibrium concentrations are measured in terms of mole fraction, x_B^{eq}, and the *chemical equilibrium constant*, $K^C(T, x_B^{eq})$, is calculated from the measured concentrations as shown in Equation 5.73 below. Therefore, it is important to distinguish between the thermodynamic equilibrium constant, $K^\ominus(T)$, which is only temperature dependent, and the experimentally-determined chemical equilibrium constant, $K^C(T, x_B^{eq})$ which also depends on the concentrations of all species whether or not they are involved in the reaction. These equilibrium constants are related to each other by $K^\gamma(T, \gamma_B^{eq})$ the activity coefficient quotient at equilibrium, which can be calculated, for example, using G^E models (Question 4.6.6) and equations of state (for example the ePC-SAFT equation Question 2.9.2),

$$K^\ominus(T) \equiv K^C K^\gamma \quad \text{with} \quad K^C = \prod_B (x_B^{eq})^{\nu_B} \quad \text{and} \quad K^\gamma = \prod_B (\gamma_B^{eq})^{\nu_B}. \quad (5.73)$$

For any pure substance, $x_B^{eq} = 1$ and $\gamma_B^{eq} = 1$ and so we see that pure solid or liquid substances have no effect upon the numerical value of the equilibrium constant. This also applies (with restrictions) to the solvent in which (bio-)chemical reactions take place, as long as the reactants are dilute enough. In many reaction systems (e.g. in biotechnology but not only there), the solvent contains so many substances that it is impractical to work with mole fractions. In this case, the equilibrium constant refers to the molality (Question 1.3.15) and consequently a different activity coefficient is needed, which approaches unity for the infinite dilution state (see Question 5.4.2.1).

TABLE 5.1
Values of equilibrium constants and standard thermodynamic functions of reaction for the dehydration reactions of Equations 5.74 and 5.75

Product	T /K	K^C	K^γ	$K^\ominus$	$\Delta_r H^\ominus$ /(kJ mol^{-1})	$\Delta_r S^\ominus$ /(kJ mol^{-1}K^{-1})	$\Delta_r G^\ominus$ /(kJ mol^{-1})
EOO	410	17.4	2.7	47.0			
					−18.9±1.3	−13.6±3.0	−14.8±2.2
	463	8.6	2.3	19.8			
EOE	410	7.1	2.1	14.9			
					−12.1±0.9	−7.6±2.1	−9.9±1.5
	463	3.9	2.4	9.4			

As an example of using Equation 5.73, we examine the liquid phase dehydration reactions between 1-octanol and ethanol to produce 1-ethoxyoctane (EOO)

$$C_8H_{17}OH + C_2H_5OH \rightleftharpoons C_2H_5 - O - C_8H_{17} + H_2O, \tag{5.74}$$

and of ethanol to produce ethoxyethane(EOE)

$$2C_2H_5OH \rightleftharpoons C_2H_5 - O - C_2H_5 + H_2O. \tag{5.75}$$

EOO has the potential for improving the quality of diesel blends due to its high cetane blending number (Pecci et al. 1991) and also as a biofuel, with reduced carbon dioxide net emissions since it can be manufactured from bioethanol. These equilibria were studied (Guilera et al. 2013) in a stirred batch (closed-system) reactor, using an ion-exchange resin as a catalyst at 2.5 MPa and gas-liquid chromatography to analyze the reacting mixture, to obtain the mole fractions and hence K^C. A sample of the results is set out in Table 5.1.

The activity coefficients used in Equation 5.57 to calculate K^γ were obtained using the UNIFAC-DORTMUND group contribution predictive method (Wittig et al. 2003). The values of $\Delta_r H^\ominus$, $\Delta_r S^\ominus$ and $\Delta_r G^\ominus$ were determined from the temperature variation of $K^\ominus$, for five temperatures over the range 410 K to 463 K, using the equations of Questions 5.3.2 and 5.3.3. It can be seen that the contributions from non-ideality through K^γ are significant so need to be included to calculate accurate conversions and thermodynamic quantities. Both reactions are exergonic (negative $\Delta_r G^\ominus$) and exothermic, which means they strongly favor the dehydrated products with high conversions but that product yields will decrease with increasing temperature.

5.4.1 What Is the Equilibrium Constant for Reacting Multiphase Mixtures?

Often there are also reversible reactions in which the species occur both in the condensed phase (solid and/or liquid) and in the gas phase in equilibrium with it. An example is the thermal decomposition of solid calcium carbonate to solid calcium oxide with the elimination of gaseous carbon dioxide

$$CaCO_3(s) \rightleftharpoons CaO(s) + CO_2(g). \tag{5.76}$$

Again, the starting points are Equation 5.11 in Question 5.2.2 and Equation 5.19 in Question 5.3.1, describing the equilibrium criterion, but here chemical potentials with different reference states for the condensed (c) and gas (g) phases have to be used. This leads to

$$\Delta_r G = 0 = \sum_{B,c} \nu_{B,c}\mu_{B,c}^{eq} + \sum_{B,g} \nu_{B,g}\mu_{B,g}^{eq}. \tag{5.77}$$

For reactions involving solids, liquids and gases, the equilibrium constant is defined as (see Equations 5.43 and 5.72)

$$K^{\ominus}(T) = \prod_{B,s} a_{B,s}^{\nu_{B,s}} \prod_{B,l} a_{B,l}^{\nu_{B,l}} \prod_{B,g} \left(\frac{y_B^{eq} p}{p^{\ominus}}\right)^{\nu_{B,g}} \exp\left(\sum_{B,g} \nu_{B,g} \int_0^p \left(\frac{V_{B,g}(T,p,y_C^{eq})}{RT} - \frac{1}{p}\right) dp\right), \tag{5.78}$$

which is related to the standard Gibbs energy, using Equation 5.34, by

$$K^{\ominus}(T) = \exp\left(-\frac{\Delta_r G_m^{\ominus}}{RT}\right) = \exp\left(-\frac{\sum_{B,s}\nu_{B,s}\mu_B^{\ominus} + \sum_{B,l}\nu_{B,l}\mu_B^{\ominus} + \sum_{B,g}\nu_{B,g}\mu_B^{\ominus}}{RT}\right). \tag{5.79}$$

Note that the same approximations described in Question 5.4 for reactions of liquids and solids and in Question 5.3.4 for reactions of gas mixtures are also applicable here. We can evaluate the equilibrium constant for the reaction of Equation 5.73 from the standard values (enthalpies of formation, Gibbs energies of formation, molar entropies, molar heat capacities) given in Table 5.2 using Equation 5.14 for $\Delta_r G^{\ominus}$, Equation 5.6 for $\Delta_r H^{\ominus}$ and Equation 5.17 for $\Delta_r S^{\ominus}$ and $\Delta_r C_p^{\ominus}$. $\Delta_r S^{\ominus}$ is for instance given by the equation

$$\Delta_r S^{\ominus} = \sum_{B} \nu_B S_B^{\ominus}. \tag{5.80}$$

Following this procedure, we obtain $\Delta_r H^{\ominus} = 179.4$ kJ mol^{-1}, $\Delta_r S^{\ominus} = 163.9$ J mol^{-1} K^{-1}, $\Delta_r G^{\ominus} = 130.5$ kJ mol^{-1} and $\Delta_r C_p^{\ominus} = -3.2$ J mol^{-1} K^{-1}. Applying the van't Hoff isotherm, Equation 5.25, we find that

$$\begin{aligned} K^{\ominus}(T = 298\text{K}) &= \exp\left(\frac{-130500}{8.314 \times 298.15}\right) = 1.37 \times 10^{-23} \\ &= \frac{x_{CaO}}{x_{CaC}} \frac{p_{CO_2}}{p^{\ominus}} \end{aligned} \tag{5.81}$$

TABLE 5.2
$\Delta_f G^{\ominus}$, $\Delta_f H^{\ominus}$, $S^{\ominus}$ and $C_p^{\ominus}$ values for the reactants and products of the calcium carbonate thermal decomposition reaction, Equation 5.76

	Calcium Carbonate (s) [CaC]	Calcium Oxide (s) [CaO]	Carbon Dioxide (g)
$\Delta_f G^{\ominus}$ /(kJ mol^{-1})	−1128.2	−603.3	−394.4
$\Delta_f H^{\ominus}$/(kJ mol^{-1})	−1207.8	−634.9	−393.5
$S^{\ominus}$/(J mol^{-1}K^{-1})	88.0	38.1	213.8
$C_p^{\ominus}$ /(J mol^{-1} K^{-1})	82.3	42.0	37.1

Since calcium carbonate and calcium oxide are present as pure phases, $x_{CaO} = x_{CaC} = 1$ (since $\gamma_{CaO} = \gamma_{CaC} = 1$, we can use mole fractions rather than activities): recognizing that $p^{\ominus} = 0.1$ MPa we have

$$p_{CO_2} = K^{\ominus}(T = 298\text{K})\ p^{\ominus} = 1.37 \times 10^{-24}\ \text{MPa}. \tag{5.82}$$

We see that at 298.15 K the partial pressure of CO_2 is negligible and thus calcium carbonate is stable at room temperature. However, if we evaluate $p_{CO_2} = K^{\ominus}(T)p^{\ominus}$ at 1,273 K, using Equation 5.41 with the first Ulich approximation, we obtain p_{CO_2}= 1.61 MPa and with the second Ulich approximation 0.92 MPa. Unlike the Haber-Bosch process, since this reaction is endothermic $K^{\ominus}$ increases with increasing temperature. An equilibrium that is almost completely on the reactant side at room temperature turns to the product side at high temperatures. That is the reason why limestone decomposition (or "calcining") is carried out in the temperature range between 1,173 K and 1,573 K (see calciner conditions for calcium looping carbon capture process, Question 6.3.5). Furthermore, the comparison of the results from the first and second Ulich approximations indicates how inaccurate simple extrapolations over a wide temperature range can be and how important it is to take into account the temperature dependence of the reaction enthalpy.

5.4.2 How Do We Define the Standard States for Solutions Containing Only Partially Soluble Solutes and Calculate Equilibrium Constants for Their Reactions?

As long as pure B is a liquid substance, pure liquid B is an appropriate choice of standard state, because mixtures of liquids often behave as shown in Figure 5.3. When x_B is equal to 1, the two fugacities in Equation 5.68 become identical, the activity coefficient must also be equal to 1 and therefore according to Equation 5.69

$$\mu_{B,l} = \mu^*_{B,l}, \tag{5.83}$$

i.e. the chemical potential of B is equal to that of pure B.

However, if a solute is only partially soluble in the liquid mixture, e.g. because it is a solid or a gas in the pure state, the standard state used so far does not exist physically. It is then possible to use an *infinitely dilute standard state* based on the observation that under such conditions the fugacities of solutes are usually proportional to their (small) mole fractions (Sandler 1989). The proportionality constant $\tilde{p}^{\infty}_{B,l}$ can be interpreted as the fugacity of B at infinite dilution but extrapolated to pure B

$$\tilde{p}_{B,l}(x_B \to 0) \propto x_B; \quad \tilde{p}_{B,l}(x_B \to 0) = x_B\tilde{p}^{\infty}_{B,l}. \tag{5.84}$$

The typical behavior of the liquid fugacity $\tilde{p}_{B,l}$ as a function of the mole fraction of B is shown as the solid curve in Figure 5.4 and compared with Equation 5.84 (which yields the upper dashed line). If the fugacity coefficients ϕ_B can be assumed to be unity, this equation reduces to Henry's law, with the standard state fugacity becoming the Henry's law constant. In order to describe the behavior of a real reacting mixture, a new activity coefficient has to be defined as follows

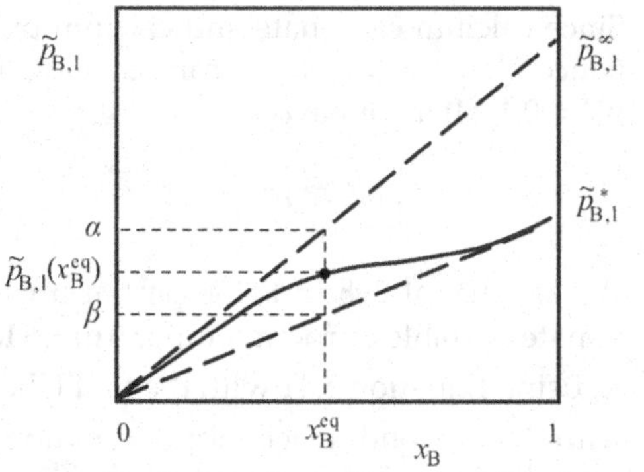

FIGURE 5.4
Fugacity of a solute as a function of its mole fraction. Solid line: real behavior, assuming that solute B exists as a liquid in pure state; lower dashed line: Lewis fugacity rule, Equation 4.112; upper dashed line: Equation 5.84.

$$\gamma^{\infty}_B = \frac{\tilde{p}_{B,l}}{x_B \tilde{p}^{\infty}_{B,l}}. \tag{5.85}$$

Adapting Equation 5.69 to this situation yields

$$\mu_{B,l} - \mu^{*}_{B,l} = RT \ln\left(\frac{x_B \tilde{p}^{\infty}_{B,l}}{\tilde{p}^{*}_{B,l}} \gamma^{\infty}_B\right). \tag{5.86}$$

The new activity coefficient γ^{∞}_B is numerically different from γ_B because it now indicates the deviation of the fugacity with respect to Eq. 5.84 instead of the Lewis fugacity rule (Equations 4.112 and 5.68).

The term $(\tilde{p}^{\infty}_{B,l}/\tilde{p}^{*}_{B,l})$ in the logarithm of Eq. 5.86 can be combined with the chemical potential of the pure liquid B in order to define a new standard state where the chemical potential of B is $\mu^{\infty}_{B,l}$ (McQuarrie and Simon 1999)

$$\mu^{*}_{B,l} + RT \ln \frac{\tilde{p}^{\infty}_{B,l}}{\tilde{p}^{*}_{B,l}} = \mu^{\infty}_{B,l}. \tag{5.87}$$

Combining Equations 5.86 and 5.87 yields

$$\mu_{B,l} - \mu^{\infty}_{B,l} = RT \ln x_B \gamma^{\infty}_B. \tag{5.88}$$

This equation describes the chemical potential of a partially soluble solute B in a liquid mixture as a function of its mole fraction. As the mole fraction of B tends to zero, the activity coefficient tends to unity owing to Eq. 5.84 and 5.85. If we extrapolate this condition to $x_B = 1$, the chemical potential of B is equal to $\mu^{\infty}_{B,l}$. This new standard state is therefore the hypothetical state of pure liquid B in which the environment of each molecule is the same as it would be at infinite dilution, whose fugacity is obtained by extrapolation of the infinite dilution behavior as shown in Figure 5.4. Another way to envisage this is that the reference state where the measurement of $(d\tilde{p}/dx)$ is actually made is at infinite dilution and the standard state is this extrapolated to $x_B = 1$ using this infinite dilution value of $(d\tilde{p}/dx)$.

The two activity coefficients γ_B^∞ and γ_B may be compared by noting that they both describe the same fugacity (Sandler 1989). Solving both Equations 5.85 and 5.68 for $\tilde{p}_{B,1}$ we obtain:

$$\tilde{p}_{B,1} = \gamma_B^\infty x_B \tilde{p}_{B,1}^\infty = \gamma_B x_B \tilde{p}_{B,1}^*. \tag{5.89}$$

Solving for γ_B^∞ and noting that $\tilde{p}_{B,1}^\infty$ can be expressed as $\gamma_B(x_B = 0)\tilde{p}_B^*$ (which can be seen using Figure 5.4 by equating $\tilde{p}_{B,1}^\infty$ to the slope of the real fugacity curve as $x_B \to 0$, which is also given by the slope of the "ideal" Lewis fugacity curve $\tilde{p}_{B,1}^*$ corrected by γ_B at this mole fraction, $x_B \to 0$) we find

$$\gamma_B^\infty(x_B) = \frac{\gamma_B(x_B)}{\gamma_B(x_B = 0)}. \tag{5.90}$$

We can see the numerical difference between the activity coefficient used in Question 5.4 based on pure liquid compound B, γ_B^{eq}, and the one used here, $\gamma_B^{\infty,eq}$, by considering the filled circle point on the real fugacity – mole fraction curve in Figure 5.4, where for an equilibrium mole fraction x_B^{eq} the fugacity is $\tilde{p}_{B,1}(x_B^{eq})$. For this reaction mixture, $\gamma_B^{eq} = \tilde{p}_{B,1}(x_B^{eq})/\beta$ which is greater than 1, whereas $\gamma_B^{\infty,eq} = \tilde{p}_{B,1}(x_B^{eq})/\alpha$, which is less than 1. The values of α $(=x_B^{eq}\tilde{p}_{B,1}^\infty)$ and β $(=x_B^{eq}\tilde{p}_{B,1}^*)$ are illustrated in Figure 5.4.

Equation 5.88 shows us that we can replace $\mu_B^\ominus$ in previous equations for the driving force (such as Equation 5.70) by $\mu_{B,1}^\infty$ and the activity a_B by $x_B\gamma_B^\infty$. Equilibrium constants can be evaluated using Equation 5.73 with $\gamma_B^{\infty,eq}$ replacing γ_B^{eq} in K^γ. We give an example of how we do this in Question 5.4.2.2.

5.4.2.1 How Do We Define Solute Standard States Using Molality?

In many reaction systems (e.g. in biotechnology but also elsewhere), the solvent contains so many substances that it is impractical to work with mole fractions. In particular, when working with reasonably diluted chemical systems (*solutions* rather than *liquid mixtures*, to use the terminology of Question 1.3.15), mole fraction is not a very convenient concentration measure for the solutes because the values are all <<1, in contrast to the solvent for which $x \sim 1$. On a linear scale that includes the solvent concentration, the solute values cluster close to $x = 0$ and cannot be easily distinguished; yet for solutions we are usually more focused on the behavior and reactions of the solutes than of the solvent. We saw in Question 4.11 that for dilute solutions the *molality*, m_B, is often used instead, indicating the number of moles of B per kg of solvent A. The relationship between mole fractions and molality is given in Equations 4.236 and 4.237. Given that in dilute systems $\sum_B x_B << 1$, these show that to a good approximation $m_B \sim x_B/M_A$, where M_A is the molar mass of the solvent A in kg. Hence, for aqueous solutions, $m_B \sim x_B/(18 \times 10^{-3}) \sim 55\, x_B$. In this case, the equilibrium constant is based on molality (see Question 1.3.15) and consequently a different activity coefficient is needed, which also approaches unity for the infinite dilution state but in this case as $m_B \to 0$.

Using molalities, Equation 5.84 has to be reformulated as

$$\tilde{p}_{B,1}(m_B \to 0) \propto m_B; \qquad \tilde{p}_{B,1}(m_B \to 0) = m_B \tilde{p}_{B,1}^{\infty,m}/m^\ominus. \tag{5.91}$$

The standard molality, $m^\ominus = 1\text{mol kg}^{-1}$ (see Equation 4.248), is included in Equation 5.91 so that the units of the fugacity of B at infinite dilution $\tilde{p}_{B,1}^{\infty,m}$ continue to be Pascals. In Figure 5.4, $\tilde{p}_{B,1}^{\infty,m}$ replaces $\tilde{p}_{B,1}^{\infty}$ on the right-hand axis and m_B replaces x_B on the x-axis. Modifying Equation 5.86 in the same way yields

$$\mu_{B,1} - \mu_{B,1}^* = RT \ln \frac{m_B \tilde{p}_{B,1}^{\infty,m}}{\tilde{p}_{B,1}^*} \gamma_B^{\infty,m}. \tag{5.92}$$

The chemical potential of B at infinite dilution is then written as

$$\mu_{B,1}^{\infty,m} = \mu_{B,1}^* + RT \ln \frac{\tilde{p}_{B,1}^{\infty,m}}{\tilde{p}_{B,1}^*}. \tag{5.93}$$

Combining Equations 5.92 and 5.93, and by analogy with Equations 5.87 and 5.88, the chemical potential of B in the liquid mixture as a function of molality then reads

$$\mu_{B,1} - \mu_{B,1}^{\infty,m} = RT \ln \frac{m_B \gamma_B^{\infty,m}}{m^\ominus}, \tag{5.94}$$

which is equivalent to Equation 4.248 when the reference chemical potential is that for infinite dilution.

5.4.2.2 How Do We Define Solute Standard States Using Molar Concentrations?

Another very common concentration measure is the *molarity* c_B, indicating the number of moles of B per dm^3 (or liter) of mixture (although it should be noted that molarity is a poor measure of concentration from a thermodynamic point of view because the volume of the mixture depends on both temperature and composition). If we use molarity, Equation 5.91 has to be reformulated as

$$\tilde{p}_{B,1}(c_B \to 0) \propto c_B; \quad \tilde{p}_{B,1}(c_B \to 0) = c_B \tilde{p}_{B,1}^{\infty,c} / c_B^\ominus, \tag{5.95}$$

where c_B is the molarity of B in the mixture in mol dm^{-3}. A standard molarity, $c^\ominus = 1$ mol dm^{-3}, needs to be included in Equation 5.95, so that the units of the fugacity of B at infinite dilution $\tilde{p}_{B,1}^{\infty,c}$ can be expressed in Pascals. Again $\tilde{p}_{B,1}^{\infty,c}$ replaces $\tilde{p}_{B,1}^{\infty}$ on the right-hand axis of Figure 5.4, with c_B replacing x_B as the concentration measure. Modifying Equation 5.86 in a similar way yields

$$\mu_{B.1} - \mu_{B,1}^* = RT \ln \frac{c_B \tilde{p}_{B,1}^{\infty,c}}{c^\ominus \tilde{p}_{B,1}^*} \gamma_B^{\infty,c}. \tag{5.96}$$

The chemical potential of B at infinite dilution is then written as

$$\mu_{B.1}^{\infty,c} = \mu_{B,1}^* + RT \ln \frac{\tilde{p}_{B,1}^{\infty,c}}{\tilde{p}_{B,1}^*}. \tag{5.97}$$

so that the chemical potential of B in the liquid mixture as a function of molarity then reads

$$\mu_{B,l} - \mu_{B,l}^{\infty,c} = RT \ln \frac{c_B \gamma_B^{\infty,c}}{c^{\ominus}}. \tag{5.98}$$

In both Equations 5.94 and 5.98, the activity coefficient will tend to unity if the concentration of the solute becomes very low because of the proportionality between the fugacity and the concentration measures in Equations 5.91 and 5.95. If at the same time the concentration becomes 1 mol kg^{-1} or 1 mol dm^{-3}, the chemical potential of B becomes equal to $\mu_{B,l}^{\infty,m}$ or $\mu_{B,l}^{\infty,c}$, respectively. The standard states are therefore defined as the solute B at infinite dilution, but extrapolated to the hypothetical states at 1 mol kg^{-1} or 1 mol dm^{-3}, respectively, in which the environment of each molecule is the same as it would be at infinite dilution.

Figure 5.5 illustrates for the 1-butanol-water system the difference between the fugacity curves plotted as a function of mole fraction and molality and between the resulting activity coefficients for butanol obtained using Equation 5.85 and its molality equivalent. 1-butanol has limited solubility in water, with $(x_B)_{max} = 0.015$; yet in molality terms this corresponds to $m_B = 0.83$ mol kg^{-1}, using $m_B \sim x_B/M_A$ from Question 5.4.2.1. Figures 5.5(a) and (b) show the

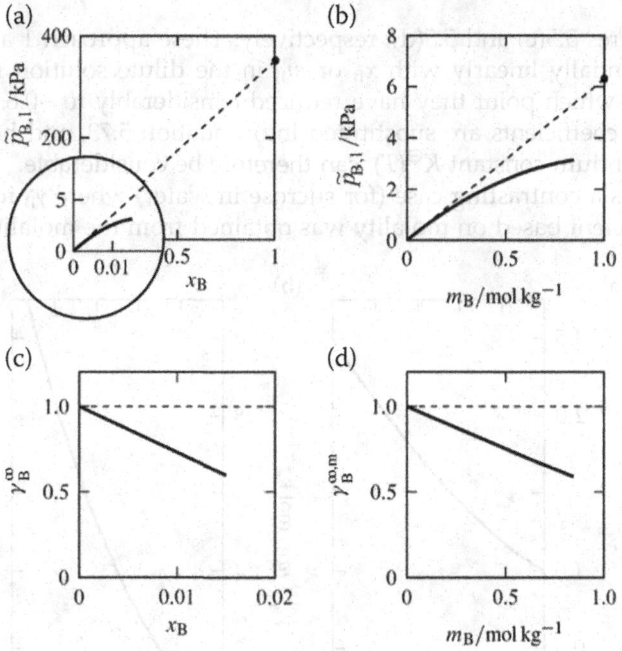

FIGURE 5.5
Fugacities and activity coefficients of dilute aqueous solutions of 1-butanol at 50.08 °C and 0.1 MPa (using the data of Fischer and Gmehling 1994). (a) shows the fugacity $\tilde{p}_{B,l}$ in an equilibrated gas phase measured as a function of 1-butanol mole fraction x_B, with the dilute region magnified to show the solubility limit of $x_B = 0.015$; the dashed line is extrapolated with the $x_B \to 0$ (Henry's law) slope to the 1-butanol solute standard state based on mole fraction, indicated by by the solid circle at $x_B = 1$. (b) shows similar data plotted as a function of 1-butanol molality, m_B, up to the solubility limit of 0.83 mol kg^{-1}; the dashed line is extrapolated with the $m_B \to 0$ slope to the 1-butanol solute standard state based on molality, indicated by the solid circle at $m_B = 1$ mol kg^{-1} = m$^{\ominus}$. (c) activity coefficients evaluated on a mole fraction basis using Equation 5.99. (d) Activity coefficients evaluated on a molality basis using Equation 5.100.

fugacity over this limited solubility region plotted as a function of both mole fraction and molality, respectively. By treating 1-butanol as a solute, we locate the solute standard state by linear extrapolation of the infinite dilution slope to the standard composition of $x_B = 1$ and $m_B = 1$ mol kg^{-1}, respectively. We see that the fugacities for the two standard states, $\tilde{p}^{\infty}_{B,1}$ (x or m), are very different. In Figure 5.5(a) for mole fractions, $\tilde{p}^{\infty}_{B,1} = 350$ kPa, which is equal to the Henry's law constant, H_{xB}, for 1-butanol-water at 50°C and 0.1 MPa. In Figure 5.5(b) for molalities, $\tilde{p}^{\infty,m}_{B,1} = 6.3$ kPa, equal to the molality Henry's law constant, H_{mB}, $= M_A H_{xB} = (18 \times 350/1{,}000)$ kPa (where M_A is the molar mass of the solvent (water); see Equation 1.23). The activity coefficients determined using Equation 5.85 in the forms

$$\gamma^{\infty}_{B} = \frac{\tilde{p}_{B,1}}{x_B \tilde{p}^{\infty}_{B,1}}, \tag{5.99}$$

and

$$\gamma^{\infty,m}_{B} = \frac{m^{\ominus} \tilde{p}_{B,1}}{m_B \tilde{p}^{\infty,m}_{B,1}}, \tag{5.100}$$

are shown in Figures 5.5(c) and 5.5(d), respectively. These approach 1 at infinite dilution and decrease essentially linearly with x_B or m_B in the dilute solution regime up to the solubility limit, at which point they have reduced considerably to ~0.6. The effect on K^{γ} when the activity coefficients are substituted into Equation 5.73, and hence on the thermodynamic equilibrium constant $K^{\ominus}(T)$, can therefore be considerable.

Figure 5.6 shows a contrasting case (for sucrose in water) where γ_B increases with m_B. The activity coefficient based on molality was obtained from the molality dependence of

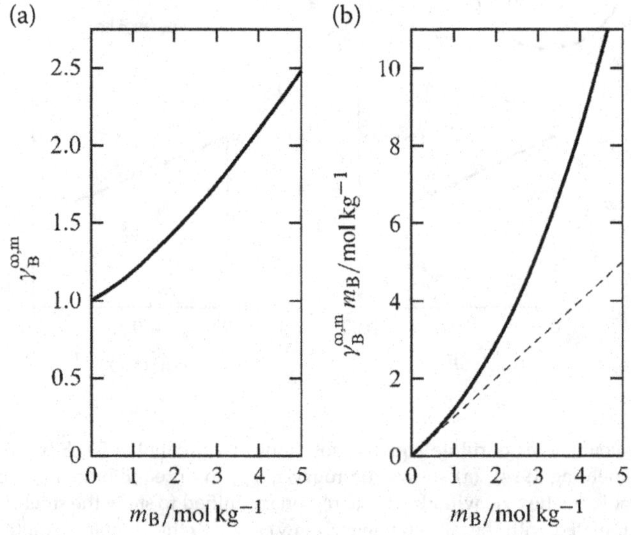

FIGURE 5.6
Molality-based activity coefficients for aqueous sucrose solutions at 25°C, 0.1 MPa (based on Robinson and Stokes 2002). (a) sucrose activity coefficient $\gamma^{\infty,m}_{B}$, evaluated from osmotic coefficients using Equation 4.247. (b) effective molality of sucrose (product of molality and $\gamma^{\infty,m}_{B}$) as a function of molality, showing the deviation from the ideal dilute behavior (dashed line).

the osmotic coefficients for the solution using Equation 4.247. It can be seen that the activity coefficient increases by more than a factor of 2 as the sucrose reaches a molality of 2.5 mol kg^{-1}, enhancing the effective concentration (Figure 5.6(b)) of the solute considerably. In this case, the dependence of $\gamma_B^{\infty,m}$ on m_B is significantly non-linear once m_B increases beyond the ideal dilute region. These departures from ideality will have a significant effect on reactions involving sucrose in aqueous solution, such as its hydrolysis to glucose and fructose in the presence of dilute acids (Goldberg et al. 1989), and are typical of many polysaccharide systems.

5.4.2.3 How Do We Determine Equilibrium Constants for Solution Reactions Using Different Measures of Solute Concentration?

To determine the thermodynamic equilibrium constants $K^{\ominus}(T)$ using molality or molarity as the measures of concentration and the standard states described in Questions 5.4.2.1 and 5.4.2.2, $\gamma_B^{\infty,eq,m}$ or $\gamma_B^{\infty,eq,c}$ must replace γ_B^{eq} in the K^{γ} term of Equation 5.73 and $m_B{}^{eq}$ and $c_B{}^{eq}$ replace x^{eq} in the K^C term. Experimental vapor pressure/fugacity vs m_B or c_B curves for each solute reactant or product can be used to obtain $\gamma_B^{\infty,eq,m}$ or $\gamma_B^{\infty,eq,c}$ using the construct of Figure 5.4 described in Question 5.4.2 in the section following Equation 5.90 and replacing Equation 5.85/5.99 for $\gamma_B^{\infty,eq}$ with Equation 5.100 for $\gamma_B^{\infty,eq,m}$ or its equivalent for $\gamma_B^{\infty,eq,c}$. Alternatively, the activity coefficients may be evaluated using thermodynamic software packages, such as GE models (Question 4.6.8) and equations of state (Questions 2.9.2 and 5.4) – see also Chapter 9. These then replace γ_B^{eq} in the K^{γ} term, enabling $K^{\ominus}(T)$ to be evaluated using the appropriate concentration measure for x_B in K^C.

$\Delta_r G^{\ominus}$ for the reaction may be evaluated directly from $K^{\ominus}(T)$ using Equation 5.26 and $\Delta_r G$ for non-standard state and non-equilibrium conditions can be determined using the equivalents of Equation 5.70 for molality or molarity

$$\Delta_r G = \sum_B \nu_B \mu_B = \sum_B \nu_B \left(\mu_{B,l}^{\infty,m} + RT \ln \frac{m_B \gamma_B^{\infty,m}}{m_B^{\ominus}} \right) = \Delta_r G^{\ominus} + RT \ln \prod_B \left(\frac{m_B \gamma_B^{\infty,m}}{m_B^{\ominus}} \right)^{\nu_B}, \quad (5.101)$$

$$\Delta_r G = \sum_B \nu_B \mu_B = \sum_B \nu_B \left(\mu_{B,l}^{\infty,c} + RT \ln \frac{c_B \gamma_B^{\infty,c}}{c_B^{\ominus}} \right) = \Delta_r G^{\ominus} + RT \ln \prod_B \left(\frac{c_B \gamma_B^{\infty,c}}{c_B^{\ominus}} \right)^{\nu_B}. \quad (5.102)$$

The example given in Question 5.4 for the dehydration of alcohols using mole fraction as the liquid mixture concentration measure should be useful, by analogy, to illustrate the above procedure for solutions where molality or molarity are more appropriate, or the only data available.

5.5 What Are Galvanic Cells?

When metallic zinc Zn(s), which is a silver-colored material, is placed in an aqueous solution of copper sulfate with a chemical formula $CuSO_4$(aq), the color of the zinc will in time change to brown. The color change is the result of Cu(s) depositing on the outer surface of the Zn(s) owing to an electrochemical reaction taking place in the solution

$$Zn(s) + Cu^{2+}(aq) \rightarrow Zn^{2+}(aq) + Cu(s). \tag{5.103}$$

Clearly, after a short time, the Zn(s) is plated with Cu(s) and the reaction ceases. This cessation can be prevented by separating the Zn(s) from the copper ions in solution, Cu^{2+}(aq), by placing a high molality aqueous solution of copper sulfate in the bottom of a beaker and carefully pouring on top of it an aqueous solution of zinc sulfate of relatively low molality to minimize the mixing of the two solutions. The $ZnSO_4$(aq) floats on the $CuSO_4$(aq) because of the difference in density of the two solutions. This separation of the two solutions also enables us to harness the electrons flowing between them when they are connected via an external circuit. A Cu(s) electrode is then placed in the bottom of the beaker in contact with Cu^{2+}(aq) while a Zn(s) electrode is suspended in the upper layer of $ZnSO_4$(aq) in contact with Zn^{2+}(aq). This arrangement forms a battery and was used to provide electricity for early telephone systems.

A more convenient method of separating the two aqueous solutions is shown in Figure 5.7 and is known as a galvanic cell. The left-hand beaker of Figure 5.7 contains $ZnSO_4$(aq) of molality about 1 mol kg^{-1} in contact with metallic Zn(s), while the right-hand beaker contains $CuSO_4$(aq) also of molality of about 1 mol kg^{-1} in contact with metallic Cu(s). In the absence of a connection between the two beakers nothing happens. However, when the metallic electrodes are, as shown in Figure 5.7, interconnected by a cable and the solutions by a salt bridge, in this case a high concentration potassium chloride KCl solution or gel contained in a tube sealed at each end by a porous plug, the reduction-oxidation (or redox) reaction of Equation 5.103 takes place as electrons flow from the left-hand side to the right-hand side. In the left-hand beaker (the anode) the reaction

$$Zn(s) \rightarrow Zn^{2+}(aq) + 2e^{-}, \tag{5.104}$$

occurs, while in the right hand beaker (the cathode) the reaction

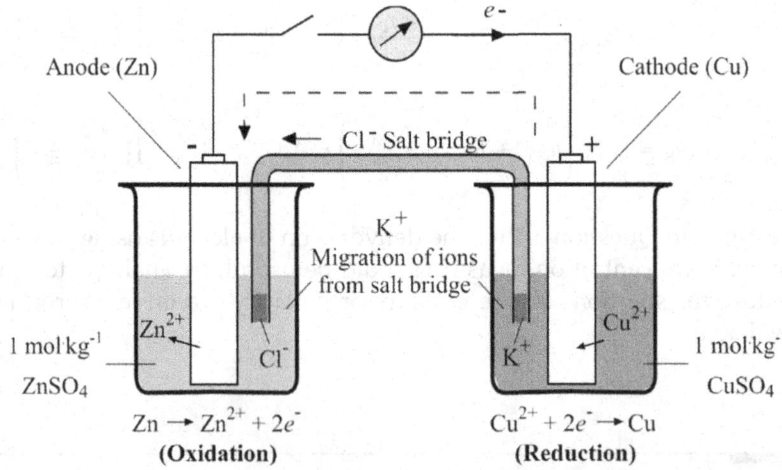

FIGURE 5.7
Schematic of a galvanic cell. LEFT: A beaker containing $ZnSO_4$(aq) and Zn(s). RIGHT: A beaker containing $CuSO_4$(aq) and Cu(s). The two beakers are interconnected by an electrically conducting wire with in this case an on/off switch, shown with an open circuit, and a galvanometer. A KCl salt bridge also interconnects the two beakers. When the circuit is closed the reaction given by Equation 5.104 occurs in the left-hand beaker and the reaction of Equation 5.105 occurs in the right-hand beaker resulting in the flow of electrons.

$$Cu^{2+}(aq) + 2e^- \rightarrow Cu(s) \tag{5.105}$$

takes place. The salt bridge completes the electrical circuit so that current can flow but the solutions cannot mix and contaminate the Zn(s) with Cu(s). The *electromotive force* (the potential difference obtained as the current tends to zero), denoted by emf E, can be used in thermodynamics to provide a method of determining the chemical potential difference for the cell reaction and it is to this that we now turn.

5.5.1 How Is Electromotive Force Related to the Cell Gibbs Energy of Reaction?

Adding one mole of charges, each of magnitude $z_B e$, where e is the electronic charge, to a system where the electrical potential is V involves an additional amount of work $z_B e\ VL$, where L is the Avogadro constant. So the chemical potential of the charged species, μ_B^e, is related to that of the same uncharged species, μ_B, by

$$\mu_B^e = \mu_B + z_B FV, \tag{5.106}$$

where μ_B^e is called the electrochemical potential and $F = eL$ is the Faraday constant, one mole of electronic charge (1 F = 96,490 C mol^{-1}).

The reaction taking place at the right-hand positive electrode (cathode) in Figure 5.7 (the "half-cell" reaction) is the reduction-oxidation (redox) reaction 5.105, where the oxidized ionic species Cu^{2+} is converted by the addition of electrons to the reduced metallic species Cu. This can be generalized for any electrode redox reaction (which by convention is always written as a reduction process) to

$$M^{z+}(\text{or Ox}) + ze^- \rightarrow M\ (\text{or Re}). \tag{5.107}$$

Using Equation 5.106

$$\mu_{M^{z+}} = \mu_M + zF\Delta V, \tag{5.108}$$

where ΔV is the potential difference between the electrode M and the M^{z+} ions in solution, called the electrode potential, E_P. Here, z is the number of electrons involved in the conversion of 1 mole of Ox into Re. Now, the Gibbs reaction energy of Equation 5.107, $\Delta_r G$, is $\mu_M - \mu_{Mz+}$; so from Equation 5.108 we see that for the half-cell reaction

$$\mu_M - \mu_{M^{z+}} = \Delta_r G = -zFE_P. \tag{5.109}$$

For the electrode potential of the right-hand half cell (R), Reaction 5.105, for which $E_P = E_R$ and z = 2, this gives

$$E_R = -\frac{\Delta_r G_R}{2F}. \tag{5.110}$$

If we also write the reaction taking place at the left-hand electrode (for which $E_P = E_L$) as a reduction reaction, the reverse of Reaction 5.104

$$Zn^{z+}(aq) + 2e^- \rightarrow Zn\,(s), \tag{5.111}$$

then

$$E_L = -\frac{\Delta_r G_L}{2F}. \tag{5.112}$$

To obtain the overall cell Reaction 5.103, we must subtract half-cell Equation 5.111 from half-cell Equation 5.105, so for the overall galvanic cell emf we obtain

$$E = E_R - E_L = \frac{[(-\Delta_r G_R) - (-\Delta_r G_L)]}{2F} = \frac{-\Delta_r G}{2F}, \tag{5.113}$$

where $\Delta_r G$ is the Gibbs energy change of the overall cell reaction. For the generalized case, corresponding to Equation 5.107 applied to the two half-cells

$$E = E_R - E_L = \frac{-\Delta_r G}{zF}. \tag{5.114}$$

Hence, by measuring the emf of a galvanic cell, under conditions as close as possible to reversible (i.e. as the current drawn tends to zero), it is possible to obtain the Gibbs energy change for the cell reaction. This could also be obtained from the electrode potential of the two half-cell reactions but they cannot be measured in isolation, only as part of an electrochemical cell. Instead, by convention, the standard electrode potential $E^\ominus$ for any individual electrode reaction is measured in a cell relative to a standard hydrogen electrode (SHE) where the activities of all the reactants are unity. This SHE is formed by bubbling hydrogen at 0.1 MPa (i.e. fugacity $\tilde{p} = 1$) and 298.15 K over an inert platinum electrode in an acidic solution (e.g. HCl) with hydrogen ion activity $a_{H+} = 1$; it is depicted as Pt|H_2|HCl, with a half-cell reaction $H^+ + e^- \rightarrow \frac{1}{2}H_2$.

A positive electrode potential indicates that reduction ($M^{z+} + ze^- \rightarrow M$) is favored relative to the SHE; for Cu^{2+}/Cu $E^\ominus = 0.34$ V so $Cu^{2+} + 2e^- \rightarrow Cu$ is favored. Conversely, a negative electrode potentials shows that oxidation is preferred: $E^\ominus$ (Zn^{2+}/Zn) is −0.76 V so that the metallic zinc will tend to lose its valence electrons and the electrode becomes negatively charged, $Zn \rightarrow Zn^{2+} + 2e^-$. Systems with higher electrode potentials are reduced by systems with lower electrode potentials. This explains why zinc is the anode in Figure 5.7 with copper as the cathode, and why Cu(s) is deposited on Zn(s) when the latter is placed in a solution of $CuSO_4$ (aq). The *standard emf* of a galvanic cell, which can be measured using a potentiometer or voltmeter with no current flowing with both reactants and products at unit activity, can also be calculated from the standard electrode potentials of its two half cells by applying Equation 5.114

$$E^\ominus = E_R^\ominus - E_L^\ominus, \tag{5.115}$$

so the standard emf of the cell shown in Figure 5.7 is

$$E^\ominus = E^\ominus_{Cu^{2+}/Cu} - E^\ominus_{Zn^{2+}/Zn} = 0.34 - (-0.76) = 1.10\text{ V}. \tag{5.116}$$

Since $E^{\ominus} > 0$, the overall cell reaction has a tendency to proceed from left to right, in the direction that it is written in Equation 5.103.

5.5.2 How Does the Cell emf Depend on Concentration?

In Question 5.2.2, we defined the Gibbs energy of reaction using Equation 5.11. Using this and the expressions for the chemical potential of liquid mixtures explained in Questions 4.5 and used in Question 5.4, Equation 5.69, we find that, in general

$$\Delta_r G = \sum_B \nu_B \mu_B = \sum_B \nu_B\left(\mu_B^{\ominus} + RT \ln a_B\right) = \Delta_r G^{\ominus} + RT \ln \prod_B a_B^{\nu_B}. \tag{5.117}$$

Using Equation 5.114, where $\Delta_r G = -zFE$, this leads to

$$E = E^{\ominus} - \frac{RT}{zF} \ln \prod_B a_B^{\nu_B}. \tag{5.118}$$

This is called the *Nernst equation* and enables the cell emf to be calculated for ion concentrations away from the equilibrium situation. At equilibrium, $\Delta_r G = -zFE = 0$, so

$$E^{\ominus} = -\frac{\Delta_r G^{\ominus}}{zF} = \frac{RT}{zF} \ln \prod_B \left(a_B^{\mathrm{eq}}\right)^{\nu_B} = \frac{RT}{zF} \ln K^{\ominus}, \tag{5.119}$$

where $K^{\ominus}$ is the standard equilibrium constant of the cell reaction, as defined in Equations 5.25 and 5.43. Equation 5.119 is the electrochemical equivalent of the *van't Hoff isotherm*, Equation 5.26.

5.5.3 How Can Irreversibility at Finite Currents Be Included to Evaluate the Cell emf?

In general, a discussion of galvanic cells should treat, for example, the speed with which ions move in a gradient of electric field. It thus involves transport phenomena that are beyond the scope of this book. Furthermore, galvanic cells are not initially at equilibrium because the gradients of chemical potential that exist within the cell always ensure diffusion occurs. Once current flows in the external circuit, the cell reaction (e.g Equation 5.103) occurs within the cell and the ionic concentrations change from their initial values until the cell reaction reaches equilibrium, at which point no current can flow. As we have seen (Equation 5.119), to relate the operation of the cell to the thermodynamics of the cell reaction, we need to avoid this irreversible behavior and draw current from the cell essentially infinitely slowly so that it operates under reversible conditions. This is usually done by using a potentiometer to balance the cell (zero current) against a constant dc supply or by using an electronic voltmeter which operates without drawing significant current. However, under certain conditions it is possible to measure the reversible cell emf when an external current is flowing and the cell is operating under irreversible conditions. To explore this, we will include the irreversible ion transport phenomena in the analysis and identify the conditions under which they may be ignored.

A galvanic cell, such as that shown in Figure 5.7, is usually replaced by a simplified representation that indicates how the solid metal electrodes, the electrolyte solutions and salt bridge are connected. For example, a general galvanic cell containing Cu(s) electrodes, unspecified solutions in which reduction and oxidation of Cu^{z+} ions can occur and a salt bridge solution is typically written as

$$\mathrm{Cu}\left|\begin{matrix}\mathrm{Re}\\ \mathrm{Ox}\end{matrix}\right|\text{bridging solution}\left|\begin{matrix}\mathrm{Re}\\ \mathrm{Ox}\end{matrix}\right|\mathrm{Cu}. \tag{5.120}$$

Each half-cell reaction is written as a reduction process as in Equation 5.105 or 5.107. The vertical lines, |, represent phase boundaries and the term "bridging solution" is often omitted so that the elimination of the liquid junction by such a device is indicated by a double vertical line | |. The emf of the cell in Equation 5.120 can be represented exactly, when current is being drawn, by a generalization of Equations 5.114 and 5.117 that includes a term that describes the transport of ions across the chemical potential gradient between the two half cells (Kjelstrup and Bedeaux 2017)

$$-FE = \frac{[\mu(\mathrm{Re}^{\mathrm{R}}) - \mu(\mathrm{Ox}^{\mathrm{R}}) - \mu(\mathrm{Re}^{\mathrm{L}}) + \mu(\mathrm{Ox}^{\mathrm{L}})]}{z} + \sum_i \int_{\mu_i^{\mathrm{L}}}^{\mu_i^{R}} (t_i / z_i)\mathrm{d}\mu_i. \tag{5.121}$$

The final term in Equation 5.121 represents the change in chemical potential as one mole of each ion migrates from one half-cell to the other, t_i is the transport number and z_i the charge number of ion i, with z the number of electrons transferred in the overall balanced cell reaction. The transport number of an ion is the fraction of the electric current arising from the flow of that ion and can be determined, for example, from the changes in concentration of the ion in the two cell compartments when a known current is passed for a measured time (Robinson and Stokes 2002, Chapter 5). Equation 5.121 can be re-written as

$$-FE = \frac{[\mu(\mathrm{Re}^{\mathrm{R}}) - \mu(\mathrm{Ox}^{\mathrm{R}}) - \mu(\mathrm{Re}^{\mathrm{L}}) + \mu(\mathrm{Ox}^{\mathrm{L}})]}{z} + \frac{\mu_i^{\mathrm{R}}}{z_i} - \frac{\mu_j^{\mathrm{L}}}{z_j} + \sum_{i\neq j} \int_{\mu_i^{\mathrm{L}}}^{\mu_i^{\mathrm{R}}} t_i\left(\frac{\mathrm{d}\mu_i}{z_i} - \frac{\mathrm{d}\mu_j}{z_j}\right), \tag{5.122}$$

which is more useful when one of the ions j is present in each part of the cell.

To illustrate the use of Equation 5.122, a specific example of the galvanic cell is considered, given by

$$\mathrm{Pt}\,|\,\mathrm{Ag}\left|\begin{matrix}\text{solution}\\ \text{of } \mathrm{AgNO_3}\end{matrix}\right|\begin{matrix}\text{bridging solution}\\ \text{of } \mathrm{AgNO_3},\ \mathrm{Fe(NO_3)_2}\\ \text{and } \mathrm{Fe(NO_3)_3}\end{matrix}\left|\begin{matrix}\text{solution}\\ \text{of } \mathrm{Fe(NO_3)_2}\\ \text{and } \mathrm{Fe(NO_3)_3}\end{matrix}\right|\mathrm{Pt}, \tag{5.123}$$

for which NO_3^- is found in each part of the system and is thus chosen as ion j with $z_j = -1$. Note that Pt is added to the left-hand electrode so that the cell emf is measured between identical electrode metal terminals to avoid the contact potential difference that occurs between two unlike metals. The half cell reactions are

$$\text{Right half-cell:}\quad \mathrm{Fe^{3+}(aq)} + e^- \rightarrow \mathrm{Fe^{2+}(aq)}, \tag{5.124}$$

$$\text{Left half-cell:}\quad \mathrm{Ag(s)} \rightarrow \mathrm{Ag^{+}(aq)} + e^-. \tag{5.125}$$

Hence, $z = 1$ and Equation 5.122 can be written for the cell of Equation 5.123 as

$$
\begin{aligned}
-FE = {} & \mu\{Fe(NO_3)_2, R\} - \mu\{Fe(NO_3)_3, R\} - \mu\{Ag, L\} + \mu\{AgNO_3, L\} \\
& + \int_{\mu^{L}_{AgNO_3}}^{\mu^{R}_{AgNO_3}} t(Ag^{+})\mathrm{d}\mu\{AgNO_3\} + \int_{\mu^{L}_{Fe(NO_3)2}}^{\mu^{R}_{Fe(NO_3)2}} \frac{t(Fe^{2+})}{2}\mathrm{d}\mu\{Fe(NO_3)_2\} \\
& + \int_{\mu^{L}_{Fe(NO_3)3}}^{\mu^{R}_{Fe(NO_3)3}} \frac{t(Fe^{3+})}{3}\mathrm{d}\mu\{Fe(NO_3)_3\},
\end{aligned} \quad (5.126)
$$

with $a^{R}_{NO3-} = a^{L}_{NO3-} = a^{bridge}_{NO3-}$ so that $\mu^{R}_{NO3-} = \mu^{L}_{NO3-}$. The first four terms on the right-hand side of Equation 5.126 involving chemical potentials constitute the Gibbs energy for the cell reaction, $\Delta_r G$, of Equation 5.117. If we wish to obtain this from measuring E under irreversible conditions when current is being drawn from the cell, we would need to know the concentration dependence all the transport numbers over the required ranges in order to evaluate the integrals in Equation 5.126. This does enable the cell emf to be calculated under current-drawing irreversible conditions when we need that, but is clearly demanding and unrealistic as a route to determining the reversible cell emf.

However, using a so-called "swamping" technique we can add to the system an inert or supporting electrolyte, such as KNO_3, with a high molality in comparison to those of the cell reactants, for the sole purpose of increasing the solution conductivity. This approach, in principle, adds an extra integral involving KNO_3, of the same form as the last three in Equation 5.126. However, this is eliminated because $\mu\{KNO_3\}$ is essentially constant so $\mathrm{d}\mu\{KNO_3\} = 0$. In addition, most of the current is carried now by the KNO_3 so the transport numbers t for the ions Ag^{+}, Fe^{2+} and Fe^{3+} and the integrals in Equation 5.126 containing them are then made sufficiently small that they may be neglected. Equation 5.126 can then be written as

$$
-FE = \mu\{Fe(NO_3)_2\} - \mu\{Fe(NO_3)_3\} - \mu\{Ag\} + \mu\{AgNO_3\}. \quad (5.127)
$$

The right-hand side of Equation 5.127 is now solely the $\Delta_r G$ of Equation 5.117 and so we can recover the reversible cell emf, without the need to draw current from the cell extremely slowly in order to ensure reversible conditions. It is also equivalent to the affinity, A, of the cell reaction, which is defined as the negative value of the Gibbs reaction energy (see Question 3.3.4)

$$
\boldsymbol{A} = -\sum_{B} \nu_B \mu_B. \quad (5.128)
$$

In this case, the galvanic cell provides E independently of the bridging solution and the thermodynamic quantities

$$
-FE = -\boldsymbol{A} = \left(\frac{\partial G}{\partial \xi}\right)_{T,p} = \Delta_r G, \quad (5.129)
$$

for the electron transfer reaction

$$
Fe(NO_3)_3 + Ag(s) \rightarrow Fe(NO_3)_2 + AgNO_3. \quad (5.130)
$$

5.5.4 How Do We Use Electrodes Which Are Reversible to Ions of Non-Metals?

The hydrogen gas/inert metal (Pt) electrode, $Pt|H_2(g)|H^+$ (which we saw in Question 5.5.1 is used under standard state conditions as the reference standard hydrogen electrode against which all standard electrode potentials are measured) is reversible to protons, H^+, with a half-cell reaction

$$H^+ + e^- \rightarrow \tfrac{1}{2}H_2(g). \tag{5.131}$$

By contrast, when chlorine dissolves in water it forms Cl^- ions so here we can form a gas/Pt electrode, $Pt|Cl_2(g)|Cl^-$, which is reversible to anions

$$\tfrac{1}{2}Cl_2(g) + e^- \rightarrow Cl^-. \tag{5.132}$$

Another type of electrode whose potential depends on the anion concentration is the metal/insoluble salt electrode; for instance, the silver/silver chloride electrode, $Ag|AgCl(s)|Cl^-$, is also reversible to chloride ions

$$AgCl + e^- \rightarrow Ag + Cl^-, \tag{5.133}$$

as is the commonly used calomel electrode $Hg|Hg_2Cl_2(s)|Cl^-$.

We can illustrate the use of these different electrodes by combining the half-cell reactions 5.131 and 5.133 to form the electron transfer reaction

$$AgCl(s) + \tfrac{1}{2}H_2(g) \rightarrow Ag(s) + H^+ + Cl^-, \tag{5.134}$$

for which the galvanic cell can be represented by

$$Pt|H_2(g)\left|\begin{matrix}\text{sol. of } H^+ \text{ and } Cl^-\\ \text{sat. with } H_2\end{matrix}\right|\begin{matrix}\text{sol. of } H^+\\ \text{and } Cl^-\end{matrix}\left|\begin{matrix}\text{sol. of } H^+ \text{ and } Cl^-\\ \text{sat. with AgCl}\end{matrix}\right|AgCl(s)\,|\,Ag\,|\,Pt. \tag{5.135}$$

For this cell, $t(Ag^+)$ is very small and the HCl is uniform throughout the cell so there is no problem running the cell under reversible conditions where the integrals involving transport numbers in Equation 5.122 vanish, so that the emf is given by

$$-FE = \mu(Ag, s) - \mu(AgCl, s) + \mu(HCl, \text{solute}) - \tfrac{1}{2}\mu(H_2, g). \tag{5.136}$$

The expressions given for the standard chemical potential of solids, solutions and ideal gases in Questions 3.7, 4.3–4.6 and 4.11 can be substituted into Equation 5.136 to give

$$\begin{aligned}-FE &= \mu^\ominus(Ag, s) - \mu^\ominus(AgCl, s) + \mu^\ominus(H^+, \text{solute}) + \mu^\ominus(Cl^-, \text{solute})\\ &\quad - \tfrac{1}{2}\mu^\ominus(H_2, g) + 2RT\ln\left(\frac{m\gamma_\pm}{m^\ominus}\right) - \tfrac{1}{2}RT\ln\left(\frac{x(H_2, g)p}{p^\ominus}\right)\\ &= FE^\ominus - \tfrac{1}{2}\mu^\ominus(H_2, g) + 2RT\ln\left(\frac{m\gamma_\pm}{m^\ominus}\right) - \tfrac{1}{2}RT\ln\left(\frac{x(H_2, g)p}{p^\ominus}\right).\end{aligned} \tag{5.137}$$

In Equation 5.137 $m^\ominus$ is the standard molality of 1 mol kg^{-1}, $p^\ominus = 0.1$ MPa and $\gamma_\pm$ is the activity coefficient of the HCl electrolyte (see Question 5.6, Equation 5.145). For substances that are solids or liquids, the differences in pressure between p and $p^\ominus$ can be ignored. The first four terms in Equation 5.137 taken together are equal to $FE^\ominus$ where $E^\ominus$ is the standard electromotive force (see Equation 5.119); for this electron transfer reaction at 298.15 K

$$E^\ominus\left\{\mathrm{AgCl(s)} + \tfrac{1}{2}\mathrm{H_2(g)} \rightarrow \mathrm{H^+} + \mathrm{Cl^-}\right\} = 0.222\ \mathrm{V}. \tag{5.138}$$

If the measured emfs, E, of the cell represented by Equation 5.135 for two molalities of HCl, m_1 and m_2, are E_1 and E_2, respectively, then subtraction of Equation 5.136 for the two measurements gives

$$\mu(\mathrm{HCl}, m_2) - \mu(\mathrm{HCl}, m_1) = \frac{E_1 - E_2}{F}, \tag{5.139}$$

which provides another route to determining the chemical potential difference for solutions.

5.6 How Can I Estimate Activity Coefficients for Ions in Electrolyte Solutions?

Ions in solution can be considered as separate components of the system subject to the requirement for electrical neutrality, given by

$$\sum_i m_i z_i = 0, \tag{5.140}$$

where m_i is the molality and z_i the charge of the ion i. The Gibbs-Duhem Equation (Equation 3.50) also applies to solutions of electrolytes subject to compliance with Equation 5.140 and, at constant temperature and pressure, is given by

$$\mathrm{d}\left\{(1-\phi_m)\sum_i m_i\right\} + \sum_i m_i\,\mathrm{d}\ln\gamma_i = 0, \tag{5.141}$$

where ϕ_m is the osmotic coefficient (see Equation 4.238). For an electrolyte $A_{\nu_+}B_{\nu_-}$ the molality of each ion is given by

$$m_+ = \nu_+ m \tag{5.142}$$

and

$$m_- = \nu_- m. \tag{5.143}$$

In view of Equations 5.142 and 5.143, Equation 5.141 can be written as

$$(\nu_+ + \nu_-)\mathrm{d}\{(1-\phi_\mathrm{m})m\} + \nu_+ m\,\mathrm{d}\ln\gamma_+ + \nu_- m\,\mathrm{d}\ln\gamma_- = 0. \quad (5.144)$$

If we define the mean activity coefficient $\gamma_\pm$ of the electrolyte as

$$(\nu_+ + \nu_-)\ln\gamma_\pm = \nu_+\ln\gamma_+ + \nu_-\ln\gamma_-, \quad (5.145)$$

then Equation 5.144 becomes

$$\mathrm{d}\{(1-\phi_\mathrm{m})m\} + m\,\mathrm{d}\ln\gamma_\pm = 0. \quad (5.146)$$

Equation 5.146 is important because it is only the mean activity coefficient $\gamma_\pm$ of the ion pair that complies with Equation 5.140 that can be measured in practice, not the individual ion activities γ_i.

In Question 4.11.2, in the context of deriving Henry's law, it was possible to state that $\phi = 1$ and $\gamma_\mathrm{B} = 1$ for a solution that was dilute, that is in practical terms for which $\Sigma_\mathrm{B}\, m_\mathrm{B} < 1$ mol kg^{-1}. This was because the pair interaction energy of nonelectrolytes in solution decreases approximately as (molality)^2. In electrolyte solutions, the pair interaction energy decreases only as the cube root of concentration and so it is not possible to make the same assumptions. However, the Debye-Hückel theory (Robinson and Stokes 2002) provides the form of $\gamma_\pm$ in both the dilute limit $\lim_{m_\mathrm{i}\to 0}\Sigma_\mathrm{i}\, m_\mathrm{i}$ and at finite m.

In the dilute limit $\lim_{m_\mathrm{i}\to 0}\Sigma_\mathrm{i}\, m_\mathrm{i}$ the Debye-Hückel law states

$$\ln\gamma_\pm = -(2\pi L\rho_\mathrm{A}^*)^{\frac{1}{2}}\left(\frac{e^2}{4\pi\varepsilon_\mathrm{A}^* k_\mathrm{B}T}\right)^{\frac{3}{2}} |z_+z_-| \left(\frac{1}{2}\sum_i m_i z_i^2\right)^{\frac{1}{2}}. \quad (5.147)$$

In Equation 5.147, the term $\frac{1}{2}\Sigma_i\, m_i z_i^2$ is called the *ionic strength* and is often given the symbol I; ρ_A^* is the density of the pure solvent, e the charge on a proton, and ε_A^* the electric permittivity of the solvent; $\varepsilon^* = \varepsilon_\mathrm{r}\varepsilon_0$ where ε_r is the relative electric permittivity and ε_0 (= 8.854 × 10^{-7} m^{-3} kg^{-1} s^4 A^2) the vacuum permittivity. We can abbreviate Equation 5.147 to

$$\ln\gamma_\pm = -\alpha|z_+z_-|(I/\mathrm{mol\ kg^{-1}})^{1/2}. \quad (5.148)$$

For water at 298.15 K, the Debye-Hückel constant α is 1.175.

For solutions of electrolytes of finite molalities (~0.1 mol kg^{-1}), the Debye-Hückel approximation is given by

$$\ln\gamma_\pm = (2\pi L\rho_\mathrm{A}^*)^{\frac{1}{2}}\left(\frac{e^2}{4\pi\varepsilon_\mathrm{A}^* k_\mathrm{B}T}\right)^{\frac{3}{2}} |z_+z_-| \frac{\left(\frac{1}{2}\Sigma_i\, m_i z_i^2\right)^{\frac{1}{2}}}{1 + d\left(\frac{1}{2}\Sigma_i\, m_i z_i^2\right)^{\frac{1}{2}} 2(2\pi L\rho_\mathrm{A}^*)^{\frac{1}{2}}\left(\frac{e^2}{4\pi\varepsilon_\mathrm{A}^* k_\mathrm{B}T}\right)^{\frac{1}{2}}}. \quad (5.149)$$

In Equation 5.149, d is an adjustable parameter called the "mean diameter of the ions". The term in the denominator $2(2\pi L\rho_\mathrm{A}^*)^{1/2}\{e^2/(4\pi\varepsilon_\mathrm{A}^* k_\mathrm{B}T)\}^{1/2}$ is approximately 3.3 × 10^9 m^{-1}kg$^{1/2}$mol$^{-1/2}$ and d is about 0.3 nm so that the product of these two quantities is about unity and Equation 5.149 can be written as

$$\ln \gamma_\pm \approx -(2\pi L\rho_A^*)^{\frac{1}{2}}\left(\frac{e^2}{4\pi\varepsilon_A^* k_B T}\right)^{\frac{3}{2}} |z_+z_-| \frac{\left(\frac{1}{2}\sum_i m_i z_i^2\right)^{\frac{1}{2}}}{1+\left(\frac{1}{2}\sum_i m_i z_i^2\right)^{\frac{1}{2}}}. \quad (5.150)$$

Equation 5.150 can be extended empirically to even higher m by adopting the form

$$\ln \gamma_\pm \approx -(2\pi L\rho_A^*)^{\frac{1}{2}}\left(\frac{e^2}{4\pi\varepsilon_A^* k_B T}\right)^{\frac{3}{2}} |z_+z_-| \frac{\left(\frac{1}{2}\sum_i m_i z_i^2\right)^{\frac{1}{2}}}{1+\left(\frac{1}{2}\sum_i m_i z_i^2\right)^{\frac{1}{2}}+\frac{1}{2}\sum_i m_i z_i^2}. \quad (5.151)$$

The use of the definition of *ionic strength* I

$$I = \frac{1}{2}\sum_i m_i z_i^2, \quad (5.152)$$

and of

$$\alpha = (2\pi L\rho_A^*)^{1/2}\left(\frac{e^2}{4\pi\varepsilon_A^* k_B T}\right)^{3/2}, \quad (5.153)$$

as well as

$$\beta = 2(2\pi L\rho_A^*)^{1/2}\left(\frac{e^2}{4\pi\varepsilon_A^* k_B T}\right), \quad (5.154)$$

permits Equation 5.149 to be written as

$$\ln \gamma_\pm = -\alpha |z_+z_-| \frac{I^{\frac{1}{2}}}{1+d\, I^{\frac{1}{2}}\beta}. \quad (5.155)$$

Equations 5.147 and 5.155 are applied to biological system in Question 8.1.2.

We can also obtain the osmotic coefficient by integration of Equation 5.146 to give

$$1-\phi_m = -\frac{1}{m}\int_1^{\ln \gamma_\pm} m\, d\ln \gamma_\pm. \quad (5.156)$$

If we use Equation 5.148 for $\ln \gamma_\pm$, Equation 5.152 to give I for ions of charge z_+ and z_- and transform Equation 5.156 to an integral over dm, performing the integration in Equation 5.156 leads, for the dilute limit, to

$$1-\phi_m = \frac{1}{3}\alpha |z_+z_-| I^{1/2} = -\frac{1}{3}\ln \gamma_\pm. \quad (5.157)$$

As for the mean ionic activity coefficient, we can extend this limiting law to finite molalities. The equivalent of Equation 5.155 with additional terms that make the equation valid to molalities of order 1 mol kg^{-1} is

$$1 - \phi_m = \frac{1}{3}\alpha|z_+z_-|\frac{I^{1/2}}{1 + d\,I^{1/2}\beta} + \frac{m}{m^\ominus}(\beta' + \beta''\exp(-cI^{1/2})). \tag{5.158}$$

where $\beta \sim 1$, $m^\ominus$ is the standard molality 1 mol kg^{-1} and α, β' and β'' are empirical constants.

At $(z_+z_-)^2I \leq 0.01$ mol kg^{-1} Equation 5.148 provides estimates of $\gamma_\pm$ that are within about ±5% of experimental determinations, while at $(z_+z_-)^2I \leq 0.1$ mol kg^{-1} Equation 5.150 provides estimates of $\gamma_\pm$ that differ from measurements also by about ±5%. The reader interested in electrolyte solutions should consult the work of Robinson and Stokes (2002).

5.7 How Can Chemical Reactions Not at Equilibrium Be Characterized?

The equilibrium of chemical reactions was discussed in Questions 5.1 to 5.4. If a chemical reaction has not reached equilibrium, there is a continuous change of the amount of substance of both reactants and products with respect to time. Thermodynamics can describe whether the potential products of a chemical reaction are more energetically stable than the reactants and quantify the relative energies of alternative potential products, but it makes no attempt to describe the stages through which the reactants pass on their way to reach the final products, nor does it calculate the rate at which equilibrium is attained. It is the subject of *chemical kinetics* that provides information about the rate of approach to equilibrium and the mechanisms for the conversion of reactants to products.

5.7.1 What Are Rate Constants and Orders of Reaction?

To discuss the principles of chemical kinetics, we will consider the general reaction of Equation 5.1 expressed in the form

$$-\nu_A A - \nu_B B - \ldots - \nu_L L \rightleftharpoons \nu_M M + \nu_N N + \ldots + \nu_W W. \tag{5.159}$$

In a closed batch reactor, the rate of consumption of reactant A, r_A, can be derived from the molar balance (Equation 1.36) as

$$r_A = -\frac{dc_A}{dt}, \tag{5.160}$$

where c_A is the amount-of-substance concentration, or simply the concentration, of species A, usually expressed as molarity (mol m^{-3} or more usually mol dm^{-3}). The rate at which product M is produced, r_M, is given by

$$r_M = \frac{dc_M}{dt}. \tag{5.161}$$

The rate of consumption of a reactant A can be expressed empirically by an equation of the form

$$r_A = k_A c_A^\alpha c_B^\beta \cdots c_L^\lambda. \tag{5.162}$$

Similarly, the rate of production of a product M may be expressed as

$$r_M = k_M c_A^\alpha c_B^\beta \cdots c_L^\lambda, \tag{5.163}$$

where the quantities k_A, k_M, α, β and λ are independent of amount-of-substance concentration and time. In Equations 5.162 and 5.163, k_A and k_M are known as the *rate constants or rate coefficients* and α, β and λ are called the *orders of reaction* ψ_B with respect to species A, B and L, respectively. The overall order of reaction, ψ, is given by $\Sigma_B\ \psi_B$, the sum of the individual orders. From the stoichiometry of the reaction given by Equation 5.159, it is evident that $k_A = -\ k_M \nu_A/\nu_M$. Rate equations such as 5.162 and 5.163 are of practical importance because they are required to predict the course of the reaction and to determine the times necessary to achieve a target extent of reaction, required product yields and the optimum economic conditions for the reaction to be carried out.

The differential rates of reactions given in Equations 5.160 and 5.161 are usually integrated before being used to describe experimental data. In this context, examples of different types of rate equations, for a first order and a second order reaction, will now be discussed.

Consider a first-order reaction (typically involving only one reactant where the rate depends linearly on the reactant concentration) given by

$$A \rightarrow B + C, \tag{5.164}$$

where the initial amount-of-substance concentration of A is c_A^o. After a time t, the remaining concentration of A has fallen to $c_A^o - c_X$ and the concentrations of B and C are both c_X. Thus, near the beginning of the reaction when A is present in very large excess compared with the amounts of products B and C, assuming no appreciable reverse reaction and that the order with respect to reactant A is $\alpha = 1$, with overall order ψ also equal to 1, application of Equations 5.160, 5.161 and 5.162 yields

$$\frac{dc_X}{dt} = -\frac{dc_A}{dt} = k_A c_A. \tag{5.165}$$

Integrating between $t = 0$, $c_A = c_A^o$ to $t = t$, $c_A = c_A^o - c_X$, the variation of c_X with time is obtained as

$$\ln\left(\frac{c_A^o}{c_A^o - c_X}\right) = k_A t, \tag{5.166}$$

or

$$c_X = c_A^o(1 - \exp(-k_A t)), \tag{5.167}$$

or

$$c_A = c_A^o \exp(-k_A t). \tag{5.168}$$

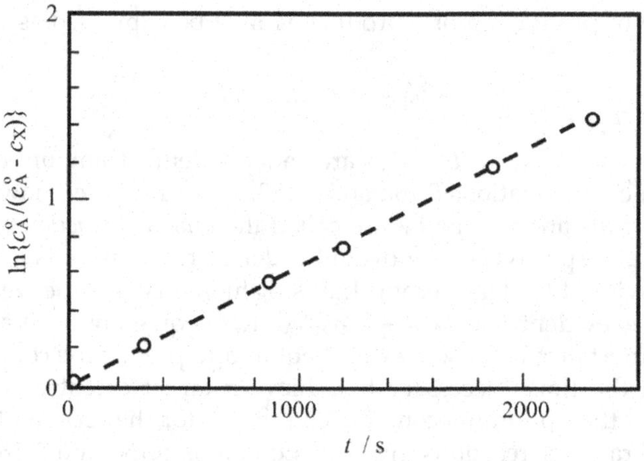

FIGURE 5.8
Variation of the amount-of-substance concentration of N_2O_5 as a function of time, according to Equation 5.143, showing the decomposition of N_2O_5 in CCl_4 (Equation 5.169) is a first-order reaction. The ratio $c_A^o/(c_A^o - c_X) = c_A(t = 0)/c_A(t)$ increases exponentially with time as c_A decreases; the slope gives the rate constant k_A directly.

As an example, consider the decomposition of N_2O_5 in CCl_4 according to the reaction

$$2N_2O_5 \overset{CCl_4}{\rightarrow} 4NO_2 + O_2, \tag{5.169}$$

for which results reported by Maskill (2006) were, as Figure 5.8 shows, well represented by Equation 5.166.

The results shown in Figure 5.8 confirm that the reaction of Equation 5.169 is first order with respect to N_2O_5 with a rate constant from the slope of $k_A = 6.22 \times 10^{-4}\ s^{-1}$. This rate constant when used with Equation 5.167 gives the amount of N_2O_5 decomposed as a function of time. For example, 99% decomposition of N_2O_5 is obtained after a time of about 7,400 s. The rate of reaction is used to determine the size of the reactor necessary in a chemical engineering process for an appropriate product specification, as well as the necessary heat transfer rate, to or from the reactor, to control the temperature.

Similarly, we can consider a second-order reaction ($\psi = 2$) given by

$$A + B \rightarrow C + D, \tag{5.170}$$

for which the initial concentrations of A and B are c_A^o and c_B^o, respectively. After a time t, a concentration c_X of both A and B have reacted, forming C and D with concentrations of c_C and c_D, respectively. For Equation 5.170, we assume as we did for Equation 5.169 that there is no appreciable reverse reaction. If the rate of the reaction, measured as in Equation 5.165 by the rate at which the reactant concentrations deplete, is proportional to the product of the two reactant concentrations, then it is said to be first order with respect to reactant A ($\alpha = 1$), first order with respect to reactant B ($\beta = 1$) and second-order overall ($\psi = 2$). The rate of reaction is given by

$$\frac{dc_X}{dt} = k_{AB}(c_A^o - c_X)(c_B^o - c_X), \tag{5.171}$$

where k_{AB} is the second-order rate constant. Separating the variables and integrating the partial fractions gives the following expression

$$\frac{1}{(c_A^o - c_B^o)} \ln\left(\frac{c_B^o(c_A^o - c_X)}{c_A^o(c_B^o - c_X)}\right) = k_{AB}t, \tag{5.172}$$

A linear plot of the measured values of the left-hand side of Equation 5.172 against t would confirm a reaction as second order. Some chemical reactions with a single reactant can also be second order. For instance, for

$$2A \rightarrow B + C, \tag{5.173}$$

the rate may be written

$$-\frac{dc_A}{dt} = k_{2A}c_A^2, \tag{5.174}$$

which on integration between the limits (c_A^o, $t = 0$) to (c_A, $t = t$) gives

$$[c_A]^{-1} - [c_A^o]^{-1} = k_{2A}t. \tag{5.175}$$

5.7.2 How Do We Include the Reversibility of Chemical Reactions in Characterizing Their Rates?

In reality, once the amount-of-substance concentration of the products becomes appreciable the rate of the reverse reaction also becomes significant and must also be taken into account. For example, consider the first-order reaction

$$A \rightleftharpoons B, \tag{5.176}$$

If the forward and reverse reactions are both first order with respect to A and B, respectively, then the rate of change of the concentration of A is

$$\frac{dc_A}{dt} = -k_A c_A + k_B c_B. \tag{5.177}$$

If initially $c_B = 0$, then $c_A = c_A^o - c_X = c_A^o - c_B$ since by stoichiometry $c_B = c_X$. Hence

$$\begin{aligned}\frac{dc_A}{dt} &= -k_A c_A + k_B(c_A^o - c_A)\\ &= -(k_A + k_B)c_A + k_B c_A^o.\end{aligned} \tag{5.178}$$

Integration of Equation 5.178 gives

$$c_A(t) = c_A^o \frac{k_B + k_A \exp(-(k_A + k_B)t)}{(k_A + k_B)}. \tag{5.179}$$

This equation reduces to Equation 5.168 if there is no reverse reaction, so $k_B = 0$. If we examine the behavior as $t \to \infty$, then

$$c_A(t \to \infty) = \frac{k_B c_A^o}{(k_A + k_B)}, \tag{5.180}$$

and

$$c_B(t \to \infty) = c_A^o - c_A(t \to \infty) = \frac{k_A c_A^o}{(k_A + k_B)}. \tag{5.181}$$

At long times, we expect the reaction to have reached equilibrium so we see that

$$\left(\frac{c_B}{c_A}\right)^{\infty} = \frac{k_A}{k_B} = \left(\frac{c_B}{c_A}\right)^{eq} = K. \tag{5.182}$$

Hence, for relatively simple reactions such as Equation 5.176, the equilibrium constant is equal to the ratio of the forward and reverse rate constants, a useful link between the thermodynamics of a chemical reaction and its kinetics. Such a relationship holds for other simple one-step reactions, such as the bimolecular second-order reactions of Equations 5.169 and 5.173 but is likely to break down for more complex, multiple step reactions.

5.7.3 How Do Reaction Rates Vary with Temperature and What Is the Gibbs Energy of Activation, ΔG^*?

The plots in Figure 5.9 show how the Gibbs energy of the molecules involved in a chemical reaction changes as they move from the initial state (the reactants) to the final state (the products). As we saw in Question 5.2, the Gibbs free energy of reaction $\Delta_r G^\ominus$ is the change in energy between the products and the reactants and characterizes the equilibrium between them. It is related to the standard state equilibrium constant $K^\ominus$ by Equation 5.26

$$\Delta_r G^\ominus = -RT \ln(K^\ominus(T)). \tag{5.183}$$

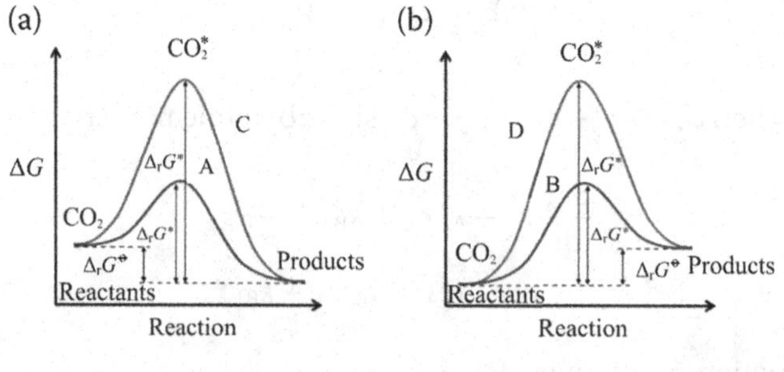

FIGURE 5.9
Gibbs energy profiles for conversion of CO_2 to (a) products which are more stable, such as carbonates and (b) products which are less thermodynamically stable, such as methane (see Question 7.5). CO_2^* is the transition state. For reactions A and B, the activation energy $\Delta_r G^*$ is quite low and the reaction may proceed at reasonable temperatures without the need for a catalyst, spontaneously in case A and with the input of energy in case B. For case C, a catalyst is likely to be needed for the reaction to proceed at an acceptable rate by reducing $\Delta_r G^*$ even though $\Delta_r G^\ominus$ is negative. Case D would require a catalyst to reduce the activation barrier $\Delta_r G^*$ as well as the input of energy to provide the increase in Gibbs energy $\Delta_r G^\ominus$.

If $\Delta_r G^{\ominus}(T)$ is negative, then the reactants are more stable than the products, $K^{\ominus}(T) > 1$, and the concentration of the products *at equilibrium* exceeds that of the reactants. However, in order to transform to the products, the reactants must undergo molecular rearrangements which usually involve chemical bonds being broken and reformed in a different configuration. The molecular configurations that the atoms involved must go through in order to rearrange into the products usually have an energy that is greater than that of both the reactants and the products. As the distances and angles between the atoms and molecular fragments change during this rearrangement process, they move across a Gibbs energy landscape, along what is termed the *reaction coordinate*, which is what is plotted as the abscissa in Figure 5.9.

The plots show schematically how the Gibbs energy G of the molecular system changes as the rearrangement takes place; the molecular configuration that has the highest energy during this change is called the *transition state*, positioned at the peak of the Gibbs energy curve as the reactants transition to the products. This means that even if the products have a lower Gibbs energy than the reactants, so that thermodynamics tells us that the reactants should spontaneously transform into the products, in order to make that transition the reactants must acquire sufficient energy to pass through the transition state before that energy is released, as the molecular reconfiguration continues towards the products. The height of this energy barrier above the energy of the reactants, $\Delta_r G^*$, which is the minimum energy required to activate the reaction, is termed the *activation energy*. $\Delta_r G^{\ominus}$ determines which final energy or product state the reactants end up in, but $\Delta_r G^*$ determines the rate at which they get there i.e. the reaction kinetics which we discussed in the previous Question 5.7. If $\Delta_r G^*$ is sufficiently large compared with the thermal energy of the reactant molecules, so that the probability of many molecules making the transition to products is very small, then although the reactants may be thermodynamically unstable with respect to the products, they will be kinetically stable because at that temperature it will take essentially an infinite time to make the transition.

The more thermal energy the reactant molecules have (i.e. the higher their temperature T), then the greater the probability that a significant number of them can rearrange via the transition state and transform to the products. We can quantify this process by the so-called *Arrhenius relationship*

$$k_r = A(T)\exp\left(-\frac{\Delta_r G^*(T)}{RT}\right), \tag{5.184}$$

where k_r is the rate constant, as defined and characterized for different types of chemical reaction in Question 5.7. In this equation, the constant $A(T)$ characterizes the rate at which the molecules come close enough to one another (through molecular collisions) to make the atomic rearrangements required for the reaction to take place; it is sometimes called the *collision number or collision factor* and can either be inferred experimentally from observations of the rate or calculated using gas kinetic theory (Atkins 2014). The exponential term (Boltzmann factor) in Equation 5.184 gives the probability that if the molecules collide, they have sufficient energy to overcome the transition state energy barrier to complete the reaction.

The activation energy $\Delta_r G^*$ is therefore a key quantity in determining the time it takes for a reactive chemical system to reach an equilibrium between the reactants and products, as determined by $\Delta_r G^{\ominus}$ and $K^{\ominus}$. Given sufficient energy to overcome $\Delta_r G^*$, a system will not necessarily transform completely to the products state, unless $\Delta_r G^\ominus$ is sufficiently large and negative and $K^\ominus \gg 1$. In many cases, the equilibrium state will involve a mixture of product and reactant molecules and achieving a product stream free from unreacted reactant molecules may prove difficult, giving rise to the need for separation and purification stages after the reactor. Manipulating temperature, pressure and product removal to drive the equilibrium strongly towards the product state is an important part of designing processes for efficient chemical conversion.

A reaction with a negative $\Delta_r G^\ominus$ will *proceed spontaneously* to the reaction products under ambient conditions if $|\Delta_r G^*| << RT = 8.314 \times 298$ J/mol = 2.5 kJ/mol. The conversion of CO_2 to carbonates by reaction with solid oxides or alkaline earth containing minerals, as described in Question 7.5.1, is a good example of this. Otherwise, steps must be taken to enhance the rate. There are two main ways to accelerate chemical reactions towards their equilibrium state

i. Increase the system temperature to enhance the rate according to Equation 5.184 so that the reaction time is low compared with the residence time of reactants in the reactor; this sometimes involves a compromise for exothermic reactions where increasing temperature shifts the equilibrium back towards the reactants.

ii. Design a catalyst that will facilitate molecules interacting and rearranging via a lower energy transition state, so reducing $\Delta_r G^*$, increasing the rate and giving the option to run the reactor at a lower temperature. The catalyst can be homogeneous, acting in solution by complexation with the reactants to facilitate molecular rearrangement and reaction, or heterogeneous, whereby the reactants adsorb on the surface of a solid catalyst, which activates the required chemical bond rearrangements before desorption of the reaction products. Design of these heterogeneous catalysts requires an understanding of the gas-solid equilibrium and the factors which encourage reactants to adsorb and products to desorb from the surface. This is considered in Question 5.8.

5.8 How Do Gases Adsorb on Solid Surfaces?

A thermodynamic understanding of gas-solid adsorption equilibrium behavior is important not only for heterogeneous catalysis but also for a host of membrane and adsorption-based gas separation, capture and storage processes, some examples of which are given in Questions 6.3 and 7.5.

How the amount of gas adsorbed on a solid surface changes with the partial pressure of the gas above the surface, at constant temperature, is called the *adsorption isotherm*. We can derive an expression for this isotherm using an approach developed by Langmuir in 1916. This assumes that all adsorption sites on the solid surface are equivalent and that adsorption is restricted to all the available sites being filled to give a molecular monolayer. This turns out to be a reasonable approximation for many gas-solid systems, especially at low pressures. It is further assumed that the rate of adsorption is proportional to the gas partial pressure, which governs the number of collisions the gas molecules have with the surface, and to the fraction of the surface sites that remain unoccupied. The rate of desorption is assumed to be proportional to the fraction of the surface covered by adsorbed molecules, θ. At equilibrium, the rates of adsorption and desorption will be equal. So if p is the gas partial pressure and the solid has Z available adsorption sites

$$k_a p Z(1-\theta) = k_d Z\,\theta, \tag{5.185}$$

where k_a is the rate constant for adsorption and k_d is that for desorption. If we ascribe $K_{ad} = k_a/k_d$ as the equilibrium constant for the gas-solid adsorption process (*cf* Equation 5.182), then

$$\theta = \frac{k_a p}{k_d + k_a p} = \frac{K_{ad}p}{1 + K_{ad}p}. \tag{5.186}$$

Equation 5.186 is known as *the Langmuir isotherm*. A useful approximation is that for low pressures $K_{ad}p \ll 1$, so that here $\theta \sim K_{ad}p$. Conversely, at high pressures, $K_{ad}p \gg 1$ and as $\theta \to 1$, $(1-\theta) = (1 + K_{ad}p)^{-1} \to (K_{ad}p)^{-1}$.

Figure 5.10 shows typical adsorption isotherms for CO_2 on a solid sorbent, fitted to the Langmuir isotherm, Equation 5.186.

We can express the surface coverage θ as $V(p)/V_\infty$, where $V(p)$ is the volume of gas adsorbed, adjusted to a standard pressure of 0.1 MPa (1 bar), and V_∞ is the adsorption capacity of the sorbent (the maximum volume that can be adsorbed, assumed in this model to be a close-packed monolayer of adsorbed molecules). Using Equation 5.186 we have

$$\frac{1}{\varsigma(p)} = \frac{1}{K_{ad}\varsigma_\infty p} + \frac{1}{\varsigma_\infty}. \tag{5.187}$$

A plot of experimental values of $V(p)$ versus $1/p$ gives V_∞ from the intercept and K_{ad} from the slope. Furthermore, from V_∞ we can estimate the number of molecules of gas adsorbed, N_{ad}, which, assuming perfect gas behavior, is

$$N_{ad} = \frac{Lp^\ominus \varsigma_\infty}{RT}, \tag{5.188}$$

where L is the Avogadro constant. If we know the molecular diameter, typically taken to be the Lennard-Jones separation parameter σ (Question 2.5.1), then we can estimate the surface area of the solid adsorbent as $0.25\,N_{ad}\pi\,\sigma^2$.

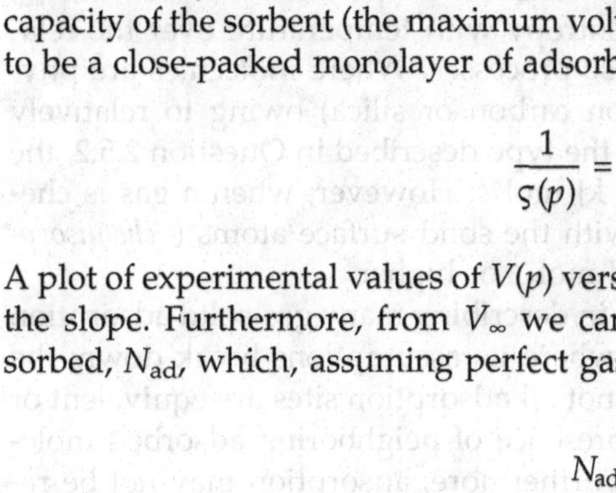

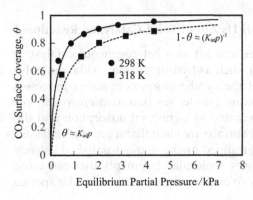

FIGURE 5.10
Adsorption isotherms for CO_2 on a typical solid capture sorbent (points), fitted to Langmuir isotherms Equation 5.186) (lines). Note the decrease in adsorption at a given partial pressure as temperature increases, in line with a negative (exothermic) enthalpy of adsorption $\Delta_{ad}H^\ominus$.

5.8.1 How Can We Obtain Thermodynamic Quantities Characterizing Gas-Solid Adsorption?

We can also use the adsorption data to obtain the thermodynamic parameters of the gas-solid equilibrium. The adsorption equilibrium constant K_{ad} is related to the standard Gibbs energy of adsorption through Equation 5.26, the van't Hoff isotherm

$$\Delta_{ad}G^{\ominus} = -RT \ln K_{ad}, \tag{5.189}$$

and to the standard enthalpy of adsorption by Equation 5.35, the van't Hoff isochore

$$\frac{d \ln K_{ad}}{dT} = \frac{\Delta_{ad}H^{\ominus}}{RT^2}. \tag{5.190}$$

Equation 5.189 may be re-written as

$$\ln K_{ad} = -\frac{\Delta_{ad}H^{\ominus}}{RT} + \frac{\Delta_{ad}S^{\ominus}}{R}, \tag{5.191}$$

which enables $\Delta_{ad}H^{\ominus}$ and $\Delta_{ad}S^{\ominus}$ to be determined from the temperature variation of K_{ad}, neglecting any small changes in enthalpy and entropy with temperature over the relatively small temperature ranges involved in these processes. Where molecules are physically adsorbed (*"physisorption"* e.g. nitrogen on carbon or silica) owing to relatively weak electrostatic and dispersion interactions of the type described in Question 2.5.2, the enthalpies of adsorption $\Delta_{ad}H^{\ominus}$ are of order 20 kJ mol^{-1}. However, when a gas is chemically adsorbed by forming a covalent bond with the solid surface atoms (*"chemisorption"* e.g. ethylene on iron), $\Delta_{ad}H^{\ominus}$ can be 200 kJ mol^{-1} or higher.

The Langmuir isotherm is remarkably accurate in describing many gas-solid adsorption processes. However, there are many situations where its assumptions break down: the solid surface can be quite heterogeneous so that not all adsorption sites are equivalent or the adsorption energy may be affected by the presence of neighboring adsorbed molecules, making it dependent on the coverage θ. Furthermore, adsorption may not be restricted to a monolayer, with further adsorbed layers developing where the main interaction is with adsorbed gas rather than directly with the solid surface. All these factors can modify the shape of the adsorption isotherm; more sophisticated models have been developed to cover these situations and the interested reader is referred to (Toth 2015) for further details.

5.8.2 What Can Gas Adsorption Isotherms Tell Us About Solid-Catalyzed Reactions?

Some solids, for example metals or metal oxides, can act as a heterogeneous catalyst to promote a chemical reaction (see Question 5.7.3), such as hydrogenation, oxidation, polymerization, hydrocarbon cracking and many of the synthesis gas conversion processes described in Questions 7.3.2–7.3.5. This occurs because under selected reaction conditions, the Gibbs free energy of adsorption $\Delta_{ad}G$ is favorable to significant adsorption and the chemistry of the gaseous reactants and the solid surface are such that a gas molecule undergoing chemisorption can be dissociated or rearranged owing to the unsatisfied valency requirements of the surface atoms. This produces molecular fragments that can more easily react with other gas molecules as they adsorb or other dissociated molecular species as they migrate across the solid surface. After completion of reaction on the surface, product molecules are likely to desorb because their partial pressure in the gas phase is low. Processes such as these provide more favorable pathways for the reaction, via lower energy molecular transition states, which lower the activation energy $\Delta_r G^*$, and hence increase the rate of reaction at a given temperature or enable lower temperatures to be used to reach the required equilibrium product yield. However, in some cases, strongly adsorbed product molecules, or unwanted by-products, can hinder the reaction by occupying potential adsorption sites and poisoning the catalyst surface, so decreasing the catalyst lifetime and requiring its replenishment.

We can use the Langmuir isotherm (or more sophisticated forms if necessary) to understand how such gas-solid catalyzed reactions proceed. The simplest case is when a gas molecule is decomposed into products on the catalyst surface in a unimolecular process e.g. the decomposition of ammonia to hydrogen and nitrogen using transition metal catalysts such as ruthenium, nickel or even iron. We can represent such a reaction by

$$\mathrm{A(g)} \underset{k_d}{\overset{k_a}{\rightleftharpoons}} \mathrm{A(ads)} \xrightarrow{k_s} \mathrm{Products}, \tag{5.192}$$

where k_s is the rate constant for the (accelerated) surface reaction. Provided that the surface reaction takes place before a gas molecule desorbs (i.e. $k_s \gg k_d$) then, using the Langmuir isotherm (Equation 5.186) for the surface concentration of species A, θ_A, the rate of the surface characterized reaction is given by

$$\mathrm{Rate} = -\frac{\mathrm{d}p}{\mathrm{d}t} = k_s\theta_A = \frac{k_s K_{ad} p}{1 + K_{ad} p}, \tag{5.193}$$

where p is the gas phase pressure of A. At low pressures, when $K_{ad}\, p << 1$, the rate is $k_s K_{ad}\, p$ so we have a first-order reaction with, using Equation 5.168, the pressure of A decreasing as

$$p = p(t = 0)(\exp(-k_s K_{ad} t)). \tag{5.194}$$

At high pressures, however, when $K_{ad}\, p \gg 1$, $\theta = 1$ and the rate becomes equal to k_s. The surface is always saturated with adsorbed gas and the rate is independent of the gas phase pressure (concentration of A) i.e. it is zero order, $\alpha = 0$ (see Question 5.7.1), with

$$\mathrm{Rate} = k_s p^0 = k_s. \tag{5.195}$$

Another common type of surface catalyzed reaction is where a gas phase molecule is able to interact with an activated molecular fragment adsorbed on the surface. Here we would expect the reaction rate to depend on both the surface coverage of the adsorbed gas A, θ_A, and the pressure of the non-adsorbing gas B, p_B

$$\mathrm{A(g)} + \mathrm{B(g)} \rightleftharpoons \mathrm{A(ads)} + \mathrm{B(g)} \rightarrow \mathrm{Products}, \tag{5.196}$$

$$\mathrm{Rate} = -\frac{\mathrm{d}p_A}{\mathrm{d}t} = k_s\theta_A p_B = \frac{k_s K_{ad} p_A p_B}{1 + K_{ad} p_A}, \tag{5.197}$$

where we have again used Equation 5.186 for θ_A. In this case, when the pressure of gas A is high, as in the initial stages of carrying out the reaction in a closed vessel, then $K_{ad}\,p_A \gg 1$, $\theta_A \sim 1$ and the rate is equal to $k_s\,p_B$, being first order with respect to gas B. The rate-determining step is the interaction of the incoming B molecules to react on the fully saturated surface with activated adsorbed A molecules. As the pressure of A in the gas phase falls, the rate becomes $k_s K_{ad} p_A p_B$, which, in this low-pressure limit, is equal to $k_s \theta_A p_B$. Hence, the reaction is now second order, depending on the partial pressures (concentrations) of both reactants, with the surface coverage θ_A, linked to p_A, ultimately becoming the controlling factor.

Some reactions will depend on the interaction of more than one activated reactant species on the solid catalyst surface, for instance when two different reactant molecules interact and rearrange on a solid surface

$$\mathrm{A(g) + B(g) \rightleftharpoons A(ads) + B(ads) \rightarrow Products}, \tag{5.198}$$

$$\text{Rate} = -\frac{\mathrm{d}p_A}{\mathrm{d}t} = k_s \theta_A \theta_B = \frac{k_s K_{ad,A} K_{ad,B} p_A p_B}{(1 + K_{ad,A} p_A + K_{ad,B} p_B)^2}. \tag{5.199}$$

Here, pressure dependence (and the temperature dependence through that of k_s, $K_{ad,A}$ and $K_{ad,B}$) can become quite complex and because the two reactants (and potentially also the products) are now in some cases competing for adsorption sites, designing the catalyst and process conditions such that the surface concentrations of both reactants are optimal is not straightforward.

References

Atkins P.W., and de Paula J., 2014, *Atkins' Physical Chemistry*, 10th ed., Oxford University Press, Oxford

Flavell-While C., 2010, "Fritz Haber and Carl Bosch – Feed the World", *The Chemical Engineer*, Institution of Chemical Engineers, March 2010 issue.

Fischer K., and Gmehling J., 1994, "P-x and γ^∞ data for the different binary butanol-water systems at 50oC", *J. Chem. Eng. Data* **39**: 309–315

Goldberg R.N., Tewari Y.B., and Ahluwalia J.C., 1989, "Thermodynamics of the hydrolysis of sucrose", *J. Biol. Chem.* **264**: 9901–9904.

Guilera J., Ramirez, E., Iborra M., Tejero L., and Cunhill F., 2013, "Experimental study of chemical equilibria in the liquid-phase reaction between 1-octanol and ethanol to 1-ethoxyoctane", *J. Chem. Eng. Data* **58**: 2076–2082.

Kjelstrup S., and Bedeaux D., 2010, *Applied Non-Equilibrium Thermodynamics, Chapter 14, in Applied Thermodynamics of Fluids*, eds. Goodwin A.R.H., Sengers J.V., and Peters C.J., for IUPAC, RSC, Cambridge.

Kjelstrup S., and Bedeaux D., 2017, *Thermodynamics of Electrochemical Systems, Chapter 4, in Springer Handbook of Electrochemical Energy*, eds. Breitkopf C. and Swider-Lyons K, Springer, Berlin Heidelberg.

Leigh G.J., 2004, "Haber-Bosch and Other Industrial Processes", in *Catalysts for Nitrogen Fixation. Nitrogen Fixation: Origins, Applications, and Research Progress, vol 1*, eds., Smith B.E., Richards R.L., Newton W.E., Springer, Dordrecht.

McQuarrie D.A. and Simon J.D., 1999, *Molecular Thermodynamics*, University Science Books, Sausalito CA, USA. 672 pp, Chapter 10.

Maskill H., Ed., 2006, *The Investigation of Organic Reactions and Their Mechanisms*, Blackwell Publishing Ltd., Oxford, UK.

Mohr P.J., Taylor B.N., and Newell D.B., 2008, "CODATA recommended values of the fundamental physical constants: 2006", *J. Phys. Chem. Ref. Data* **37**: 1187–1284.

Pecci G.C., Clerici M.G., Giavazzi F., Ancillotti F., Marchionna M., and Patrini R., 1991, "Oxygenated diesel fuels. Part 1: Structure and properties correlation", Proceedings of the *Ninth International Symposium on Alcohol Fuels*, ISAF, Firenze, pp. 321–326.

Robinson R.A., and Stokes R.H., 2002, *Electrolyte Solutions*, 2nd rev. ed., Dover Publications, New York.

Sandler S.I., 1989, "*Chemical and Engineering Thermodynamics*," 2nd ed., John Wiley & Sons Inc., New York, Chapter 7.8.

Smil V., 2001, *Enriching the Earth: Fritz Haber, Carl Bosch, and the Transformation of World Food Production*, MIT Press, Cambridge, MA.

Thermodynamic Research Center (TRC), 1942–2007, *Thermodynamic Tables Hydrocarbons*, ed. Frenkel M., National Institute of Standards and Technology Boulder, CO, Standard Reference Data Program Publication Series NSRDS-NIST-75, Gaithersburg, MD.

Thermodynamic Research Center (TRC), 1955–2007, *Thermodynamic Tables Non-Hydrocarbons*, ed. Frenkel M., National Institute of Standards and Technology Boulder, CO, Standard Reference Data Program Publication Series NSRDS-NIST- 74, Gathersburg, MD.

Toth J., 2015, "Adsorption Isotherms", in *Encyclopedia of Surface and Colloid Science*, eds. Somasundaran P., Deo N., Farinato R., Grassian V., Lu M., Malmsten M., Mittal K.L., Nagarajan R., Partha P., Sukhorukov G., and Wasan D., 3rd ed., CRC Press, Boca Raton, USA.

6

Power Generation and Refrigeration

6.1 What Is a Cyclic Process and Its Use?

As the name implies, a cyclic process (see Question 1.3.5 for a definition of a thermodynamic process) consists of a series of steps that result in a closed cycle. Thus, in all characteristic thermodynamic diagrams, such as pressure as a function of specific volume denoted (p, v), temperature as a function of specific entropy (T, s), and specific enthalpy as a function of specific entropy (h, s), the lines describing the individual process steps form a closed loop. The term "cyclic process" may refer either to a closed system or to a series of open systems.

For a closed system, the fluid within the system undergoes a series of processes so that at the end of the cycle the fluid and the system are returned to the initial state. In an open system a fluid flows through a series of mechanical components and at the end of the cycle the fluid is returned to its initial thermodynamic state, for example, the fluid in a steam-driven power plant. In some processes, the surroundings are also regarded as one of the system components. This extension applies, for example, to an open-cycle gas-turbine engine, where air initially at ambient temperature flows through the engine components and the air temperature increases, the air is then discharged from the engine to the surroundings, where it mixes with ambient air and the temperature returns to the original ambient value. In this example, the heat exchanger for heat rejection is the air surrounding the engine.

Two types of cyclic processes may be distinguished, and detailed information on the respective processes may be found in general literature on engineering thermodynamics or specialized books, e.g. Invernizzi (2013), Dincer (2017) and Grassi (2017). These are illustrated in Figure 6.1. In the first process, shown in Figure 6.1a, heat is provided to the system to obtain a net output of work or power. These power cycles constitute the origin of engineering thermodynamics and are the subject of Question 6.2. The lines describing the process steps constitute a closed clockwise loop. The second type of process arises when net work input is required to realize the uptake of heat at a temperature and its rejection at a higher temperature. A refrigeration cycle, where heat is removed from a space that is to be cooled and the heat is finally rejected to the surroundings, is one example that is addressed in Question 6.5. Another example is a heat pump in which heat from the surroundings is brought into the space and is used for heating. In the characteristic diagram, shown in Figure 6.1b, the individual steps in the process are expressed by lines that form a closed counter-clockwise loop.

DOI: 10.1201/9780429329524-6

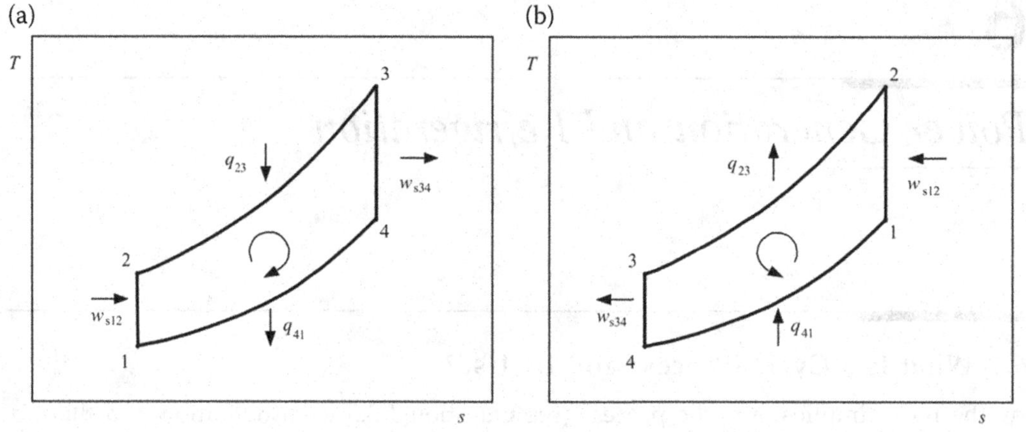

FIGURE 6.1
Schematic of the (T, s) diagram for the following: (a) for a power cycle, the process steps are performed clockwise; heat q_{23} is required to withdraw work w_{s34} from the process. (b) for either refrigeration or heat pump cycle the process steps are performed counterclockwise; work w_{s12} is required to transport heat from a lower temperature level to a higher temperature level where it is rejected from the process. For a refrigeration cycle, the purpose of the process is to cool a space by transferring heat q_{41} while for a heat pump the purpose is to heat a space by transferring heat q_{23}.

6.2 What Are the Characteristics of Power Cycles?

Because it is a defining characteristic of any state variable Z that its value is independent of the history of the process by which a specific state is reached, it is then obvious for a cyclic process that $Z_{\text{final}} = Z_{\text{initial}}$, that is the variable has the same value at the beginning and the end of the process, and the cycle integral over all changes dZ in this process becomes zero

$$\oint dZ = 0. \tag{6.1}$$

Considering the First Law for a series of open systems, we may write

$$\dot{Q}_{12} + P_{12} = \dot{m}\left\{(h_2 - h_1) + \frac{1}{2}(u_{f2}^2 - u_{f1}^2) + g(z_2 - z_1)\right\}, \tag{6.2}$$

$$\dot{Q}_{23} + P_{23} = \dot{m}\left\{(h_3 - h_2) + \frac{1}{2}(u_{f3}^2 - u_{f2}^2) + g(z_3 - z_2)\right\}. \tag{6.3}$$

In Equations 6.2 and 6.3, the subscript numerals refer to the process step. Equations 6.2 and 6.3 are for two steps in the process, which have additional steps with similar equations, that vary only by the subscript numerals defining the process step. These equations are not given here. The final step in the process returns the fluid from state n to state 1 and is given by

$$\dot{Q}_{n1} + P_{n1} = \dot{m}\left[(h_1 - h_n) + \frac{1}{2}(u_{f1}^2 - u_{fn}^2) + g(z_1 - z_n)\right]. \tag{6.4}$$

Here we use a form of Equation 1.77 in which we employ the symbol P for power, which is equivalent to the time derivative of shaft work, $P = \dot{W}_s$, and $\dot{m}$ is the mass flowrate, which is evidently conserved. Summing all forms of Equations 6.2 through 6.4 we obtain

$$\sum_{i=1}^{n-1} \dot{Q}_{i(i+1)} + \sum_{i=1}^{n-1} P_{i(i+1)} + \dot{Q}_{n1} + P_{n1} = 0, \tag{6.5}$$

where the sum of all heat and work fluxes (remembering that a flux has a positive sign when entering into the system, a negative sign when leaving it) is zero. The *net power* produced is equal in magnitude to the sum of all heat fluxes crossing the system boundaries and given by $P = \sum_{i=1}^{n-1} P_{i(i+1)} + P_{n1}$. Usually, the heat fluxes are grouped into the following: (1), $\dot{Q}$, the sum of all heat fluxes entering the system (heat provided); and (2), $\dot{Q}_0$, the sum of all heat fluxes leaving the system (heat rejected). In view of this definition, we can now rewrite Equation 6.5 as

$$P = -\dot{Q} - \dot{Q}_0, \tag{6.6}$$

and the net (shaft) work produced may be written as

$$w_s = \frac{P}{\dot{m}} = \sum_{i=1}^{n-1} w_{s,i(i+1)} = -\sum_{i=1}^{n-1} q_{i(i+1)} \tag{6.7}$$

or

$$w_s = -q - q_0. \tag{6.8}$$

Power cycles may make use of closed systems (such as in an internal combustion engine) or a series of open systems (such as in a gas turbine); the working fluid may be a gas (e.g. in a Stirling engine) or a fluid undergoing phase transitions (e.g. in a steam cycle). However, at least the basic and idealized variants of many practical processes may be described by four process steps that are characterized by an alternating sequence of steps involving the transfer of solely (or sometimes mainly) work or heat, respectively, as shown in Figure 6.2.

The power cycle illustrated in Figure 6.2 includes the following steps:

step 1–2: Work input as w_{s12} (increasing system pressure)
step 2–3: Heat input as q_{23} (heat provided)
step 3–4: Work output as w_{s34} (that is to be maximized)
step 4–1: Heat rejected as q_{41} (waste heat)

The net work produced is these steps is given by

$$w_s = w_{s12} + w_{s34}, \tag{6.9}$$

where $w_{s12} > 0$, $w_{s34} < 0$ and $w_s < 0$ and $|w_s|$ is to be maximized.

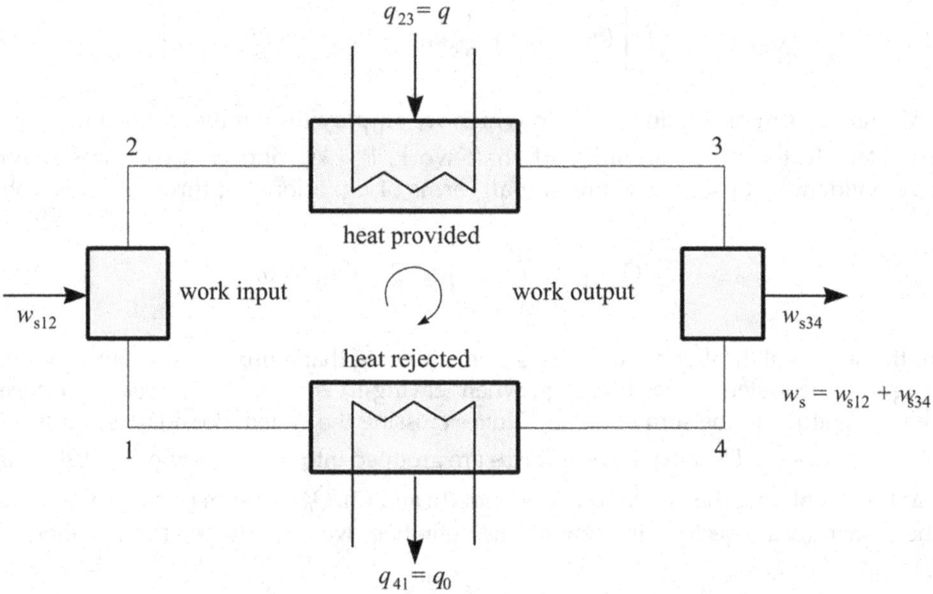

FIGURE 6.2
Typical scheme of a basic, idealized power cycle: The cycle consists of a series of steps where only work or heat are transferred, respectively.

Before turning to some specific processes in the questions following this, we examine some generic and characteristic features of processes by reference to an example four-step process. However, there is no requirement for exactly these four steps to occur, and in an actual process there are often many more than four.

We begin by examining two characteristic diagrams: (p, v) and (T, s). For simplicity, we assume that all process steps are reversible, a restriction that if lifted does not alter the conclusions reached.

Figure 6.3 shows both a (p, v) and a (T, s) diagram. Figure 6.3a illustrates the shaft work, which is the central quantity of a power cycle, given by (*cf.* Question 1.3.6)

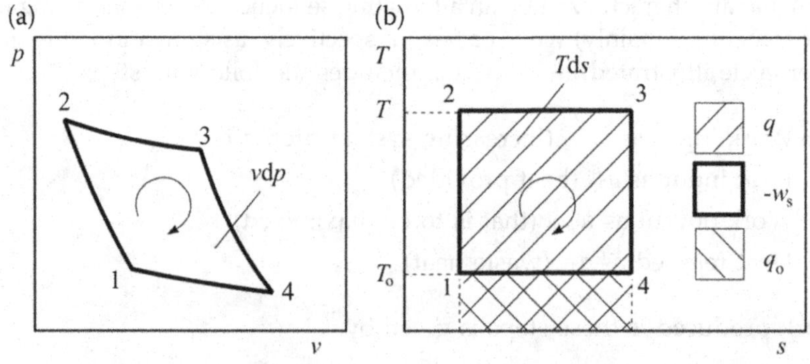

FIGURE 6.3
A power cycle: (a) (p, v) diagram and (b) a (T, s) diagram. For a reversible process, the area enclosed by the cycle represents the net work output; (b) depicts a Carnot cycle, where heat is provided at constant temperature T, and rejected at constant temperature T_0.

$$(w_{sij})_{rev} = \int_i^j v\mathrm{d}p. \tag{6.10}$$

The total work output given by

$$|w_s| = q - |q_0| \tag{6.11}$$

may therefore also be written, omitting the index for the reversible process in Equation 6.10, as

$$w_s = \oint v\mathrm{d}p. \tag{6.12}$$

Equation 6.12 implies that a change of volume during the process is a prerequisite for net work output. The work output increases with the area enclosed by the process as shown in Figure 6.3. In the case when there are irreversible steps, the work output is diminished by dissipation, but the result does not change in principle.

Considering the (T, s) diagram shown in Figure 6.3b, we have for the total differential of the specific enthalpy (*cf.* Equation 3.37)

$$\mathrm{d}h = T\mathrm{d}s + v\mathrm{d}p, \tag{6.13}$$

and because enthalpy is a state variable, it must assume the initial value at the end of the cycle so that

$$\oint \mathrm{d}h = 0. \tag{6.14}$$

As a consequence of Equation 6.14, Equation 6.13 becomes

$$0 = \oint \mathrm{d}h = \oint T\mathrm{d}s + \oint v\mathrm{d}p \tag{6.15}$$

so that

$$\oint T\mathrm{d}s = -\oint v\mathrm{d}p = -w_s. \tag{6.16}$$

It follows from Equation 6.16 that a temperature change is required if a cycle process is to deliver a net work output. These two results rather vividly illustrate how very general and far-reaching results can be derived from thermodynamic analysis.

The particular (T, s) diagram shown in Figure 6.3b depicts a special process that consists of two (reversible) isothermal and two (reversible) adiabatic (i.e. isentropic) steps. This is the *Carnot cycle* (Carnot 1872), *cf.* Question 3.2.1, where heat is provided at constant temperature T, and rejected at constant temperature T_0. The elegant rectangular shape shown in Figure 6.3b is a result of the specific process steps adopted. The two isentropic steps are also adiabatic (i.e. no heat is transferred) and because the whole process is assumed to be reversible, the change of entropy within the two isothermal steps is solely connected with heat transfer and not due to any irreversibilities. The heat provided to the process is thus

$$q = \int_2^3 T\mathrm{d}s = T(s_3 - s_2), \tag{6.17}$$

and the heat rejected is given by

$$q_0 = \int_4^1 T\mathrm{d}s = T_0(s_1 - s_4). \tag{6.18}$$

We now consider the efficiency of a power cycle in which the central goal is to maximize the output (net work w_s) for a given input (heat, q, provided). The *thermal efficiency*, η_{th}, characterizing the overall quality of the process, is defined by

$$\eta_{th} = \frac{-w_s}{q} = \frac{-P}{\dot{Q}}. \tag{6.19}$$

The minus sign in the numerator of Equation 6.19 arises solely from the desire for the thermal efficiency to be positive; the work delivered is negative. The η_{th} ranges between zero and one. The definition of w_s is the net work output summed over all work steps in the process. All steps in a process that require work input diminish the total work output. The net work argument is particularly useful because of the direct connection between the respective components of the process. For example, in a gas turbine shown in Figure 6.4 the compressor stage of the engine is driven by the same shaft as the gas turbine. Thus, the work generated by the gases expanding through the turbine is partially offset by the work done in compressing the gases.

In contrast, q only refers to the sum of all heat provided; q_0, the sum over all heat rejected, does not reduce the expended effort. Heat is rejected at a lower temperature. If a process is poorly designed so that too much heat is rejected, this does not reduce the heat provided to the process, for example, from combustion of coal or gas.

As the Carnot cycle is an ideal and reversible process, it constitutes a reference process in engineering thermodynamics. It provides an upper limit for the thermal efficiency that can be obtained in a power cycle. Equation 6.19 can be written in the general form as

$$\eta_{th} = \frac{q - |q_0|}{q} = 1 - \frac{|q_0|}{q}, \tag{6.20}$$

and it is a consequence of the Second Law (as discussed in Question 3.2) that the heat provided cannot be transformed completely into useful work and that, necessarily, part of the heat must be discarded at a lower temperature.

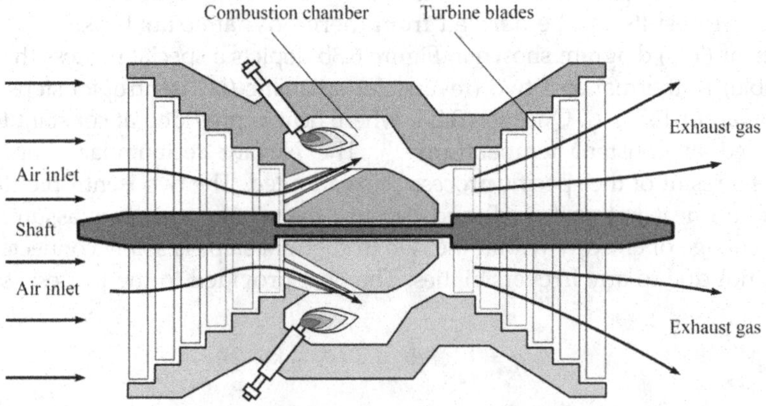

FIGURE 6.4
A schematic diagram of a gas turbine. A common shaft is used for both compressor and turbine blades.

The heat and the net power output in the Carnot cycle, as illustrated in Figure 6.3, exhibit the property that heat is provided and rejected at constant temperature, which makes the evaluation of Equation 6.20 particularly easy. The fundamental characteristic of this process, however, is that all steps are performed in a reversible manner. The thermal efficiency of the Carnot cycle, which is also the maximum thermal efficiency of a power cycle, only obtained in a reversible process, is given by

$$\eta_{\text{th,C}} = \eta_{\text{th,rev}} = 1 - \frac{|q_0|}{q} = 1 - \frac{T_0(s_4 - s_1)}{T(s_3 - s_2)} = 1 - \frac{T_0}{T}, \tag{6.21}$$

where the subscript C denotes the Carnot cycle and the subscript 'rev' denotes that the steps in the process are reversible. The fundamental consequence of this formula is that all power cycles should be designed so that heat is provided at a temperature as high as possible and heat is rejected at a temperature as close to the ambient temperature as possible to yield the highest efficiency.

As a specific example, if heat was provided at a temperature of T = 773 K (t = 500°C) and rejected at ambient temperature of T_0 = 298 K (t_0 = 25°C), a maximum thermal efficiency of $\eta_{\text{th,rev}}$ = 0.61 would result. This value of efficiency is the base line to which an engineer has to compare his or her design, recognizing that the real values for the thermal efficiency will be considerably lower (often by almost a factor of two) because of the inevitable losses and irreversibility within a process. However, if one could increase the base temperature T, by choice of materials that withstand higher temperatures, the efficiency would increase. A temperature of T = 873 K (t = 600°C) would result in a "reference" thermal efficiency of $\eta_{\text{th,rev}}$ = 0.69.

6.2.1 Why Do Power Plants Have Several Steam Turbines?

We begin our discussion with an idealized scheme for a simple steam power plant that, as Figure 6.5 shows, consists of a series of process steps alternately involving transfer of work and heat, respectively. The working fluid water undergoes the following processes:

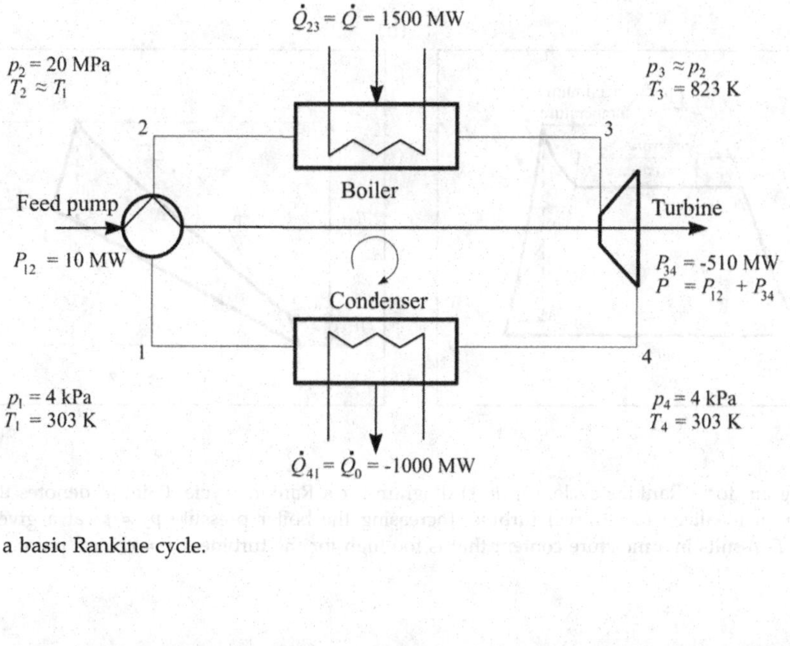

FIGURE 6.5
Schematic of a basic Rankine cycle.

step 1–2: Adiabatic compression (pumping) of liquid water to the boiler pressure

step 2–3: Constant-pressure addition of heat in the boiler through the heating of subcooled water to its vaporization temperature, complete vaporization and then superheating of the water vapor

step 3–4: Adiabatic expansion of the vapor in a steam turbine usually into the two-phase region close to the saturated vapor line

step 4–1: Heat rejection and complete condensation at constant pressure

These four processes are characteristic of a basic steam power plant, which is also called a Rankine (or sometimes Clausius-Rankine) cycle.

Figure 6.5 shows a Rankine cycle and includes typical values for T, p, P and $\dot{Q}$ in each process. A general requirement for the thermal efficiency discussed previously is that heat should be provided at the highest possible temperature and rejected at the lowest possible temperature. On the basis of the upper temperature limits imposed by materials used to construct the machinery the highest practical temperature is about 550°C. The lower temperature where heat is rejected is determined by the temperature of the surroundings where the power plant is located. For the purpose of this example we assume a condensation temperature of 30°C, which corresponds to a water vapor pressure of about 4 kPa. The heat and power fluxes listed in Figure 6.5 are for a power plant with a net output power of 500 MW, where a part (albeit small) of the power available at the turbine (shown in Figure 6.4) is consumed by the feed pump. As a rule of thumb, we may assume an overall thermal efficiency of 1/3 (state-of-the-art power plants achieve a thermal efficiency of >0.4). The efficiency of 1/3 means that a heat flux of 1,500 MW must be provided, of which a fraction of two-thirds is discharged at low temperature, mainly as a consequence of the Second Law, but also because of inevitable irreversibilities and losses within the process.

One of the major losses within a Rankine cycle is the necessarily nonideal operation of the steam turbine. In a perfect turbine, process step 3–4 would be reversible and, thus, isentropic, resulting in an ideal state denoted by 4^s in Figure 6.6. In a real process, the entropy of the fluid is increased, yielding fluid at a higher temperature and enthalpy as shown in Figure 6.6. Therefore, not all of the available energy (equal to the exergy

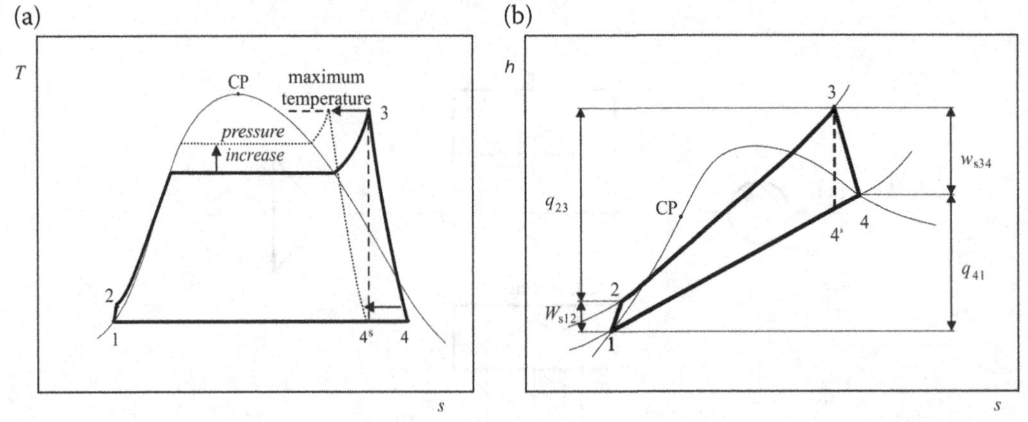

FIGURE 6.6
(a) (T, s) diagram for a Rankine cycle. (b) (h, s) diagram for a Rankine cycle. Point 4^s denotes the state after expansion in an idealized (isentropic) turbine. Increasing the boiler pressure $p_2 = p_3$ at a given maximum temperature T_3 results in a moisture content that is too high for the turbine. CP = critical point.

discussed in Question 3.8) of the fluid at state 3 is exploited in the real process. Similar considerations hold for the feed pump that operates in the step 1–2 of the process. To illustrate the salient points, Figure 6.6 is not drawn to scale in this region of the diagram. The rise in both temperature and enthalpy are relatively small; the process is operated close to the saturation line and the differences would be practically indistinguishable on the overall scale of Figure 6.6. For an ideal process, the step involving work would be reversible (and thus isentropic), giving a vertical line; in reality, the entropy of the fluid increases and the line is not vertical.

The thermal efficiency of the Rankine cycle can be obtained from the (h, s) diagram of Figure 6.6 as

$$\eta_{\mathrm{th,R}} = \frac{-w_{\mathrm{s}}}{q} = \frac{-(w_{\mathrm{s34}} - w_{\mathrm{s12}})}{q_{23}} = \frac{(h_3 - h_4) - (h_2 - h_1)}{h_3 - h_2} \approx \frac{h_3 - h_4}{h_3 - h_2}, \tag{6.22}$$

where the last step follows because of the relatively small enthalpy change that accompanies the liquid compression. From a fundamental thermodynamic point of view, the obvious measure to improve the efficiency is to increase the spread of temperatures between the levels where heat is provided and where heat is discharged. Because the upper temperature is determined by the materials used for construction of the power plant and the lower temperature by the ambient value the margin for efficiency improvement from this source is small. There are, of course, always efforts to develop materials that could enhance the upper temperature.

In the remaining discussions, we provide reasons why particular design features are incorporated in power plants. It is important to recognize it is the average temperature of heat provision that is of paramount importance. On the basis of this fact, it is therefore desirable to obtain a high temperature in the two-phase region for water and this can be realized by increasing the boiler pressure. Recent developments in steam turbine plants also use pressures >22 MPa that are supercritical and result in step 2–3 of the process extending outside the two-phase region. However, increasing the $p_2 = p_3$ at a maximum temperature T_3 will shift point 3 to the left of Figure 6.6 in both the (T, s) and (h, s) diagrams. As a consequence, after expansion of the vapor, point 4 lies further into the two-phase region with a higher fraction of liquid water present and this leads to the formation of larger water droplets that lead to increased erosion of the turbine blades. Indeed, it is because of erosion that the steam quality x (*cf.* Question 4.7.1) is maintained >0.9 at the end of expansion.

Combining the requirements for a high boiler pressure and temperature with the need for a state after expansion near to that of the saturated vapor leads to what is termed the *reheat power plant design* for which the schematic is shown in Figure 6.7 and the corresponding (T, s) diagram in Figure 6.8. After the steam is expanded to a medium pressure in a high-pressure turbine, it is reheated to about the original maximum temperature. In a second step, the steam is expanded again, this time to the condenser pressure in a low-pressure turbine. However, additional turbines increase the complexity of a plant and a large number of turbines are neither beneficial nor economical. Consequently, a second reheat step and thus a third turbine operating at an additional intermediate pressure level are normally introduced only in the case of boiler pressures close to or above the critical pressure of water of 22 MPa. For all steam turbines, increasing the boiler pressure and temperature in a reheat process is one of the most important thermodynamic methods used to increase the efficiency of a steam power plant.

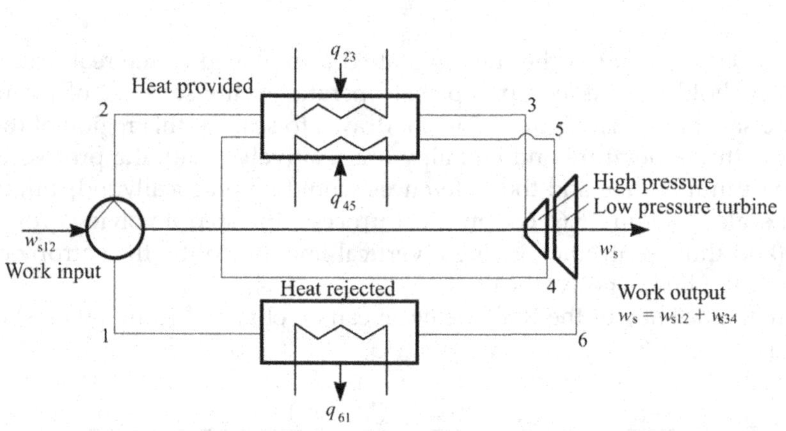

FIGURE 6.7
Schematic of a reheat Rankine cycle. After expansion in a high-pressure turbine, the steam is reheated and expanded again in a second turbine.

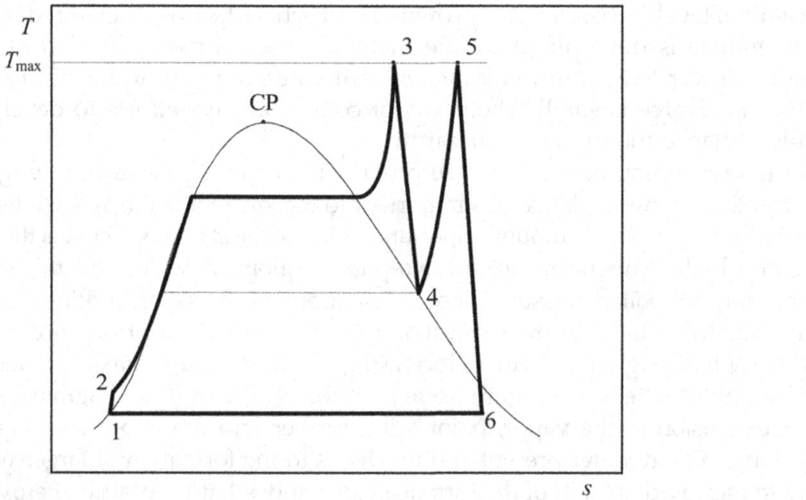

FIGURE 6.8
(T, s) diagram of a reheat Rankine cycle. The maximum temperature is determined by the materials of construction. The average temperature of heat provision can be increased, while the quality x after the final expansion is large.

For the sake of completeness, we mention another important variation that also increases the average temperature of heat provision. In the Rankine cycle discussed so far, liquid water at low temperatures is fed into the boiler after compression. However, it is advantageous to heat the water in a regenerative scheme. In this case, steam from a turbine is extracted and is either directly mixed with the feedwater or used for preheating via a heat exchanger. Steam power plants use a series of feedwater heaters each at a different temperature that use steam bleed at appropriate points of the turbine stages.

6.2.2 What Is a Combined Cycle?

The term "combined cycle" commonly refers to a combination of a gas-turbine cycle and a steam-power cycle; the introduction of the combination is intended to increase the

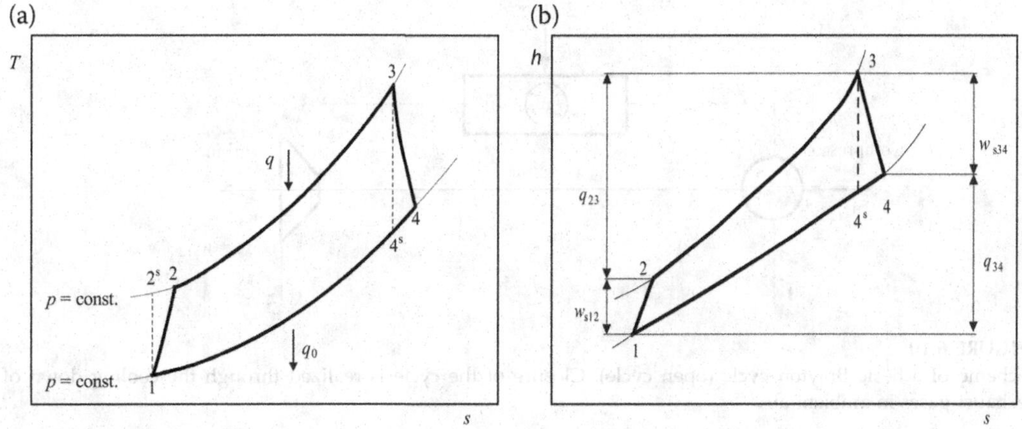

FIGURE 6.9
The Brayton cycle. (a) (T, s) diagram. (b) (h, s) diagram. The superscript s denotes the idealized (isentropic) processes.

overall efficiency. The key feature of this approach is the use of the waste heat from the gas turbine as a partial replacement for the heat that must be provided to a steam cycle, normally from the combustion of fossil fuel.

We start with the operation scheme of the gas-turbine cycle, which consists of three steps, as shown in Figure 6.9:

step 1–2: Adiabatic compression of ambient air to a pressure of up to 2 MPa (through a common shaft, the compressor is directly driven by the turbine, as shown in Figure 6.4)

step 2–3: Combustion of gas in the chamber modeled as a constant-pressure heat addition

step 3–4: Adiabatic expansion of hot compressed gas in a turbine

The gas turbine cycle is commonly referred to as either a Brayton or a Joule cycle. The Joule cycle is normally used only for the particular case when both compression and expansion are performed reversibly.

This "open-cycle" arrangement is normally utilized within a gas turbine for the generation of electricity and, as illustrated in Figure 6.10, at first sight, overlooks the closure of the cycle. There are closed cycles where a fourth process is used to reject heat with a heat exchanger (and the combustion chamber is replaced by an additional heat exchanger). This closed-loop system often uses helium as the working fluid, and only finds limited application, because it is impractical and uneconomic for large-scale power generation from the combustion of gas. In an open-cycle gas turbine, the final step of heat rejection at constant pressure is omitted, resulting in the elimination of an additional mechanical component. Practically, closure is obtained by heat rejection to ambient air, and the prerequisite for a thermodynamic cycle (the thermodynamic properties before and after the cycle must be identical) is accomplished by the surroundings: Air at ambient conditions occurs at the beginning and end of the cycle.

From the (T, s) and (h, s) diagrams for this cycle (shown in Figure 6.9), the thermal efficiency of the system can be determined. In both diagrams, we have already accounted for the irreversibilities in the operation of both compressor and turbine. In both cases, the

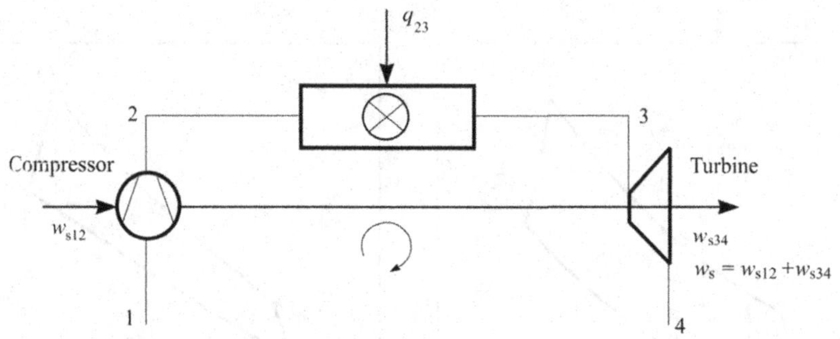

FIGURE 6.10
Scheme of a basic Brayton cycle (open cycle). Closure of the cycle is realized through the cooling down of exhaust gases in ambient air.

pressure change is connected with an increase in entropy. In an ideal Brayton (or Joule) cycle, both compression and expansion are isentropic and are represented by vertical lines in the diagrams.

By analogy to the Rankine cycle, the thermal efficiency of the Brayton cycle is given by

$$\eta_{\text{th,B}} = \frac{-w_s}{q} = \frac{-(w_{s34} - w_{s12})}{q_{23}} = \frac{(h_3 - h_4) - (h_2 - h_1)}{h_3 - h_2} = 1 - \left(\frac{h_4 - h_1}{h_3 - h_2}\right). \tag{6.23}$$

For simplicity, we assume the working fluid is an ideal gas for which the heat capacity at constant pressure c_p is constant and the enthalpy differences are given by

$$h_y - h_x = c_p(T_y - T_x), \tag{6.24}$$

so that Equation 6.23 becomes

$$\eta_{\text{th, B}} = 1 - \left(\frac{T_4 - T_1}{T_3 - T_2}\right). \tag{6.25}$$

Again, the assumption of an ideal gas does not affect the conclusions obtained from the analysis.

For an ideal Brayton cycle with isentropic compression and expansion (and a constant heat capacity ratio γ), Equation 6.25 can be simplified utilizing the pressure ratio $\Pi = p_2/p_1 = p_3/p_4$ and

$$\frac{T_2}{T_1} = \left(\frac{p_2}{p_1}\right)^{\frac{\gamma-1}{\gamma}} = \Pi^{\frac{\gamma-1}{\gamma}} = \left(\frac{p_3}{p_4}\right)^{\frac{\gamma-1}{\gamma}} = \frac{T_3}{T_4}, \tag{6.26}$$

to obtain

$$\eta_{\text{th,B,id.}} = 1 - \frac{T_1\{(T_4/T_1) - 1\}}{T_2\{(T_3/T_2) - 1\}} = 1 - \frac{T_1}{T_2} = 1 - \frac{1}{\Pi^{(\gamma-1)/\gamma}}. \tag{6.27}$$

For the idealized process, Equation 6.27 implies that the efficiency increases with increasing pressure ratio Π. Other parameters that influence the performance of a gas-turbine cycle are the temperature T_3, which is the maximum temperature of the process, where the gas enters the turbine and the temperature ratio $\tau = T_3/T_1$. In view of the materials used to construct the turbine, the inlet temperature is limited to about 1,500°C (1,800 K); operation at these high temperatures is only possible with the use of air-cooled turbine blades. It is uneconomical to raise the pressure ratio >20 for a given maximum temperature T_3 because the net work output $-w_s = -(w_{s34} - w_{s12})$ has a maximum and a further increase of the pressure ratio Π results in a decrease in the net work output. This observation can be rationalized by considering that for fixed temperatures T_1 and T_3 (and thus a fixed enthalpy difference $h_3 - h_1$) an increase in Π results in an increase in the compressor work w_{s12} and reduction of the heat q_{23}. Closer inspection shows that the compressor work w_{s12} increases at a rate greater than the turbine work output $-w_{s41}$, resulting in a maximum for the net work output $-w_s$. From the condition $\mathrm{d}w_s/\mathrm{d}\Pi = 0$, the optimum pressure ratio is given by

$$\Pi_{\text{opt}} = \tau^{\gamma/\{2(\gamma-1)\}}, \tag{6.28}$$

which is equivalent to the condition $T_2 = T_4$. Table 6.1 lists the variation of w_s and q as a function of Π. The derivation of Equation 6.28 is discussed in detail in the literature, for example, by Burghardt and Harbach (1993). The results listed in Table 6.1 reveal that the thermal efficiency gradually increases with the increasing pressure ratio, with a maximum (albeit shallow) for the net work output at a pressure ratio $\Pi_{\text{opt}} = 19.75$.

The thermal efficiency of the gas-turbine cycle can be increased by several methods and two significant ones are as follows: (1) multistage compression with repeated intercooling and reheating between stages; and (2) utilization of the exhaust gas to preheat the air before entering the combustion chamber in a counterflow heat exchanger. Item 1 reduces the overall work required for compression because the process approaches isothermal compression and requires less work than adiabatic compression (compare Question 1.6.5). Item 2 is used when the compression ratio Π and thus the compressor exit temperature T_2 are not too high and the exhaust gas at temperature T_4 preheats the air before entering the combustion chamber.

It is the high outlet temperatures of up to 900 K for a modern gas turbine that leads to the combined cycle. The exhaust gases of the gas turbine may be used either to preheat the water for the steam cycle or to act as the sole heat source for the steam turbine

TABLE 6.1

The work w_s and heat q for each step of a Brayton cycle as a function of the pressure ratio Π. The values are based on fixed intake temperature $T_1 = 290$ K and maximum temperature $T_3 = 1{,}595$ K ($\tau = T_3/T_1 = 5.5$) and air the working fluid assumed to be a perfect gas

Π	w_{s12}/(kJ kg^{-1})	q_{23}/(kJ kg^{-1})	w_{s34}/(kJ kg^{-1})	w_s/(kJ kg^{-1})	η_{th}
14.00	328	983	−848	−520	0.53
16.00	352	958	−876	−524	0.55
18.00	374	936	−900	−526	0.56
19.75	392	919	−919	−527	0.57
20.00	394	916	−921	−527	0.58
22.00	413	897	−939	−526	0.59
24.00	431	879	−956	−525	0.60

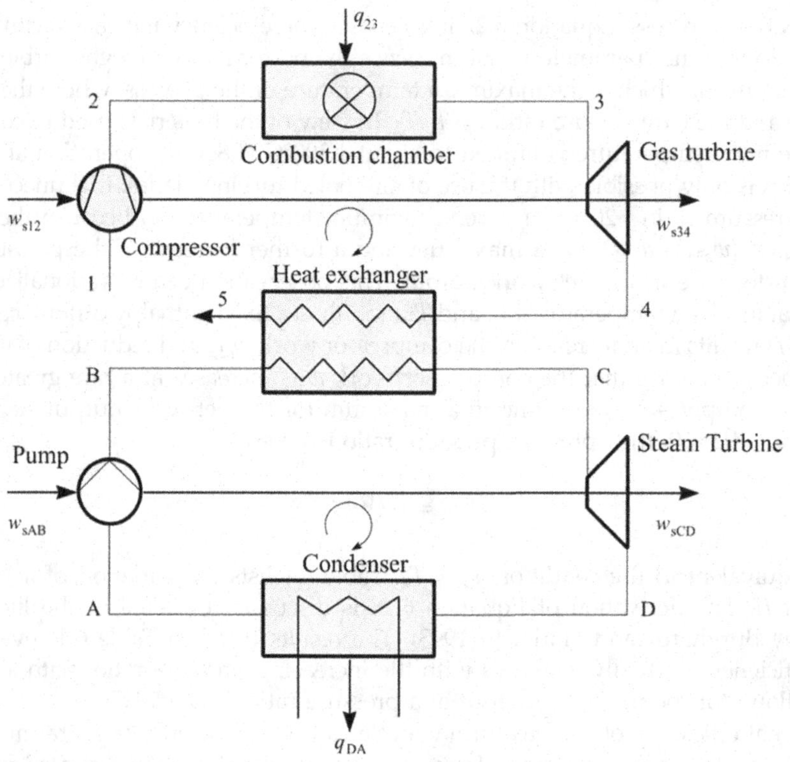

FIGURE 6.11
Scheme of a combined cycle. In this configuration, the hot exhaust of the gas turbine is used as the sole heat source for the steam process.

through a heat exchanger (boiler). One example for such a combination is depicted in Figures 6.11 and 6.12 for which the heat rejected from the gas turbine and given by

$$q_{45} = \int_4^5 T ds \tag{6.29}$$

is used completely to provide the heat for the steam cycle, which is given by

$$q_{BC} = \int_B^C T ds. \tag{6.30}$$

Practically, of course, there are some losses owing to imperfect heat transfer; also, to avoid corrosion, the exhaust gases are not completely cooled to ambient temperature. However, for our current purpose, we can assume that the exhaust gases are completely utilized and that there is no additional heat for the steam cycle. In that case, the maximum thermal efficiency of the combined cycle is given by

$$\eta_{th,c,max} = \frac{-w_{s,B} - w_{s,R}}{q_{23}} = \eta_{th,B} + \eta_{th,R}(1 - \eta_{th,B}). \tag{6.31}$$

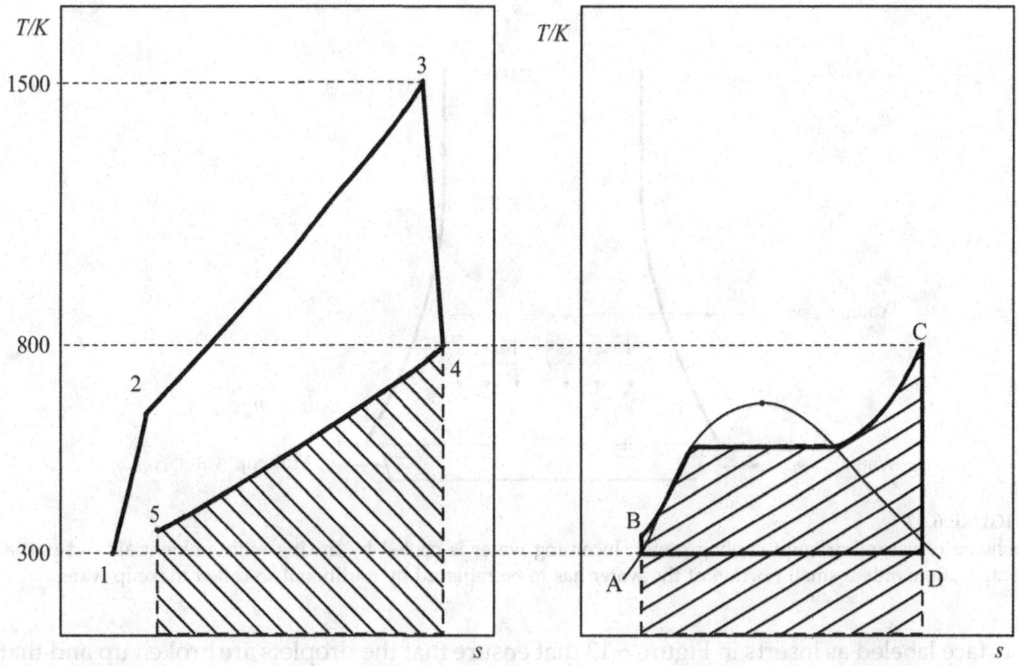

FIGURE 6.12
(T, s) diagrams of a combined cycle that incorporates a Brayton cycle and a Rankine cycle. The heat q_{45} rejected from the gas turbine at a comparatively high temperature provides the heat input q_{BC} for the steam cycle.

For the typical conditions used in Table 6.1 for a gas Brayton cycle with a pressure ratio Π of 20.0, $\eta_{th,B}$ is 0.58. If we couple this to a simple steam Rankine cycle as shown in Figures 6.5 and 6.6, Equation 6.22 gives $\eta_{th,R} = 0.33$. Using these values, Equation 6.31 gives $\eta_{th,c,max} = 0.72$. Typical equipment efficiencies for the steam boiler (0.8) and the turbines (0.95) reduce η_c. In practice, combined cycles using a dual supercritical reheat cycle may attain an overall thermal efficiency of over 60%.

6.2.3 How Does a Cooling Tower Work?

A consequence of the Second Law of Thermodynamics is that a large power plant inevitably has to discharge energy of the order of 1 GW, because only a portion of the energy provided can be used as work to generate electricity. In a steam-powered electricity generation plant, the steam exiting the turbine must be condensed with cooling water and, owing to the volume required, the water is often extracted from lakes or rivers, passed through heat exchangers and discharged to the source of the water at a temperature greater than the source; this action in principle has an environmental consequence that will not be considered further here. To limit the temperature increment, wet cooling towers are used which rely on the enthalpy of vaporization to cool the water. In the case of water, the enthalpy of vaporization is relatively high and requires a relatively low mass to evaporate as a function of time to decrease the water temperature. Cooling towers are used in other industrial applications or with large air conditioning systems where single-phase energy exchangers utilizing air or water are insufficient to dissipate the energy.

Cooling towers may operate with either forced or natural convection, and a schematic of the latter is shown in Figure 6.13. In this case, a stream of warm water is sprayed onto a solid

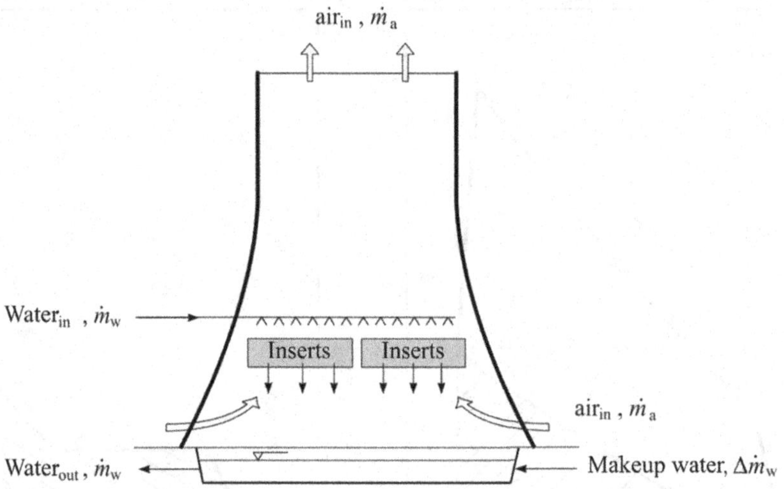

FIGURE 6.13
Scheme of natural-draught cooling tower: Incoming water is cooled by mixing with ambient air and partial evaporation; only a small portion of the water has to be replaced by additional so-called makeup water.

surface labeled as inserts in Figure 6.13 that ensure that the droplets are broken up and that there is intense mixing of them with atmospheric air that is drawn in from below and takes up moisture as water vapor leaving at the top of the tower with a higher humidity. When this exhaust air mixes with colder ambient air, a plume of fog may become visible, in a similar fashion to that discussed in Question 4.7.2.

The energy required for the evaporation of some of the water is mostly taken from the warm water that leaves with a lower temperature at the bottom of the tower and may be returned to the coolant stream. Because some cooling water is evaporated as a part of this process, it is necessary to add water to replace that which is lost, but because water has a relatively large enthalpy of evaporation the amount of water required is, as we will now show, rather small by comparison. To illustrate this point, we will consider the mass flowrate of water required for a steam-powered electricity generating plant that must dissipate a heat flux of $\dot{Q}$ = 1 GW. In the first case, we assume that cooling occurs solely by water obtained from a river that enters the cooling system at a temperature of 10°C (283 K) and then exits at a temperature of 35°C (308 K) where the specific heat capacity of water is approximated as 4.2×10^3 J kg^{-1} K^{-1}. From

$$\dot{Q} = \Delta \dot{H} = \dot{m} c_p \Delta T, \tag{6.32}$$

we have

$$\dot{m} = \frac{\dot{Q}}{c_p \Delta T} = \frac{10^9\ \mathrm{W}}{4.2 \times 10^3\ \mathrm{J\,kg^{-1}\,K^{-1}} \times (35 - 10)\ \mathrm{K}} = 9.5 \times 10^3\ \mathrm{kg\,s^{-1}}. \tag{6.33}$$

Alternatively, the cooling water may be circulated and chilled with a cooling tower. If we assume that the water in its circulation through the cooling tower is only cooled down to a temperature of 20°C, providing a temperature difference of only 15 K instead of 25 K, a higher mass flowrate of $\dot{m}_w = 16 \times 10^3$ kg s^{-1} results. For the cooling tower we assume that ambient air enters at a temperature of t = 10°C and a relative humidity Φ = 0.7 and the air exits the cooling tower saturated with water vapor at a temperature of 25°C. The moisture content ω of the air, defined by Equation 4.186, can be calculated from Equation 4.188 with $p_v^{sat}(10°C)$ = 1.23 kPa and $p_v^{sat}(25°C)$ = 3.17 kPa to give ω_{in} = 0.0054 and ω_{out} = 0.020. From an energy balance over the whole cooling tower, a mass flow of dry air $\dot{m}_a = 19 \times 10^3$ kg s^{-1} is obtained. Because water evaporates during the cooling process, it must be replaced by what is termed "additional makeup water" at an assumed temperature of 10°C. The mass flowrate of the additional water $\Delta\dot{m}_w$ required is obtained from

$$\Delta\dot{m}_w = \dot{m}_a(\omega_{out} - \omega_{in}) = 0.28 \times 10^3 \text{ kg s}^{-1}, \tag{6.34}$$

which is less than 2% of the total water mass flowrate.

6.2.4 Why Are Cooling Towers Shaped Like an Hour Glass?

Natural draught wet cooling towers, whose operation is described in the previous question, are often shaped like an hour glass, or more accurately as a hyperboloid – the shape generated by rotation of a hyperbola around its central vertical axis. They are designed in this way for two principal reasons: To optimize the heat and mass transfer between the rising air and the falling water inside the tower, and to give the tower structural stability.

The towers can be up to 200 m high with a diameter of typically 150 m at the base and 90 m at the top. As the warm cooling water from the power plant is pumped into the central region of the tower, it is sprayed through nozzles to form small droplets which fall down the tower, contacting the upward moving air in countercurrent flow. The process is designed to enhance the rates of heat transfer between the water and the upflowing air and the accompanying mass transfer of water to the air through evaporative cooling described in Question 6.2.3. This is to ensure that the transport of heat and mass (liquid water to vapor) are both sufficiently fast that the thermodynamic equilibrium conditions described in the previous question can be achieved by the time the air reaches the top of the tower. Converting the water into small droplets increases the air-water contact surface area significantly – the mean droplet size in this region is typically 2–4 mm and up to 15% of the heat transfer can occur in this region. As they move down, the droplets come into contact with the solid inserts (or "fills") whose role is to further increase the contact area as well as the contact time, either by causing the droplets to spread into thin films or break up further through splashing. The majority of the heat and mass transfer occurs in this fill zone. Then the water falls from the inserts to the bottom of the tower in slightly larger droplets in the so-called rain zone, where the final ~10% of the cooling takes place. The cooled water accumulates in a pond at the base of the tower, from where it is pumped back to the power plant steam condenser where the cooling cycle starts all over again.

The feed air is drawn naturally into the bottom of the tower and travels upwards due to the buoyancy effect of the gradient of decreasing density of the warming air as it moves up the tower and contacts the descending hot water. Mean air velocities are typically a few m s^{-1}. The hyperboloid shape of the tower influences another important factor in optimizing the convective air-water heat and mass transfer: the relative velocity of the air over the surface of the water droplets and films. The decreasing tower diameter towards the throat section, which is typically 55% of the base diameter, causes an acceleration of the air flow (the so-called "Venturi effect"; Chabra and Sankar 2017) by a factor of over three, while the air remains in laminar flow as it passes upwards through the inserts and column packings, picking up moisture on the way. After the tower throat, its diameter expands again and pressure decreases, causing the moist air to expand rapidly with further cooling. The continuing increase in tower diameter towards the exit increases the mixing contact area between the emitting hot, damp air and the cross-flow cooler external atmospheric air, supporting efficient turbulent mixing. On meeting this cooler atmosphere, the temperature of the water saturated air from the tower decreases sharply, causing the vapor to condense and produce the mist cloud plume often seen on cool days above these towers. Some very fine, micron-sized water droplets entrained in the upward air flow are also emitted into the atmosphere (drift emissions). Since these can contain polluting suspended and dissolved impurities (such as anti-scaling or corrosion additives), a key part of the design of the tower internals is to minimize the emission of these droplets by control of droplet size and filtration systems. The largest hyperboloid towers are capable of circulating about 80,000 cubic meters of water per hour.

There are also good civil engineering reasons (Reid 1984) for choosing the hyperboloid shape for large cooling towers. The bi-curved shape gives superior strength to the structure and resistance to external forces, particularly high winds. The towers have to be sufficiently tall to release the water vapor safely into the atmosphere and to give enough time and contact for air-water equilibrium to be reached, as described previously. In order to support such tall, heavy structures, the base must be consolidated and spread over a large area. Hence, the tower requires a large circular base. The stress distribution in the thin shell structure is such that a relatively small wall thickness is required, typically less than 2 cm, so requiring relatively small amounts of reinforced concrete and minimizing the weight of the structure. The wide base not only adds strength but also provides a large space for the installation of the water injection and cooling system.

6.3 What Is the Energy Penalty We Must Pay to Capture CO_2 from Fossil Fuel Power Plants?

A major concern since the last quarter of the twentieth century has been the potentially very damaging environmental, economic and social effects of climate change (previously referred to as "global warming") caused by the release of CO_2 and other so-called greenhouse gases (GHGs) from the use of fossil fuels, not only for power but also for heating and major industrial processes such as petrochemicals and cement manufacture. The "carbon footprint" of using fossil fuels can be minimized by using a process called carbon capture and storage (CCS). This is a generic approach that can be applied to remove CO_2 from any process using fossil fuels in a centralized facility (see Figure 6.14).

A CCS process consists of a number of stages:

- *Capture* of CO_2 (and sometimes other GHGs such H_2S, SO_2, NO_x) by passing the exhaust gas stream through a capture system that typically involves dissolution in a solvent, adsorption on a high surface area solid, reaction with a suitable reagent such as a metal oxide or physical separation using, for example, membranes.
- *Release* (or stripping) of the captured CO_2 from the capture system.
- *Compression* of the released captured CO_2 to the supercritical state (above 304.13 K, 7.38 MPa) where it has the (high) density of a liquid, suitable for efficient transport and storage and the (low) viscosity of a gas, optimizing the ease of injection into porous natural mineral systems in subsurface storage sites.

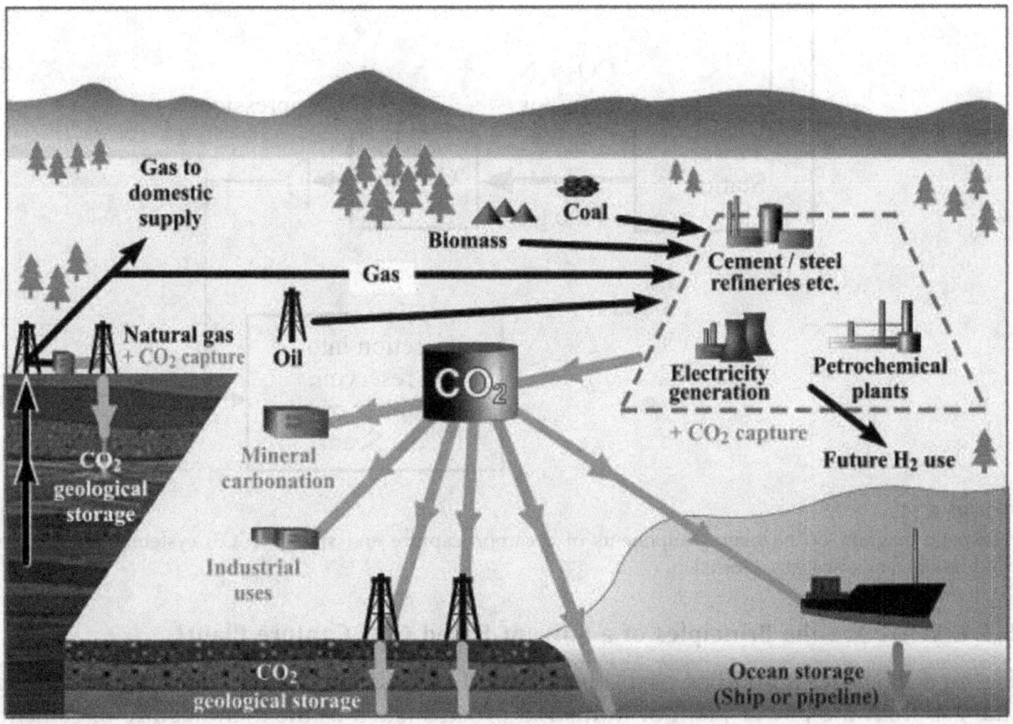

FIGURE 6.14
Schematic diagram illustrating a range of possible carbon capture and storage (CCS) systems, involving capture of CO_2 from a centralized facility using fossil fuels, compression to the supercritical state and transport by pipeline or tanker to a suitable underground storage site (depleted oil and gas reservoirs, deep saline aquifers or uneconomic coal seams) where it is injected and sealed in for long-term (indefinite) storage.

- *Transport* of the compressed CO_2 to the geological storage site by pipeline or tanker.
- *Injection* and sealing into subsurface geological sites for long-term storage; the main types of site are depleted oil and gas reservoirs, deep saline aquifers or uneconomic coal seams.

This overall system is illustrated schematically in Figure 6.15. Each of these stages involves the use of energy, so the total net energy used by the CCS system is the *energy penalty* (EP) that must be paid for preventing the CO_2 produced in the process from being released into the atmosphere. If the process is a power station, for instance, its electricity output will be reduced because some of the steam normally used to drive the turbines is diverted to power the CO_2 capture process, having an impact on both the power station output and its economics. The additional cost of electricity as a result of using CCS will arise from the cost of the capture process itself and this energy penalty (see also Table 7.2 in Question 7.2.2).

As an example of incorporating CCS into a power generation system, we will consider a case study where a solvent-based CO_2 capture process is used to extract CO_2 from the post-combustion flue gas stream of a combined-cycle gas turbine (CCGT) plant, described in Question 6.2.2 and shown in Figures 6.11 and 6.12.

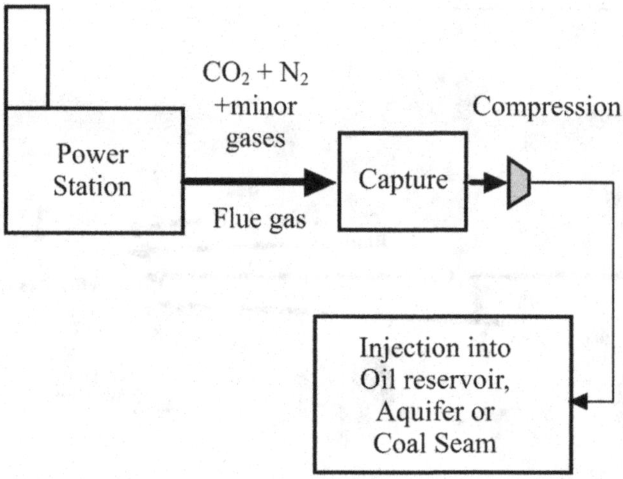

FIGURE 6.15
Schematic diagram of the main components of a carbon capture and storage (CCS) system: CO_2 capture, compression and subsurface injection.

6.3.1 What Are the Principles of a Solvent-Based CO_2 Capture Plant?

Figure 6.16 shows an outline of a solvent-based CO_2 capture plant. The CO_2 containing flue gas from the power plant or industrial process (such as the manufacture of cement, iron or steel) is sent to an *absorption column* where it is contacted in countercurrent flow

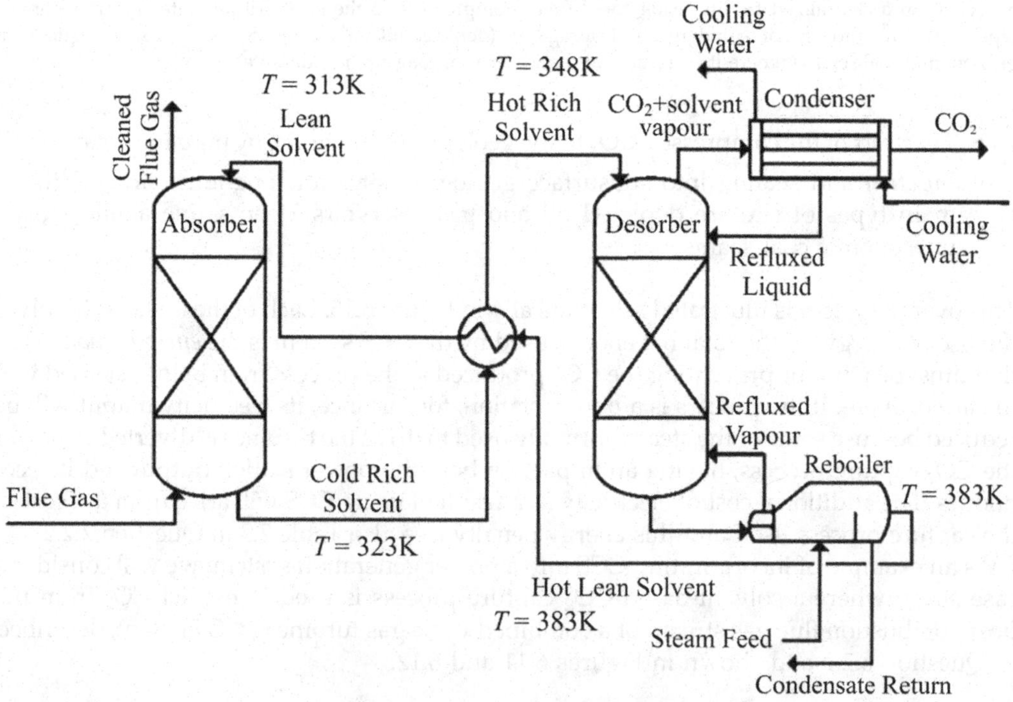

FIGURE 6.16
Process flow diagram for a CO_2 solvent absorption capture plant.

with a solvent. The cleaned gas leaves this column at the top whereas the CO_2-rich solvent leaves at the bottom. The latter is passed through a heat exchanger where it is heated by hot recycled CO_2-lean solvent and enters the top of a *stripper* or *desorber column*, where it is heated to remove the CO_2. A pure CO_2 stream leaves the top of the stripper column and the hot CO_2-lean solvent is recycled to the top of the absorber column via the heat exchanger heating the CO_2-rich stream leaving the bottom of that column.

The thermodynamic properties involved in this process are:

- the solubility s of CO_2 in the solvent
- the heat of dissolution $\Delta_{soln}H$ of CO_2 in the solvent
- the heat of desorption $\Delta_{des}H$ of CO_2 from the solvent = $-\Delta_{soln}H$ if $T_{soln} = T_{des}$

6.3.2 What Are the Energy Requirements of a Solvent-Based CO_2 Capture Plant?

The design of a full solvent-based CO_2 capture plant involves a complete consideration of the heat and mass transfer processes in the columns, heat exchanger, condenser and solvent reboiler and the liquid-vapor phase behavior as well as the gas-solvent dissolution thermodynamics. This is beyond the scope of this text. However, we can obtain a reasonable estimate of the energy requirements of the process by simply considering the key solution thermodynamic quantities given above. As an example, we choose the most common CO_2 capture solvent, monoethanolamine (MEA).

6.3.2.1 Absorption stage

The absorption column of Figure 6.16 typically operates at about 40°C (313 K). The incoming flue gas stream from a CCGT plant will be at ~0.1 MPa with a CO_2 content of typically 3–4%, although with exhaust gas recycling this can rise to 5–9%. For this exercise we will take 5% as a suitable CO_2 concentration, equivalent to a CO_2 partial pressure, p_{CO_2}, of 5 kPa. Under these conditions, for a typical 30% MEA by weight aqueous solution, the MEA equilibrium loading is 0.5 mol CO_2/mol MEA.

If we assume that natural gas is composed entirely of methane (typically the content is 95–98%, the rest being mainly ethane), then combustion in the CCGT plant proceeds as

$$CH_4 + 2O_2 \rightarrow 2H_2O + CO_2. \tag{6.35}$$

Hence, burning 1 kg of methane produces an amount of n_{CO_2} = 1,000/16 mol = 62.5 mol CO_2, which requires 125.0 mol MEA (M_{MEA} = 61 kg mol^{-1}) for complete dissolution. Now, 1 kg of the 30% aqueous MEA solvent contains 300/61 mol = 4.9 mol MEA, so the mass of solvent required is m_{solv} = 125.0/4.9 kg = 25.5 kg.

The enthalpy of dissolution (or absorption) under the column conditions, $\Delta_{soln}H$, is –85 kJ/mol CO_2, so if the column is run adiabatically an energy balance (Chapter 1) yields

$$\Delta_{soln}H \; n_{CO_2} = m_{solv} \; c_{p,solv} \; (T_{out} - T_{in}), \tag{6.36}$$

where T_{out} and T_{in} are the outlet and inlet temperatures, respectively, of the absorber column solvent stream and $c_{p,solv}$ is the heat capacity of the solvent under the column conditions, for which a typical mean value is 3.8 kJ kg^{-1} K^{-1}. Hence, Equation 6.36 yields 85.0 × 62.5 = 25.5 × 3.8 × (T_{out} / K – 313), from which T_{out} would be 368 K (95°C). However, heat losses from the absorber and the cooling effect of water evaporation mean that T_{out} is typically ~323 K (50°C). Moreover, an increase in temperature leads to a decrease in the solvent absorption capacity, so additional cooling of the absorber solvent as it passes down the column is sometimes performed to enhance the CO_2 loading.

6.3.2.2 Stripping stage

Let us now examine the energy requirement for removing CO_2 from the CO_2-rich solvent as it passes down the stripping column. This CO_2 "regeneration energy", $\Delta_R H$, is the steam energy input to the reboiler of Figure 6.16, sometimes referred to as the *specific reboiler duty* (SRD). This will consist of three components:

- the heat of desorption of CO_2 from the solvent, $-\Delta_{soln}H$.
- the heat Q_{sens} involved in heating the CO_2-rich solvent in the stripper from the feed temperature at the top, T_{feed}, to its exit temperature from the reboiler, T_{reb}, at the bottom

$$Q_{sens} = m_{solv}\, c_{p,solv}\,(T_{reb} - T_{feed}).$$

  This heat, used simply to increase the temperature of a material with no phase change and so increase its stored enthalpy (or internal energy) is often called *"sensible heat"*.

- the heat Q_{vap} going into forming the excess steam leaving the stripper, needed to maintain the partial pressure driving force for CO_2 desorption from the solvent,

$$Q_{vap} = m_{vap}\,\Delta_l^g h_w$$

  where m_{vap} is the mass of water vaporized and $\Delta_l^g h_w$ is the specific latent heat of water vaporization.

The aim is to generate enough steam to maintain CO_2 desorption while minimizing the large energy loss in the condenser. Typical conditions in the stripping column to achieve this are:

- T_{feed} = 348 K: as seen in Figure 6.16, the CO_2-rich solvent stream from the bottom of the absorber, typically at ~323 K, is usually heated in a heat exchanger on its way to the top of the stripping column by the hotter CO_2-lean solvent from the reboiler (~383 K), which is on its way to the top of the absorber column, where it enters at about 313 K.
- T_{reb} = 383 K.
- p_{strip} = 150 kPa.
- rich solvent loading = 0.5 mol CO_2/mol MEA (top feed to stripping column). Conditions are usually chosen so that Q_{vap} is kept as low as possible. One factor is that at 150 kPa, the water bubble point (Question 4.7.4) is 111°C or 384 K (T_{bp}); if T_{reb} is increased above this temperature, more and more water evaporates, so keeping T_{reb} close to $T_{bp}(p)$ helps minimize Q_{vap}. Calculating m_{vap} requires a full consideration of the vapor-liquid equilibria along the temperature gradient within the stripper column. This shows that as the CO_2-rich solvent loadings

decrease from 0.5 to 0.3 mol CO_2/mol MEA, the CO_2 equilibrium partial pressure falls from ~150 kPa to ~20 kPa so that stripping from CO_2 loadings of 0.5 mol CO_2/mol MEA requires significantly less water vapor, reducing Q_{vap}.

- lean solvent loading = 0.2 mol CO_2/mol MEA; stripping to lower loading values requires large amounts of water vapor, which increases the regeneration enthalpy $\Delta_R H$ to unacceptably high values.

A full analysis of the behavior of both the absorber and stripper columns of a solvent carbon capture process requires a consideration of the continuous counterflow of the liquid and gas streams in both columns, the temperature gradients in each vessel, the rates of heat and mass transfer between the different phases and the thermodynamics of the gas-liquid phase behavior of water, MEA and CO_2-MEA. However, for simplicity, we can calculate the regeneration energy $\Delta_R H$ for the batch equivalent of this continuous process, using the typical column conditions given above, as follows:

Assuming first of all that all the CO_2 absorbed in the MEA solvent, 62.5 mol for every kg of methane burned, is regenerated, then

$$\begin{aligned} \Delta_R H &= -\Delta_{soln} h \, m_{CO_2} + Q_{sens} + Q_{vap} \\ &= -\Delta_{soln} h \, m_{CO_2} + m_{solv} \, c_{p,solv} (T_{reb} - T_{feed}) + m_{vap} \, \Delta_l^g h_W. \end{aligned} \tag{6.37}$$

For the conditions chosen, very little excess steam is generated so, assuming $m_{vap} \sim 0$,

$$\Delta_R H = 85 \times 62.5 \text{ kJ} + 25.5 \times 3.8 \times (383 - 348) \text{ kJ} = (5312.5 + 2391.5) \text{ kJ} = 8.7 \text{ MJ}.$$

Thus, our simple calculation suggests we use 8.7 MJ of energy to capture the 2.75 kg of CO_2 produced from combusting 1 kg of CH_4; so, expressing the regeneration enthalpy per unit mass, we have

$$\begin{aligned} \Delta_R h &= 8.7 \text{ MJ/kg } CH_4 \text{ combusted} \\ &\text{or } 3.2 \text{ MJ/kg (141 kJ/mol) } CO_2 \text{ captured.} \end{aligned} \tag{6.38}$$

There are several reasons why for a real absorber-stripper system the energy requirement may differ from this. The main one is that design of the stripper may be such that there is a significant amount of excess steam generated, m_{vap}, so that there will be an additional energy requirement of $Q_{vap} = m_{vap} \, \Delta_l^g h_W$. Typically, this could increase the energy penalty by ~25%. Conversely, using the heat in the hot CO_2-lean MEA stream to heat only part of the cool CO_2-rich solvent stream, as well as to inject heat into the middle of the stripper column, enables more efficient use of any excess hot water vapor coming up the column and can reduce the Q_{vap} energy penalty by up to 20%. These effects may to some extent balance each other out.

6.3.3 How Much Electricity and Carbon Dioxide Do We Generate by Burning 1 Kilogram of Natural Gas in a CCGT Power Plant?

Again, we assume that natural gas is composed entirely of methane and we use the result of Question 6.2.2 that the CCGT plant has an efficiency of ~60%. If the methane is burned in air, the relevant combustion reaction is

$$\begin{aligned} &CH_4 + 2O_2 + 8N_2 \rightarrow 2H_2O(l) + CO_2 + 8N_2 \\ &\Delta_c H^{\ominus} = -890.8\ \text{kJ/mol}\ (298\text{K},\ 0.1\ \text{MPa}). \end{aligned} \tag{6.39}$$

Nitrogen is included because, although it does not take part in the combustion reaction, we will see that it has a significant effect on the energy balance and hence the thermal energy available for power generation.

For a typical CCGT plant, the typical operating temperature for the gas turbine stages (Brayton cycle) illustrated in Figures 6.9 to 6.12 are

Compressor in: T_1 = 298 K
Compressor out: T_2 = 700 K = inlet to gas combustor
Gas Turbine in: T_3 = 1,500 K = outlet from gas combustor
Gas Turbine out: T_4 = 800 K = inlet to heat exchanger (HRSG) = T_C, inlet to steam turbine
HRSG outlet: T_5 = 353 K

The heat recovery steam generator (HRSG) is the heat exchanger in the center of Figure 6.11 where the hot gas turbine exit stream is used to generate the steam used to drive the steam turbine.

Thus, to calculate the overall heat input to the combined gas turbine Brayton cycle and steam turbine Rankine cycle, we need to calculate the enthalpy change of the combusted gas stream from the inlet to the combustor at 700 K to its exit from the HRSG at 353 K, at an operating pressure of typically 2 MPa. We will

- assume that methane behaves as a perfect gas in the range 0.1–2 MPa, so that the heat of combustion does not vary with pressure at the conditions in the gas turbine;
- use the heat capacities of the gases involved in the reaction of Equation 6.39 over the temperature range 298 K and 700 K, as set out in Table 6.2;
- use the cycle in Figure 6.17 to give the enthalpy change from the turbine inlet conditions of CH_4 and O_2 to the HRSG outlet conditions for $H_2O(l)$ and CO_2;
- as is common in combustion calculations, assume a ratio of $N_2/O_2 = 0.79/0.21 = 3.76$ for air, which rounded yields 8 mol N_2 associated with the reaction of 2 mol O_2.

TABLE 6.2
Perfect gas heat capacities of gases involved in CCGT gas turbine combustion (Maitland 2021)

Substance	$C_{p,m}$ (298 K)/(J·mol^{-1}·K^{-1})	$C_{p,m}$ (353 K)/(J·mol^{-1}·K^{-1})	$C_{p,m}$ (700 K)/(J·mol^{-1}·K^{-1})
$CH_4(g)$	35.77	37.09	46.67
$O_2(g)$	29.39	29.69	31.35
$H_2O(l)$	75.33	75.33	–
$CO_2(g)$	37.15	51.33	43.95
N_2 (g)	29.06	29.13	29.80

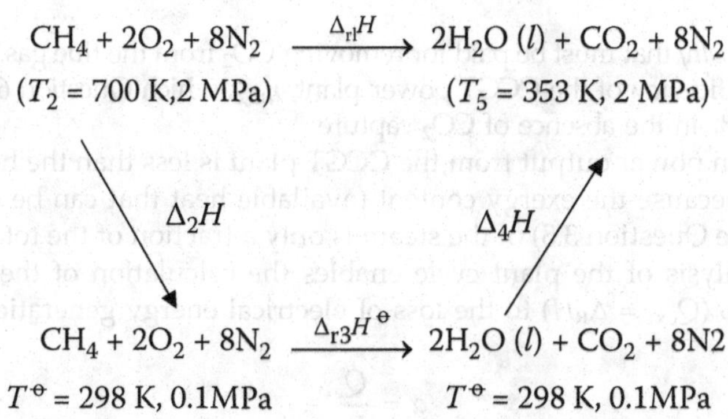

FIGURE 6.17
Energy cycle relating enthalpy change on combustion of methane in gas turbine and heat transfer to steam turbine via the heat recovery steam generator HRSG to standard enthalpy of combustion.

Applying the First Law to this cycle (see Hess's law, Question 5.2.1) gives

$$\begin{aligned}\Delta_{r1}H &= \Delta_2 H + \Delta_{r3}H^\ominus + \Delta_4 H \\ &= \bar{C}_{p2M}(T^\ominus - T_2) + 2\bar{C}_{p2O}(T^\ominus - T_2) + 8\bar{C}_{p2N}(T^\ominus - T_2) + \Delta_c H^\ominus(298K, 0.1\text{MPa}) \\ &\quad + 2\bar{C}_{p4w}(T_5 - T^\ominus) + \bar{C}_{p4C}(T_5 - T^\ominus) + 8\bar{C}_{p4N}(T_5 - T^\ominus),\end{aligned} \tag{6.40}$$

where $\bar{C}_{p2M}$ is the mean heat capacity of methane for step 2 between T_2 and $T^\ominus$, approximated by the average of the values at T_2 and $T^\ominus$, and the symbols for the other heat capacities are chosen accordingly.

Using Table 6.2 for the heat capacities in Equation 6.40 gives

$$\Delta_{r1}H = -776.7\ \text{kJ/mol or } -48.5\ \text{MJ/kg of } CH_4. \tag{6.41}$$

So if we combust 1 kilogram (or 1,000/16 mol = 62.5 mol) of methane in a CCGT plant with an efficiency of 60%, the electrical energy generated, E_{el}, is

$$E_{el} = 48.5\ \text{MJ} \times 0.60 = 29.2\ \text{MJ}. \tag{6.42}$$

Given the approximations made (e.g. perfect gases) and the variations in the complexity and temperature conditions of the different stages of practical combined cycles, these results are remarkably close to those of commercial systems. Henceforth, we take the enthalpy input from gas combustion, $\Delta_{r1}h_1$, to be typically –50.0 MJ/kg of CH_4 and the baseline electricity generation of a CCGT plant to be 30 MJ/kg of CH_4. From Equation 6.39, we see that in burning 62.5 mol (1 kg) of methane, we produce 62.5 mol of CO_2 or 62.5 mol × 0.044 kg mol^{-1} = 2.75 kg of CO_2.

6.3.4 What Is the Energy Penalty and Efficiency Decrease Incurred by Adding Amine-Solvent Post-Combustion Carbon Capture onto a CCGT Power Plant?

We now return to the CCGT power plant shown in Figures 6.11 and 6.12. The steam needed to release the CO_2 in the solvent stripper of the capture process described in Question 6.3.1 is taken from the HSRG. The steam used in the stripper is therefore no longer available to generate electricity. So the energy used, $\Delta_R H$ = 3.2 MJ per kg of CO_2 captured (Equation 6.38), is the so-called *energy penalty* that must be paid for removing CO_2 from the flue gas. This will lead to a reduction in the efficiency of the CCGT power plant, $\eta_{th,c}$, which Question 6.2.2 indicated was typically $\eta_{th,c} = 60\%$ in the absence of CO_2 capture.

The reduction in power output from the CCGT plant is less than the heat content of the removed steam because the exergy content (available heat that can be converted to mechanical work, see Question 3.8) of the steam is only a fraction of the total heat content. A full enthalpy analysis of the plant cycle enables the calculation of the ratio (α) of heat extracted for CCS ($Q_{ext} = \Delta_R H$) to the loss of electrical energy generation, E_{loss},

$$\alpha = \frac{Q_{ext}}{E_{loss}}. \tag{6.43}$$

The ratio α depends on the state of the steam when it is extracted, in particular for a reheat Rankine cycle, for example (see Figures 6.7 and 6.8), as either single phase superheated (HP) steam or as a two-phase low pressure (LP) mixture; for the typical conditions chosen here, with steam input to the stripper reboiler at ~110°C, $\alpha \sim 3$ (Nord and Bolland 2020).

We have seen in Question 6.3.3 that burning 1 kg of methane in the CCGT plant generates 50.0 MJ of heat (Q_{in}) and that this is typically converted at 60% efficiency to 30.0 MJ of electrical energy, E_{el}. From Equation 6.38, the thermal energy penalty ($Q_{ext} = \Delta_R H$) of adding the carbon capture unit is 8.70 MJ per kg of methane burned in the power plant. So the overall efficiency of the CCGT plant fitted with post-combustion carbon capture is

$$\eta_c = \frac{\text{Work output}}{\text{Heat input}} = \frac{E_{el} - E_{loss}}{Q_{in}} = \frac{E_{el} - Q_{ext}/\alpha}{Q_{in}} = \frac{30.0 - 8.70/3}{50.0} = 0.54. \tag{6.44}$$

Hence, for this typical case, there is a *loss in efficiency of 6%-points* and a *decrease in electrical output of 10%* compared to the plant without capture. The increase in operating costs arising from this decrease in efficiency of generation must be added to the cost of building and operating the solvent CO_2 capture unit in determining the additional cost of produced electricity by adding carbon capture to a CCGT power station.

The other major energy requirement for CCS is in the compression of the captured CO_2 to its supercritical state prior to injection into the subsurface reservoir. We can estimate the energy required by assuming that this takes place isentropically according to Equation (1.107).

Compressing 1 mol of CO_2 with a volume $V_1 = 22.4 \times 10^{-3}\ \text{m}^3\ \text{mol}^{-1}$ at standard conditions of $p_1 = 0.1$ MPa, 298 K, to its critical pressure of $p_2 = 7.38$ MPa, with $\gamma = 1.29$, has an energy requirement of

$$w_{s12} = 16.3\ \text{kJ mol}^{-1}\frac{\text{mol}}{0.044\ \text{kg}} = 370\ \text{MJ/t}\ CO_2. \tag{6.45}$$

This is 12% of the thermal energy (3,200 MJ/t CO_2, from Equation 6.38) required to capture CO_2 from the power plant flue gas, giving a total CCS energy requirement of approximately 3.6 GJ/t CO_2.

6.3.5 How Does the Calcium Looping CO_2 Capture Process Work and How Much Energy Does It Use?

In Questions 6.3.2 and 6.3.4 we examined the energy required to capture CO_2 from the flue gas of a gas-fired power station using a solvent absorption process. We will now consider two alternative carbon capture processes: calcium looping and chemical looping combustion.

Calcium looping is a CO_2 capture process based on

a. passing a flue gas containing CO_2 through a fluidized bed (carbonator reactor) of calcium oxide (lime), whereby the CO_2 and lime combine to form calcium carbonate

$$CaO + CO_2 \rightarrow CaCO_3 \; \Delta_r H^{\ominus} = -178 \text{ kJ/mol}, \quad (6.46)$$

b. transferring the calcium carbonate to a calciner reactor where it is heated to release CO_2, while itself reverting to calcium oxide which is recycled to the carbonator to capture more CO_2

$$CaCO_3 \rightarrow CaO + CO_2 \; \Delta_r H^{\ominus} = 178 \text{ kJ/mol}. \quad (6.47)$$

The process is illustrated schematically in Figure 6.18.

Typical conditions are:

- Carbonator: $t = 500 - 750°C$, $T = 773 - 1{,}023$ K, $p = 0.1–3$ MPa
- Calciner: $t = 850 - 1{,}000°C$, $T = 1{,}123 - 1{,}273$ K, $p = 0.1$ MPa.

The carbonator's operating temperature is chosen as a compromise between higher equilibrium capture at lower temperatures (given that reaction 6.46 is exothermic, see Question 5.3.5) accompanied by a decreased reaction rate. Similarly, a temperature of >850°C in the calciner strikes a balance between increased extent and rate of calcination at higher temperatures and reduced CaO sorbent degradation rate (through high-temperature sintering that reduces available surface area) at lower temperatures.

We will carry out the same exercise as for the MEA scrubbing capture plant in Question 6.3.2 and estimate the energy penalty in capturing the 62.5 mol, or 2.75 kg, of CO_2 produced on burning 1 kg of methane in a CCGT power plant. The inlet CO_2 gas is assumed to be at the CCGT gas turbine outlet temperature of 527°C (800 K) and the carbonator to run at 650°C (923 K). We will initially assume Reaction 6.46 goes to completion with 100% capture.

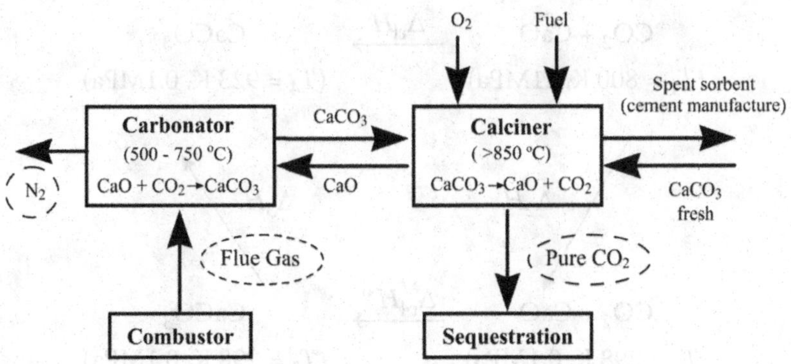

FIGURE 6.18
Schematic diagram of calcium looping CO_2 capture process.

6.3.5.1 Carbonator

The enthalpy change for reaction 6.46 at the specified conditions can be calculated using the cycle given in Figure 6.19. Applying the First Law to this cycle (and using Hess's law) in the same way as in Question 6.3.3, we find that

$$\Delta_{r1}H = -10.08 \text{ MJ} \tag{6.48}$$

per 62.5 mol (or 2.75 kg) CO_2 captured from burning 1 kg CH_4.

6.3.5.2 Calciner

The chemical change taking place in the calciner, assumed to operate at 950°C (1,223 K), is given by Equation 6.49. We can evaluate the energy required in the calciner from $\Delta_r H^{\ominus}$ for this reaction by using a Hess's law cycle similar to that in Figure 6.19 for the carbonator. The result is

$$\Delta_{r4}H^{\ominus} = 12.59 \text{ MJ} \tag{6.49}$$

per 62.5 mol (or 2.75 kg) of CO_2 captured from burning 1 kg CH_4.

6.3.5.3 Combined

Combining the two preceding calculations, we obtain for the overall process:

$$\text{Net energy requirement} = 12.59 \text{ MJ} - 10.08 \text{ MJ} = 2.51 \text{ MJ}. \tag{6.50}$$

However, because the carbonator is only at 650°C, the excess heat generated there cannot be used to heat the calciner to the higher temperature of 950°C. (It can nevertheless be used in a relatively efficient Rankine steam cycle to generate additional electricity, as we will see shortly). Hence

$$\begin{array}{l}\text{Overall heating requirement for the process} \\ \text{(calciner) or primary energy penalty} = 12.6 \text{ MJ}.\end{array} \tag{6.51}$$

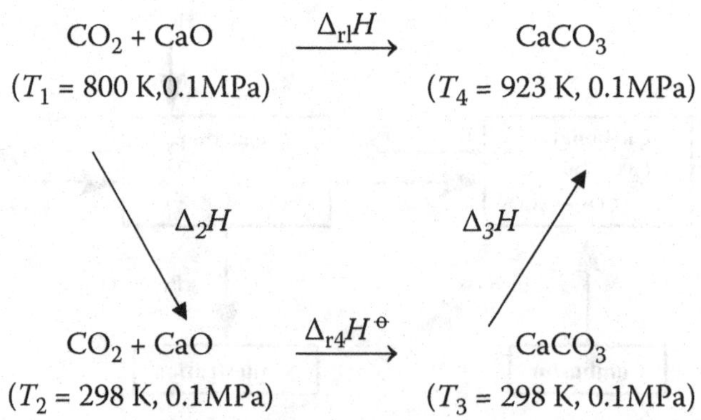

FIGURE 6.19
Energy cycle relating enthalpy change on carbonation of lime in a calcium looping carbonator to the standard enthalpy of carbonation.

So the energy requirement for this calcium looping carbon capture process is

$$\begin{aligned}&12.6\ \text{MJ/kg CH}_4\ \text{combusted}\\&\text{or } 12.6/2.75\ \text{MJ/kg} = 4.6\ \text{MJ/kg CO}_2\ \text{captured}\\&\text{or } 4600 \times 0.044\ \text{kJ/mol} = 202.4\ \text{kJ/mol CO}_2\ \text{captured.}\end{aligned} \tag{6.52}$$

This may be supplied externally from the HRSG of the power plant or internally within the calciner by, for example, feeding a fuel (coal, gas or biomass, alongside the calcium carbonate from the carbonator) and oxygen to burn and create heat within the calciner. For the latter option, oxygen is used rather than air to avoid contaminating the CO_2 exit stream, destined for underground storage, with nitrogen; there would therefore be an additional energy penalty in separating oxygen from air.

6.3.5.4 Additional power generation

This primary energy penalty may be offset by using the excess heat from both the calciner and the carbonator to raise steam to produce additional electricity. The hot CO_2 exit stream from the calciner at 100% CO_2 capture would consist of 62.5 mol or 2.75 kg for every kg of methane burned in the CCGT. This stream is at 1223 K, 0.1 MPa for which the specific enthalpy of the CO_2 is 1547.8 kJ/kg. So

$$\text{Enthalpy of CO}_2\ \text{exit stream} = 1547.8 \times 2.75\ \text{kJ} = 4{,}256.5\ \text{kJ} = Q_{\text{calciner}}. \tag{6.53}$$

The exergy content of the HT steam generated using this heat is relatively high, with an α value (Equation 6.43) of ~2.2. So the heat to power conversion is quite efficient and the net work generated in the CCGT from this, $W_{\text{calciner}}^{CO_2}$ is

$$-W_{\text{calciner}}^{CO_2} = Q_{\text{calciner}}/\alpha = 4.26\ \text{MJ}/2.2 = 2.0\ \text{MJ}. \tag{6.54}$$

The carbonator produces 10.1 MJ of high-grade heat (Equation 6.48), which may also be used to raise steam for power generation,

$$-W_{\text{carbonator}} = 10.1/\alpha = 10.1/2.2\ \text{MJ} = 4.6\ \text{MJ}. \tag{6.55}$$

A third possibility is to use the hot calciner CaO exit stream at 1,223 K to raise steam. The enthalpy of CaO at 1,223 K is 44.6 kJ/mol, therefore

$$Q_{\text{calciner}}^{\text{CaO}} = 62.5 \times 44.6\ \text{MJ} = 2.79\ \text{MJ}, \tag{6.56}$$

and if all this heat could be extracted to raise steam, then the additional electrical energy produced would be

$$-W_{\text{calciner}}^{\text{CaO}} = Q_{\text{calciner}}/\alpha = 2.79\ \text{MJ}/2.2 = 1.3\ \text{MJ}. \tag{6.57}$$

6.3.5.5 *Overall net energy penalty*

The primary energy requirement of the calciner is 12.6 MJ per kg of methane combusted in the CCGT (Equation 6.51). If there is no additional electricity generation using heat from the calcium-looping exit streams then, using Equation 6.44, the efficiency of the CCGT with calcium looping capture, η_{cc}^{Cal} is

$$\eta_{cc}^{Cal} = \frac{\text{Work output}}{\text{Heat input}} = \frac{E_{el} - E_{pen}/\alpha}{Q_{in}} = \frac{30 - 12.6/2.2}{50.0} = 0.48. \tag{6.58}$$

If we include the compensating power contributions from the carbonator heat and the hot CO_2 exit stream,

$$\eta_{cc}^{Cal} = \frac{\text{Work output}}{\text{Heat input}} = \frac{E_{el} - E_{pen}/\alpha + W_{calciner}^{CO_2} + W_{carbonator}}{Q_{in}} = \frac{30 - 5.8 + 4.6 + 2.0}{50.0} = 0.62. \tag{6.59}$$

So, under these conditions, the energy penalty for calcium looping can, in principle, be made quite small and far less than for the amine solvent process. However, a number of factors act to reduce the net efficiency below this ideal value. The operating temperatures of the carbonator and the calciner are both compromises between equilibrium yields, reaction rates and sintering degradation of the adsorbent. This means that in practice carbonators operate with 80–95% capture efficiency with CO_2 release in the calciner in excess of 90%. Overall, CO_2 capture efficiencies are therefore typically 80–90%. A reduction in capture efficiency to 80% and inclusion of heat exchange inefficiencies ($\eta \sim 0.8$) in extracting heat from the calciner CO_2 exit stream and the carbonator modifies Equation (6.59) to

$$\begin{aligned}\eta_{cc}^{Cal} &= \frac{\text{Work output}}{\text{Heat input}} = \frac{E_{el} - E_{pen}/\alpha + W_{calciner}^{CO_2} + W_{carbonator}}{Q_{in}} \\ &= \frac{30.0 - 0.8 \times 5.8 + 0.8 \times 0.8(4.6 + 2.0)}{50.0} = 0.59.\end{aligned} \tag{6.60}$$

In this analysis, there are a number of other energy requirements and process variations that have been neglected, such as calcining the original loading of limestone to form lime for the carbonator, powering the pumps moving the gases and solids around the looping system, the partial recycle of some of the calciner CO_2 exit stream back to the carbonator and the addition of steam to the flue gas stream entering the carbonator. These will lead to decreases in the overall process efficiency compared to the (close to) upper thermodynamic limits presented here. Overall, however, the calcium-looping process is less sensitive to capture inefficiencies and heat losses than the amine solvent process of Question 6.3.2 because of its capability to generate additional electricity from excess heat in the process, so reducing the energy penalty.

6.3.6 What Is the Chemical Looping CO_2 Capture Process and How Much Energy Does It Use?

A major problem for practical power cycles, exacerbated by the need to extract power to drive carbon capture systems, is the irreversibility inherent in complex systems of finite size.

Chemical looping is an approach that enables a significant reduction in the irreversibility of fuel combustion using thermochemical processes alone. Instead of carrying out the combustion of a hydrocarbon fuel, such as natural gas using air or oxygen, directly to CO_2 and steam, combustion of the primary fuel is split into two redox stages. This ***chemical looping combustion (CLC)*** involves an *oxygen carrier*, M (usually a metal), whose oxide acts as the source of oxygen for the combustion process. As we have seen in Question 6.2, the conventional way to reduce the lost work of combustion for a power plant is to increase the temperature at which the working fluid receives heat, T. The extent to which this can be done is limited by the metallurgical constraints on the materials that will sustain the required extreme temperatures. CLC is a different way of increasing the reversibility of power generation without the need to go to such high temperatures.

6.3.6.1 How Does Chemical Looping Approach a Reversible Combustion Pathway?

The lost work in a real power cycle is linked to the extent to which the real temperature path, T_{real}, deviates from that corresponding to equilibrium or reversible behavior, T_{rev}. A good indication of the latter for an isothermal process involving a combustible or oxidizable material is the *combustion (or equilibrium) temperature*, T_c, defined as the temperature at which the reaction takes place reversibly under standard conditions (McGlashan 2008). Since

$$\Delta_r G \approx \Delta_r H^{\ominus} - T\,\Delta_r S^{\ominus}, \tag{6.61}$$

and under reversible conditions $\Delta_r G = 0$ and $T = T_c$, we have

$$T_c = \frac{\Delta_r H^{\ominus}}{\Delta_r S^{\ominus}} = \frac{T^{\ominus}\Delta_r H^{\ominus}}{(\Delta_r H^{\ominus} - \Delta_r G^{\ominus})}, \tag{6.62}$$

where $T^{\ominus}$ is the standard temperature (298.15 K) (see Question 1.9). Since $\Delta_r H^{\ominus}$ is negative for a combustion reaction, $\Delta_r S^{\ominus}$ must also be negative for Equation 6.62 to give a physically realistic value; as we shall see, this is almost always the case for chemical-looping reactions. When both $\Delta_r H^{\ominus}$ and $\Delta_r S^{\ominus}$ are negative, then the reaction only proceeds if $T < T_c$ because $\Delta_r G$ must be negative for the reaction to proceed (*cf.* Equation 6.62). So T_c is the maximum temperature at which the combustion or oxidation can proceed, and it will do so reversibly. T_c also corresponds to the maximum temperature that could be obtained in a perfectly reversible heat engine (Carnot cycle) driven by combustion or oxidation of that material. To proceed perfectly reversibly, these processes would have to take place very (in fact infinitely) slowly, which of course imposes huge practical constraints. The aim in CLC is to operate the combustion of a fuel such as methane using redox reactions, operating as closely as possible to the relevant T_c's for the reactions involved, to take advantage of the large efficiency advantages that this imparts (see Question 6.3.6.2), while operating sufficiently away from exact equilibrium conditions to give acceptable reaction and heat transfer rates and so keep process equipment size and costs under control.

A schematic diagram of a chemical looping combustion system is given in Figure 6.20. An oxygen carrier (typically a metal) is first oxidized with air in the *oxidizer reactor*. This highly exothermic reaction produces a nitrogen-rich gaseous exit stream, released to atmosphere, which is readily separated from the other solid or liquid oxide stream. This hot stream moves to the second reaction vessel, the *reducer*, where the metal oxide is converted back to metal by reaction with a fuel, here taken to be methane. The reduction reaction is endothermic and also produces two easily separated product streams: reduced

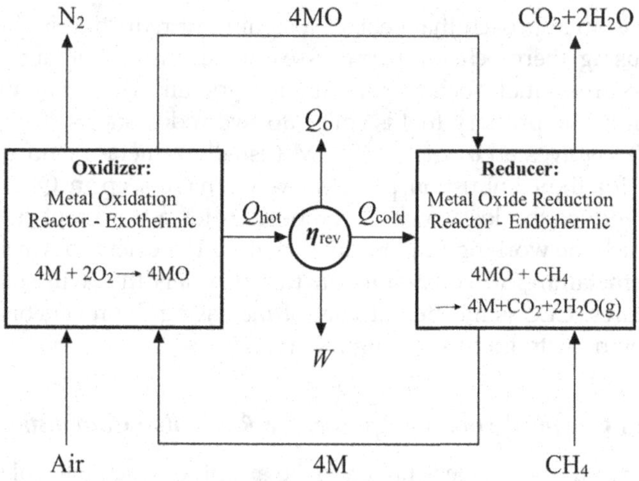

FIGURE 6.20
Schematic diagram of chemical looping combustion process for power production with integrated carbon capture.

metal which can be condensed and a *flue gas* containing CO_2 and steam. The metal stream is returned to the oxidizer vessel for re-oxidation while water condenses from the flue gas stream as it cools, leaving an essentially pure stream of CO_2, which is then compressed to a supercritical state ready for transport and storage. Any remaining water condenses in the intercoolers of the CO_2 compression system.

As the metal/metal oxide system is pumped around the process loop, heat (Q_{hot}) is transferred from the exothermic reaction in the oxidizer to a heat engine (power cycle). The available heat is converted into shaft work in a turbine and hence to electrical power. The remaining heat (Q_{cold}), rejected from the heat engine, passes to the reducer vessel where it provides the heat required to drive the endothermic reduction process. Additional waste heat (Q_o) may be rejected; if this is significant, it requires a second heat engine to ensure that heat rejection takes place at $T^{\ominus}$ so that after extraction of further power, no available energy leaves the process by this route.

If the two redox reactions are each carried out at their respective T_c's, the process will be essentially reversible. Hence, the metal and metal oxide inlet streams must be heated and cooled to their required temperatures by recovering heat from the product streams, leaving the two reactors using adiabatic heat exchangers.

6.3.6.2 How Do We Calculate the Thermodynamic Efficiency of a Chemical Looping Combustion Power Cycle?

We will now carry out a similar exercise as for the amine-scrubbing and calcium looping capture plants in earlier questions and estimate the energy penalty in capturing the 62.5 mol, or 2.75 kg, of CO_2 produced on burning 1 kg of methane fuel. For this reaction

$$
\begin{aligned}
&CH_4 + 2O_2 \rightarrow 2H_2O(g) + CO_2 \\
&\Delta_c H^{\ominus} = -802.6\ \text{kJ/mol}\ (298K,\ 0.1\ \text{MPa}) \\
&\Delta_c G^{\ominus} = -801.1\ \text{kJ/mol}\ (298K,\ 0.1\ \text{MPa}) \\
&\Delta_c S^{\ominus} = -4.9\ \text{J/mol}\ (298K,\ 0.1\ \text{MPa}).
\end{aligned}
\tag{6.63}
$$

We see that $\Delta_c S^\ominus \sim 0$; hence, we assume that the waste heat Q_o (= $T^\ominus \Delta_c S^\ominus$ for this reversible process, see Questions 3.2.3.1 and 3.2.4) is zero and so there is no need for a second heat engine to utilize any rejected unavailable heat. Now the overall CLC cycle efficiency is

$$\eta_{CLC} = \frac{-W_{net}}{-(\Delta_c H^\ominus)_{fuel}}. \tag{6.64}$$

The heat input to the heat engine, Q_{hot}, is the enthalpy generated in the oxidizer $-(\Delta_r H^\ominus)_{ox}$; hence, the work output from the heat engine is given by

$$-W_{net} = Q_{hot}\, \eta_{th} = -(\Delta_r H^\ominus)_{ox}\, \eta_{th}, \tag{6.65}$$

where η_{th} is the reversible thermal efficiency of the heat engine. So, combining Equations 6.64 and 6.65, the efficiency of the overall CLC plant using methane, $\eta_{CLC,M}$, is

$$\eta_{CLC,M} = \frac{(\Delta_r H^\ominus)_{ox}}{(\Delta_r H^\ominus)_{fuel}} \eta_{th} = \beta\, \eta_{th}, \tag{6.66}$$

where

$$\beta = \frac{(\Delta_r H^\ominus)_{ox}}{(\Delta_r H^\ominus)_{fuel}}. \tag{6.67}$$

Now if we do run the reactors at their corresponding optimum T_c temperatures (as given by Equation 6.62) so that they run under essentially isothermal, reversible conditions then we can assume that the thermal efficiency of the heat engine is given by the corresponding Carnot efficiency (see Equation 3.14), so that

$$\eta_{th} = 1 - \frac{T_{c,red}}{T_{c,ox}} = 1 - \frac{(\Delta_r H^\ominus / \Delta_r S^\ominus)_{red}}{(\Delta_r H^\ominus / \Delta_r S^\ominus)_{ox}}. \tag{6.68}$$

We can relate $(\Delta_r H^\ominus)_{ox}$, $(\Delta_r H^\ominus)_{red}$ and $(\Delta_c H^\ominus)_{fuel}$ by the Hess's law cycle, shown in Figure 6.21. This shows that $(\Delta_r H^\ominus)_{fuel} = (\Delta_c H^\ominus)_{ox} + (\Delta_r H^\ominus)_{red}$; therefore, rearranging Equation 6.68, we obtain

$$\eta_{th} = 1 - \frac{(\Delta_r S^\ominus)_{ox}[(\Delta_c H^\ominus)_{fuel} - (\Delta_r H^\ominus)_{ox}]}{(\Delta_r H^\ominus)_{ox}[(\Delta_c S^\ominus)_{fuel} - (\Delta_c S^\ominus)_{ox}]}, \tag{6.69}$$

$$\eta_{th} = \frac{\beta - \chi}{\chi(1-\beta)} \text{ where } \chi = \frac{(\Delta_r S^\ominus)_{ox}}{(\Delta_r S^\ominus)_{fuel}}. \tag{6.70}$$

6.3.6.3 What Is a Practical Example of a Chemical Looping Combustion Cycle and How Efficient Can It Be?

We will use zinc/zinc oxide as the redox system to produce power by the combustion of methane to CO_2 and steam.

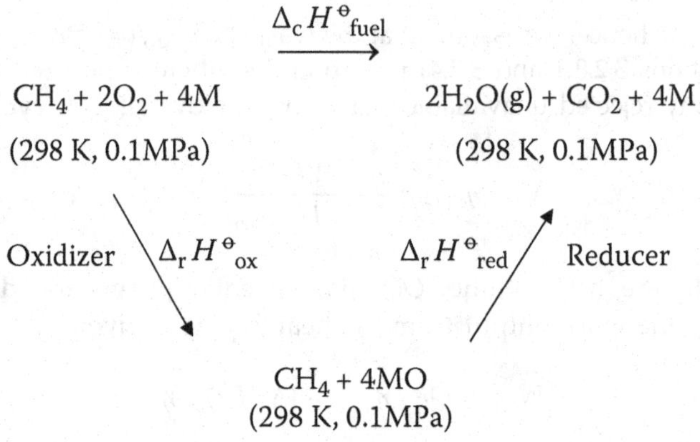

FIGURE 6.21
Cycle linking the overall combustion of CH_4 to CO_2 and H_2O to the reactions taking place in the oxidizer reactor (M to MO) and the reducer reactor (MO to M).

a Oxidizer

The oxidizer reaction is

$$\begin{aligned} &4Zn + 2O_2 \rightarrow 4ZnO(s) \\ &\Delta_r H^\ominus = -1923.5 \text{ kJ/mol } (298K,\ 0.1 \text{ MPa}) \\ &\Delta_r G^\ominus = -1611.7 \text{ kJ/mol } (298K,\ 0.1 \text{ MPa}) \\ &\Delta_r S^\ominus = -1.046 \text{ kJ/mol } (298K,\ 0.1 \text{ MPa}). \end{aligned} \tag{6.71}$$

We can use these data to calculate the optimum (reversible) temperature for this reaction, $T_{c,ox}$, using Equation 6.62

$$T_{c,ox} = T^\ominus \frac{\Delta_r H^\ominus}{\Delta_r H^\ominus - \Delta_r G^\ominus} = 298\text{K}\frac{-1923.5}{-1923.5 + 1611.7} = 1838 \text{ K}. \tag{6.72}$$

The operating temperature must be lower than 1,838 K for the reaction to proceed spontaneously, taking place reversibly at this optimum temperature.

b Reducer

The reducer reaction is

$$\begin{aligned} &CH_4(g) + 4ZnO(s) \rightarrow CO_2(g) + 2H_2O(g) + 4Zn(s) \\ &\Delta_r H^\ominus = 1120.9 \text{ kJ/mol } (298\text{K},\ 0.1 \text{ MPa}) \\ &\Delta_r G^\ominus = 810.6 \text{ kJ/mol } (298\text{K},\ 0.1 \text{ MPa}) \\ &\Delta_r S^\ominus = 1.041 \text{ kJ/mol } (298\text{K},\ 0.1 \text{ MPa}). \end{aligned} \tag{6.73}$$

These data give the optimum combustion temperature, $T_{c,red}$

$$T_{c,red} = T^\ominus \frac{\Delta_r H^\ominus}{\Delta_r H^\ominus - \Delta_r G^\ominus} = 298\text{K}\frac{1120.9}{1120.9 - 810.6} = 1076 \text{ K}. \tag{6.74}$$

Here, the temperature must be higher than 1,076 K for the reaction to proceed spontaneously, again with reversible operation near this temperature.

The net overall reaction for the CLC process, the summation of Equations 6.71 and 6.73 is, of course, the combustion of methane, Equation 6.63, with the source of oxygen being the metal oxide

$$CH_4 + 2O_2 \rightarrow 2H_2O(g) + CO_2 \; \Delta_c H^o = -802.6 \text{ kJ/mol (298K, 0.1 MPa)}. \quad (6.75)$$

c Thermodynamic efficiency

Using the reaction thermodynamic data above, and data from Equation 6.63 for methane combustion, to evaluate β and χ, we find

$$\beta = \frac{-1923.5}{-802.6} = 2.397 \text{ and } \chi = \frac{-1046}{-4.9} = 213.5. \quad (6.76)$$

So, the thermal efficiency of the reversible heat engine in the CLC process is (Equation 6.70)

$$\eta_{th} = \frac{\beta - \chi}{\beta(1-\chi)} = \frac{2.397 - 213.5}{2.397(1 - 213.5)} = 0.41. \quad (6.77)$$

Note that because $\chi \gg 1$ and $\chi \gg \beta$, to a good approximation $\eta_{th} = 1/\beta$, so β is the key parameter in determining thermal efficiency; the lower the ratio $(\Delta_r H^{\ominus})_{ox}/(\Delta_c H^{\ominus})_{fuel}$, the higher the heat engine cycle efficiency.

From Equation 6.66, $\eta_{CLC,M} = \beta\,\eta_{th}$, we see that because $\eta_{th} \approx 1/\beta$, a CLC plant operating under totally reversible conditions can be 100% efficient. The ideal overall power efficiency of this zinc oxide CLC is very close to this

$$\eta_{CLC,M} = \beta\,\eta_{th} = 2.4 \times 0.41 = \mathbf{0.98}. \quad (6.78)$$

This much enhanced reversible efficiency, achievable at much lower temperatures than for conventional gas-steam cycles, is the main attraction of this approach. The Zn/ZnO chemical looping system enables methane to be oxidized reversibly with a peak system temperature, $T_{c,ox}$, of 1,838 K, which is challenging but manageable for the metallurgical constraints on the equipment components. It is two orders of magnitude smaller than the T_c temperature of 164×10^3K required (insert the values of Equation 6.63 into Equation 6.62) to achieve reversible behavior for methane combustion in a conventional single-stage burner! The efficiency at the Carnot limit calculated here is similar to the gas turbine Brayton cycle for the CCGT gas turbine and significantly higher than the ~0.3 efficiency of a single stage basic steam cycle on that plant (see Question 6.2.1).

Apart from this respectable power generation efficiency, chemical looping has a number of significant advantages over several other carbon capture technologies:

- It avoids a high-energy separation process involving the separation of the difficult to condense CO_2 at low partial pressure from a mixture of other non-condensable gases, especially nitrogen.
- This ease of separation of CO_2 from the sole co-product, steam, is a similar advantage to that gained using oxy-combustion of methane, using pure oxygen rather than air (Cabral and Mac Dowell 2019). However, in CLC, the oxygen is chemically extracted

from air in the process, thereby avoiding the very large energy costs of cryogenic distillation separation of O_2 from air required as a pre-cursor for oxy-combustion.

- CLC reduces the equilibrium temperature of the two "combustion" reactors to within metallurgical limits and approaches reversible heat engine efficiencies without the need for unrealistically high temperatures.

However, there are also some significant practical challenges. Identifying optimal and practically viable redox systems for chemical looping combustion is one of the biggest challenges for this technology. The main constraints are the metallurgical limits of materials operating at the high temperatures set by $T_{c,ox}$ and the thermal stability and volatility of the oxides. In general, $T_{c,ox}$ increases with the boiling point of the elements, so that systems that best meet the metallurgical constraints are quite volatile with less stable oxides. Cadmium was an early contender for a viable system but its toxicity and high cost make it an unlikely practical system. Zinc by contrast is non-toxic and much cheaper, with a reasonable balance of $T_{c,ox}$ and $T_{c,red}$, so it is a more attractive option.

In considering practical options for chemical looping processes, they need to operate at acceptable rates, using process vessels that are not unacceptably large or costly. Therefore, the constraints that the system is entirely reversible and that $\Delta_c S^{\ominus}$ for the fuel is essentially zero need to be relaxed. The oxidizer and reducer may need to operate at temperatures below their optimum T_c values, which in fact makes the metallurgical constraints less demanding. Larger $\Delta_c S^{\ominus}$ values mean that it is necessary to introduce a second cycle to enable the waste heat from the CLC cycle (Q_o) to be discharged to the surroundings at $T^{\ominus}$ and ensure that all available work has been extracted from this stream. A liquid metal like sodium can be used as the working fluid in the primary cycle owing to the very high temperatures involved, whereas steam could be used in the secondary cycle. The efficiency of this double cycle system becomes, by extension of Equation 6.64,

$$\eta_{CLC} = \frac{|W_{net}^{A} + W_{net}^{B}|}{-(\Delta_c H^{\ominus})_{fuel}}, \tag{6.79}$$

where A is the primary, now irreversible, heat engine and B is the secondary engine installed to take the rejected heat. The two Rankine cycles operate as a combined cycle in much the same way as the CCGT system of Question 6.2.2 and in this case give an overall efficiency of

$$\eta_{CLC} = \beta\left(\eta_{th}^{A}\left(1 - \eta_{th}^{B}\right) + \eta_{th}^{C}\eta_{th}^{B}\right), \tag{6.80}$$

where η_{th}^{C} is the efficiency of the Carnot reversible cycle given by Equation 6.68. For a Zn/ZnO CLC-sodium/steam(l,v) dual turbine system, the two cycles A and B have efficiencies lower than the reversible value given in Equation 6.77, both between 0.30 and 0.35. It follows from Equation 6.80 that η_{CLC} in this irreversible case is about 0.85.

For CLC, there is no intrinsic energy penalty for capturing CO_2 since CO_2 separation is an integral part of the combustion process. Its efficiency as a carbon capture process should therefore be judged by a comparison of its power generation efficiency with that of a conventional CCGT plant without carbon capture. If a CLC plant could be run with an overall efficiency of 85%, then there is a net electricity generation benefit of (0.85 – 0.60) × 50 MJ/kg = 7.5 MJ/kg of methane burned compared with unabated CCGT plants (see Question 6.2.2) and (0.85 – 0.54) × 50 MJ/kg = 15.5 MJ/kg compared with amine-solvent capture (see Question 6.3.4). Indications are, however (Schnellmann et al. 2019), that when compressor and pump work, as well as heat exchanger irreversibilities and current metallurgical constraints on the oxidizer reactor temperature, are taken into account, the efficiency of CLC processes in practice may only be about 55%. This corresponds to a 2.5 MJ/kg methane burned energy penalty for CLC compared to the best CCGT plants without carbon capture, comparable to that for amine-solvent CO_2 capture. So the method has many novel features and interesting potential to combine a near-zero energy penalty carbon capture system with highly efficient power production. However, there remain many challenges to achieving its full thermodynamic potential in practice.

6.4 Why Does a Diesel Car Have a Better Fuel Efficiency Than a Gasoline Car?

The balance between the use of Diesel engines or gasoline engines to power freight vehicles or passenger cars has varied considerably over the lifetime of fossil-fueled vehicles. The incentives have been fashion, climate change arguments and performance. However, in the context of this book, we will concern ourselves solely with an examination of the relative fuel efficiency of the two hydrocarbon sources of energy to power the car, diesel and gasoline (also commonly known as petrol). These and other aspects are treated in detail in the specialized literature, e.g. Kirkpatrick (2020), Lakshminarayanan and Agarwal (2020) and Szybist et al. (2020). Often, car manufacturer specifications cite average fuel consumption for a car required to travel a distance of 100 km, which has been determined under well-defined test conditions. As an example, consider a Diesel engine with a power of about 100 kW that consumes 4.3 dm^3 (or 4.3 liters) of fuel to travel a distance of 100 km (i.e. equivalent to about 55 miles per U.S. gallon [mpg(US)] or 66 miles per U.K. gallon [mpg(UK)]). For a car with a gasoline engine also with a power of about 100 kW, the fuel consumption is about 5.7 dm^3 of fuel to travel a distance of 100 km (i.e. equivalent to about 41 mpg (US) or 50 mpg (UK)), and about 30% higher than for a car powered with diesel. The question to pose is then as follows: What is the thermodynamic reason for this considerable difference?

The specific energy content of gasoline and diesel fuels is similar at about 43 MJ kg^{-1} while the mass densities are 0.74 kg dm^{-3} for gasoline and 0.82 kg dm^{-3} for diesel that result in volumetric energy content of 32 MJ dm^{-3} for gasoline and 35 MJ dm^{-3} for diesel. Thus, the energy content by volume is about 10% of the observed difference in fuel economy between a car powered by diesel compared with a petrol version. The additional 20% difference arises from the thermal efficiencies of the two engine types that we will now consider.

In a car, both types of engines operate in a four-stroke manner originally involving the following process steps (shown in Figure 6.22):

step 0–1: Intake of the mixture of air and fuel (1st stroke)

step 1–2: Compression of the gas mixture (2nd stroke)

step 2–3: Ignition through either a spark plug in a gasoline engine or autoignition in a Diesel engine and then combustion

step 3–4: Expansion of the gas mixture (3rd stroke)

step 4–0: Expulsion of the burnt gas (4th stroke)

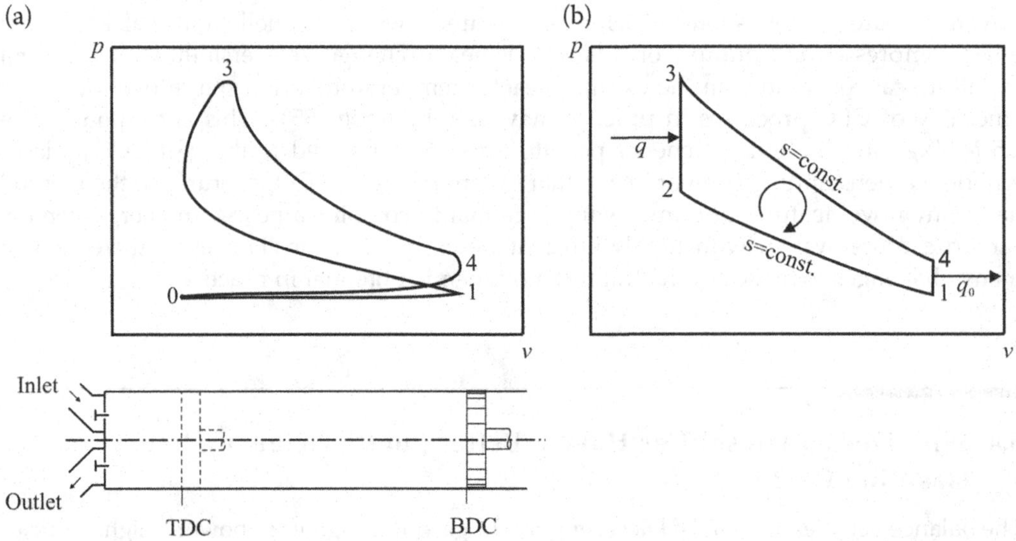

FIGURE 6.22
Schematic of a gasoline engine (Otto-cycle) in a (p, v) diagram. The exhaust stroke $(4 \to 0)$ and the intake stroke $(0 \to 1)$ of the real cycle (a) are replaced by an isochoric heat discharge in the idealized cycle (b). The acronyms TDC and BDC stand for top dead center and bottom dead center, respectively.

A general modification of this four-step process is that in modern cars no air-fuel mixture is sucked into the combustion chamber because fuel is injected directly. Direct injection provided a tremendous improvement in the performance of Diesel engines, and the fuel is injected at pressures of about 200 MPa, or even above, to ensure proper mixture formation. Direct injection is standard today also for gasoline engines albeit at lower pressure levels.

To model these processes thermodynamically, there are a number of assumptions that will be introduced to permit the simplified treatment given here. However, these assumptions do not change the overall outcome of the arguments for either engine.

First, to avoid the changes in the chemical composition of the working fluid we assume that the fluid is air (which as a first step is justified because of the relatively large mass fraction of nitrogen). Second, despite pressures of up to an order of 10 MPa, we assume that the air behaves as an ideal gas. Third, the heat released by combustion and the energy removal by the discharge of burnt gases are replaced by heat transfer across the system boundaries, so that steps 0–1 and 4–0 are omitted, and the system is now considered as a piston-cylinder closed system. With this transition to a closed system the primary quantity describing the work in the system is now the volume work and not the shaft work as for an open system (*cf.* Question 1.3.6). This formal problem, however, is resolved when we take into account that the total flow work for the process is zero, $\oint d(pv) = 0$, rendering the volume work and the shaft work identical. Finally, to find a mathematical description for the process the actual steps, with rounded shapes between them, are replaced by idealized, well-defined steps (see Figure 6.22(b)).

With these assumptions we may now provide the thermodynamic analyses of both diesel and gasoline cycles and answer the question posed regarding their thermal efficiencies. In this context, we define the characteristic property that is called the *compression ratio*, given by

$$\varepsilon = \frac{v_2}{v_1} = \frac{V_{TDC}}{V_{BDC}}. \tag{6.81}$$

Equation 6.81 is the ratio between the volumes when the piston is at top dead center (TDC), V_{TDC}, where the volume enclosed in the cylinder is a minimum, and often termed the clearance volume, and when it is at bottom dead center (BDC), V_{BDC}, where the volume is a maximum. The difference between V_{TDC} and V_{BDC} is the displacement volume.

We begin with a closer look at the idealized cycle for a gasoline spark-ignition engine, which is often called an *Otto cycle* after Nikolaus Otto, who in 1876 built the first engine of this type. Parenthetically, it is worth remarking that those early versions had little in common with modern Otto engines, apart from the basic working principle. The idealized Otto cycle consists of the following processes:

step 1–2: Reversible adiabatic (i.e. isentropic) compression
step 2–3: Isochoric addition of heat
step 3–4: Reversible adiabatic (i.e. isentropic) expansion
step 4–1: Isochoric rejection of heat.

The thermal efficiency of this process is given by

$$\eta_{\text{th,O}} = 1 - \frac{|q_0|}{q} = 1 - \frac{|q_{41}|}{q_{23}}. \tag{6.82}$$

Because

$$q_{23} = c_v(T_3 - T_2) \tag{6.83}$$

and

$$q_{41} = c_v(T_1 - T_4), \tag{6.84}$$

Equation 6.82 becomes, when we assume a constant heat capacity, c_v,

$$\eta_{\text{th, O}} = 1 - \left[\frac{T_4 - T_1}{T_3 - T_2}\right] = 1 - \frac{T_1\left(\frac{T_4}{T_1} - 1\right)}{T_2\left(\frac{T_3}{T_2} - 1\right)}. \tag{6.85}$$

From Chapter 1 Question 1.6.5, the expression for a reversible adiabatic process is given by Equation 1.100 for an ideal gas and when applied to the Otto engine gives

$$\frac{T_3}{T_4} = \left(\frac{v_4}{v_3}\right)^{\gamma-1} = \left(\frac{v_1}{v_2}\right)^{\gamma-1} = \frac{T_2}{T_1}. \tag{6.86}$$

In Equation 6.86, γ is the ratio of specific heat capacities at constant pressure to that at constant volume and, as in Chapter 1, it is given by $\gamma = c_p/c_v$. Because $v_1 = v_4$ and $v_2 = v_3$, Equation 6.86 becomes

$$\frac{T_4}{T_1} = \frac{T_3}{T_2}. \tag{6.87}$$

Thus, the thermal efficiency of an Otto engine of Equation 6.85 is then

$$\eta_{\text{th, O}} = 1 - \frac{T_1}{T_2}, \tag{6.88}$$

or, when the compression ratio defined by Equation 6.81 is used, Equation 6.88 becomes

$$\eta_{\text{th, O}} = 1 - \frac{1}{\varepsilon^{\gamma-1}}. \tag{6.89}$$

Examination of Figure 6.23 reveals that the efficiency of the ideal Otto cycle increases steeply at first with increasing compression ratio ε and then flattens off. Nevertheless, from a thermodynamic viewpoint alone, it would be desirable to increase the compression ratio.

However, increasing the compression ratio leads to a marked increase in the temperature at the end of the compression stroke. This results in auto ignition of the fuel and uncontrolled combustion (engine knock) that can damage the engine. Use of higher octane gasoline and fuel injection permits Otto engines to reach compression ratios often between 10 and 12. The desire to increase the compression ratio while also avoiding uncontrolled combustion and the resultant engine knock will be used in our discussion of the Diesel engine, in which fuel vapor auto-ignites.

One of our assumptions was that the fluid contained in the process was air, for which as an ideal gas $\gamma = 1.4$. We can now determine how variations of chemical composition alter the thermal efficiency η_{th}. The gases resulting from combustion are mainly water vapor and carbon dioxide, with nitrogen as an "almost" inert gas. At room temperature (T = 298 K) and low pressure (p = 0.1 MPa) water vapor and carbon dioxide have a heat capacity ratio of $\gamma \approx 1.3$. As Figure 6.23 shows, η_{th} varies with the heat capacity ratio and thus with chemical composition, but because the mole fraction of nitrogen is the largest in the whole process the practical effect is small.

The Diesel cycle, named after Rudolf Diesel, who presented the first prototype of his engine in 1897, permits the use of higher compression ratios (and thus pressures). The (p, v) diagram for the Diesel cycle, shown in Figure 6.24, is similar to the (p, v) diagram of the Otto cycle shown in Figure 6.23. The major difference between Figure 6.24 and Figure 6.23 is that for Figure 6.24 the process of heat addition is now modeled as one at constant pressure. A more refined model of either the Otto or the Diesel process splits the combustion phase into two processes, namely a constant-volume and a constant-pressure process, where the precise segmentation depends on the cycle and is called a *dual* cycle or *Seiliger* cycle. The basic model of the Diesel cycle, however, follows the process depicted in Figure 6.24, where there are two isentropic processes, one isochoric process and one isobaric process that makes the treatment more complicated than that for the Otto cycle.

Because the heat addition, q_{23}, in the Diesel cycle is at constant pressure, the thermal efficiency of this cycle, $\eta_{\text{th,D}}$, is given by

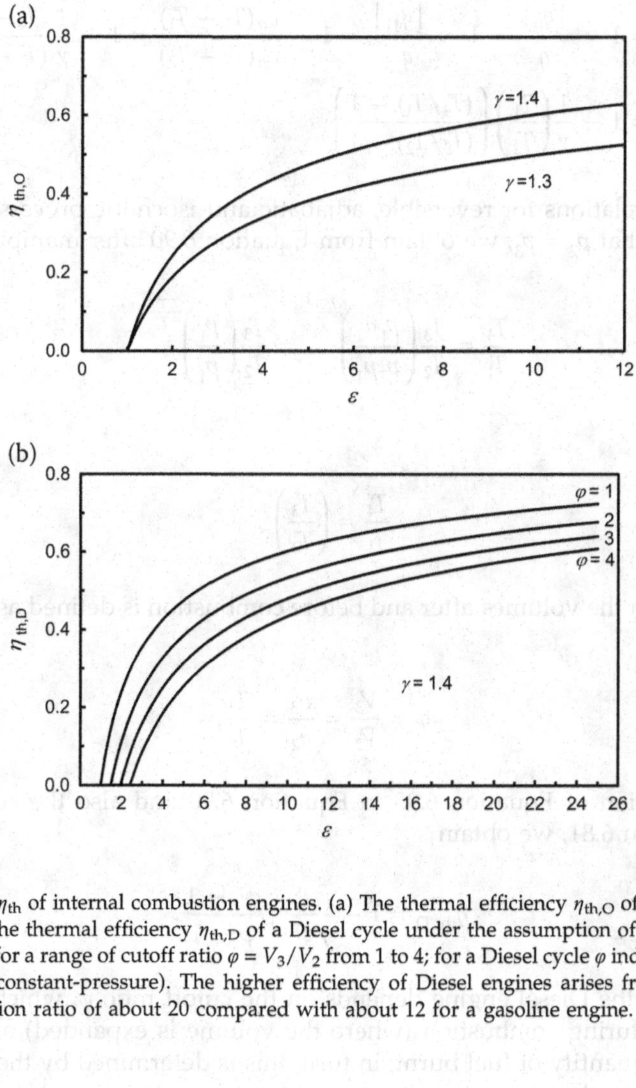

FIGURE 6.23
Thermal efficiencies η_{th} of internal combustion engines. (a) The thermal efficiency $\eta_{th,O}$ of an Otto cycle for γ = 1.3 and γ = 1.4. (b) The thermal efficiency $\eta_{th,D}$ of a Diesel cycle under the assumption of a constant isentropic exponent γ = 1.4 but for a range of cutoff ratio $\varphi = V_3/V_2$ from 1 to 4; for a Diesel cycle φ indicates the duration of the heat release (at constant-pressure). The higher efficiency of Diesel engines arises from the possibility of realizing a compression ratio of about 20 compared with about 12 for a gasoline engine.

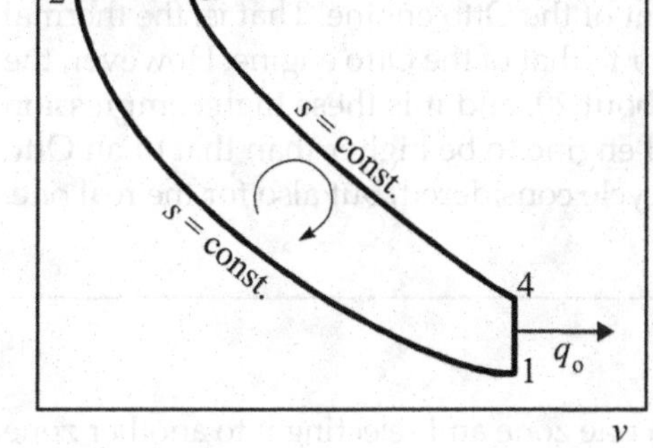

FIGURE 6.24
Schematic of a Diesel engine as a (p, v) diagram. In contrast to the Otto cycle the heat released by combustion, modeled as heat input (2 to 3), is now realized by an isobaric process instead of an isochoric one.

$$\eta_{\text{th, D}} = 1 - \frac{|q_0|}{q} = 1 - \frac{|q_{41}|}{q_{23}} = 1 - \frac{c_v(T_4 - T_1)}{c_p(T_3 - T_2)} = 1 - \frac{1}{\gamma}\frac{(T_4 - T_1)}{(T_3 - T_2)}$$
$$= 1 - \frac{1}{\gamma}\left(\frac{T_1}{T_2}\right)\left\{\frac{(T_4/T_1) - 1}{(T_3/T_2) - 1}\right\}. \tag{6.90}$$

Using the (p, T) relations for reversible, adiabatic and isochoric processes with an ideal gas and the fact that $p_2 = p_3$, we obtain from Equation 6.90 after manipulation

$$\frac{T_4}{T_1} = \frac{T_3}{T_2}\left(\frac{p_4 p_2}{p_3 p_1}\right)^{\frac{\gamma-1}{\gamma}} = \frac{T_3}{T_2}\left(\frac{p_4}{p_1}\right)^{\frac{\gamma-1}{\gamma}}, \tag{6.91}$$

or

$$\frac{T_4}{T_1} = \left(\frac{T_3}{T_2}\right)^{\gamma}. \tag{6.92}$$

The ratio between the volumes after and before combustion is defined as the *cutoff ratio* φ and is given as

$$\varphi = \frac{V_3}{V_2} = \frac{v_3}{v_2} = \frac{T_3}{T_2}. \tag{6.93}$$

Using the definition of Equation 6.93 in Equation 6.90 and also the compression ratio given by Equation 6.81, we obtain

$$\eta_{\text{th, D}} = 1 - \frac{1}{\gamma\,\varepsilon^{\gamma-1}}\frac{\varphi^{\gamma} - 1}{\varphi - 1}. \tag{6.94}$$

The efficiency of the Diesel engine depends on the cutoff ratio φ, which is given by the volume change during combustion (where the volume is expanded) and therefore depends upon the quantity of fuel burnt; in turn, this is determined by the quantity of fuel injected (accelerator depression).

Figure 6.23 shows the thermal efficiency of the Diesel engine as a function of the compression ratio with the cutoff ratio as a parameter and reveals that $\eta_{\text{th,D}}$ decreases with increasing φ. Using L'Hopital's rule to examine the limit as $\varphi \to 1$, we find the efficiency of the Diesel engine approaches as a limiting case that of the Otto engine. That is, the thermal efficiency of the Diesel engine (with $\varphi > 1$) is inferior to that of the Otto engine. However, the Diesel engine permits compression ratios up to about 20, and it is these high compression ratios that enable the overall efficiency of a Diesel engine to be higher than that of an Otto engine. This is the case not only for the idealized cycle considered, but also for the real one.

6.5 What Is a Refrigeration Cycle?

Refrigeration is the process of removing heat from one zone and rejecting it to another zone of higher temperature. The primary purpose of refrigeration is to lower the temperature of one zone and then to maintain it at that temperature. In this case, heat is transferred from a low to a higher temperature, requiring a machine and a thermodynamic cycle, which are called refrigerators and refrigeration cycles, respectively.

In Chapter 1, within Question 1.8.6, we discussed the temperature drop of a working fluid after its flow through a constriction in an isenthalpic process. We now consider how that phenomenon can be exploited in a closed thermodynamic cycle to produce continuous cooling. Refrigerators and heat pumps are essentially the same devices that differ only in their specific objective. We discuss two types of refrigeration cycles in the remainder of this question: the vapor-compression cycle and the (ammonia) absorption cycle.

6.5.1 What Is a Vapor-Compression Cycle?

The refrigeration cycle is a closed loop of four processes. A schematic diagram of this refrigeration cycle is given in Figure 6.25. Figure 6.25a shows a cycle where the refrigerant expands through a turbine which is, owing to cost, unusual and occurs solely in large installations and Figure 6.25b shows a cycle where the refrigerant expands through a

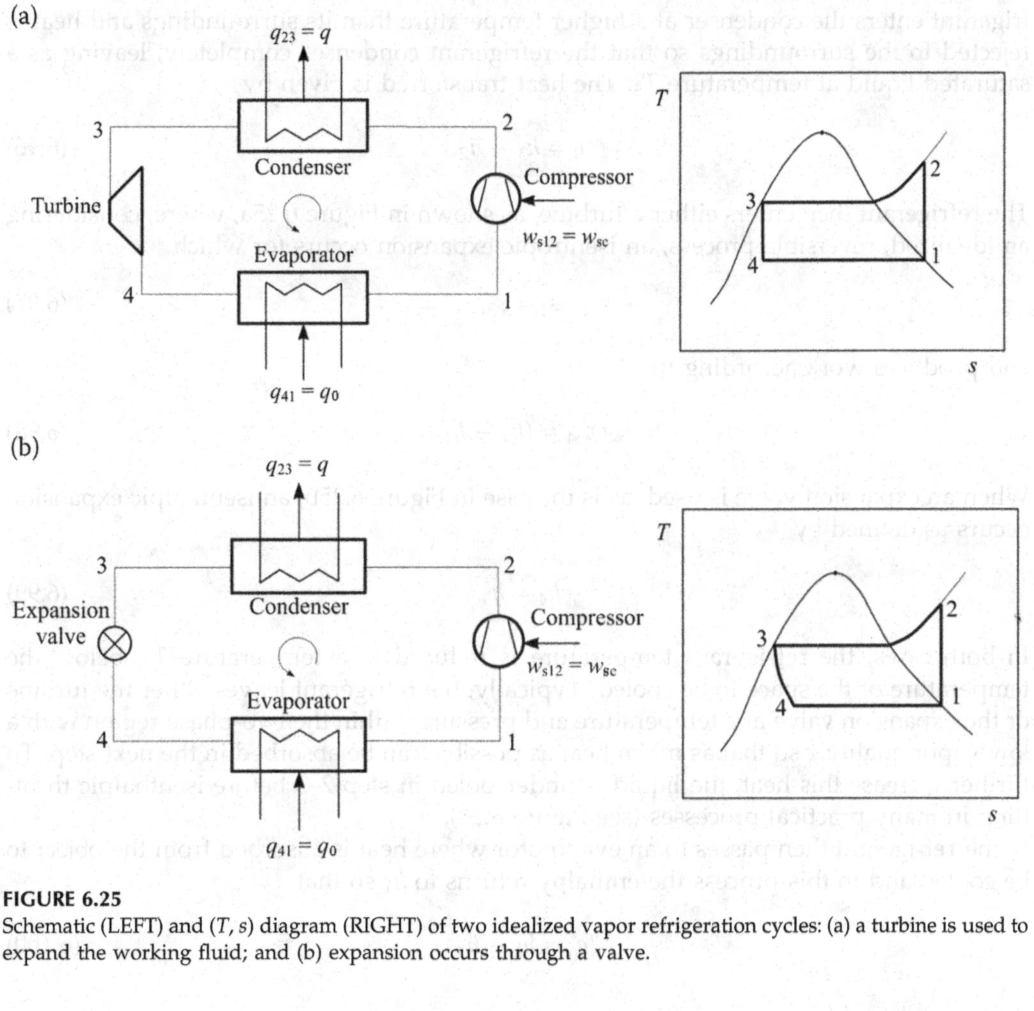

FIGURE 6.25
Schematic (LEFT) and (T, s) diagram (RIGHT) of two idealized vapor refrigeration cycles: (a) a turbine is used to expand the working fluid; and (b) expansion occurs through a valve.

valve; it is the most widely used approach for refrigerators, air conditioning and heat pumps.

The four stages of a simplified process illustrated in Figure 6.25 are as follows:

- step 1–2: The refrigerant is adiabatically compressed, raising the pressure so that the corresponding saturation temperature is above ambient temperature
- step 2–3: The refrigerant rejects heat to the environment through a heat exchanger
- step 3–4: The refrigerant is expanded through either a turbine at constant entropy (as shown in Figure 6.25a) or through a throttling valve at constant enthalpy (as shown in Figure 6.25b) and condenses
- step 4–1: The fluid evaporates as it absorbs heat from the space to be cooled

In the idealized refrigerator, the compression in the first stage is isentropic and the work required is given by

$$w_{s12} = (h_2 - h_1), \tag{6.95}$$

and the fluid temperature is raised from temperature T_1 to temperature T_2. The refrigerant enters the condenser at a higher temperature than its surroundings and heat is rejected to the surroundings so that the refrigerant condenses completely, leaving as a saturated liquid at temperature T_3. The heat transferred is given by

$$q = h_3 - h_2. \tag{6.96}$$

The refrigerant then enters either a turbine, as shown in Figure 6.25a, where, considering an idealized, reversible process, an isentropic expansion occurs for which

$$s_4 = s_3, \tag{6.97}$$

and produces work according to

$$w_{s34} = (h_4 - h_3). \tag{6.98}$$

When an expansion valve is used, as is the case in Figure 6.25b, an isenthalpic expansion occurs as defined by

$$h_4 = h_3. \tag{6.99}$$

In both cases, the refrigerant temperature is reduced to a temperature T_4, below the temperature of the space to be cooled. Typically, the refrigerant leaves either the turbine or the expansion valve at a temperature and pressure within the two-phase region with a low vapor quality x so that as much heat as possible can be absorbed in the next step. To further increase this heat, the liquid is undercooled in step 2–3 before isenthalpic throttling in many practical processes (see Figure 6.26).

The refrigerant then passes to an evaporator where heat is absorbed from the object to be cooled and in this process the enthalpy returns to h_1 so that

$$q_0 = h_1 - h_4. \tag{6.100}$$

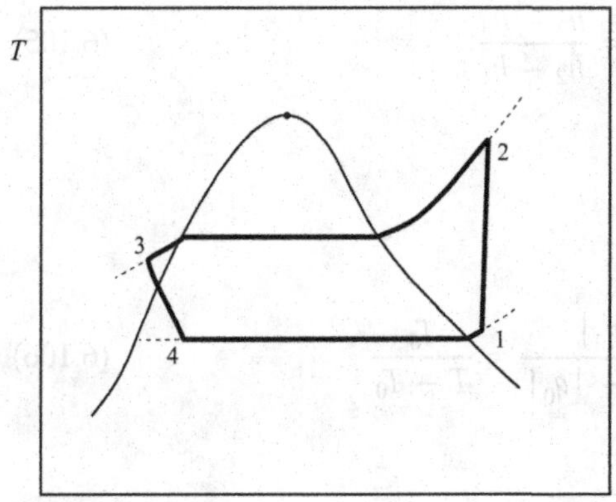

FIGURE 6.26
(*T*, *s*) diagram of a real vapor refrigeration cycle.

In the ideal case, the refrigerant leaves the evaporator as a saturated vapor; however, in the actual cycle, the vapor is superheated to prevent liquid droplets entering the compressor and causing damage to it and it leaves the condenser subcooled so as to provide greater cooling capacity. The (*T*, *s*) diagram for a real refrigeration cycle is shown in Figure 6.26 and should be compared with Figure 6.25b.

We now consider the coefficient of performance (COP) for these cycles. If the purpose of using these cycles is to cool a space, then the COP, in this case often also termed energy efficiency ratio, EER, is defined as the ratio of cooling effect to the required net work and receives an index c. The net work required in any cycle is found by the application of the First Law

$$|w_{s,\,net}| = |q| - |q_0|. \tag{6.101}$$

Thus, for the cycle shown in Figure 6.17a, the COP is given by

$$COP_c = \frac{|q_0|}{|w_{s,net}|} = \frac{h_1 - h_4}{(h_2 - h_3) - (h_1 - h_4)}, \tag{6.102}$$

while for the idealized cycle shown in Figure 6.25b and the real cycle shown in Figure 6.26 (no work production), the COP is given by

$$COP_c = \frac{|q_0|}{|w_{s,net}|} = \frac{h_1 - h_4}{h_2 - h_1}. \tag{6.103}$$

If these cycles are used to heat a space (with a heat pump), then the COP is defined as the ratio of heating effect to the required net work and given the index h. Thus, for the cycle of Figure 6.25a, the COP is given by

$$COP_h = \frac{|q|}{|w_{s,net}|} = \frac{h_2 - h_3}{(h_2 - h_3) - (h_1 - h_4)}, \tag{6.104}$$

while for the idealized cycle shown in Figure 6.25b and the real cycle shown in Figure 6.26, the COP is given by

$$\mathrm{COP_h} = \frac{|q|}{|w_{\mathrm{s,net}}|} = \frac{h_2 - h_3}{h_2 - h_1}. \tag{6.105}$$

For a Carnot cycle, the COP is

$$\mathrm{COP_c} = \frac{|q_0|}{|w_{\mathrm{s,net}}|} = \frac{|q_0|}{|q| - |q_0|} = \frac{T_0}{T - T_0} \tag{6.106}$$

for cooling, while the COP for heating is

$$\mathrm{COP_h} = \frac{|q|}{|w_{\mathrm{s,net}}|} = \frac{|q|}{|q| - |q_0|} = \frac{T}{T - T_0}. \tag{6.107}$$

The coefficients of performance given by Equations 6.102 through 6.105 are less than those of the ideal reversible Carnot cycle given by Equations 6.106 and 6.107. In these equations, T and T_0 indicate the higher and lower temperature levels of the Carnot cycle, respectively.

The cooling capacity of the refrigerator cycle is given by

$$\dot{Q}_\mathrm{c} = \dot{m}_{\mathrm{ref}}\, q_0, \tag{6.108}$$

where $\dot{m}_{\mathrm{ref}}$ is the rate at which the mass of the refrigerant is circulated around the refrigerator cycle and q_0 is the heat absorbed in the evaporator per mass of refrigerant.

The power required to move and compress the refrigerant is given by

$$P = \dot{m}_{\mathrm{ref}} w_{\mathrm{s12}}, \tag{6.109}$$

where w_{s12} is the work required for compression per unit mass of circulating refrigerant.

6.5.2 How Do I Choose the Right Refrigerant for an Application?

The evaporation and condenser temperatures are fixed for given refrigeration tasks by the temperatures of the space to be cooled and the temperature of the surroundings. The choice between the several refrigerants available depends on many factors and generally the most important are as follows: (1) the vapor pressure in the evaporator and condenser; (2) the specific enthalpy of vaporization, which should be as high as possible to obtain the greatest cooling effect per mass of fluid circulated; (3) the specific volume of the gaseous refrigerant, which should be as low as possible to minimize the specific work required in the compressor; (4) chemical stability; (5) toxicity; (6) cost and (7) environmental factors. For item 1, the vapor pressure in the evaporator should not be lower than atmospheric pressure to avoid air leaking in, while the vapor pressure at the condenser should not be too high (typically below 2 MPa) to avoid high equipment cost associated with handling high pressures. The working fluids used in refrigeration cycles have been the subject of intense public and scientific interest almost since the introduction of the first processes. Initially, the concerns were solely around the threat that the refrigerants employed posed to the users, either because they were hazardous to human health if released (for example ammonia) or because they were combustible (for example hydrocarbons). A series of materials known as fluorocarbons were developed specifically for the purpose of refrigeration with fewer safety concerns and properties tuned to the applications. However, in the last decades of the twentieth century, it was revealed that those compounds when released into the atmosphere (as is inevitable when in large scale use) damaged the ozone layer surrounding the Earth, which provides protection from some harmful components of the sun's radiation. As a result, there was international agreement (the Montreal Protocol 1987) to phase out these materials and replace them with hydrofluorocarbons, not harmful to the ozone layer. Yet these materials too have been shown to have negative effects on the Earth's climate (greenhouse effect). The Kyoto Protocol in 1997, later amended by a number of other agreements, aimed at a reduction of the emission of greenhouse gases, and so increasingly stricter regulations on the use of conventional refrigerants have come into effect in various regions of the world. There is now a continual search for new materials with no harmful effects on the environment or health. Besides the development of completely new refrigerants, "classical" working fluids such as hydrocarbons or carbon dioxide have attained increased attention again in spite of their disadvantages, namely flammability or the operation at high pressures, respectively. As a consequence of this continuous quest for new refrigerants, we have refrained from making concrete reference to examples of refrigerants from just one period of time but focus on the general principles of refrigeration cycle design. There are numerous reviews on refrigerants, including their thermodynamic and environmental properties, for example those by McLinden et al. (1998), Abas et al. (2018) and Bobbo et al. (2018).

As a general rule, the compression ratio should be as low as possible to minimize the energy required for compression, and the choice of the refrigerant has a major impact on this number. There are a number of cases, however, where a large compression is required to obtain higher temperature in the condenser or lower temperature in the evaporator. The performance of compressors, of course also depending on the compressor type, usually deteriorates beyond a pressure ratio of about 5–7. Larger pressure ratios then require the use of two or more compressor stages accompanied by intercooling between the stages to reduce the refrigerant volume and, thus, reduce the required compression power as illustrated in the (T, s) diagram of Figure 6.27.

A high pressure ratio between the two relevant temperature levels can also be handled by means of a cascade cycle system that consists of two cycles completely independent of each other except that the evaporator of the higher-temperature cycle acts as the condenser for the low temperature cycle. When multiple compressor stages are used, intermediate pressure levels are normally chosen in a way that pressure ratios between two consecutive levels are similar.

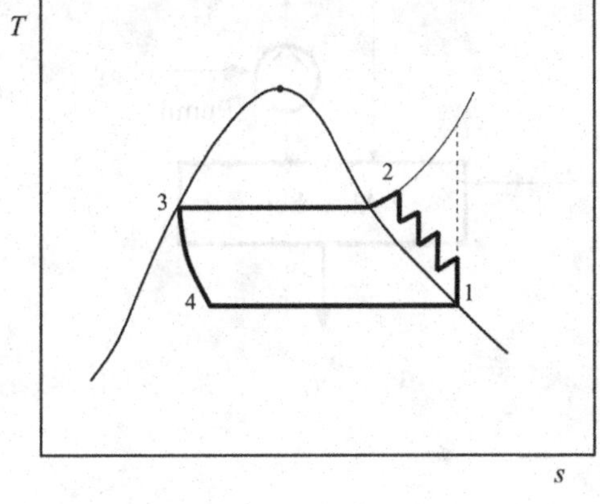

FIGURE 6.27
T-s diagram of a vapor refrigeration cycle with a multistage compressor.

6.5.3 What Is an Absorption Refrigerator Cycle?

The absorption refrigeration cycle, shown schematically in Figure 6.28, is similar to the vapor compression cycle shown in Figure 6.25 with the major difference being the method used for refrigerant compression. In the vapor compression refrigerator, the working fluid is compressed to a high pressure by a compressor while in the absorption refrigerator cycle the refrigerant is first absorbed into water and then the liquid solution is compressed to a high pressure by a pump from which the absorbed gas is subsequently released by heating.

The most common refrigerant for absorption refrigeration cycles is ammonia. It has a high solubility in water (at a temperature of 298 K and a pressure of 0.1 MPa about 320 g of $NH_3(g)$ are soluble in 1 dm^3 of $H_2O(l)$, that is a molality of 18.8 $mol \cdot kg^{-1}$) and the ammonia reacts with water in a reversible reaction of

$$NH_3(g) + H_2O(l) \rightleftharpoons NH_4^+(aq) + OH^-(aq) \tag{6.110}$$

to give a basic solution. The steps in the process shown in Figure 6.28 are as follows:

step 1–2: Ammonia vapor at a temperature T_1 is dissolved or absorbed in liquid water

step 2–3: The refrigerant rejects heat to the environment through a heat exchanger

step 3–4: The refrigerant is expanded in the throttling process at constant enthalpy and condenses

step 4–1: The fluid evaporates as it absorbs heat from the environment to be cooled

In step 1–2, heat is rejected to the environment to maintain the temperature as low as possible so as to increase the amount of substance of NH_3 that can dissolve in water;

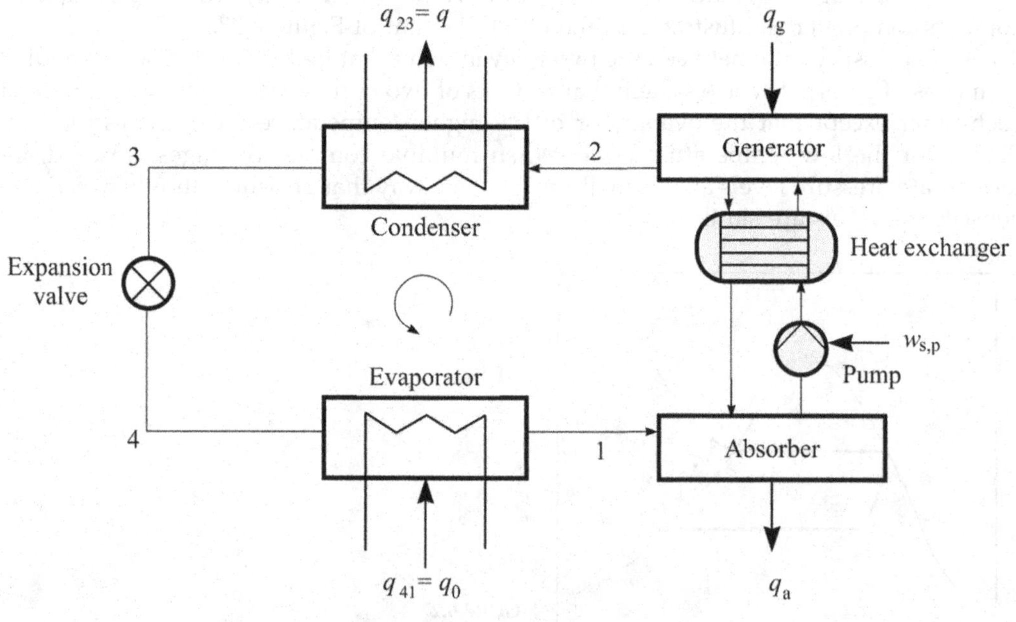

FIGURE 6.28
Schematic diagram of the absorption refrigeration cycle.

the solubility increases with decreasing temperature. For example, at a temperature of 273 K 900 g of $NH_3(g)$ dissolve in 1 dm^3 of $H_2O(l)$, that is, a molality of about 53 mol kg^{-1} assuming a mass density of 1 kg dm^{-3}. The liquid solution is then pumped to the high pressure of the generator and heat transferred to produce $NH_3(g)$ from the water solution. In the absorption refrigerator, the work required to pump the liquid solution (which is essentially an incompressible fluid) is much less than for compression if the ammonia were gaseous, as it would be in a vapor-compression cycle. The work for this liquid pumping is approximately given by

$$w_{s,p} = v(p_2 - p_1), \tag{6.111}$$

where v is the specific volume of the liquid solution (assumed constant) and p_1 and p_2 are the pressures of the evaporator and condenser, respectively. The energy required to operate the pump is much less than the energy required to evolve $NH_3(g)$ from the $NH_3(aq)$.

The steps of the absorption refrigeration cycle are the same as those of a vapor-compression cycle except for step 1–2. The heat transferred in the condenser is given by Equation 6.96 and that of the evaporator is given by Equation 6.100. There is isenthalpic expansion through a throttling valve. Absorption refrigeration becomes economically attractive when there is a source of inexpensive heat energy, q_g, to evolve $NH_3(g)$ from the $NH_3(aq)$.

The COP for absorption refrigeration is defined as the ratio of the cooling effect to the required energy input (heat input to the generator plus pump work); this differs from the definition of COP used for a vapor-compression cycle given by Equation 6.103. The work input to the pump is relatively small so that the COP is given by

$$\mathrm{COP_c} = \frac{|q_0|}{|q_g| + |w_{s,p}|} \approx \frac{|q_0|}{|q_g|} = \frac{h_1 - h_4}{|q_g|}. \tag{6.112}$$

6.5.4 Can I Use Solar Power for Cooling?

The maximum radiant flux on the surface of the Earth of about 1 kW m^{-2} that arises from the sun can only be achieved on a clear summer day at noon. The resultant mean radiant flux over a time of 1 d is about 250 W m^{-2}. This energy can be used in a number of ways, and here we consider how it may be used in refrigeration by partial substitution for the heat provided to a refrigeration cycle (for example, the ammonia absorption described in Question 6.5.3) that would otherwise be obtained from another source. Solar energy is particularly appropriate for cooling buildings because the demand for cooling during a day is essentially in phase with the energy available from the sun.

For solar-powered absorption-refrigeration, water is used as the working fluid and a solution of an alkali metal halide, for example LiBr, as the absorbent that relies on the solubility of LiBr in H_2O of 1.67 kg in 1 dm^3 of water at a temperature of 298 K and a pressure of 0.1 MPa; this is a molality of about 18.8 mol kg^{-1} and is similar to that of a saturated solution of ammonia in water. From safety and environmental perspectives, water is an extremely advantageous refrigerant. The heat required to separate the water from the aqueous solution of lithium bromide requires a high temperature in the generator that is provided by the sun with special solar energy collectors. One form of solar collector uses evacuated tubes made of glass, where the round profile favors the near-perpendicular incidence of the sun rays on the tube during the whole day. In addition, the vacuum within the tubes reduces convection and conduction heat loses and thus achieves

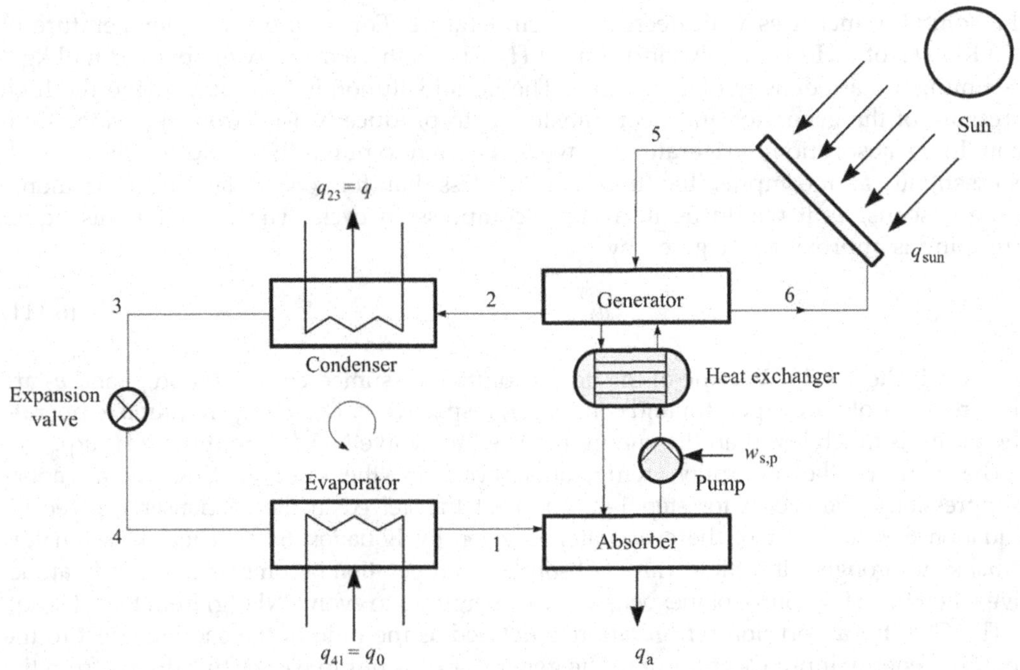

FIGURE 6.29
Schematic diagram of a solar cooling cycle.

high thermal efficiency and temperature. Another form of solar collector relies on mirrors or lenses to focus the energy and to obtain the temperatures required. Figure 6.29 shows the schematic diagram of the basic elements of this absorption cycle.

An analytical description of the absorption cycle was given in Question 6.5.3. Heat is provided to the generator by another working fluid circulating between the solar collector and the generator. Thus, the energy from the sun is used to reduce the energy required from other sources and thus the cost of operating the cooling system.

The COP for this cooling system is defined as the ratio of the cooling effect to the required energy input (heat plus pump work). Since the work input to the pump is usually small the COP is given by

$$\mathrm{COP_c} = \frac{|q_0|}{|q_{sun}| + |w_{s,p}|} \approx \frac{|q_0|}{|q_{sun}|} \approx \frac{h_1 - h_4}{|q_{sun}|}. \tag{6.113}$$

The COP for the solarpowered absorption refrigeration cycle is usually between 0.6 and 0.75. Manufacturers give an average value of specific collector surface area of between 3 and 4.5 m^2 for each kW of cooling capacity.

6.6 What Is a Liquefaction Process?

Liquefied gases are used in many practical situations: For example, liquid oxygen is shipped and stored in many hospitals in chilled tanks until required, and then allowed to boil to release oxygen gas for patients; liquid chlorine is shipped and stored for sterilization of water; and liquefied natural gas is shipped and stored to be used as fuel. The reduction in volume per unit amount of substance from the gaseous to the liquid phase is significant and further, because of the reduction in volume, the cost of transportation is also reduced substantially. As an example, let us consider liquefied natural gas for which the major chemical component is methane so that we can assume, for the purpose of this discussion, that liquefied natural gas (LNG) is entirely methane. Gaseous methane at a temperature of 298 K and pressure of 0.1 MPa has a molar volume of 24.7 $dm^3 \cdot mol^{-1}$, while at a temperature of 111 K and pressure of 0.1 MPa the molar volume of the liquid is about 0.038 $dm^3 \cdot mol^{-1}$, that is, about 650 times less than for the same amount of substance in the gas phase. Liquefaction is the process whereby a material in the gas phase is converted to the liquid phase.

A gas can be liquefied only at temperatures below the critical temperature (see Question 4.2 and Figure 4.1). At temperatures above the critical temperature, a substance will remain in the gaseous state irrespective of the applied pressure. There are certain substances commonly used as liquefied gases that have a very low critical temperature and examples of these substances with their critical temperatures T_c are as follows: Hydrogen (T_c = 33.145 K), oxygen (T_c = 154.58 K), helium (T_c = 5.1953 K), nitrogen (T_c = 126.19 K) and methane (T_c = 190.56 K). Liquefaction of these gases can be achieved only at temperatures below T_c and these temperatures cannot be obtained with ordinary refrigeration techniques because of the low efficiency and high power (energy) consumption. The most widely used liquefaction cycle is known as the Linde process and it is to this that we now turn.

The Linde process, shown schematically in Figure 6.30, is used to liquefy a gas. During this cycle, the following steps occur:

step 1–2: Gas supplied is mixed with the gas (at state 8) that was not liquefied during its pass through the Linde process, and then enters a multistage compressor with intercooling to avoid heating of the gas and therefore to minimize the power required for compression

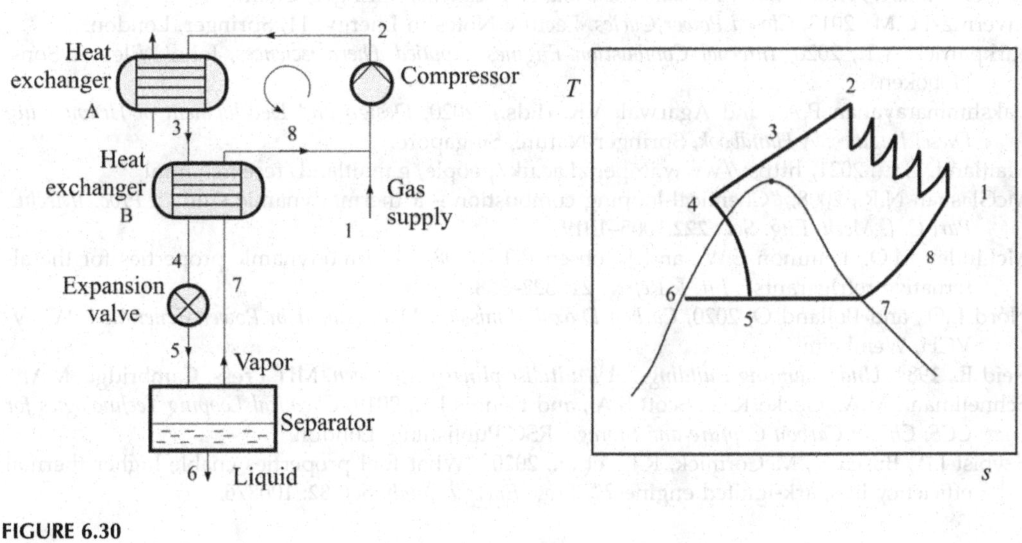

FIGURE 6.30
Linde liquefaction process.

step 2–3: The high-pressure gas is cooled passing through a heat exchanger from state 2 to state 3

step 3–4: The gas is cooled further by passage through another heat exchanger cooled with low temperature gas discharged from steps 5–7 and 7–8

step 4–5: The high-pressure gas is throttled through an expansion valve to low pressure and low temperature as a saturated liquid + gas mixture

step 5–6 and 7: The (liquid + gas) mixture is separated to give a saturated gas (at state 7) and saturated liquid (at state 6) that is removed from the process as the required product

step 7–8: The low-temperature saturated gas is returned to the start of the Linde process after passing through a heat exchanger that cools the high-pressure gas stream in step 3–4

References

Abas N., Kalair A.R., Khan N., Haider A., Saleem Z., and Saleem M.S., 2018, "Natural and synthetic refrigerants, global warming: A review", *Renew Sustain. Energy Rev.* **90**: 557–569.

Bobbo S., Di Nicola G., Zilio C., Brown, J.S., and Fedele L., 2018, "Low GWP halocarbon refrigerants: A review of thermophysical properties", *Int. J. Refrig.* **90**: 181–201.

Burghardt M.D., and Harbach J.A., 1993, *Engineering Thermodynamics*, Harper Collins College Publishers, New York.

Cabral R.P., and Mac Dowell N., 2019, *Oxy-fuel Combustion Capture Technologies; Ch6 in Carbon Capture and Storage*, RSC Publishing, London.

Carnot S., 1872, "Reflections sur la puissance motrice du feu et sur les machines propres à développer cette puissance", *Annales scientifiques de l'École Normale Supérieure ser. 2* **1**: 393–457.

Chabra R.P. and Sankar V. (Eds.), 2017, *Coulson and Richardson's Chemical Engineering, 7th Edition, Volume 1A: Fluid Flow: Fundamentals and Applications*, Butterworth-Heinemann Elsevier, Oxford.

Dincer I., 2017, *Refrigeration Systems and Applications*, John Wiley & Sons, Chichester.

Grassi W., 2017, *Heat Pumps: Fundamentals and Applications*, Springer, Cham.

Invernizzi C.M., 2013, *Closed Power Cycles*, Lecture Notes in Energy, 11, Springer, London.

Kirkpatrick A.T., 2020, *Internal Combustion Engines: Applied Thermosciences*, John Wiley & Sons, Hoboken.

Lakshminarayanan P.A., and Agarwal A.K. (Eds.), 2020, *Design and Development of Heavy Duty Diesel Engines: A Handbook*, Springer Nature, Singapore.

Maitland, G.C., 2021, https://www.imperial.ac.uk/people/g.maitland/research.html

McGlashan N.R., 2008, "Chemical-looping combustion – a thermodynamic study", *Proc. IMechE, Part C: J. Mech. Eng. Sci.*, **222**:1005–1019.

McLinden M.O., Lemmon E.W., and Jacobsen R.T., 1998, "Thermodynamic properties for the alternative refrigerants", *Int. J. Refrig.* **21**: 322–338.

Nord L.O., and Bolland O, 2020, *Carbon Dioxide Emission Management in Power Generation*, Wiley-VCH, Weinheim.

Reid E., 1984, *Understanding Buildings: A Multidisciplinary Approach*, MIT Press, Cambridge, MA.

Schnellmann M.A., Gorke R.H., Scott S.A., and Dennis J.S., 2019, *Chemical Looping Technologies for CCS; Ch7 in Carbon Capture and Storage*, RSC Publishing, London.

Szybist J.P., Busch S., McCormick, R.L., et al., 2020, "What fuel properties enable higher thermal efficiency in spark-ignited engines?", *Prog. Energy Comb. Sci.* **82**: 100876.

7

Energy and the Environment

7.1 Why Do We Need Different Energy Vectors and How Do They Impact the Environment?

In this chapter we continue our consideration of the applications of the basic thermodynamic principles we introduced in Chapters 1 to 5 with an examination of two areas of great significance for the twenty-first century: energy and the environment. Although thermodynamics cannot claim to provide solutions to either of these issues on its own, we hope to illustrate, by means of examples based on key issues, how thermodynamics can provide insight into the energy requirements involved, which areas it is sensible to pursue to give viable, practical approaches and those where it is not. It is impossible to cover all the areas that are relevant to the debate, but our approach is to provide, through specific examples, illustrations of the methodology that has universal applicability.

Since the mid-nineteenth century, the major source of energy for domestic use, travel and industrial processes has been through the combustion of fossil fuels. Initially, the main means of delivering and transporting energy (often referred to as an energy vector) was coal and the Industrial Revolution was driven by using the heat generated by its combustion to power steam engines through the cyclic vaporization of water to steam and its condensation back to water. In these reciprocating expanding engines, the volume expansion on producing steam drove the movement of pistons whose reciprocal motion was mechanically coupled to power machinery and devices carrying out work. The chemical energy of combustion was hence converted, with some loss as heat, into useful mechanical work (see Questions 1.6.3, 3.8 and 3.9). Coal could also be burned directly, simply to produce heat.

With the advent of electricity as a major energy vector, coal-fired generation of high temperature and pressure steam has long been used to drive electricity-generating steam turbines, with the condensed steam being recycled and reheated to enable continuous generation of electrical power. Here the chemical energy of combustion is converted into the enthalpy of high-pressure steam that is then converted to mechanical energy by rotating the turbine blades, whose shaft in turn drives an electromagnetic generator to produce electrical energy. The changes in thermodynamic state of the steam working fluid that drives these power-producing devices, both heat engines and turbines, can be described by thermodynamic cycles, such as the purely hypothetical reversible Carnot cycle and the Rankine cycle. These cycles can be used to estimate the efficiency with which the original chemical energy can be converted into mechanical work or electrical energy and are closely linked to the Second Law of Thermodynamics. The principles are set out in Chapter 3 and covered in detail for different types of cycle in Questions 6.2 and 6.3.

During the twentieth century, the discovery and widespread exploitation of oil and gas led to a redistribution of the use of fossil fuel–based energy for different sectors. Coal

DOI: 10.1201/9780429329524-7

continued to be used for electricity with an increasing contribution from natural gas, aided by the design of more efficient combined power cycles (see Questions 6.2.2 and 6.3.3) and nuclear energy, whereas space heating of buildings converted from coal to oil, gas or electricity and land, sea and air transport has been transformed and powered largely by liquid hydrocarbons with some contribution from electricity. During the latter quarter of the last century, there was increasing evidence that the continued use of fossil fuels, and the associated release of carbon dioxide, CO_2, was causing significant climate change owing to the radiative forcing effects caused by CO_2 and other gases, notably methane, released during oil and gas production and use, which are now called collectively "greenhouse gases" because of the analogy with greenhouse trapping of solar radiation they cause by their presence in the upper atmosphere.

The concerns about the continued use of fossil fuels for energy (and for producing petrochemicals and plastic materials on which a large part of modern society depends) causing catastrophic climate change has led to a major drive to transition to decarbonized energy vectors, such as electricity generated using renewable sources (solar and wind energy) or hydrogen, which when combusted to produce heat converts solely to water. In the meantime, it has become clear that during that transition, any continued use of fossil fuels must be accompanied by the capture of as much of the released CO_2 as possible (feasible mainly for large, centralized facilities for power production and industrial manufacture) and to store it (most practically in underground geological sites such as depleted oil and gas reservoirs or deep saline aquifers) to prevent its release into the atmosphere. This process is referred to as carbon capture and storage (CCS). Climate change indicates that energy provision and care for the environment are inextricably linked.

This chapter will consider what thermodynamics can tell us about how alternative energy sources compare with fossil fuels regarding their energy delivery capacity, the efficiency of their energy conversion processes, how energy from these often-intermittent sources can be stored and how the release of CO_2 from the continued use of fossil fuels can be minimized. The answers will enable different energy sources to be compared and guide the different choices that need to be made in selecting energy options and decarbonization processes for different applications, given a variety of constraints and environments. We start with questions about the thermodynamics behind using hydrogen as an energy vector.

7.2 Does a Hydrogen Energy Economy Make Thermodynamic Sense?

Hydrogen is a carbon-free energy vector that can be used to generate electrical power by combustion in gas turbines (large-scale grid power) or as an energy source for fuel cells (short to medium-range transport). It can also be used as a direct replacement for natural gas for domestic or commercial-scale heating, although substitution above the 20% level requires major changes to burners because of the vastly different combustion properties of hydrogen as compared to natural gas. There are two main routes for producing hydrogen at the large commercial scale required for such applications:

a. ***Steam methane reforming (SMR)*** whereby natural gas is catalytically converted to hydrogen via a two-stage process:

Steam reforming to carbon monoxide and hydrogen ("Syngas") – *strongly endothermic*

$$CH_4 + H_2O \rightleftarrows CO + 3H_2 \quad \Delta_r H^\ominus = 206\ \text{kJ/mol} \quad \Delta_r G^\ominus = 141\ \text{kJ/mol}. \tag{7.1}$$

Water gas shift (WGS) reaction to produce further hydrogen from the carbon monoxide

$$CO + H_2O \rightleftarrows CO_2 + H_2 \quad \Delta_r H^\ominus = -41\ \text{kJ/mol} \quad \Delta_r G^\ominus = -28.5\ \text{kJ/mol}. \tag{7.2}$$

The end by-product is CO_2, which must be captured and permanently stored using CCS to make hydrogen produced by this route ("blue hydrogen") a carbon-free energy vector. Although there are alternative routes to convert natural gas to hydrogen, such as auto-thermal reforming and partial oxidation (Kalamaras and Efstathiou 2013), both of which require a supply of pure oxygen, SMR is the most cost-effective route to hydrogen for power applications and we will focus on this here. The net enthalpy requirement to convert 1 mol of methane to 4 mol of hydrogen is, adding Equations 7.1 and 7.2, 165 kJ/mol CH_4 or 41.3 kJ/mol H_2. This is equivalent to 41.3/0.002 kJ/kg H_2 or 20.7 MJ/kg H_2. The thermal efficiency of commercial SMR processes is in the range 70%–85% so the energy requirement is ~27 MJ/kg H_2.

b. ***Electrolysis of water:***

The direct decomposition of water to hydrogen and oxygen is very unfavorable thermodynamically, requiring 237 kJ/mol of energy:

$$H_2O(l) \rightarrow \tfrac{1}{2}O_2(g) + H_2(g) \quad \Delta_r G^\ominus = 237\ \text{kJ/mol}. \tag{7.3}$$

When this is carried out in an electrochemical cell (in this case an *electrolyzer*), the two half-cell reactions (see Question 5.5.1) are

$$\text{Anode (oxidation):} \quad H_2O(l) \rightarrow \tfrac{1}{2}O_2(g) + 2H^+(aq) + 2e^- \quad E^\ominus = +1.23\ \text{V}, \tag{7.4}$$

$$\text{Cathode (reduction):} \quad 2H^+(aq) + 2e^- \rightarrow H_2(g) \quad E^\ominus = 0.00\ \text{V}, \tag{7.5}$$

where $E^\ominus$ is the standard electrode potential. So the standard potential of a water electrolysis cell at 25°C is

$$E^\ominus_{\text{cell}} = E^\ominus_{\text{cathode}} - E^\ominus_{\text{anode}} = -1.23\ \text{V}. \tag{7.6}$$

Using the Nernst equation (see Equation 5.113)

$$\Delta_r G^\ominus = -z\,F\,E^\ominus, \tag{7.7}$$

we verify that this is equivalent to a standard Gibbs energy change of 237 kJ/mol. If the hydrogen produced by this route is to be considered carbon-free ("green hydrogen"), then the electrolyzer must be powered by renewable electrons, using renewable or decarbonized electricity.

So there are energy penalties in producing hydrogen from either natural gas or water, meaning that the energy content of hydrogen is significantly less than in the primary energy vectors (gas or electricity) used to produce it. Its main advantage is its clean burning to produce only water, which is easily separated by condensation. In order to determine the thermodynamic viability of using hydrogen as an energy vector, we need to ask two more specific questions.

7.2.1 How Can We Compare the Energy Content of Hydrogen with Other Energy Carriers?

The *energy density of a fuel or energy vector* is the amount of thermal energy (enthalpy) stored within that substance per unit volume. This should be contrasted with the *specific energy* of a substance, which is the amount of energy per unit mass. These measures of fuel capacity are usually defined using the standard enthalpy of combustion since they reflect the intrinsic energy content of a substance as described by the First Law (see Questions 1.5.3 and 1.8.4). In some practical contexts, these enthalpies are referred to as the *higher heating value* (HHV) of a substance, corresponding to all reactants and products being at their standard state (usually 298 K) so that vapor products such as water condense to liquids and their enthalpy of condensation is released on combustion of the fuel. This is in contrast to the *lower heating value* (LHV) where it is assumed that all combustion products remain as vapors. Table 7.1 gives values for both energy density and specific energy for a range of fuels or energy vectors.

TABLE 7.1

Energy density and specific energy for a range of fuels and energy vectors†

Energy Vector (Fuel)	Specific Energy MJ/kg	Energy Density MJ/L
Hydrogen:		
• Liquid	141.9	10.0
• Gas 700 bar, 25°C	141.9	5.3
• Gas 1 bar, 25°C	141.9	0.012
Methane 1 bar, 25°C	55.6	0.038
LNG (−160°C)	53.6	22.2
Natural Gas	53.6	0.036
LPG (C_3H_8)	49.6	25.3
Petrol (Gasoline)	46.4	34.2
Heating Oil	46.2	37.3
Diesel	45.6	38.6
Kerosene	43.0	35.0
Crude Oil	41.9	37.0
Dimethyl Ether (DME)	31.7	21.2
Coal (bituminous)	20.0	25.0
Methanol	19.7	15.6
Hydrazine	19.5	19.3
Wood	18.0	14.5
LOHC*	8.8	8.2
Domestic Waste	8.0	3.0

Notes

† Based on higher heating values (HHV)

* LOHC = Liquid Organic Hydrogen Carriers; values for typical fluid (perhydrodibenzyltoluene).

The critical issue in using this table to choose viable energy vectors for particular applications is identifying the limiting factor in the design of the process or device which is being powered. If the mass of the fuel were the sole consideration, then the specific energy of fuels ranging from dense solids to low pressure gases only varies by a factor of three to five: Compare coal (~30 MJ/kg) with oil (~42 MJ/kg), natural gas (~54 MJ/kg) and hydrogen (142 MJ/kg) for instance; on this basis, hydrogen looks attractive. However, the cost, weight and size of energy storage devices, be that fuel stock tanks, batteries or other forms of energy storage (see Question 7.5), are critical factors in determining the suitability of specific energy vectors for particular applications, especially for transportation. It can be seen that several fuels have higher specific energies than petroleum and diesel but none have higher energy densities, which is critical where storage volume is at a premium, as it often is. Gaseous hydrogen is particularly unfavorable in this respect. On the other hand, for applications involving pipeline or bulk supply to centralized facilities (for example industrial or domestic heating), its high specific energy can be better exploited.

Figure 7.1 compares this mass versus volumetric energy capacity balance for different fuels, as well as devices for storing and transporting energy such as batteries and fuel cells. Hydrogen, even as a liquid, is a factor of three lower on energy density compared

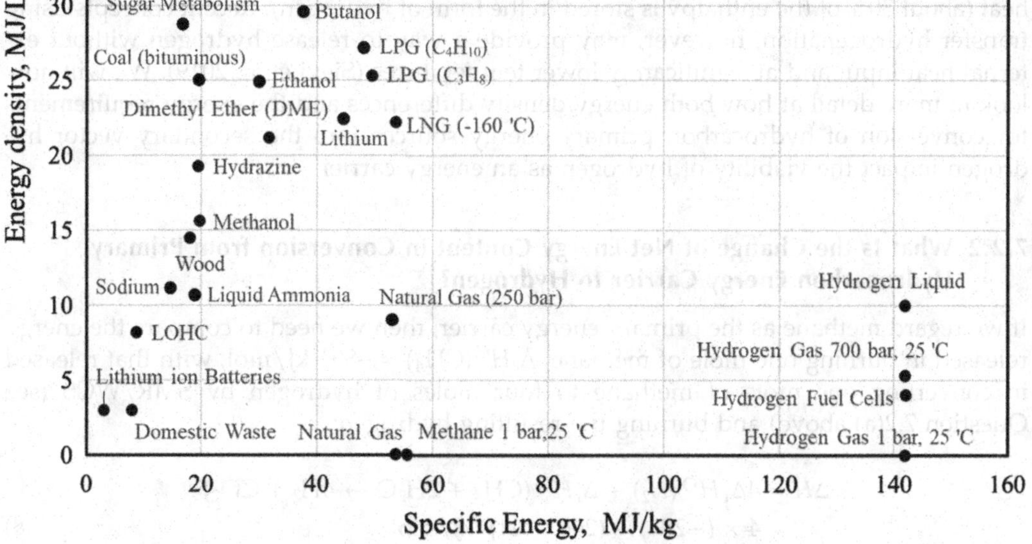

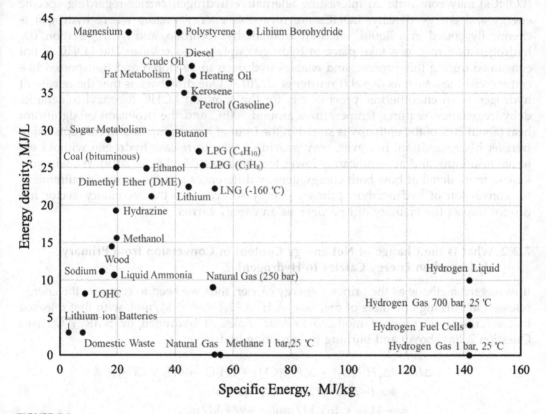

FIGURE 7.1

Plot of energy density vs specific energy for a range of fuels and energy vectors. Of energy storage devices, only Li ion batteries and hydrogen fuel cells are visible on this plot. Most other storage systems (see Question 7.5) have energy storage densities and specific energies at least a factor of ten smaller than Li ion batteries.

with other liquid fuels, which places hydrogen fuel cells, despite their compact nature, down in the bottom right-hand corner of the plot, limiting their ability to compete with fossil fuels for transport on the basis of range capability for a single tank of fuel. Since many renewable energy sources are intermittent, the need to provide energy storage processes and devices at all scales, not just for portable devices and transport, is essential. The specific energy and energy density of these storage systems are critical characteristics in determining how low-carbon energy sources can replace high energy density fossil fuels for different applications. A particularly challenging area is long-distance transport which requires either low-carbon footprint liquid fuels or batteries/fuel cells with sufficient energy density. To compensate for the carbon emissions of continuing to use fossil fuels in these areas until these are in place, it will be necessary to remove some CO_2 from the atmosphere. So as well as examining the thermodynamics of carbon capture and storage processes for *preventing* CO_2 from fossil fuels entering the atmosphere (as we did in Question 6.3), in this Chapter we will also examine the thermodynamic feasibility of direct air capture methods for *removing* CO_2 from the atmosphere.

Ease and safety of storage are other critical factors and here hydrogen, despite being very attractive from a low/zero carbon standpoint, is at a significant disadvantage compared to conventional liquid fuels, needing to be either compressed or cooled to start to compete in terms of energy density. However, liquid organic hydrogen carriers (LOHCs) may constitute an interesting alternative hydrogen carrier regarding specific energy and energy density. LOHCs are reversible energy vectors where hydrogen is chemically bound in a liquid, often an aromatic hydrocarbon, and hydrogenation/dehydrogenation reactions take place in hydrogen uptake and release. The LOHC is not consumed during this process, and enables hydrogen to be stored and transported like commercial fuels such as diesel (Preuster et al. 2017). One drawback is that the release of hydrogen is an endothermic process, e.g. for the common LOHC dibenzyl toluene its dehydrogenation requires temperatures around 300°C and the provision of significant heat (about 30% of the enthalpy is stored in the form of hydrogen). Recent concepts using transfer hydrogenation, however, may provide a way to release hydrogen without external heat input and at significantly lower temperatures (Sievi et al. 2019). We will now look in more detail at how both energy density differences and the energy requirements for conversion of hydrocarbon primary energy sources into the secondary vector hydrogen impact the viability of hydrogen as an energy carrier.

7.2.2 What Is the Change of Net Energy Content in Conversion from Primary Hydrocarbon Energy Carrier to Hydrogen?

If we regard methane as the primary energy carrier, then we need to compare the energy released in burning one mole of methane, $\Delta_c H^{\ominus}(CH_4) = -891$ kJ/mol, with that released in converting one mole of methane to four moles of hydrogen by SMR/WGS (see Question 7.2(a) above) and burning the resulting hydrogen:

$$\begin{aligned}\Delta H &= 4\Delta_c H^{\ominus}(H_2) + \Delta_r H^{\ominus}(CH_4 + 2H_2O \rightarrow 4H_2 + CO_2)\\ &= 4 \times (-286) + (206 - 41)\ \text{kJ/mol}\\ &= -1144 + 165\ \text{kJ/mol} = -979\ \text{kJ/mol}.\end{aligned} \tag{7.8}$$

So, on a molar basis, converting methane and water to hydrogen via SMR/WGS results in a net increase in energy content of 88 kJ per mol of original methane, or ~10%. Here we see the value in converting hydrogen bound in liquid water into free hydrogen gas which can be used as a fuel. On a mass basis, combustion of 1 kg of CH_4 releases 891/0.016 kJ = 55.7 MJ (its specific energy), whereas the net energy release in converting that 1 kg of methane to 0.5 kg of hydrogen via SMR/WGS and then burning it is 979/0.016 kJ = 61.2 MJ. However, there is an additional energy penalty in delivering the SMR/WGS hydrogen as carbon-free and that is the CCS energy requirement to capture and store the CO_2 co-produced with the hydrogen; see Question 7.2.5 and Tables 7.2 and 7.3.

If we neglect the hydrogen produced in the WGS reaction, then the energy released by burning the 3 moles of hydrogen produced in Equation 7.1 is –652 kJ/mol, which means that the energy conversion efficiency of converting methane to hydrogen by SMR alone is 652/891 = 73%. A similar calculation for the steam reforming to hydrogen of a liquid hydrocarbon like octane (representative of the alkane fractions from C_n = 5–12 typically found in petrol or gasoline, which has a standard heat of combustion $\Delta_c H^{\ominus}$ (C_8H_{18}) = –5,460 kJ/mol), for which

$$C_8H_{18} + 8H_2O \rightarrow 8CO + 17H_2 \quad \Delta_r H^{\ominus} = 1275 \text{ kJ/mol}, \tag{7.9}$$

gives an energy efficiency of only 66%. There are, therefore, major decreases in net energy content from the primary crude oil and natural gas energy sources to hydrogen as an energy carrier as we move down the energy conversion chain (see also Question 7.8).

Hydrogen as an energy carrier compares even less favorably with hydrocarbons once we consider their energy capacity on a volumetric basis, which is what really matters in assessing

TABLE 7.2

Comparison of carbon emissions and electricity costs for range of power plants and for SMR hydrogen (Metz et al. 2005; IEAGHG 2017)

Process	GHG Emissions w/o CCS kg CO_2/MWh	GHG Emissions with CCS kg CO_2/MWh	Electricity Cost w/o CCS US$/MWh	Electricity Cost with CCS US$/MWh
PC	760	112	43–52	63–99
IGCC	775	108	41–61	55–91
NGCC	370	52	31–50	43–77
	GHG emissions w/o CCS kg CO_2/kg H_2	**GHG emissions with CCS kg CO_2/kg H_2**	**H_2 Cost w/o CCS US$/kg H_2**	**H_2 Cost with CCS US$/kg H_2**
SMR	20	3	1.2–1.5	1.5–2.0

TABLE 7.3

Comparison of carbon emissions and electricity costs for production of 1 kg hydrogen using electricity from a range of power plants with SMR hydrogen process costs

Process	GHG Emissions w/o CCS kg CO_2/kg H_2	GHG Emissions with CCS kg CO_2/kg H_2	Cost w/o CCS US$/kg H_2	Cost with CCS US$/kg H_2	Approx % Cost Increase with CCS
PC	40.1	5.9	2.3–2.7	3.3–5.2	65
IGCC	40.9	5.7	2.2–3.2	2.9–4.8	50
NGCC	19.5	2.7	1.6–2.6	2.3–4.1	60
SMR	19.7	2.5	1.2–1.5	1.5–2.0	30

its viability as a liquid fuel. One liter of methane with a density of 0.72 kg/m^3 at STP (0°C, 1 bar) has a volumetric higher heating value (HHV) (or energy density) of 55.7 × 0.72 MJ/L = 40.0 MJ/L. By contrast the SMR/WGS hydrogen generated from it has, under the same conditions where it has a density of 0.09 kg/m^3, a net HHV of 61.2 × 0.09 MJ/L = 5.5 MJ/L. The loss of net energy content per unit volume of fuel is therefore 34.5 × 100/40.0 = 88%. For hydrogen produced from SMR only, the net HHV is (652/16) × 0.09 = 3.7 MJ/L, a net energy decrease per unit of volume of 36.3 × 100/40.0 = 91%.

Hence, the low density of the resulting fuel compared to its hydrocarbon pre-cursor results in a decrease in energy density of about 90% for low-pressure hydrogen gas compared with natural gas. Further, energy losses occur if there is a need to compress hydrogen to the dense gas state or to liquify it to enhance its energy density, which at its highest is still only about 20% of liquid hydrocarbon values. This significant disadvantage needs to be outweighed by the potential benefits of zero carbon emissions using hydrogen and maybe using devices more efficient than traditional combustion technologies for extracting the energy stored within hydrogen. It is to these issues that we next turn.

7.2.3 How Does the Efficiency of a Hydrogen Fuel Cell Compare with a Petrol Internal Combustion Engine?

A fuel cell is an electrochemical energy conversion device which can capture and use the power of hydrogen very efficiently. It directly converts the chemical energy inherent in hydrogen into electricity and is essentially pollution-free, producing only water and heat, which can also be used for other applications such as combined heat and power CHP (Breeze 2017). As we shall see, fuel cells combine this clean electricity generation with efficiencies two to three times those of traditional fossil fuel power plants and engines. The first fuel cell was, in fact, invented a long time ago, in 1839 by William Grove who developed what he called a "gas voltaic battery" (Grove 1839; Sella 2020). It was Ludwig Mond and Charles Langer who first used the term "fuel cell" some 50 years later (Mond and Langer 1889; Appleby 1990).

A fuel cell has many of the features found in an electrochemical battery: Two electrodes, an anode and a cathode, in contact with an electrolyte (see Figure 7.2). The most common fuel cells convert hydrogen and oxygen into water and in the process produce electricity. While both devices convert chemical into electrical energy, the chemicals enclosed within

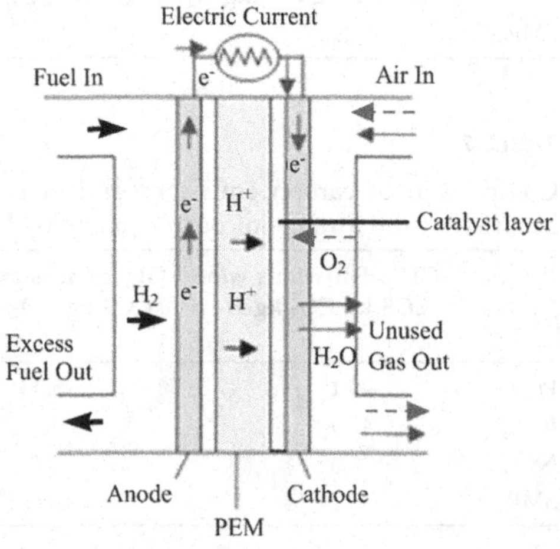

FIGURE 7.2
Schematic diagram of polymer exchange membrane fuel cell, showing the flow of reactants, products, electrons and protons. Only the latter can pass through the thin film solid electrolyte membrane.

a battery eventually deplete and either its life ends or it needs to be recharged. By contrast, the chemical feedstocks are fed continuously into a fuel cell and as long as they flow, electricity continues to be generated. There are a variety of different types of fuel cell using different electrolytes and operating in different temperature ranges. Some are better suited to transportation or small-scale device or portable applications, such as polymer exchange membrane (PEMFC), direct methanol (DMFC) and phosphoric acid (PAFC) fuel cells (Dicks and Rand 2018). Others may in time be used for large-scale stationary power generation; solid oxide (SOFC) and molten carbonate (MCFC) fuel cells fall into this category. These operate at very high temperatures, which can make reliability a problem, but the steam generated by the electrochemical conversion of hydrogen and oxygen to water can be usefully used to drive turbines to generate additional electricity which improves the overall efficiency of the process. Here we will focus on PEMFCs to illustrate the thermodynamic principles of all fuel cells. These operate at relatively low temperatures (60–80°C), which means they can warm up to their operating temperature quickly, an advantage for vehicle power applications.

As seen in Figure 7.2, a solid thin-film electrolyte, typically containing perfluorosulphonic acid or conducting aromatic polymers, is sandwiched between an anode and a cathode, with a thin high-surface area layer of finely dispersed catalyst (typically platinum nanoparticles) placed between each electrode and the electrolyte layer. Hydrogen gas is pumped into the fuel cell on the anode side; the electrodes have fine channels to allow the gas to have intimate contact with the catalyst, through which it flows.

On contacting the catalyst, a hydrogen molecule decomposes into two protons H^+ and two electrons e^-. Only the protons can pass through the electrolyte membrane; the electrons are conducted through the anode where they follow an external circuit, doing useful work such as turning a motor before returning to the cathode. On the cathode side of the cell, oxygen gas (O_2), often in the form of air, flows from the electrode channels into the catalyst forming two oxygen atoms. One of these combines with two H^+ ions, which have passed through the membrane and two electrons from the external circuit to form a molecule of water (H_2O). This process is exothermic, generating heat that can be used outside the cell.

This reaction in a single fuel cell produces only about 0.7 volts. To get this voltage up to a reasonable level, many separate fuel cells must be combined to form a fuel-cell stack. The individual cells are coupled using bipolar plates, typically made of lightweight metals or conducting carbon-thermoset composites. This scalability makes fuel cells suitable for a wide range of applications, for example small computers (50–100 watts), domestic use (1–5 kW), road and marine transport (50–120 kW) and grid-scale power (1–200 MW or higher).

A thermodynamic analysis will enable us to explore the power generated by a hydrogen fuel cell unit and compare its efficiency with other power generation systems. Figure 7.3 shows a simplified schematic of a fuel cell with its chemistry, reactants and products.

Figure 7.3 defines the thermodynamic system (or control volume) to which we can apply the First Law (see Question 1.5),

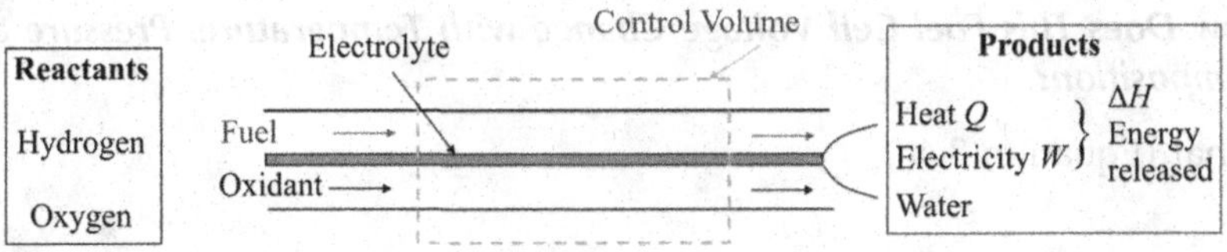

FIGURE 7.3
Schematic diagram of the mass and energy flows in a hydrogen fuel cell.

$$\Delta H = Q + W_{el}, \tag{7.10}$$

where the work done by the system, $-W_{el}$, is that due to transport of electrons from the anode to the cathode across a potential difference, E, generating a current i for a time Δt

$$W_{el} = -E\, i\, \Delta t = -Eq, \tag{7.11}$$

where q is the charge (units A s) transferred when the cell reaction occurs, also given by

$$q = z\, F, \tag{7.12}$$

where z is the number of moles of electrons involved, each mole having a charge of F = 96,493 Coulombs, F being Faraday's constant. So

$$\text{Electrical work } W_{el} = -z\, F\, E. \tag{7.13}$$

The maximum work, assuming no irreversibilities, is also given by the change in free energy $\Delta G_r^{\ominus}$ associated with the chemical energy released during the fuel cell reaction (see Questions 5.5.1 and 5.5.2, Equations 5.101–5.108). The half cell reactions are

$$\text{Anode:} \quad H_2 \rightarrow 2H^+ + 2e^- \tag{7.14}$$

$$\text{Cathode:} \quad \tfrac{1}{2}O_2 + 2H^+ + 2e^- \rightarrow H_2O \tag{7.15}$$

$$\text{Overall reaction:} \quad H_2 + \tfrac{1}{2}O_2 \rightarrow H_2O. \tag{7.16}$$

$$\begin{aligned} W_{el} &= -z\, F\, E^{\ominus} = \Delta_r G^{\ominus} \\ &= G_f^{\ominus}(\text{products}) - G_f^{\ominus}(\text{reactants}) \\ &= G_f^{\ominus}(H_2O) - G_f^{\ominus}(H_2) - 0.5G_f^{\ominus}(O_2), \end{aligned} \tag{7.17}$$

$$\text{so} \quad \Delta_r G^{\ominus} = [-306.7 - (-39.0) - 0.5(-61.1)]\ \text{kJ/mol} = -237.2\ \text{kJ/mol}, \tag{7.18}$$

and using Equation 7.18, we can calculate the open-circuit (reversible) voltage generated by the H_2–O_2 fuel cell at standard conditions (0.1MPa, 298.15 K) as

$$E^{\ominus} = -\frac{\Delta_r G^{\ominus}}{z\, F} = \frac{237.2\ \text{V}}{2 \times 96.493} = +1.23\ \text{V}. \tag{7.19}$$

7.2.3.1 How Does This Fuel Cell Voltage Change with Temperature, Pressure and Composition?

Recalling that (Equation 3.40)

$$dG = V\,dP - S\,dT + \sum_B \mu_B\, dn_B \tag{7.20}$$

and Equation 3.41

$$\left(\frac{\partial G}{\partial T}\right)_p = -S, \tag{7.21}$$

using Equation 7.17 we find

$$\left(\frac{\partial E}{\partial T}\right)_p = -\frac{\Delta S}{zF}, \tag{7.22}$$

so that the cell voltage at the temperature T is

$$E_T = E^\ominus + \frac{\Delta S^\ominus}{zF}(T - T^\ominus), \tag{7.23}$$

where $T^\ominus$ is the standard temperature 298.15K. $\Delta_r S^\ominus$ for the cell reaction is –44.3 J/(mol K) so we can see that the reversible cell voltage decreases with increasing temperature. If we assume that $\Delta S^\ominus$ in Equation 7.23 is independent of temperature, we can estimate the voltage at the normal operating temperature of a PEMFC, around 70°C or 343 K:

$$E_{343} = 1.23\text{ V} - 44.3 \times (343 - 298)/(2 \times 96,493)\text{ V} = 1.22\text{ V}, \tag{7.24}$$

Here, the effect is very small. However, for a SOFC working typically at 1,000 K, the cell voltage drops to 1.07 V. For the pressure dependence, again using Equation 3.41,

$$\left(\frac{\partial G}{\partial p}\right)_T = V, \tag{7.25}$$

so using Equation 7.17 we have

$$\left(\frac{\partial E}{\partial p}\right)_T = -\frac{\Delta V}{zF}, \tag{7.26}$$

which leads, assuming perfect gas behavior for the hydrogen and oxygen to evaluate ΔV, to

$$\left(\frac{\partial E}{\partial p}\right)_T = -\frac{\Delta n\, RT}{zFp}, \tag{7.27}$$

where Δn is the change in the number of moles of products compared with the reactants, which for the hydrogen fuel cell is –0.5 mol (see Equation 7.16). Integrating Equation 7.27 leads to

$$E_p = E^\ominus - \frac{\Delta n\, RT}{zF}\int_{p_0}^{p}\frac{dp}{p} = E^\ominus - \frac{\Delta n\, RT}{zF}\ln\left(\frac{p}{p_0}\right). \tag{7.28}$$

If we apply Equation 7.28 for p = 100 bar at 298 K, E increases slightly to E_{100} = 1.26 V.

Finally, let us examine the effect of changing the concentration of the reactants. To do this, we use Equation 4.145 for the chemical potential of each species in the fuel cell,

$$\mu_i = \mu_i^{\ominus} + RT \ln a_i, \tag{7.29}$$

where for a gas the relative activity, a_i, (Equation 4.146) is strictly its fugacity (see Question 4.3.1, Equation 4.31) but we will equate it here to its partial pressure p_i (as in Equation 4.55).

For a general reaction (with stoichiometrc coefficients $\nu_i = -a, -b, c, d$), $a\text{A} + b\text{B} \rightleftharpoons c\text{C} + d\text{D}$, the change in molar Gibbs energy (see Equations 5.11 and 7.29) is

$$\begin{aligned} \Delta_r G &= \left(c\mu_C^{\ominus} + d\mu_D^{\ominus}\right) - \left(a\mu_A^{\ominus} + b\mu_B^{\ominus}\right) + RT \ln \frac{a_C^c a_D^d}{a_A^a a_B^b} \\ &= \Delta_r G^{\ominus} + RT \ln \frac{a_C^c a_D^d}{a_A^a a_B^b}. \end{aligned} \tag{7.30}$$

We recognize this as the generalized form of the *van't Hoff isotherm*; see Equations 5.34 and 5.58. If we use Equation 7.17 for $\Delta_r G$ and $\Delta_r G^{\ominus}$, we obtain the electrochemical equivalent of the van't Hoff equation, the *Nernst equation* (see Equation 5.113):

$$E(p, T) = E^{\ominus} - \frac{RT}{zF} \ln \frac{a_C^c a_D^d}{a_A^a a_B^b}. \tag{7.31}$$

Here, $E = E_{rev}$, the maximum open-circuit voltage generated by the cell. Applying this equation to the hydrogen fuel cell reaction gives

$$E(p, T) = E^{\ominus} - \frac{RT}{zF} \ln \frac{a_{H_2O}}{a_{H_2} a_{O_2}^{0.5}} = E^{\ominus} - \frac{RT}{2F} \ln \frac{1}{p_{H_2} p_{O_2}^{0.5}}, \tag{7.32}$$

since water remains liquid for operating temperatures below 100°C; hence, $a_{H2O} = 1$.

So if we operate the fuel cell at 298 K with H_2 at 5 bar and using air also at 5 bar (i.e. with O_2 at a partial pressure of 5 × 0.2 bar = 1 bar):

$$E = 1.23\text{ V} - \frac{8.314 \times 298\text{ V}}{2 \times 96493} \ln \frac{1}{5 \times 1} = (1.23 + 0.02)\text{ V} = 1.25\text{ V}. \tag{7.33}$$

7.2.3.2 What Is the Thermodynamic Efficiency of a Fuel Cell?

The maximum thermodynamic efficiency η of the cell occurs under reversible, open-circuit (zero-current) conditions (see Question 5.5.1) and is given by the ratio of the useful work output it produces to the total chemical (thermal) energy in the H_2–O_2 system,

$$\eta = -\frac{W}{Q_{in}} = \frac{\Delta_r G^{\ominus}}{\Delta_r H^{\ominus}}, \tag{7.34}$$

where $\Delta_r G^{\ominus}$ and $\Delta_r H^{\ominus}$ (equivalent to the higher heating value, HHV) are the standard changes in Gibbs free energy and enthalpy, respectively, for the cell reaction $H_2 + ½ O_2 \rightarrow H_2O$. So at 298 K and 1 bar

$$\eta_{FC} = \frac{237}{286} = 0.83. \tag{7.35}$$

This is a very high efficiency compared with most devices (an internal combustion energy has a maximum theoretical efficiency of only about 0.6, depending on the engine design and the fuel; see Question 6.4) and is a major advantage of fuel cells. It will go down slightly with increasing temperature, because ΔG decreases as T increases; for instance, the efficiency is 0.75 at 500 K.

It is instructive to compare this with the efficiency of a Carnot heat engine (see Question 6.2). This is the most efficient heat engine and although it cannot be applied practically in real power plants, it gives the (theoretical) maximum efficiency that practical power plants based on Rankine or other power cycles might achieve. A steam turbine operating at 400°C (corresponding to T_h = 673 K) with the water exhausted through a condenser at 40°C (T_l = 313 K), even if it could operate at the efficiency of a Carnot cycle (which practical plants fall well short of; see Question 6.3.3), it would only have an efficiency of (using Equation 6.21)

$$\eta_{Carnot} = \frac{T_h - T_l}{T_h} = \frac{673 - 313}{673} = 0.53. \tag{7.36}$$

This is similar to the efficiency of an ideal internal combustion engine (Otto cycle) with a compression ratio of about 10 (see Equation 6.89 and Figure 6.23) and is well below that of a typical fuel cell. In fact, conventional combustion-based power plants typically operate at efficiencies of about 35–40% and while fuel cells cannot achieve the ideal reversible efficiencies calculated previously, because of various irreversibilities and losses (see later), they can typically generate electricity at efficiencies of about 60%, or even higher if the heat generated can be exploited in cogeneration. Fuel cells are therefore not constrained by the Carnot efficiency limit and are inherently less irreversible than combustion-based power cycles, operating isothermally and at much lower temperatures. In fact, a Carnot engine would have to operate between 1,750 K and 300 K to achieve an efficiency of 83%.

We can use Equations 7.34 and 7.36 to map out how the reversible efficiencies of a hydrogen fuel cell and a Carnot heat engine change with temperature. This is illustrated in Figure 7.4. We can see that a PEMFC operating at 80°C will have an efficiency vastly in excess of a Carnot engine, whereas for a SOFC operating at 600°C, the fuel cell advantage is just a few percent. Above 680°C in fact the fuel cell efficiency is lower than that of an equivalent Carnot engine. However, as pointed out previously, no practical power cycle can operate as efficiently as a Carnot heat engine, so the fuel cell remains more efficient than practical power cycles at all temperatures of real interest.

Just as the Carnot cycle gives the maximum reversible efficiency for a heat engine, Equation 7.34 gives the maximum efficiency of a fuel cell, achieved at open circuit with no load. As the current drawn increases, so the output voltage achieved falls off and the efficiency decreases. These losses arise from a number of irreversible polarization processes. The rapid decrease in cell voltage with current drawn initially is due to activation losses, caused by the slowness of the reaction taking place at the electrode surfaces; some voltage is

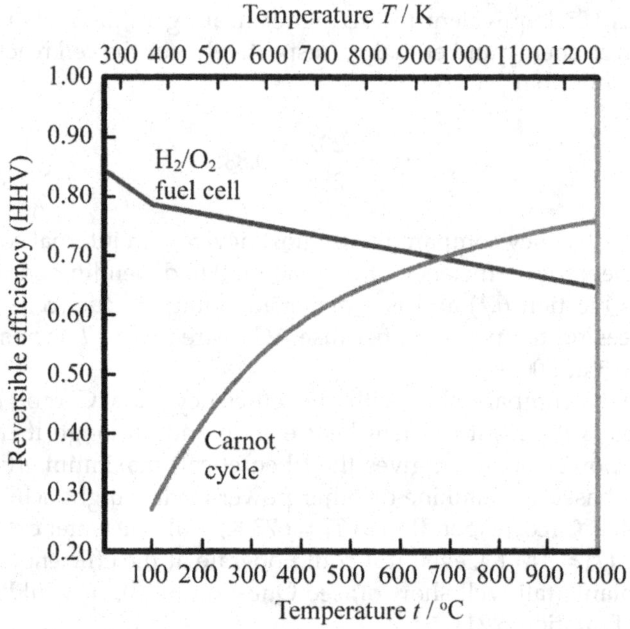

FIGURE 7.4
Variation of reversible efficiency of a hydrogen fuel cell and a Carnot heat engine with temperature.

also lost in driving the redox reactions that transfer electrons to and from the electrodes. The rate of voltage decrease then slows down as ohmic polarization takes over, arising from the resistance to electron flow through the electrode material; these losses vary linearly with current density. At high current densities, the rate of decrease of cell voltage increases again as concentration losses become important, arising from changes in concentration of the reactants at the electrode surfaces as the fuel is consumed more quickly than it can be supplied. This drop-off in efficiency as current is drawn is illustrated in Figure 7.5.

In practice, these and other losses lead to fuel cell vehicles operating at significantly lower efficiencies than their thermodynamic maximum value of 83%. If we compare fuel cells for transport with petroleum internal combustion engines, which typically are only 20–30% efficient in converting the chemical energy contained in petroleum into mechanical power

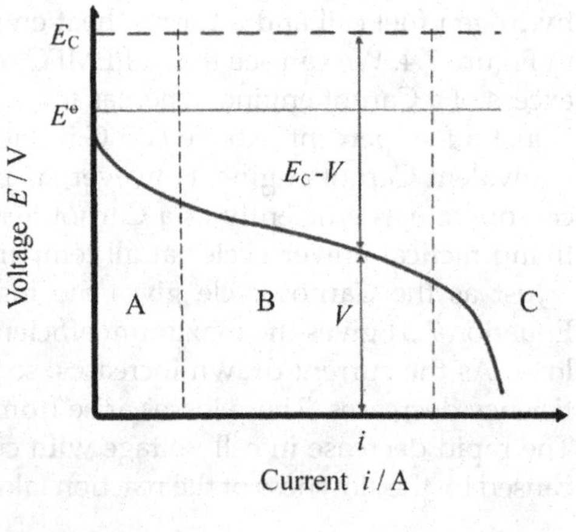

FIGURE 7.5
Polarization curve for a fuel cell (schematic). E_c is the hypothetical voltage of the fuel cell if all the molar enthalpy $H^{\ominus}$ were converted into electricity; $E^{\ominus}/E_c = \Delta G^{\ominus}/\Delta H^{\ominus} = \eta = 0.83$. $E_c - V$ represents the energy loss in the fuel cell liberated as heat, compared with the energy converted to electricity, V. This loss increases from $E_c - E^{\ominus}$ at open-circuit conditions (only 17% loss for a hydrogen fuel cell), following the polarization curve which tracks activation losses in region A, ohmic losses in region B and reactant concentration losses in region C.

(at best ~35%), hydrogen fuel cells powering electric vehicles operating at typically 50–60% efficiency are two to three times more efficient.

7.2.4 What Is the Energy Requirement for Making Hydrogen by Water Electrolysis?

At the start of Question 7.2, we introduced the basic equations for the electrolytic decomposition of water to hydrogen and oxygen. Equation 7.6 gave the standard potential of a water electrolysis cell at 25°C as

$$E^{\ominus}_{\text{cell}} = E^{\ominus}_{\text{cathode}} - E^{\ominus}_{\text{anode}} = -1.23\text{ V}. \tag{7.37}$$

In Question 7.2(b), using Equation 7.7, $\Delta G^{\ominus} = -z\,F\,E^{\ominus}$, we verified that this is equivalent to the standard Gibbs free energy, $\Delta_r G^{\ominus}$, of the overall decomposition reaction, 237 kJ/mol.

$\Delta_r G^{\ominus}$ is positive; hence, the decomposition is non-spontaneous and electrical work must be done for the reaction to proceed. The cell therefore operates at a negative voltage, in contrast to the corresponding hydrogen fuel cell, where for the reverse reaction forming water $\Delta_r G^{\ominus}$ is negative, the reactions release net Gibbs free energy and the cell operates at a positive voltage. $\Delta_r G^{\ominus}$ is directly related to the equilibrium constant for the cell reaction under standard conditions. To calculate the cell potential under non-standard state conditions, we use a version of Equation 5.58,

$$\Delta_r G = \Delta_r G^{\ominus} - RT \ln Q_r = -z\,F\,E, \tag{7.38}$$

where the *reaction quotient* Q_r measures the relative amounts of products and reactants, defined as

$$Q_r = \prod_B (\lambda_B(T))^{\nu_B} \tag{7.39}$$

When the system is at equilibrium, Q_r becomes the equilibrium constant $K^{\ominus}(T)$ (see Equation 5.60). So for a half-cell reaction, such as the cathode electrolysis reaction (see Question 7.2(b))

$$E = E^{\ominus} - \frac{RT}{zF}\ln\left(\frac{\prod_B (\lambda_B(T))^{\nu_B}_{\text{red}}}{\prod_B (\lambda_B(T))^{\nu_B}_{\text{ox}}}\right) \approx E^{\ominus} - \frac{RT}{zF}\ln\left(\frac{[\text{Red}]}{[\text{Ox}]}\right), \tag{7.40}$$

where $\lambda_B(T)_{\text{red}}$ and $\lambda_B(T)_{\text{ox}}$ are the activities of the reduced and oxidized species, respectively, often approximated by their concentrations [Red] and [Ox].

For the full cell reaction (7.3), this becomes another version of the Nernst Equation 7.31

$$E_{\text{cell}} = E^{\ominus}_{\text{cell}} - \frac{RT}{zF}\ln\left(\frac{[\text{Products}]}{[\text{Reactants}]}\right) = E^{\ominus}_{\text{cell}} - \frac{RT}{zF}\ln Q_r. \tag{7.41}$$

Although the reversible cell voltage is −1.23 V, we need to pay particular attention to the entropic part of the free energy when we look at what controls the water electrolysis process:

$$\Delta_r G^{\ominus} = \Delta_r H^{\ominus} - T\,\Delta_r S^{\ominus}$$
$$237\text{ kJ mol}^{-1}\text{K}^{-1} = (286 - 48.6)\text{ kJ mol}^{-1}\text{K}^{-1}. \tag{7.42}$$

The splitting of the water molecules makes the entropic energy $T\Delta_r S^\ominus$ term favorable for the electrolysis and may be contributed from the surroundings. However, under ambient conditions the kinetics are such that it is faster to obtain energy for the $T\Delta_r S^\ominus$ term from electrical energy than by transfer of heat from the surroundings. Therefore, practical rates of water electrolysis require that it is the enthalpy of formation, $\Delta_f H$ (the higher heating value, that corresponds to the water reactant being liquid), that should be used to indicate the required operating voltage rather than $\Delta_f G$. $\Delta H_r = -\Delta H_f = 286$ kJ/mol corresponds to −1.48 V and represents the minimum (negative) electric potential required for water electrolysis at the standard state, referred to as V_{TN}, the thermo-neutral voltage:

$$-V_{TN} = \frac{\Delta_r H^\ominus}{zF} = \frac{\Delta_r G^\ominus}{zF} + T\frac{\Delta_r S^\ominus}{zF} = 1.48\ \text{V}. \tag{7.43}$$

The efficiency of a water electrolysis cell is usually expressed relative to $\Delta_r H^\ominus$ and hence to this minimum electrical potential V_{TN} as

$$\eta = \frac{V_{TN}}{V_m}, \tag{7.44}$$

where V_m is the measured cell potential. The cell efficiency decreases at higher current densities. A practical measure of efficiency is the energy required to produce 1 kg of hydrogen. The ideal efficiency is determined by the standard enthalpy of decomposition $-\Delta_f H^\ominus$ of 286 kJ/mol, which means that 286 kJ produce 0.002 kg of H_2, which is 143 MJ/kg H_2 or 39.7 kWh/kg H_2.

In fact, this voltage of −1.48 V, based on equilibrium electrode potentials, represents the minimum required (negative) voltage for electrolysis to occur. In practice, when a reasonable potential is applied to drive the electrolytic decomposition to completion, the process becomes kinetically controlled and several processes, such as ion conduction, mass transfer, electron and ion transfer at the electrodes and surface contact hindrance including bubble formation, cause irreversibilities, contributing to losses in cell efficiency through added resistances, and a numerically higher applied (negative) potential is required to overcome them: the so-called *overpotential*. Capacitive effects can also occur at the electrodes as charge builds up. When solution resistance dominates the kinetics, a linear increase in applied negative voltage results in a linear increase in current.

Because pure water is an effective insulator owing to its low autoionization ($K_w = 1 \times 10^{-14}$), it has a very low conductance. Thus, electrolysis proceeds very slowly unless a very large potential is applied to enhance the autoionization. Therefore, electrolytes are added to increase conductivity, prevent the buildup of charge by protons and hydroxyl ions at the electrodes and enable a good current to flow. There are a range of electrolyzer types depending on the electrolyte, electrode materials and operating temperature, higher temperatures enabling the required heat to be supplied externally rather than relying on the internal resistive joule heating. The main systems are alkaline water, operating at low temperatures (30–80°C); solid oxide, operating at high pressures and temperatures (500–850°C); microbial, using organic matter such as biomass and wastewater; and polymer exchange membranes (PEMs), similar to those used in fuel cells. A schematic diagram of a PEM water electrolyzer is given in Figure 7.6.

The membranes are commonly made of perfluorosulphonic acid polymers such as Nafion (Kim and Pivovar 2007; Kusoglu and Weber 2017). Key elements in cell design for

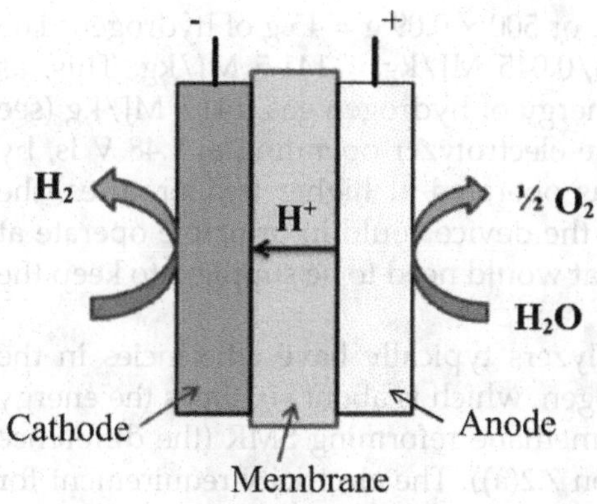

Anode: $H_2O \rightarrow 2H^+ + \frac{1}{2}O_2 + 2e^-$

Cathode: $2H^+ + 2e^- \rightarrow H_2$

Overall cell: $H_2O \rightarrow H_2 + \frac{1}{2}O_2$

FIGURE 7.6
Schematic diagram of polymer exchange membrane (PEM) water electrolyzer.

both technical performance and cost are the choice of electrodes and of electrocatalysts that coat the membrane and whose role is to promote the charge transfer kinetics and so decrease the activation energy by enhancing proton transport from the electrode to the membrane.

The electrode reactions for PEM electrolyzers are as set out earlier in this section. Let us examine the current and energy required to produce 500 L per hour of hydrogen at STP, starting with the reversible case i.e. efficiency $\eta = 1$ so that, from Equation 7.44, $V_m = V_{TN}$. Assuming hydrogen behaves as a perfect gas under the relatively low pressure conditions of a PEM cell, 500 L corresponds at STP to 500 mol/22.4 = 22.3 mol of hydrogen. The cell reaction tells us that hydrogen gas is produced at the cathode, requiring two electrons

$$\text{Cathode (reduction):} \quad 2\,H^+(aq) + 2e^- \rightarrow H_2(g). \tag{7.45}$$

The reaction therefore consumes 22.3 mol × 2 = 44.6 mol e^-, which corresponds to a charge of

$$Q = 44.6\,F = 44.6 \times 96{,}495\ C = 4.30 \times 10^6\,C. \tag{7.46}$$

The corresponding current i is

$$i = Q/t = 4.30 \times 10^6\,C/3600\,s = 1{,}195\ A. \tag{7.47}$$

Assuming that the cell is operated at the minimum required negative potential of −1.48 V, the power consumption P is

$$P = i\,V = 1{,}195 \times 1.48\ W = 1{,}769\ W. \tag{7.48}$$

The energy consumption is therefore

$$E = P\,t = 1.769\ kWh = 1.769 \times 3600\ kJ = 6{,}368\ kJ. \tag{7.49}$$

The energy used corresponds to producing 500 L or 500 × 0.09 g = 45 g of hydrogen. The energy consumption per kg is, therefore, 6,368/0.045 MJ/kg = 141.5 MJ/kg. This, as expected, is essentially identical to the specific energy of hydrogen gas, 141.8 MJ/kg (see Table 7.1). Under these reversible conditions, the electrolyzer operating at 1.48 V is, by definition, 100% efficient. If the electrolyzer was operated at higher temperatures, the minimum negative voltage would decrease and the device could in principle operate at more than 100% efficiency. However, external heat would need to be supplied to keep the system operating.

The best industrial PEM and alkaline electrolyzers typically have efficiencies in the range 70–80%, requiring about 190 MJ/kg hydrogen, which is about six times the energy requirement of forming hydrogen using steam methane reforming SMR (the difference coming from the energy of methane, see Question 7.2(a)). The electricity requirement for the electrolysis route is 50–55 kWh, corresponding to about $7/kg at 2020 mean global electricity prices, some two to three times the actual cost of hydrogen produced by SMR.

7.2.5 What Is the Carbon Footprint of Making Hydrogen by Water Electrolysis Using Electricity from Different Sources?

The carbon footprint of hydrogen production by electrolysis is very sensitive to the source of electricity used. Let us compare the greenhouse gas (GHG) emissions (expressed in kg CO_2 evolved in in producing 1 MWh of electricity) from (a) a pulverized coal power plant (PC), (b) an integrated gas combined cycle plant (IGCC), (c) a natural gas combined cycle plant (NGCC) and (d) an SMR hydrogen plant (in this case CO_2 emissions per kg of hydrogen produced). We can use the methods given in the answers to Question 6.3 to provide the CO_2 emissions from these different processes and also the additional electricity cost (or additional cost of hydrogen production in the case of SMR) arising from the CCS capture energy penalty, as set out in Table 7.2.

For an industrial electrolyzer requiring 190 MJ/kg of hydrogen produced or 190 × 1,000/3,600 = 52.8 kWh per kg H_2, then the carbon footprint in producing 1 kg of hydrogen by electrolysis using electricity from these three different types of power plants compared with that of steam methane reforming is set out in Table 7.3 (see Question 7.2.2).

We see from Table 7.3 that without capturing and storing (CCS) the CO_2 generated as a consequence of powering electrolysis or SMR hydrogen production processes, SMR has both the lowest carbon footprint (apart from the most efficient NGCC plants) and the lowest cost. The use of CCS technology has a dramatic effect in reducing the CO_2 emissions by an order of magnitude, not to zero but to low levels. However, the additional energy required in CCS to release the captured CO_2 from the capture solvent or adsorbate and then to compress it to supercritical conditions for injection into subsurface storage sites results in an increase in the cost of hydrogen production of order 50%, sometimes higher, depending on the precise CO_2 capture technology used (see Questions 6.3.4–6.3.6). Hydrogen from SMR is likely to remain the most effective "near-zero" emissions option as CCS becomes more effective and cheaper (so-called "blue hydrogen").

Truly "green hydrogen" requires the use of a carbon-free process, such as electrolysis powered by solar energy. Such a process is shown schematically in Figure 7.7, which illustrates a photovoltaic (PV) electrolysis system consisting of a multi-junction solar cell connected to two PEM electrolyzers in series. The general configuration for such a system is a sequence of electrolyzers powered by a bank of multi-junction solar panels. For simplicity, Figure 7.7 shows two PEM electrolyzers in series powered by a single triple-junction PV solar cell.

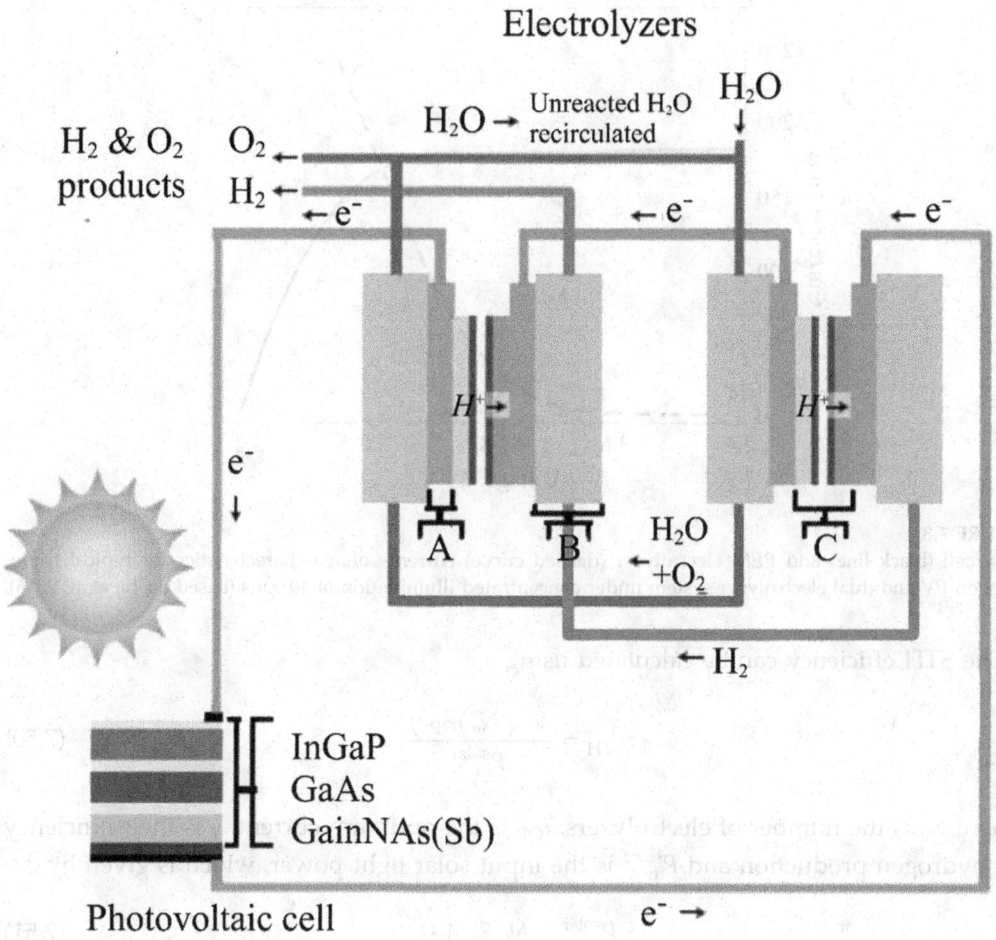

FIGURE 7.7
Schematic of a PV-electrolysis device consisting of PEM electrolyzers in series, powered by a multi-junction solar cell. A = current conductor; B = H_2 gas flow; C = PEM cell as in Figure 7.6.

A key factor in maximizing the solar-to-hydrogen (STH) efficiency of such a unit is to match the current-voltage (*i-V*) characteristics of the multi-junction PVs with those of the electrolyzers. We have seen that the thermodynamic minimum voltage V_{TN} required to electrolyze water is −1.48 V at 300 K, with practical operating voltages down to −1.9 V. However, the maximum power point voltage ($-V_{MPP}$) of typical commercial triple-junction solar cells under concentrated illumination (up to 1,000 times the sun's power incident on the Earth) is 2.0–3.5 V. Using a negative voltage for electrolysis in excess of $-V_{TN}$ results in energy wasted as heat and reduces the efficiency of the hydrogen production. The aim should be, therefore, to match V_{MPP} and V_{TN} as closely as possible.

This is illustrated in Figure 7.8, which gives the i-V characteristics of a typical dual electrolyzer and a triple-junction solar cell (see e.g. Jia et al. 2016). The intersection of these two curves gives the optimum operating voltage V_{OP} and current i_{OP} for the system, ensuring that the solar cell and the electrolyzer stack are well matched near the V_{MPP} of the solar cell.

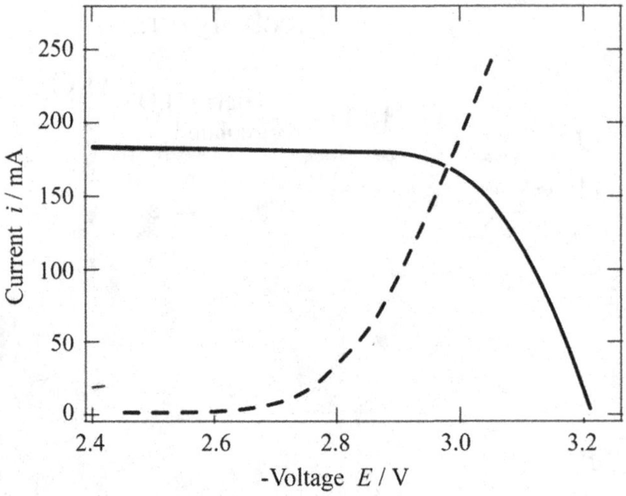

FIGURE 7.8
Solar cell (black line) and PEM electrolyzer (dashed curve) current-voltage characteristics for typical triple-junction PV and dual electrolyzer system under concentrated illumination of 40 suns (based on Jia et al. 2016).

The STH efficiency can be calculated using

$$\eta_{STH} = \frac{V_{TN}\, N_e\, i_{OP}\, \eta}{P_{in}^{solar}}, \qquad (7.50)$$

where N_e is the number of electrolyzers, i_{OP} is the optimum current, η is their efficiency for hydrogen production and P_{in}^{solar} is the input solar light power, which is given by

$$P_{in}^{solar} = N_s\, E_e\, A_{PV}. \qquad (7.51)$$

Here, N_s is the solar energy concentration factor (number of suns), E_e is the maximum irradiance (solar power incident on the Earth's surface, 100 mW cm^{-2}) and A_{PV} is the surface area of the solar cell. For a typical small-scale laboratory system such as that in Figure 7.7 (e.g. Jia et al. 2016), N_e might be 2, i_{OP} ~160 mA, $\eta \sim 1$, N_s ~40, E_e = 100 mW cm^{-2} and A_{PV} ~0.35 cm^2. Hence

$$\eta_{STH} = (1.48 \times 2 \times 160 \times 1)/(40 \times 100 \times 0.35) = 473.6/1400 = 0.34 \text{ or } 34\%. \qquad (7.52)$$

For a commercial electrolyzer operating with $\eta \sim 0.75$ and typically at 80°C where V_{TN} is –1.40V, the STH efficiency would be lower, ~25%. For comparison, the solar-to-electricity efficiency of the PV under these laboratory operating conditions is ~40%, slightly higher than the ideal STH efficiency.

So if a solar cell is linked directly to an electrolyzer system, the maximum conversion of incident solar energy to hydrogen as an energy carrier is about 35%, due mainly to the generation of heat in the PV. (For a detailed consideration of the thermodynamics of a solar cell, see for example Parrott 1992). Since solar input energy is essentially "free", the cost of solar-based electricity lies mainly in the nature and quantity of PV materials required for the energy conversion, and of subsequent storage in batteries if not used

TABLE 7.4

CO_2 emissions for different hydrogen production methods based on life cycle analysis (Agarwal et al. 2015)

Hydrogen Production Method	kg CO_2/kg H_2	mol CO_2/mol H_2
"Green" Hydrogen		
Biomass	2.9	0.13
Hydroelectric/electrolysis	2	0.09
Solar photovoltaic / electrolysis	6	0.27
Wind/electrolysis	1.5	0.07
Nuclear/high-temperature electrolysis	4	0.18
Nuclear/thermochemical	2.4	0.11
"Blue" Hydrogen		
Steam reforming + CCS	2.5	0.11
"Grey" Hydrogen		
Steam reforming without CCS	12	0.55

directly. Table 7.3 indicates that once the cost of solar-based electricity becomes comparable with the most efficient fossil-fuel power plants (NGCC), solar-electrolyzers will displace SMR with CCS as the preferred hydrogen production route on the basis of both lowest energy cost and lowest carbon footprint.

In fact, even for renewable sources of electricity, the carbon footprint is not zero as there are inevitably GHG emissions due to extraction of raw materials, construction of plant and manufacture of components such as PV cells. It is interesting to see that when a life cycle ("cradle-to-grave") analysis is carried out there is significant variability in CO_2 emissions even amongst so-called low emission routes, as illustrated in Table 7.4.

"Blue" hydrogen from SMR with CCS has a comparable carbon footprint to electrolysis using nuclear and renewable electricity. It is, therefore, likely to remain a competitive route for producing low-carbon hydrogen until there are major reductions in the cost of electrolyzers and renewable electricity.

7.3 Rather Than Bury CO_2 in CCS Processes, Why Not Convert the CO_2 to Useful Chemical Products and Fuels?

Whilst capturing CO_2 and storing it in geological reservoirs is effective as a means of preventing large volumes of CO_2 waste product entering the atmosphere and so decarbonizing the process that produces it, we have seen in Question 6.3 that the energy required to do this can be considerable. Except in the case where the captured CO_2 is used for enhanced oil recovery from a reservoir before it is eventually stored there, there is no saleable product from CCS processes to cover the cost. On the other hand, if it is possible to use CO_2 as a chemical feedstock and convert it to valuable materials or chemicals in which it remains sequestered, then it may be possible to prevent its release to the atmosphere while adding significant monetary value to a cheap waste by-product of hydrocarbon combustion. This entire process is called ***carbon dioxide capture and utilization*** (CCU).

The thermodynamic potential for such processes is illustrated in Figure 7.9.

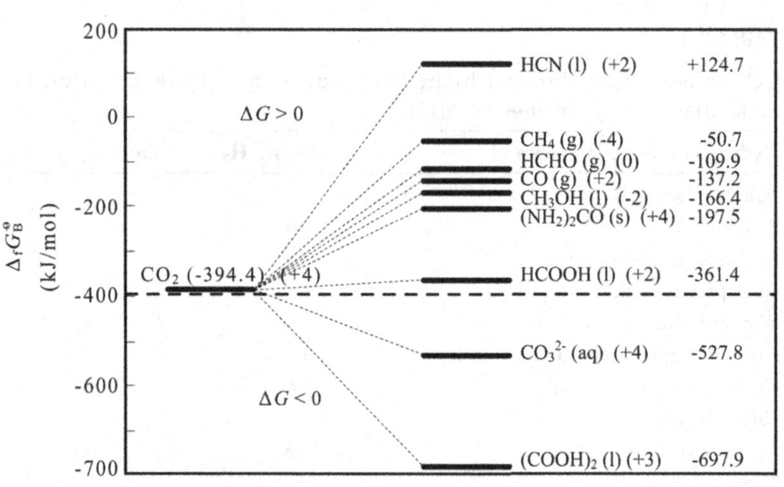

FIGURE 7.9
Gibbs free energy of formation of a range of C1 compounds compared with CO_2. The oxidation state of carbon in each compound is indicated in brackets after the chemical formula, e.g. (+4).

This figure shows that CO_2, with carbon in an oxidation state of +4, is one of the most thermodynamically stable carbon compounds. Only carbonates (inorganic and organic, oxidation state +4) and some carboxylic acids such as oxalic acid (oxidation state +3) have more negative values of $\Delta_f G^{\ominus}$ for which the conversion is either spontaneous or requires relatively small energy inputs to overcome the activation energy between CO_2 and the products (see Question 5.7.1, reactions of type A in Figure 5.5). Most other compounds have higher (less negative) values of $\Delta_f G^{\ominus}$ than CO_2 and conversion of CO_2 involves a reduction process to lower oxidation states, requiring the input of energy as well as in most cases the use of a catalyst to reduce the activation energy $\Delta_r G^*$ to reasonable values (reactions of types B or D in Figure 5.5). Note that while increasing the temperature will enhance the kinetics of the reaction, the rate being proportional to $\exp(-\Delta_r G^*/RT)$ (see Equation 5.179), if the reaction entropy $\Delta_r S^{\ominus}$ is negative, then a raised temperature will increase $\Delta_r G^{\ominus}$ ($=\Delta_f G^{\ominus}$ (products) $-$ $\Delta_f G^{\ominus}$ (CO_2)), which may cause it to become positive, and certainly reduce the equilibrium constant and hence the yield of product (see Question 5.3.2, the van't Hoff isotherm, and Question 5.3.3, Le Chatelier's principle).

Thermodynamics therefore enables us to determine which CCU reactions are energetically feasible, what the likely product yields are and how much energy must be expended to produce them. Energy is required not only because of the change in thermodynamic state between reactants and products but also to take the molecules involved into the higher energy non-equilibrium states that they must pass through in transitioning from reactants to products, as explained in more detail in Question 5.7.1.

7.3.1 How Easily Is CO_2 Converted into Solid Carbonates?

The conversion of CO_2 into solid carbonate materials is probably the most effective CCU process in that it is thermodynamically favored and, given the availability of large amounts of suitable natural (usually calcium or magnesium based) minerals, can convert large volumes of CO_2 into solid materials where it is permanently sequestered and which have significant commercial value, such as cements and building materials. Typical minerals used are wollastonite ($CaSiO_3$), olivine (Mg_2SiO_4), serpentine ($Mg_3Si_2O_5(OH)_4$) and brucite ($Mg(OH)_2$), which react with CO_2 to form carbonates such as calcite ($CaCO_3$), magnesite ($MgCO_3$) and dolomite ($CaMg(CO_3)_2$), often alongside quartz (SiO_2). Industrial by-products and waste materials such as fly-ash, steel slags, kiln dust and tailings which contain high concentrations of Ca and Mg can also be used. Mineral oxides like CaO exhibit the most exothermic behaviour. The relative free energies of formation for the relevant materials involved in these conversions are illustrated in Figure 7.10.

The main corresponding exothermic reactions are

$$CaSiO_3 + CO_2 \rightarrow CaCO_3 + SiO_2 \quad \Delta_r H^{\ominus} = -90\ \text{kJ/mol}, \tag{7.53}$$

$$Mg_2SiO_3 + CO_2 \rightarrow 2MgCO_3 + SiO_2 \quad \Delta_r H^{\ominus} = -89\ \text{kJ/mol}, \tag{7.54}$$

$$Mg_3Si_2O_5(OH)_4 + CO_2 \rightarrow 2MgCO_3 + 2SiO_2 + 2H_2O \quad \Delta_r H^{\ominus} = -64\ \text{kJ/mol}, \tag{7.55}$$

$$CaO + CO_2 \rightarrow CaCO_3 \quad \Delta_r H^{\ominus} = -178\ \text{kJ/mol}. \tag{7.56}$$

These reactions will take place spontaneously at ambient conditions, all having negative values of $\Delta_f G^{\ominus}$; the last one is exploited in the calcium-looping CO_2 capture process, as described in Question 6.3.5. Mineralization can take place *ex situ*, where the extracted or waste solid reactants are contacted with a CO_2-rich liquid or gas stream in a reactor, or *in situ*, where CO_2 is pumped into suitable rock formations at depth (basalt, typically 10% Ca and Mg as $CaMgSi_2O_6$, is particularly effective here). Despite the strong thermodynamic driving force (equivalent to equilibrium constants far greater than 1), reaction of CO_2 with solid oxides or minerals is generally slow because of the growth of layers of carbonate on the mineral surface, through which the CO_2 gas must then diffuse to access more unreacted mineral. Fine grinding or milling of the solid minerals to increase

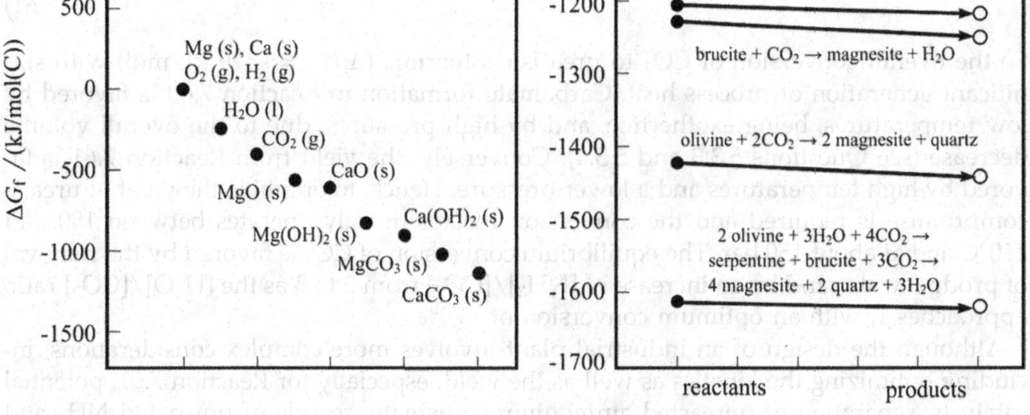

FIGURE 7.10
Standard Gibbs energies of formation and reaction for compounds involved in typical mineral carbonation processes.

reaction surface area or heating using the heat released in the reactions are ways to accelerate the mineralization or improve product yields. In fact, maximum yields can be obtained by dispersing the fine minerals in water, often with addition of weak acids or alkalis, where on dissolution Ca^{2+} and Mg^{2+} ions are released and react with the CO_3^{2-} ions present due to the dissolution of CO_2 to form carbonic acid:

$$CO_2 + H_2O \rightleftharpoons HCO_3^- + H^+ \quad (7.57)$$

$$HCO_3^- \rightleftharpoons CO_3^{2-} + H^+. \quad (7.58)$$

Another way to exploit the free energy decrease of the mineralization process is to generate electricity in a fuel cell that uses a cation exchange membrane to separate the reactants and products of the mineralization reaction

$$Ca(OH)_2 + 2CO_2 + 2NaCO_3 \rightleftharpoons 2NaHCO_3 + CaCO_3 \quad \Delta_r G^\ominus = -89.26\ \mathrm{kJ/mol}, \quad (7.59)$$

while enabling the exchange of Ca^{2+} ions between the anode and cathode compartments. Under standard conditions when the dissolved CO_2 and salt are saturated, the cell voltage is 0.68 V and the maximum electrical energy produced per tonne of CO_2 consumed is 415 kWh.

7.3.2 How Easily Is CO_2 Converted to Urea?

The other important compound whose formation from CO_2 is thermodynamically favorable is urea. This is usually produced from ammonia via a two-step process: the rapid formation of ammonium carbamate in an exothermic reaction that is highly exothermic

$$2NH_3(g) + CO_2(g) \rightleftharpoons H_2N\text{–}COONH_4(s)\ \Delta_r G^\ominus = -23.8\ \mathrm{kJ/mol},\ \ \Delta_r H^\ominus = -151\ \mathrm{kJ/mol}, \quad (7.60)$$

followed by the slow, endothermic dehydration of the carbamate to urea

$$H_2N\text{–}COONH_4(s) \rightleftharpoons (NH_2)_2CO(g) + H_2O(g)\ \Delta_r G^\ominus = 16.7\ \mathrm{kJ/mol},\ \ \Delta_r H^\ominus = 32\ \mathrm{kJ/mol}. \quad (7.61)$$

So the overall conversion of CO_2 to urea is exothermic ($\Delta_r H^\ominus = -129$ kJ/mol) with significant generation of process heat. Carbamate formation in Reaction 7.60 is favored by low temperatures, being exothermic, and by high pressures, due to the overall volume decrease (see Questions 5.3.3 and 5.3.4). Conversely, the yield from Reaction 7.61 is favored by high temperatures and a lower pressure. Hence, to optimize the yield of urea, a compromise is required and the conversion reactor usually operates between 180 and 210°C and at about 150 bar. The equilibrium conversion of CO_2 is favored by the removal of product water and by an increase of $[NH_3]/[CO_2]$ from 2 to 3 as the $[H_2O]/[CO_2]$ ratio approaches 1, with an optimum conversion of ~70%.

Although the design of an industrial plant involves more complex considerations, including optimizing the kinetics as well as the yield, especially for Reaction 7.61, potential catalysts, separation of unreacted ammonium carbamate, recycle of unreacted NH_3 and CO_2 and optimizing heat recovery, this simple thermodynamic analysis rationalizes the overall features of the process. Urea is one of the chemicals produced in the highest quantity (~180 m tons per annum). It is widely used as a fertilizer and also to form urea-formaldehyde resins, thermoset polymers that are widely used for decorative laminates, hard casings, textiles, paper, wood glue and foams (The Chemical Company 2021; Munoz and Fernandez, 2012).

7.3.3 How Can CO_2 Be Reduced to Create the Syngas Building Blocks for Synthesizing Other Chemicals and Fuels?

As shown in Figure 7.9, the bigger the decrease in the oxidation state of carbon from its maximum value of 4 in CO_2, the larger the input of energy required to convert it to other substances. In fact, most of the conversions involve the use of hydrogen as a reductant, and so require a source of hydrogen. The energy requirements and carbon footprint for different routes to hydrogen can be found in Question 7.2.5. Given the context of utilizing captured waste CO_2 to produce valuable chemicals, it makes sense to use "renewable" or "green" hydrogen, produced with the lowest carbon footprint, as we have seen in Question 7.2. This is likely to come from water electrolysis using electricity from low-carbon emission or renewable sources, although as we have seen in Table 7.4, there is significant variability in CO_2 emissions amongst these, and hydrogen from SMR with CCS has a comparable carbon footprint with significant cost advantages.

We will now use thermodynamics to look at which chemicals and fuels it is viable to produce from CO_2. One route for producing other organic chemicals is to use the *reverse water-gas shift reaction* to produce carbon monoxide

$$CO_2 + H_2 \rightleftharpoons CO + H_2O \quad \Delta_r G^\ominus = 28.6\ \text{kJ/mol}, \quad \Delta_r H^\ominus = -41.1\ \text{kJ/mol}. \quad (7.62)$$

There is, therefore, an energy penalty for this transformation and also a significant activation energy (Reaction 7.62 is of type D in Figure 5.9) so that a catalyst (typically Cu/alumina) and high temperatures are required to give significant rates and yields of CO. The equilibrium can be moved in either direction using pressure and temperature adjustments so that the H_2/CO ratio can be adjusted to match the requirements of particular syntheses, such as the Fischer-Tropsch process (Mahmoudi et al. 2017) for producing long-chain hydrocarbons, synthetic petroleum and diesel, or the production of ammonia and methanol.

An alternative process is to use CO_2 itself for the production of *synthesis gas* (H_2 + CO, commonly called "*Syngas*") or hydrogen by reaction with natural gas or other fossil fuel feedstocks. In contrast to steam methane reforming (SMR), considered in detail in Question 7.2, no water is involved in this dry methane reforming (DMR) with CO_2

$$CO_2 + CH_4 \rightleftharpoons 2CO + 2H_2 \quad \Delta_r G^\ominus = 41.7\ \text{kJ/mol}, \quad \Delta_r H^\ominus = 247\ \text{kJ/mol}. \quad (7.63)$$

The reaction is highly endothermic and also strongly endergonic (positive $\Delta_r G^\ominus$, so highly non-spontaneous) under standard conditions. Since for this reaction (Arona and Prasad 2016)

$$\Delta_r G^\ominus = (61{,}770 - 67{,}32\ T/\text{K})\ \text{kJ/mol}, \quad (7.64)$$

$\Delta_r G^\ominus$ does not decrease to zero until T = 917.5 K or 644.3°C. So very high reaction temperatures are required to give significant yields of syngas and since the reaction is of type D in Figure 5.9, catalysts (usually transition metals, originally Ni/Co, more recently Ru/Pt/Ru) are required to give viable reaction rates.

Thermodynamics also enables us to explore alternative routes to syngas and tune the $[H_2]/[CO]$ ratio so that it is a suitable for different follow-on syntheses. Whereas for SMR, this ratio is 3:1 (Reaction 7.1) and for DMR above it is 1:1, combining these into a bi-reforming process

$$CO_2 + 2H_2O + 3CH_4 \rightleftharpoons 4CO + 8H_2 \quad \Delta_r G^\ominus = 42.0\ \text{kJ/mol}, \quad \Delta_r H^\ominus = 659\ \text{kJ/mol}, \tag{7.65}$$

gives a ratio of 2:1, which is the ratio required for methanol synthesis (see Question 7.3.4) and is also better suited to the Fischer-Tropsch (FT) process (see Question 7.3.4). The presence of steam reduces the tendency to form catalyst deactivating coke.

Another very effective way to produce syngas combines the exothermic partial oxidation of methane (POM):

$$CH_4 + 0.5O_2 \rightarrow CO + 2H_2 \quad \Delta_r H^\ominus = -35.6\ \text{kJ/mol}, \tag{7.66}$$

with SMR and DMR in a single reactor, almost eliminating carbon formation and increasing catalyst lifetime and process efficiency:

$$20CH_4 + 9O_2 + CO_2 + H_2O \rightarrow 21CO + 41H_2 \quad \Delta_r H^\ominus = 12.9\ \text{kJ/mol}. \tag{7.67}$$

This process is mildly endothermic, so it does require some energy input but has greater flexibility in tuning the feed gas composition to vary the H_2/CO ratio in the product gas within the range 2–3, preferred for syngas conversion to higher hydrocarbons via the FT process.

7.3.4 Which Conversions of Syngas to Organic Chemicals and Fuels Make Thermodynamic Sense?

Synthesis gas with a $[H_2]/[CO]$ ratio of 2:1 can be converted to methanol, produced in high volume as a route to many organic molecules and also as a potential alternative fuel:

$$CO + 2H_2 \rightleftharpoons CH_3OH \quad \Delta_r G^\ominus = -25.1\ \text{kJ/mol}, \quad \Delta_r H^\ominus = -90.7\ \text{kJ/mol}. \tag{7.68}$$

This reaction is strongly exothermic and with its strong thermodynamic driving force (when $\Delta_r G^\ominus$ is negative the reaction is said to be *exergonic*), the equilibrium under most conditions lies strongly to the right. Starting from CO_2 requires the reverse water gas shift reaction to take place (Equation 7.62) to convert it to CO with a free energy penalty of 28.6 kJ/mol; adding this to Reaction 7.68 gives the direct conversion reaction

$$CO_2 + 3H_2 \rightleftharpoons CH_3OH + H_2O \quad \Delta_r G^\ominus = 3.5\ \text{kJ/mol}, \quad \Delta_r H^\ominus = -49.6\ \text{kJ/mol}. \tag{7.69}$$

Both reactions are favored by high pressures (reduction in moles of gas on reaction) and by lower temperatures. However, since there is a significant activation energy for both (Reaction 7.68 is type B in Figure 5.9, Reaction 7.69 type D), catalysts are used (typically Cu/ZnO), but the kinetics of reaction still requires temperatures in excess of 200°C for reasonable rates, meaning that the yield of methanol will be a compromise between thermodynamic conversion and rate. Reaction 7.69 using CO_2 captured from flue gas and green hydrogen produced by electrolysis using low-carbon (e.g. wind-generated) electricity has the potential to produce methanol which is essentially carbon neutral when used as a fuel in an internal combustion engine or when used in a direct methanol fuel cell. It can also be used for biodiesel manufacture and as a blended fuel with gasoline.

Methanol from CO_2 or syngas can be converted to olefins and higher hydrocarbons, including synthetic gasoline. A useful fuel can be obtained by the dehydration of methanol or directly from CO_2 and hydrogen – dimethyl ether, DME:

$$2CH_3OH \rightleftharpoons CH_3OCH_3 + H_2O \quad \Delta_rG^\ominus = -24.6\ \mathrm{kJ/mol}, \quad \Delta_rH^\ominus = -23.4\ \mathrm{kJ/mol}, \tag{7.70}$$

$$2CO_2 + 6H_2 \rightleftharpoons CH_3OCH_3 + 2H_2O \quad \Delta_rG^\ominus = -17.6\ \mathrm{kJ/mol}, \quad \Delta_rH^\ominus = -122.6\ \mathrm{kJ/mol}. \tag{7.71}$$

Reactions 7.70 and 7.71 can take place in parallel during the hydrogenation of CO_2 and the selectivity for methanol or DME adjusted by changes in catalyst, temperature, pressure and H_2/CO feed ratio. DME can be used as a replacement fuel for liquid petroleum gas (LPG), having similar physical properties, can be produced by low carbon footprint routes as above and has the additional advantages of being non-toxic and non-carcinogenic.

The most widely used process for production of long-chain alkanes from CO_2 once it has been transformed by the reverse water gas shift reaction, Equation 7.62, to syngas is the *Fischer-Tropsch process*, invented by Franz Fischer and Hans Tropsch in 1925. It involves a series of reactions that produce a distribution of mainly linear alkanes:

$$(2n+1)H_2 + nCO \rightarrow C_nH_{2n+2} + nH_2O \text{ where } n \text{ is typically 10–20}, \quad \Delta_rH^\ominus = -165\ \mathrm{kJ/mol}. \tag{7.72}$$

The Gibbs energy of reaction varies with chain length n as (Müller et al. 2014)

$$\Delta_rG^\ominus = -54.1\,n - 52.8\ \mathrm{kJ/mol}. \tag{7.73}$$

So the thermodynamic driving force per carbon atom decreases slightly with increasing chain length but reaches a limiting value of –54.1 kJ/mol carbon. Typical reaction temperatures are 150–350°C, with pressures up to about 30 bar (3 MPa), using transition metal catalysts, typically based on iron, cobalt or ruthenium. As expected for reactions that are exothermic and lead to a decrease in the number of moles of gas on reaction, higher conversions and longer chain lengths are achieved at lower temperatures and higher pressures. A large variety of products can be produced by varying the temperature, the catalyst and the $[H_2]/[CO]$ ratio by the methods described previously. Typical products include synthetic diesel (alkanes with n = 10–20), gasoline or petroleum (alkanes and cycloalkanes with n = 4–12) and paraffin wax ($n \sim$ 20–40).

7.3.5 How Can CO_2 Be Converted to Polymers and Be Sequestered in Solid Plastic Materials?

A major use of CO_2 which can result in its long-term sequestration in long-life articles manufactured from polymeric materials is its co-polymerization with epoxides to produce *polycarbonate polymers* having a range of structures and functionalities, as illustrated in Figure 7.11.

The formation of polymer competes with the formation of the cyclic carbonate as a side product. Studies of a range of epoxide monomers indicate that all polycarbonates have essentially the same enthalpies of formation, in the range 88–96 kJ/(mol monomer

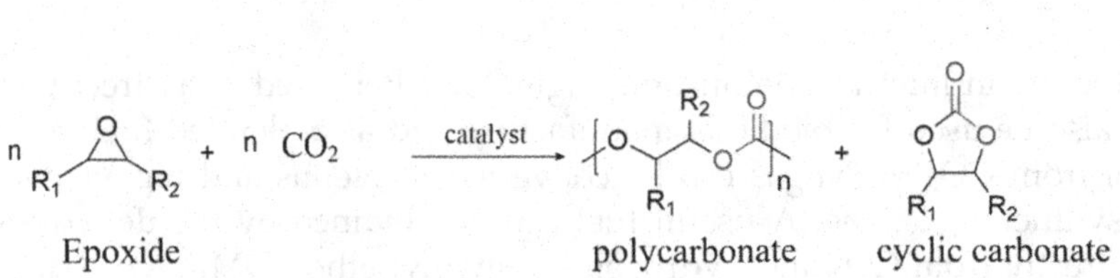

FIGURE 7.11
Coupling of CO_2 with substituted epoxides to form polycarbonate copolymers along with cyclic carbonate side product.

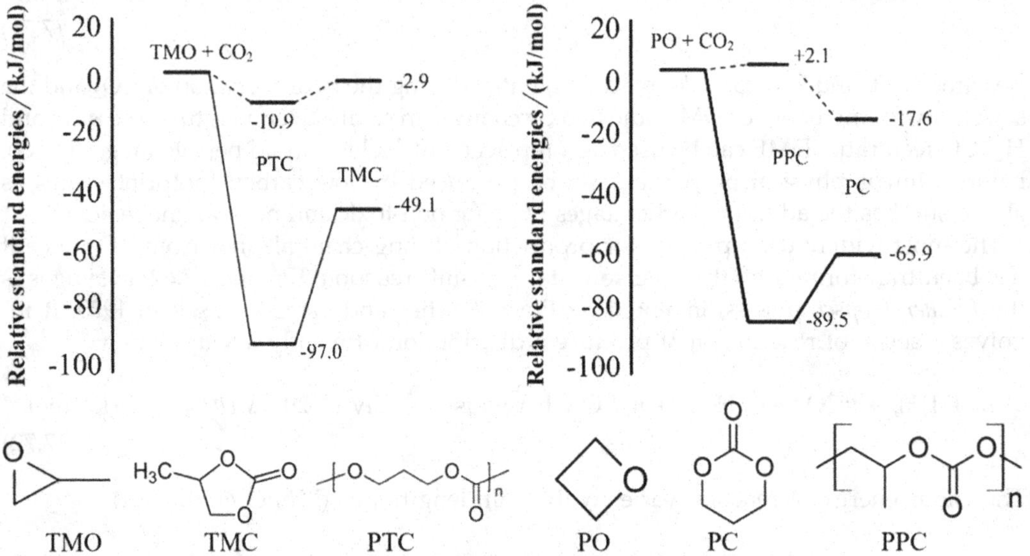

FIGURE 7.12
Enthalpies (joined by solid lines) and Gibbs energies (joined by dotted lines) for the conversion of (a) trimethylene oxide (TMO) to poly(trimethylene carbonate) (PTC) or cyclic trimethylene carbonate (TMC), and (b) propylene oxide (PO) to polypropylene carbonate (PPC) or cyclic propylene carbonate (PC) (Darensbourg and Yeung 2013).

added). The carbonate formation reaction is also exothermic but with enthalpies lower by 20–35 kJ/mol. Considering Gibbs energy changes, for trimethylene oxide (TMO or oxetane; see Figure 7.12) the formation of the linear polymer poly(trimethylene carbonate) (PTC) is exergonic, with a Gibbs energy of −10.9 kJ/(mol monomer added) compared to −2.9 kJ/mol for formation of the cyclic carbonate. So linear polymer formation is highly favored over the carbonate; indeed, the cyclic trimethylene carbonate TMC polymerizes via a ring opening reaction on heating, usually with a catalyst, to the same PTC, a biodegradable polymer with biomedical applications. Typical catalysts are aromatic organometallic complexes of transition metals such as Co or Cr. For TMO, essentially quantitative conversion to PTC polymer can be achieved using heat and a catalyst.

By contrast, for another epoxide, propylene oxide (PO), while the enthalpy of polymerization with CO_2 is still much more exothermic (−89.1 kJ/mol) than formation of the cyclic carbonate (−65.7 kJ/mol), the Gibbs energy is slightly endergonic (with a positive reaction Gibbs energy $\Delta_r G^\ominus$, +2.1 kJ/mol) while the carbonate formation reaction is exergonic (−17.6 kJ/mol). So the ring opening polymerization is even more endergonic at +19.7 kJ/mol and in this case thermodynamics favors the formation of the carbonate rather the polymer. In this case, the carbonate formation is favored because of the larger entropy decrease on formation of the polymer.

In general, for CO_2 reacting with epoxides, the thermodynamic stability of the polymer compared to the cyclic carbonate will depend on the nature of the epoxide ring structure and the substituent groups R1, R2 etc. and on the activation barrier for the growing polymer chain to extend rather than for it to fold back and break off as a carbonate fragment. ΔG^* will be determined both by the structure of the epoxide (how bulky the constituent groups are, how strained the carbonate ring is) and by the use of blocking groups on the organometallic catalyst. For these polymerization reactions, the amount of carbonate by-product produced can therefore be controlled by the temperature (makes both $\Delta_r G$ and $\exp(-\Delta G^*/RT$ lower) and the design of the catalyst (changing ΔG^*(polymerization) relative to ΔG^*(carbonate formation)) (Darensbourg and Yeung 2013).

7.3.6 Is the Direct Conversion of Captured CO_2 to Useful Chemicals Thermodynamically Viable?

There are many potential chemical transformations for converting CO_2 to chemicals, fuels and intermediates for other chemical syntheses, but because of the high oxidation state and low Gibbs energy of CO_2 compared with most C-containing compounds, most of these are endergonic and endothermic and require significant input of energy. This does not rule them out as feasible and examples of widely used compounds where the energy requirements are modest and viable commercial processes exist include:

- *Salicylic acid* $C_6H_5(OH)COOH$ (a precursor in the synthesis of aspirin, acetylsalicylic acid) by alkaline carboxylation of phenol (Stanescu et al. 2006; Müller et al. 2014);
- *Oxalic acid* $(COOH)_2$ (one of the few carbon compounds which is more thermodynamically stable than CO_2; see Figure 7.9) which requires two molecules of CO_2 to react to form a C-C bond via a reaction that is both endothermic and enderogonic but for which a viable electrochemical process exists (Fischer et al. 1981);
- *Formic acid* HCOOH, a pre-cursor for a wide range of useful chemicals, which forms easily at relatively mild conditions ($t \sim 70°C$, $p \sim 5$ MPa), sometimes aided by using water-soluble Ru or Rh organometallic catalysts, at pH ~ 8 in aqueous solution, where CO_2 dissolves to form bicarbonate ion

$$HCO_3^- + H_2 \rightleftharpoons HCOO^- + H_2O \quad \Delta_r G^\ominus = -35\ \text{kJ/mol.} \tag{7.74}$$

7.4 Does Direct Air Capture Make Thermodynamic Sense?

Direct air capture (DAC) is the process of removing CO_2 from the air and generating a concentrated stream of CO_2 that may then be stored (as in Question 6.3) or utilized (as in Question 7.3). Whereas CCS is prevention of CO_2 entering the atmosphere, DAC is a cure – extracting it back, once it has been released. If the CO_2 captured in this way is then stored, it can be deemed to be permanently removed from the atmosphere, a condition termed "negative emissions." In this regard, it has a similar end point to processes that involve bioremoval of CO_2 from the air via photosynthesis, such as bioenergy with CCS (BECCS) (Daggash et al. 2020) or reforestation (Nolan et al. 2021). A typical hardwood tree absorbs up to 22 kg per year from the atmosphere, removing a tonne over a 45-year period. Indeed, DAC devices have been referred to as "artificial trees." Removing CO_2 from air at ~400 ppm and concentrating it to >90% purity requires both more energy and much larger treated gas volumes than for CO_2 capture from concentrated point sources as described in Question 6.3. Here, we examine the energy requirements of such DAC processes.

7.4.1 What Is the Thermodynamic Minimum Energy Requirement of Separating CO_2 from the Other Components of Air?

The simplest thermodynamic analysis is to examine the Gibbs energy change required to separate the CO_2 in air from the other components, irrespective of the particular process used to carry out the separation. For the CO_2 in air at a partial pressure p_{CO_2}, assuming ideal gas behavior, the chemical potential is (see Equation 4.56)

$$\mu_{CO_2}(\text{air}) = \mu^{\ominus}_{CO_2} + RT \ln\left(\frac{p_{CO_2}}{p_0}\right), \tag{7.75}$$

where p_0 is the total air pressure. Direct air capture moves from this initial condition to a final state of pure CO_2 for which

$$\mu_{CO_2}(\text{pure, separated}) = \mu^{\ominus}_{CO_2}. \tag{7.76}$$

So, the overall Gibbs energy change is

$$\Delta G = \mu_{CO_2}(\text{DAC}) = \mu^{\ominus}_{CO_2} - \mu^{\ominus}_{CO_2} - RT \ln\left(\frac{p_{CO_2}}{p_0}\right) = RT \ln\left(\frac{p_0}{p_{CO_2}}\right). \tag{7.77}$$

Equation 7.77 is equivalent to Equation 3.193 for the minimum work required to separate air into its constituents, simplified to apply to the extraction of 1 mol of CO_2. In Question 3.9, this work was obtained from (minus) the exergy loss on mixing the air components together; we see here that this is equivalent to the Gibbs energy of demixing. It enables us to calculate the thermodynamic minimum energy required to extract 1 mole of CO_2 from the air at 100 kPa (1 bar), 25°C, at early twenty-first-century atmospheric levels of 400 ppm (p_{CO2} = 0.04 kPa). We can use the same equation to calculate the minimum energy required to capture CO_2 from flue gas emitted by a natural gas power plant (p_{CO2} ~ 5.0 kPa) and a coal power plant (p_{CO2} ~ 12.0 kPa), both at typical conditions of 0.1 MPa and 65 °C. The results are shown in Table 7.5, along with typical actual energy requirements, taking into account irreversibilities, losses and other power requirements of actual carbon capture processes (see Question 7.4.3 for realistic DAC processes and Question 6.3 for power plant carbon capture).

Table 7.5 indicates that because of the much lower CO_2 concentration in air than in a typical process flue gas (by a factor of 100–300), the energy requirement to extract 1 mole of CO_2 will inevitably be higher, by a factor of 2–4, depending on the precise processes involved. However, because of the lower thermal energy requirements (some DAC processes can be carried out at ambient conditions), it may be possible to design processes with a higher thermodynamic efficiency.

TABLE 7.5
Thermodynamic efficiencies of direct air capture compared to amine solvent carbon capture from power plants

Energy Requirement/(kJ/mol)	DAC	Natural Gas Power CC	Coal Power CC
Thermodynamic limit	19.4	8.4	5.3
Typical plant	330*	180	180
Efficiency (%)	5.8	4.7	2.9

* The DAC plant figure is less certain than the power plant data due to less deployment at scale to validate the calculated estimates of Question 7.4.3.

7.4.2 How Do the Conditions for Direct Air Capture Affect the Energy Requirements?

Because of the low concentration of CO_2 in the atmosphere, and the targets of >90% capture efficiency and high CO_2 purity (>95%) in the extracted stream, similar to those of typical power plant and industrial CCS processes, carbon capture may not be energetically viable for DAC. Besides, this low air concentration means that much larger volumes of air need to be treated to extract useful quantities of CO_2, which in turn means that the plant size, and hence process capital cost, will be higher as well. We can carry out a thermodynamic analysis to address these issues, to assess the compromises that might be made between plant size (capital costs), energy requirements (operating costs) and product quality.

Figure 7.13 show a schematic idealized DAC process to extract 1 mole of CO_2 from ambient air.

We can specify the fractional recovery of CO_2 in stream 3 as ζ and the purity of that stream as χ (=x_{CO_2}). Since the process recovers 1 mole of CO_2 in stream 3 (N_3) and stream 1 contains N_1x_1 moles of CO_2

$$\zeta = \frac{1}{N_1\, x_1}, \tag{7.78}$$

and since stream 3 contains 1 mole of CO_2, then

$$N_3 = \frac{1}{\chi}. \tag{7.79}$$

We take the same approach as in Question 7.4.1 to calculate the thermodynamic minimum work of separation, but this time we include the other components of air, mainly nitrogen

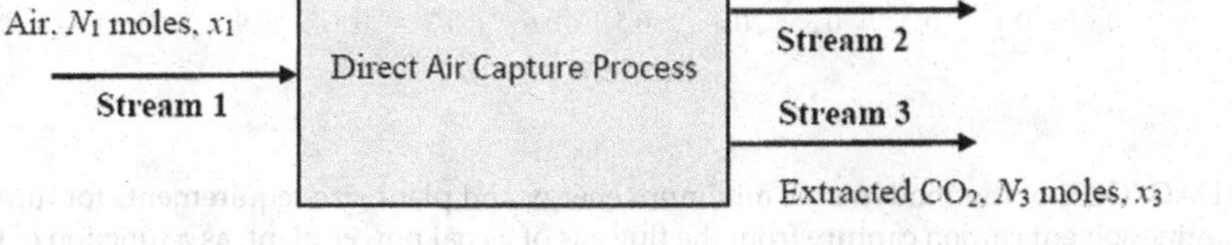

FIGURE 7.13
Schematic diagram of ideal DAC process. N_i = number of moles in stream *i*; x_i = CO_2 mole fraction in stream *i*.

and oxygen, and assume they behave as a single pseudo component. For each stream i with components j we have

$$G_i = \sum_j N_{i,j}\mu_{i,j}, \tag{7.80}$$

with, assuming the system behaves as an ideal gas mixture (see Question 4.3.4),

$$\mu_{ij} = \mu_{ij}^{\ominus}(T, p) + RT\ln(x_{ij}). \tag{7.81}$$

For an isothermal, isobaric separation at 298 K, 0.1 MPa, then the Gibbs energy of separation is the difference ΔG between the product streams 2 and 3 and the feed air stream 1

$$\frac{\Delta G}{RT} = N_2[x_2\ln x_2 + (1-x_2)\ln(1-x_2)] + N_3[x_3\ln x_3 + (1-x_3)\ln(1-x_3)] + N_1[x_1\ln x_1 + (1-x_1)\ln(1-x_1)]. \tag{7.82}$$

Since $N_1 = N_2 + N_3$, N_1 is set by Equation 7.78 and N_3 by Equation 7.79 and the mole balance on CO_2 is

$$N_1x_1 = N_2x_2 + N_3x_3 = N_2x_2 + 1, \tag{7.83}$$

then all the N_i and x_i values are specified and so Equation 7.82 can be evaluated for any set of initial conditions for stream 1. This enables us to compare DAC, where for 400 ppm CO_2 in air $x_1 = 0.0004$, with other CO_2 removal processes. Figure 7.14 compares the minimum energy requirements for this DAC with those for carbon capture from a coal-fired power

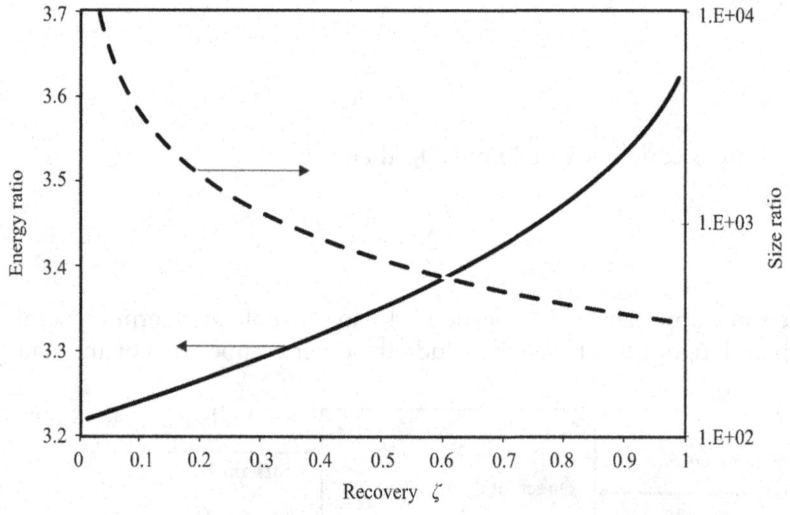

FIGURE 7.14
Comparison (DAC/CCS) of thermodynamic minimum energy and plant size requirements for direct air capture compared to amine solvent carbon capture from the flue gas of a coal power plant, as a function of CO_2 recovery ζ (based on Brandani 2012).

plant with a flue gas composition of 12% CO_2 i.e. $x_1 = 0.12$ for a situation where it has been assumed that for both processes the CO_2 purity χ is 0.95.

It can be seen that as the CO_2 recovery approaches 1, the ratio of energy requirement of the two processes is in line, as expected, with that given by the simpler analysis of Question 7.4.1, and amounts to 3.66. With decreasing recovery of CO_2 the DAC energy requirement decreases but the ratio is always >3. Assuming that the major factor determining plant size is the amount of gas treated ($N_1 = 1/(\zeta x_1)$ from Equation 7.78), then the relative size of separators for the two processes increases linearly from 300 for $\zeta = 1$ to 6,000 for $\zeta = 0.05$, as also shown in Figure 7.14. There is clearly a balance to be struck between energy requirements (operating costs) and plant size (capital cost), with an optimum in the region of $\zeta = 0.5$–0.6, but DAC will inevitably be more expensive and have a larger footprint than a post-combustion carbon capture plant. In the context of reducing CO_2 levels in the atmosphere, DAC only makes sense if low-carbon renewable energy is used to power the separation and it is used as a negative emissions remedial technology to complement carbon capture (emissions prevention) from large centralized power and industrial plants.

To give further insight into the thermodynamic requirements of the full separation process, it is useful to re-cast Equation 7.82 in terms of the total pressure in stream 1, p_0, and the partial pressures of CO_2 in the stream 1 air, p_1, and the depleted air stream 2, p_2. Using the CO_2 mole balance in Equation 7.83, we have

$$N_1 x_1 = N_2 x_2 + 1. \tag{7.84}$$

So assuming ideal gas behavior we have

$$N_1 (p_1/p_0) = N_2 (p_2/p_0) + 1 \tag{7.85}$$

or

$$N_1 p_1 - p_0 = N_2 p_2 = (N_1 - 1) p_2, \tag{7.86}$$

hence

$$N_1 = (p_0 - p_2)/(p_1 - p_2) \tag{7.87}$$

and

$$N_2 = 1 - N_1 = (p_0 - p_1)/(p_1 - p_2). \tag{7.88}$$

We can therefore re-write Equation 7.82, recognizing that $N_3 = x_3 = 1$, as

$$\begin{aligned}\frac{\Delta G}{RT} = \frac{p_0 - p_1}{p_1 - p_2}\left(\frac{p_2}{p_0}\ln\left(\frac{p_2}{p_0}\right) + \frac{p_0 - p_2}{p_0}\ln\left(\frac{p_0 - p_2}{p_0}\right)\right) \\ - \frac{p_0 - p_2}{p_1 - p_2}\left(\frac{p_1}{p_0}\ln\left(\frac{p_1}{p_0}\right) + \frac{p_0 - p_1}{p_0}\ln\left(\frac{p_0 - p_1}{p_0}\right)\right),\end{aligned} \tag{7.89}$$

which simplifies to

$$-\frac{\Delta G}{RT} = \ln\left(\frac{p_1}{p_0}\right) + \left(1 - \left(\frac{p_1}{p_0}\right)\right)\frac{p_2}{p_1 - p_2}\ln\left(\frac{p_1}{p_2}\right)$$
$$+\left(1 - \left(\frac{p_1}{p_0}\right)\right)\left(1 - \left(\frac{p_2}{p_0}\right)\right)\frac{p_0}{p_1 - p_2}\ln\left(\frac{p_0 - p_1}{p_0 - p_2}\right). \tag{7.90}$$

This form for the Gibbs energy change for the DAC process makes it easy to track the CO_2 through the process and shows that of the n_1 moles of air in the input air stream 1

- 1 mol CO_2 is separated and compressed from p_1 to p_0
- the uncaptured fraction of the CO_2 in the air stream, $\{1 - (p_1/p_0)\}\{p_2/(p_1 - p_2)\}$ moles, expands from a partial pressure p_1 in stream 1 to p_2 in stream 2
- the remainder of the air stream (O_2, N_2 = $\{1 - (p_1/p_0)\}\{1 - (p_2/p_0)\}\{p_0/(p_1 - p_2)\}$ moles) is compressed from a partial pressure of $p_0 - p_1$ in stream 1 to $p_0 - p_2$ in stream 2

In the limit that $p_2 \to 0$ so that all the CO_2 in the air is captured, then Equation 7.90 reduces to

$$\Delta G = -RT\left(\ln\left(\frac{p_1}{p_0}\right) + \left(1 - \frac{p_1}{p_0}\right)\right). \tag{7.91}$$

Since at 400 ppm $p_1 \ll p_0$, the second term is negligible and so Equation 7.91 becomes identical to Equation 7.77 in Question 7.4.1. We saw in Table 7.5 that the DAC minimum energy requirement for separation based on this is about 20 kJ/mol. This means that if we use an absorption or adsorption process to separate CO_2 from an air stream, then the Gibbs energy of absorption/adsorption must be at least 20 kJ/mol. We will now examine some practical DAC processes and compare their energy consumption with this thermodynamic limit.

7.4.3 Can We Adapt Solvent-Based Carbon Capture Approaches to Direct Air Capture?

One approach to DAC is to use a wet scrubbing technique to absorb acidic CO_2 in an aqueous alkaline solvent. Such a process is illustrated schematically in Figure 7.15.

This process has similarities with amine-based solvent and solid calcium looping methods for centralized carbon capture from industrial processes (see Questions 6.3.2 and 6.3.5) and combines features of both. As with both these methods, it is a "thermal swing" process, using heating to recover the captured CO_2 and regenerate the capture agent. In the direct capture step, CO_2 is absorbed by a sodium hydroxide solution to produce soluble sodium carbonate in a strongly exothermic reaction

$$2NaOH(aq) + CO_2(g) \rightarrow Na_2CO_3(aq) + H_2O(l)\ \Delta_rG^\ominus = -56.1\ \text{kJ/mol},\ \ \Delta_rH^\ominus = -109.4\ \text{kJ/mol}. \tag{7.92}$$

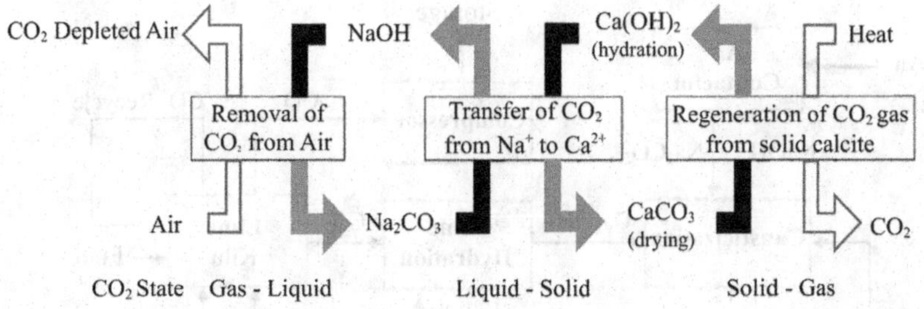

FIGURE 7.15
Schematic summary of soda-lime wet direct air capture process.

The value of $\Delta_r G^\ominus$ for this reaction is well in excess of the thermodynamic minimum free energy requirement calculated in Questions 7.4.1 and 7.4.2 and although one might argue that a system with a lower binding energy, such as sodium bicarbonate, might be adequate, there are advantages in having the additional driving force in extracting CO_2 from streams with a very low CO_2 partial pressure, like air.

The recovery of the solubilized CO_2 starts with a "causticization" reaction whereby an aqueous suspension of calcium hydroxide ("slaked lime") is added to precipitate calcium carbonate with >95% recovery of the carbonate ions from solution in another (mildly) exothermic reaction

$$Na_2CO_3(aq) + Ca(OH)_2(s) \rightarrow 2NaOH(aq) + CaCO_3(s) \quad \Delta_r H^\ominus = -5.3\ \text{kJ/mol.} \quad (7.93)$$

The calcium carbonate precipitate is filtered, dried and transferred to a lime kiln where it is heated to >900°C to produce a concentrated stream of CO_2 and calcium oxide powder

$$CaCO_3(s) \rightarrow CaO\ (s) + CO_2(g) \quad \Delta_r H^\ominus = 179.2\ \text{kJ/mol.} \quad (7.94)$$

This calcination step is a highly endothermic process and the use of pure oxygen in firing the lime kiln to avoid an additional gas separation stage (CO_2 from N_2 and other trace gases) adds additional energy requirements (~25 kJ/mol O_2). For a fuller analysis of this reversible reaction, see Question 6.3.5. The cyclic process loop is completed by the hydration of lime back to calcium hydroxide in another exothermic reaction

$$CaO(s) + H_2O(l) \rightarrow Ca(OH)_2(s) \quad \Delta_r H^\ominus = -64.5\ \text{kJ/mol.} \quad (7.95)$$

As shown in Figure 7.15, the key reagents sodium hydroxide and calcium hydroxide are both recycled in the process, although in practice some makeup will be required from time to time. The overall process is the sum of Reactions 7.92–7.95 with an overall standard enthalpy change of zero. However, the CO_2 in Reaction 7.92 is at a partial pressure of ~40 Pa (0.0004 bar) (p_1 in Question 7.4.2), whereas in Reaction 7.94 it leaves the process as a pure stream at 0.1 MPa or 1 bar ($p_0 = p_3$ in Question 7.4.2). As we have seen in Question 7.4.1, this has a thermodynamic minimum energy requirement of 19.4 kJ/mol. The actual process will require more energy than this because it is not carried out reversibly; the consequent inefficiencies of the various process steps can be offset by various means including heat recovery and utilization.

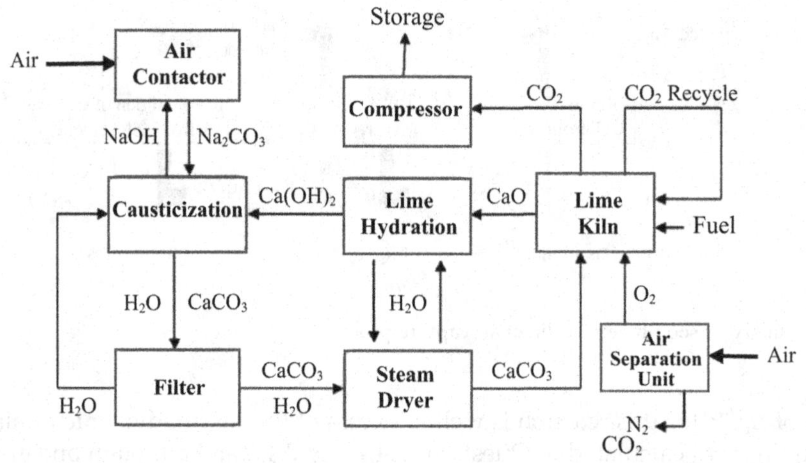

FIGURE 7.16
Typical process flow diagram for a soda-lime wet DAC process.

To achieve the outline process given in Figure 7.15 requires quite a complex practical operation, as illustrated by the process flow diagram given in Figure 7.16.

The absorption column contains packing material with a high surface to volume ratio over which the caustic soda solution spreads to give thin films of liquid with high gas-liquid contact area to enable rapid achievement of saturation by reactive dissolution of CO_2. The main energy requirements are for the air blowers (114 m^3 of ambient air must be treated to capture 1 mol of CO_2 at 400 ppm air concentration and 50% CO_2 recovery) and the solvent pumping through the column. These depend on the precise design of the column and the packings but a typical value is about 85 kJ/mol (Table 7.6, line (a)). The size and shape of the towers is a challenge if the aim is to capture very large amounts of CO_2. To capture 0.4 Mt

TABLE 7.6
Comparison of the energy requirements for direct air capture with those for flue gas carbon capture processes

Operation	Energy kJ/mol
Air pumping	85 (a)
$CaCO_3$ filtercake drying	64 (b)
O_2 cryogenic separation	15 (c)
Calcination	256 (d)
CO_2 compression*	12 (e)
Total DAC energy requirement	**432**
Heat recovery from CaO hydration	107 (f)
Possible DAC net energy requirement	325
Solvent-based flue gas carbon capture	141 (Equation 6.38)
Calcium looping flue gas carbon capture	202 (Equation 6.52)
Chemical looping flue gas carbon capture	<100 (Question 6.5.6)

Note
* The additional CO_2 generated by using gas for firing the lime kiln has not been included so that the DAC energy requirements are those needed if the entire process was powered by carbon-free (e.g. renewable) electricity.

CO_2/y, comparable to a commercial CCS project, even for a 10 m high column, this would require a diameter of about 80 m, which is physically unrealistic. The maximum practical diameter is about 10 m, which would mean that 75 such towers would be required for a DAC farm at this scale.

After capture, the conversion by causticization to calcium carbonate is almost 100%. This is then dewatered mechanically, either by pressure or vacuum filtration, which requires relatively little energy. The $CaCO_3$ filtercake has a residual moisture content of ~22% by weight, so this needs to be dried further before calcination in the kiln. If this is done by direct heating, for every mole of $CaCO_3$ in the filtercake we must evaporate (22/18)/(78/100.1) = 1.57 moles of water. So under standard conditions (see Table 7.6(b))

$$\text{Energy of evaporation} = \text{Drying energy} = \Delta_v H^{\ominus}(H_2O) \times 1.57 = 40.65 \times 1.57 = 63.8\ \text{kJ/mol}. \tag{7.96}$$

However, since Reaction 7.95 requires 1 mole of water (18 g) to hydrate 1 mole of CaO (100 g), leaving 18/118 = 0.15 or 15% residual water in the filtercake enables the steam for that hydration to be generated within the lime kiln. The additional 7% moisture may be removed by steam drying. A Hess's law calculation (analogous to that shown in Figure 6.19) using the enthalpy of vaporization of water and the heat capacities of CaO, H_2O and $Ca(OH)_2$ shows that Reaction 7.95 generates 107.1 kJ/mol of recoverable heat per mole of lime at 400°C (Table 7.6(f)); using this to raise steam for the filtercake drying is more than adequate, even with a drier efficiency of only 50%.

The captured CO_2 is regenerated from the dried calcium carbonate by heating to ~900°C in a lime kiln, which typically operates with an energy efficiency of 70% relative to the minimum heat requirement of 179.2 kJ/mol (4.1 GJ/t CO_2) under standard conditions given by Equation 7.94. If the kiln is powered by burning gas, it is fired with oxygen rather than air to avoid contamination with nitrogen; in this case, the exit CO_2 stream consists of that generated by burning the natural gas as well as that from the DAC process. We can estimate the former by using the standard heat of combustion of methane from Question 6.3.3, Equation (6.39) (–891 kJ/mol) to give the number of moles of methane that must be burned to give the required kiln calcination energy, which, using Equation (7.65) is ΔH = 179.2/0.7 kJ/mol = 256.0 kJ/mol (Table 7.6(d))

Moles of CH_4 to be burned to power kiln = 256.0/891 mol = 0.29 mol of additional CO_2 generated

Moles of cryogenic oxygen required for oxy-combustion = 2 × 0.29 mol = 0.58 mol O_2.

Therefore, 1.29 moles of CO_2 need to be stored or utilized for every mole captured. The energy required to supply the oxygen cryogenically separated from air is 0.58 × 25 kJ/mol = 14.5 kJ/mol (Table 7.6(c)). Ideally, the kiln and all the DAC equipment would be powered by renewable near zero-carbon electricity so that the additional CO_2 production is small.

There is one additional energy cost, which is that required to compress the CO_2 to above its critical point (73.8 bar, 31.1°C) for transport and storage. Typically it is compressed to 80 bar; the energy required for this is 281 kJ/kg or 12.4 kJ/mol (Table 7.6(e)). If the kiln is gas-fired, then the CO_2 compression energy would be 1.29 × 12.4 kJ/mol = 16.0 kJ/mol CO_2 extracted from the air. If the CO_2 is used directly as a chemical feedstock in CCU (see Question 7.3), then this energy may not be needed unless the CO_2 is transported to the CCU plant via a supercritical pipeline.

These energy requirements for this wet DAC process are summarized in Table 7.6, where they are compared with those for centralized CCS processes discussed in Questions 6.3.2, 6.3.5 and 6.3.6.

It can be seen that the ratio of the DAC and solvent-based flue gas process energy requirements from this more detailed process analysis is 432/141 = 3.1, which is in line with the thermodynamic minimum energy expectations from Questions 7.4.1 and 7.4.2, even though as expected the absolute energy requirement for DAC and the other systems are more than an order of magnitude higher for realistic processes. The DAC process outlined here uses 9.82 GJ/t CO_2 extracted from the air, or 7.25 GJ/t CO_2 if full heat recovery from the CaO hydration is achieved. This can be compared with the 12 GJ/t CO_2 of commercial plants using thermal swing DAC processes in the early 2020s, capturing up to 1000 t CO_2 per year.

7.4.4 How Might the Energy Requirements for Direct Air Capture Be Reduced?

Table 7.6 indicates that the key to reducing DAC energy requirements is to reduce significantly the post-capture regeneration energy. Alternative approaches need to replace the relatively high energy wet absorption thermal swing approach with a dry solid adsorption approach, reducing the need to heat large volumes of liquid solvent or carbonate precipitate (chemically bound CO_2) and using less energy-intensive methods to remove the captured CO_2 and regenerate the sorbent. A pressure swing reversible gas-solid adsorption process (PSA) might be expected to provide this. However, because of the low CO_2 partial pressure in the feed stream, it turns out that the pumping and compression requirements to obtain adequate contrast between feed and regeneration pressures make PSA an unviable option for DAC. Here we describe a more promising approach based on using the evaporation of water as the source of free energy for the CO_2 desorption: a moisture swing process.

This class of methods depends on adsorbing CO_2 onto a solid sorbent, which typically consists of a solid matrix functionalized with amine or other basic groups to which CO_2 can bind with an adsorption energy of typically 30–50 kJ/mol, just above the minimum thermodynamic requirement of 20 kJ/mol but far less than the 179 kJ/mol with which it is chemically bound to CaO in the wet thermal swing process. The sorbent takes the form of fine porous particles which can form a packed bed, or porous membrane sheets, having a high surface area (up to 2 m^2/g), through which large volumes of air can be easily drawn.

A typical solid sorbent is a polymer-resin system that is functionalized with quaternary ammonium groups, giving cationic surface sites, accompanied by hydroxide and carbonate counter-anions, to which the CO_2 binds adequately but quite weakly (enthalpy of adsorption $\Delta_a H^{\ominus} = 32$ kJ/mol) (Wang et al. 2011). Such a system has an adsorption capacity of 20–40 L/kg. The resin gives Langmuir-type adsorption isotherms (see Question 5.8) with CO_2 of the form shown in Figure 7.17. Even at CO_2 concentrations of 100 ppm, the adsorption reaches a saturation of over 95% so these resins are very effective at capturing CO_2 from very low concentration environments.

An interesting feature of this type of sorbent is that humidity has a large effect on the adsorption equilibrium. Figure 7.18(a) shows how the CO_2 saturation θ reduces dramatically with humidity at a constant ambient temperature, 23°C, and a constant p_{CO2} of 0.0004 bar (0.04 kPa). The saturation decreases to 0.5 in the presence of only 2% water vapor (relative humidity $\Phi \sim 66\%$; see Question 4.7.2, Equation 4.185). So this effect provides the basis of an alternative desorption mechanism for this class of adsorbents: moisture (lower case 'm') swing adsorption (MSA) whereby CO_2 is adsorbed from

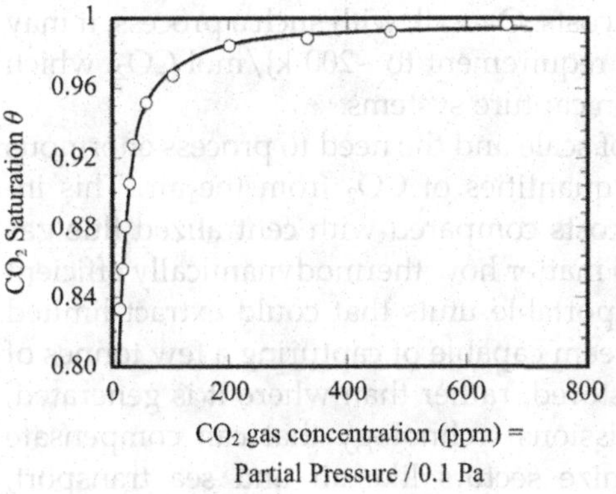

FIGURE 7.17
Typical adsorption isotherm for CO_2 on a typical DAC solid sorbent, fitted to Langmuir isotherm, Equation 5.157.

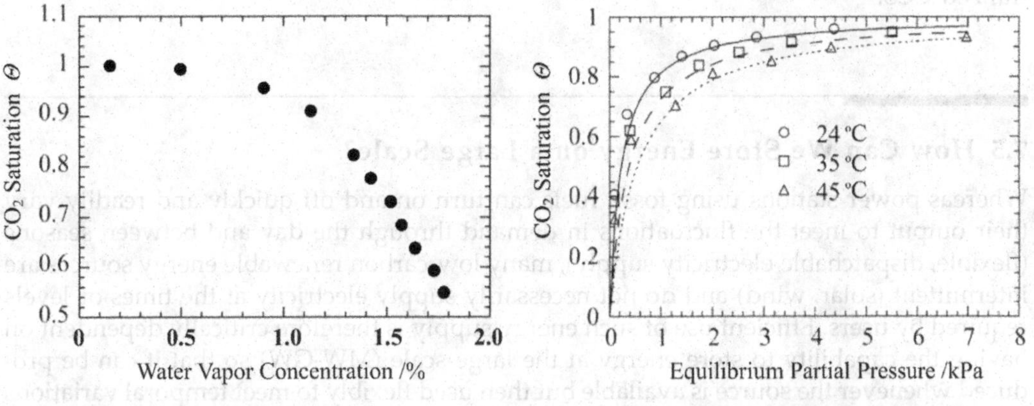

FIGURE 7.18
(a) Effect of humidity (water vapor concentration/%) on the CO_2 adsorption saturation θ of a hydroxide/carbonate functionalized cationic sorbent at 23 °C; (b) CO_2 desorption isotherms – released CO_2 partial pressure after desorption by exposure of loaded sorbent to water as a function of initial CO_2 surface saturation. (Adapted from Wang et al. 2011.)

ambient air at low humidity (<1% H_2O, Φ ~ 30%) and the sorbent is then exposed to higher moisture content air or liquid water. The water preferentially hydrates the surface and displaces the CO_2 from the adsorbent. Figure 7.18(b) illustrates that on desorption, the CO_2 partial pressure can rise to several kPa, at least a factor of 100 greater than in the ambient air from which it has been captured. It also shows how increasing temperature enhances desorption if an additional driver is needed.

Analysis of the desorption isotherms using Equations 5.184–5.186 gives $\Delta_d G^\ominus$ = 15.0 kJ/mol (adsorption is thermodynamically favored so desorption is not spontaneous), the standard enthalpy of desorption $\Delta_d H^\ominus$ as 31.8 kJ/mol and the entropy change $\Delta_d S^\ominus$ as 56.7 J/(mol K), leading to the release of ~20 mol CO_2/kg sorbent from confined adsorbed layers on the solid surface to free gas at ~10 kPa. The desorption is highly entropically favored; the additional enthalpic contribution to the free energy required for the release of the adsorbed CO_2 is provided by the energy of surface hydration by the water, present in large concentration.

This water swing process has potential energy saving advantages. It combines a sorbent with high binding capacity at low CO_2 partial pressures with a low-energy mechanism for release of the adsorbed gas and regeneration of the sorbent in a cyclic water-swing process. The operating energy requirements are mainly concerned with moving fluids around the system with additional heating and compression costs. Overall, with such a process, it may be possible to reduce the total operating energy requirement to ~200 kJ/mol CO_2, which starts to become competitive with flue-gas carbon capture systems.

The problem with all DAC approaches is one of scale and the need to process enormous volumes of air in order to extract worthwhile quantities of CO_2 from the air. This inevitably means large devices and large capital costs compared with centralized flue gas capture (CCS), as discussed in Question 7.4.2, no matter how thermodynamically efficient these plants are. However, deployed as small, portable units that could extract limited quantities of CO_2 (units of about 100 m^3 in size seem capable of capturing a few tonnes of CO_2 a day) near to where it can be utilized or stored, rather than where it is generated, and as a potential contributor to negative emissions technology that can compensate for emissions from more difficult to decarbonize sectors like air and sea transport, DAC should be viewed as a complementary rather than competing approach to centralized CCS.

7.5 How Can We Store Energy on a Large Scale?

Whereas power stations using fossil fuels can turn on and off quickly and readily vary their output to meet the fluctuations in demand through the day and between seasons (flexible, dispatchable electricity supply), many low-carbon renewable energy sources are intermittent (solar, wind) and do not necessarily supply electricity at the times or levels required by users. Efficient use of such energy supply is therefore critically dependent on having the capability to store energy at the large scale (MW-GW) so that it can be produced whenever the source is available but then used flexibly to meet temporal variations in demand. The most widespread method of storage of electrical energy is through batteries. However, although these are commonplace and efficient for powering small devices at the mW-W level, such as mobile phones (a typical phone battery has a capacity of ~10 Wh or 36kJ), and electric car batteries can store about 100 kWh or 360 kJ, batteries are not yet capable of storing energy at the GWh level required to serve large-scale power distribution systems.

In looking for large-scale energy storage systems, ideally we seek methods which combine high energy density (see Question 7.2.1) and charging rates while minimizing energy losses and leakage. In addition to chemical (batteries and hydrogen converted to liquids, such as LOHCs, methanol and ethanol, in which it can be stored at large scale for long periods), the main classes of methods are:

- mechanical (hydropumping, compressed air and flywheel)
- electrical (capacitors, superconducting magnets)
- thermal (molten salts, phase change materials, solid media).

Chemical, mechanical and electrical storage are termed "high quality" because in principle there is no thermodynamic limitation on converting the stored form of energy to another form for use, often electrical, although all practical processes will suffer from losses. Thermal storage methods are intrinsically "low quality" because of the Second Law limitation on converting stored thermal energy back into work, specifically electrical energy. Nevertheless, there is an increased interest in using such systems, often termed

"Carnot batteries", for storing electrical energy (Dumont et al. 2020). The overall efficiency can be improved by either using waste heat or by storage at high temperatures.

We will examine a selection of these methods from a thermodynamic standpoint to illustrate their potential. In order to compare them (see Question 7.5.10), we will evaluate a number of measures of their efficiency and capacity, including:

- ***The round-trip efficiency,*** η_{RT}: This is the ratio of the amount of energy released from a storage device or process on discharge compared with the energy initially stored in the charging stage. The target of course is to minimize energy losses within the storage process and during the charging and discharge energy transfer steps.
- ***The energy storage density,*** ρ_e: This measures the amount of energy stored per unit volume, mass (called specific energy in Question 7.2.1) or surface area of the device/process and determines both the portability of the stored energy and the feasibility of using the method for particular applications, as well as being another measure of the efficiency of energy storage.
- ***The total energy capacity***: This is the total amount of energy that can be delivered from a single charge-discharge cycle of the storage process, which gives an indication in particular of its ability to service large energy consuming processes or communities.

7.5.1 What Is the Basis of Thermal Energy Storage?

Thermal storage devices can be deployed to supply energy at a range of different scales, from localized processes requiring kWh of energy to heating or powering districts and towns at the GWh level. A common source of heat is thermal solar energy (concentrated solar power, where sunlight is directed using parabolic mirrors on receiver tubes at the focal point or by sun-tracking mirrors (heliostats) onto solar towers where a heat transfer fluid collects the energy). Other common sources include heat from combined heat and power (CHP) plants and waste heat from industrial processes. As well as balancing energy demand between day and night, thermal processes can store thermal energy in the summer for winter use or conversely be cooled in winter to provide air conditioning in the summer. They store energy either as *sensible heat,* where the temperature of a single phase material increases as it is heated, or additionally as *latent heat* in materials having a convenient phase change. Losses are minimized by insulating the system and the energy released by connection to a colder system for direct heating or conversion to power. For sensible heat, the energy stored is given simply by

$$Q = m\, c_p\, \Delta T \tag{7.97}$$

The storage medium can be a liquid (typically oil, water or molten salts) or a solid such as underground rock. A high specific heat capacity is an advantage, and within its limited liquid range water is an ideal fluid, having a relatively high value of c_p = 4.2 kJ/(kg K), which means that if heated from 20°C to 80°C it can store 250 kJ/kg.

Phase change materials have the advantage of having a significantly higher energy storage density since they exploit the additional latent heat associated with a phase transition

$$Q = m\, c_p\, \Delta T + m\, \Delta_\alpha^\beta h \tag{7.98}$$

where $\Delta_\alpha^\beta h$ is the enthalpy change associated with the phase transition (endothermic for solid to liquid or liquid to gas). Again, water has some advantage if its melting can be exploited since this absorbs 333 kJ/kg, which is 80 times higher than the heat required to raise the temperature of the liquid by just 1°C.

7.5.2 How Can Thermal Energy Be Stored at High Temperatures?

In order to store energy, particularly solar energy, in a way that can be readily used to generate electricity using gas and steam cycle plants such as those described in Question 6.2, it is most effective to do this at high temperatures. Molten salts are very effective for this, having a wide liquid range with high boiling points, low vapor pressure and high heat capacities per unit volume, minimizing the size of storage tank required. They can store both sensible and latent heat when starting from the solid state. Eutectic mixtures (Atkins 2014) are particularly effective as their freezing point at the eutectic composition is lower than that for the pure salts and represents the lowest temperature at which the system remains liquid, so extending the liquid range to lower temperatures. For instance, sodium nitrate melts at 308°C (581 K) and potassium nitrate at 320°C (593 K), whereas a eutectic 50 mol% mixture of the two salts (46wt% $NaNO_3$) melts almost 100°C lower at 222°C (495 K) and can be heated as a liquid up to 600°C (873 K). In fact, a near-eutectic mixture of 60wt% $NaNO_3$ (called "solar salt") is often used as this is cheaper but still has a liquid range of 230°C (503 K) to 580°C (853 K). It has a heat of fusion of 161 kJ/kg (about half that of water) and a mean heat capacity c_{Pms}, (used in Question 7.5.3) of 1.53 kJ/kg K (about a third that of water). A commonly used ternary eutectic mixture ("Hitec" – 40% $NaNO_2$-7% $NaNO_3$-53% KNO_3) has a lower freezing point, 142°C (415 K), but can only be used up to 500°C (773 K) before it starts to vaporize. To use such systems in energy storage processes requires precise characterization of their thermodynamic properties, in particular their phase diagrams, melting and boiling points and their heat capacities as a function of temperature.

A schematic diagram of a typical thermal energy storage system, here coupled to a concentrated solar power process, is shown in Figure 7.19.

Using the 60wt% $NaNO_3$, 40wt% KNO_3 molten salt mixture as an example, it is kept liquid at about 250°C (523 K) in an insulated "cold" storage tank. The concentrated solar radiation from parabolic mirrors or heliostats is focused on collector tubing heat exchangers through which flows the cold molten salt, acting as a heat transfer fluid. The salt is heated within its liquid range to about 580°C (853 K) and stored in an insulated "hot" storage tank where the energy can be stored effectively for up to a week if needed. This stored energy can then be used to generate electricity on demand by pumping the hot molten salt through another heat exchanger in a steam generator to drive a conventional turbine/generator system as described in Question 6.2. If this is a simple steam system, we saw there that the thermodynamic (Second Law) restriction on conversion of heat to work/electricity leads to an efficiency of about 33%. Alternatively, the stored thermal energy can be used directly with much higher efficiency for district heating, driving industrial processes like steam methane reforming (SMR) to produce hydrogen or in seawater desalination (Olwig et al. 2012; Belessiotis et al. 2016).

Variations on this process include replacing the two "hot" and "cold" storage tanks by a single tank "thermocline" system, where the hot and cold salt are separated by a vertical temperature gradient and mixing is minimized by buoyancy forces or by a divider plate. On charging, the salt flows out of the cold section and returns to the hot section; discharge follows the opposite route. Some plants use oil or organic heat transfer fluids such a biphenyl to collect the solar heat, which is then transferred to the molten salt via a second

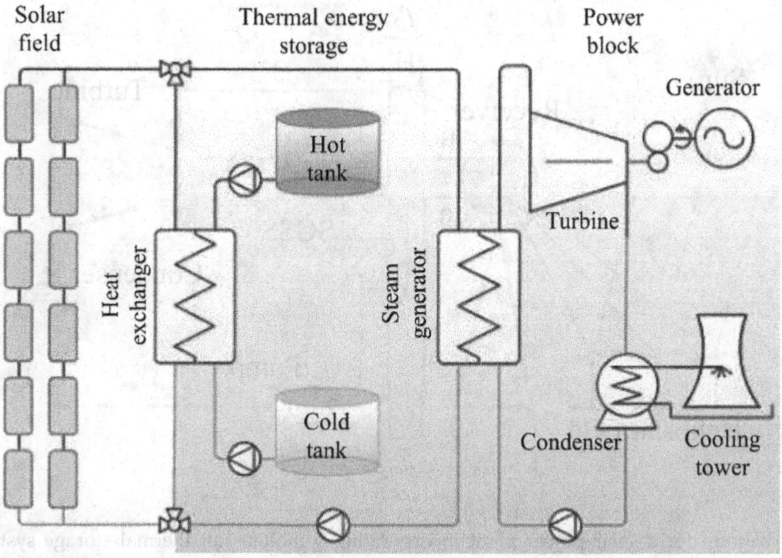

FIGURE 7.19
Schematic diagram of a concentrated solar power plant coupled to a thermal energy storage system.

heat exchanger with a small (~5%) heat loss. Another way to reduce costs is to add cheaper high heat capacity mineral filler material, such as quartz rock, to the hot tank. For electricity generation, some systems use the hot molten salt to heat air to drive an air turbine Brayton cycle, using the exhaust heat to drive a steam turbine to give a combined power cycle (see Question 6.2.2). Molten salt thermal storage has also been coupled to nuclear power systems as a method for modulating supply.

7.5.3 What Is the Efficiency and Size of a Molten Salt Thermal Energy Storage System for a 100 MW Solar Power Plant?

Let us consider a 25 MW_e (the subscript "e" indicates electrical power) solar power plant, based on harvesting 100 MW_{th} of solar energy (the subscript "th" indicates thermal), operating as in Question 7.5.2 with a thermal storage system that uses 60wt% $NaNO_3$/40wt% KNO_3 "solar salt" as the storage fluid. The main elements of the system are shown in the block diagram of Figure 7.20, which gives the mass flow rates ($\dot{m}_i$) and temperatures (T_i) of the different streams.

Because thermal energy storage involves conversion of primary energy sources such as solar or wind into stored thermal energy that is then discharged and often converted to another form of energy, usually electrical, it is useful to examine the efficiency of the system in terms of *exergy* as well as energy. The concept of exergy is explained in Question 3.8. One consequence of the Second Law is that heat cannot be completely converted into useful work, and the extent to which this is the case depends on temperature. As we have seen in Question 3.2.4, for an ideal Carnot cycle this is demonstrated by the efficiency for conversion of heat to work in a heat engine (see Equation 3.14)

$$\eta_C = \frac{|\text{Work out}|}{\text{Heat In}} = \frac{|W|}{Q} = \frac{T_2 - T_1}{T_2} = 1 - \frac{T_1}{T_2} \tag{7.99}$$

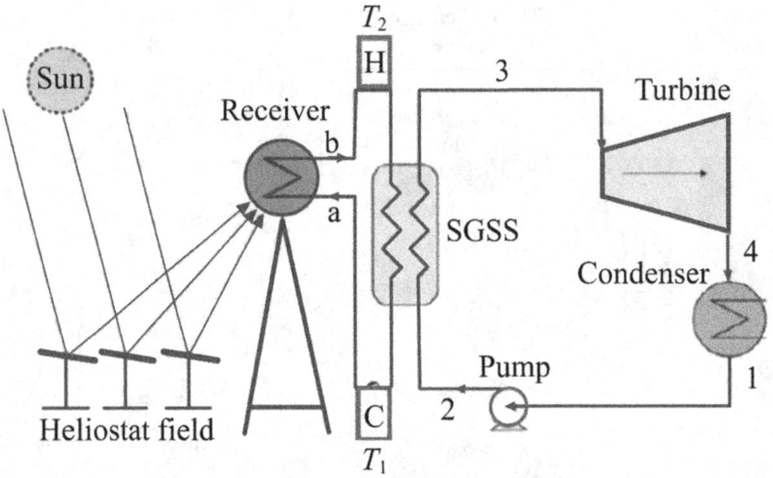

FIGURE 7.20
Process flow diagram for a solar power plant incorporating a molten salt thermal storage system: H = hot storage tank; C = cold storage tank; SGSS = steam generator subsystem; a = inlet to solar receiver heat exchanger; b = outlet from solar receiver heat exchanger.

The higher the upper cycle temperature T_2, the higher the efficiency of heat to work conversion. So temperature is a key quantity in determining how "useful" heat is. Also when a solid body or fluid stream transfers heat to another body or fluid, it cannot raise its temperature to a value higher than its own. Since exergy is a thermodynamic quantity that incorporates the temperature of a system as well as its energy content, it is particularly helpful in quantifying how effectively heat can be used for storing and transporting energy. Question 3.8 explains that exergy is that part of the total energy that can go on to be converted into useful work and so is a convenient way to quantify the extent to which heat transferred through a storage system can be used to generate mechanical or electrical work. The higher the temperature of a fluid or body is, the more effective it is for doing work. The term "quality" is sometimes used to characterize the ability of a source of thermal energy to generate useful work: a high-temperature fluid stream contains "high quality" enthalpy, which can be more effectively converted to useful work than a low-temperature, low quality stream.

Question 3.8 showed how the exergy *e* associated with the enthalpy of a stream *i* at temperature *T* is defined relative to a reference temperature T_0, usually taken as ambient temperature (we will use 298 K); see Equations 3.170 and 3.173. Using these equations, we can write the difference in specific exergy between any two streams a and b as

$$\Delta e = (h_a - h_b) - T_0(s_a - s_b) = c_P(T_a - T_b) - c_P\, T_0\, ln(T_a/T_b). \quad (7.100)$$

7.5.3.1 Solar energy capture

Consider first the light capture and concentration system whose mirrors reflect and concentrate the sun's rays onto the heat collection system (either troughs for parabolic mirrors or a central tower receiver for heliostats). The rate of solar energy harvested, $\dot{Q}_s$ (J/s), depends on the intensity of the sun's radiation and the aperture area of the collecting mirrors. Much of this will be captured and delivered to the heat storage system ($\dot{Q}_{sc}$), but some will be lost by a variety of mechanisms ($\dot{Q}_{sl}$). The energy balance is simply

$$\dot{Q}_s = \dot{Q}_{sc} + \dot{Q}_{sl} \tag{7.101}$$

and the First Law energy capture efficiency is

$$\eta_{th,cap} = \dot{Q}_{sc}/\dot{Q}_s. \tag{7.102}$$

Because from Equation 3.166 the rate of exergy accumulation is, in general, given by

$$\dot{E}_s = \dot{W}\,(T_1 \rightarrow T_0) = \dot{Q}\left(1 - \frac{T_0}{T}\right), \tag{7.103}$$

the rate of exergy accumulation (available work) associated with the heat from the solar radiation, $\dot{E}_s$, is

$$\dot{E}_s = \dot{Q}_s\left(1 - \frac{T_0}{T_{sun}}\right), \tag{7.104}$$

with a similar equation linking E_{sc}, the exergy delivered to the heat collection system, with $\dot{Q}_{sc}$. T_{sun} is the effective black body temperature of the sun as an exergy source, ~5,800 K. This means that the rate of exergy accumulated because of the incident radiation is slightly less than the energy (heat) content accumulation

$$\frac{\dot{E}_s}{\dot{Q}_s} = \frac{\dot{E}_{sc}}{\dot{Q}_{sc}} = \left(1 - \frac{T_0}{T_{sun}}\right) = (1 - 298/5800) = 0.95 \tag{7.105}$$

and, similarly, the exergy of the captured radiation $\dot{E}_{sc}$ is ~95% of its energy content, $\dot{Q}_{sc}$. However, since both are affected by the same "Carnot" factor as in Equation 7.105 then the Second Law exergy solar capture efficiency is the same as the First Law capture efficiency for energy

$$\eta_{ex,cap} = \frac{\dot{E}_{sc}}{\dot{E}_s} = \frac{\dot{Q}_{sc}}{\dot{Q}_s} = \eta_{th,cap}. \tag{7.106}$$

In practice, about 25% of the incident radiation is lost in the light capture process and so $\eta_{th,cap}$ is about 0.75. Assuming that for our solar plant the incident radiation is 100 MW, then the rate of energy delivery to the heat storage transfer system is only 75 MW while the rate of exergy delivery is 71.3 MW.

7.5.3.2 Heat storage charging

In the heat storage transfer system, the focused, captured heat $\dot{Q}_{sc}$ is directed at a heat exchange system through which flows the molten salt. There are some heat exchange losses here ($\dot{Q}_{scl}$), typically 10%, so that an energy balance gives

$$\dot{Q}_{sc} = \dot{Q}_{sst} + \dot{Q}_{scl} = \dot{m}_{ms}(h_{ms2} - h_{ms1}) + \dot{Q}_{scl}, \tag{7.107}$$

where $\dot{m}_{ms}$ is the mass flow rate of molten salt through the storage transfer heat exchanger and $\dot{Q}_{sst}$ is the rate of energy storage as heat in the molten salt as its temperature rises from T_1 (cold tank, stream a in Figure 7.20) to T_2 (hot tank, stream b). The energy efficiency of this storage stage is

$$\eta_{th,st} = \frac{\dot{Q}_{sst}}{\dot{Q}_{sc}} \sim 0.90, \tag{7.108}$$

with the rate of energy storage being

$$\dot{Q}_{sst} = \dot{m}_{ms}\, c_p (T_2 - T_1). \tag{7.109}$$

Taking typical values of 160 kg/s for $\dot{m}_{ms}$, $T_1 = 563$ K (290°C) and $T_2 = 838$ K (565°C), then

$$\dot{Q}_{sst} = 160 \times 1.53 \times (838 - 563) \text{ MW} = 67{,}320 \text{ kJ/s or } 67.3 \text{ MW, with}$$
$$\eta_{th,st} = 67.3/75.0 = 0.90, \tag{7.110}$$

where $\eta_{th,st}$ is the thermal or First law efficiency. The corresponding exergy balances for this stage are

$$\dot{E}_{sc} = \dot{E}_{sst} + \dot{E}_{l,scl} + \dot{E}_{l,irr}, \tag{7.111}$$

where $\dot{E}_{l,irr}$ are the internal losses due to irreversibility, $\dot{E}_{l.scl}$ the exergy loss associated with the heat exchange loss $\dot{Q}_{scl}$ and the exergy uptake of the molten salt storage stream $\dot{E}_{sst}$ is, using Equation 7.100,

$$\begin{aligned}\dot{E}_{sst} &= \dot{m}_{ms}[(h_{ms2} - h_{ms1}) - T_0(s_{ms2} - s_{ms1})]\\ &= \dot{m}_{ms}[C_P(T_2 - T_1) - T_0 C_P \ln(T_2/T_1)]\end{aligned} \tag{7.112}$$

$$\begin{aligned}&= [160 \times 1.53 \times (838 - 563) - 160 \times 298 \times 1.53 \times \ln(838/563)] \text{ kJ/s}\\ &= [67,320 - 29,015] \text{ kJ/s} = 38,305 \text{ kJ/s} = 38.3 \text{ MW}.\end{aligned} \tag{7.113}$$

This means that the exergetic or Second Law efficiency (see Equation 3.180) for the heat storage stage is

$$\eta_{ex,st} = \frac{\dot{E}_{sst}}{\dot{E}_{sc}} = \frac{38.3}{71.3} = 0.54. \tag{7.114}$$

So although the energy transferred in this storage ("charging") step only decreases due to losses by 10%, the exergy content decreases by 46% and now represents only 38% of the incident solar exergy.

The heated molten salt is stored in the "hot" storage tank until it is required for power generation. It may be stored for days to smooth out daily or day/night variations in solar radiation, or for much longer periods if being used for to cope with seasonal variations and to provide solar-sourced power in winter. Although the tank will be well insulated, some thermal losses may be expected (~2% per week for 100 MW_{th} tanks with a surface area of ~500 m^2). We will ignore this in the calculation here but good insulation is essential for efficient thermal storage.

7.5.3.3 Heat storage discharge

The discharge phase releases the thermal energy from the molten salt to raise steam for power production. In Figure 7.20, the molten salt is pumped from the hot storage tank through the heat exchangers in the power plant steam generator (SGSS) where, in this example, it heats water to create superheated steam to drive a turbine to produce electricity in a steam Rankine cycle, as described in Question 6.2.1. The heat transferred from the molten salt is used to heat water from point 2 in Figure 6.6, at 30°C (303 K), 20 MPa (stream 2 in Figure 7.20) to point 3 in Figure 6.6, superheated steam at 550°C (823 K), (stream 3 in Figure 7.20) to undergo almost isentropic expansion in a turbine which in turn drives a generator to produce electricity. We saw in Question 6.2.2 that the ideal efficiency of such a Rankine cycle is 0.33. The key issue for the energy storage system is its ability to supply enough energy to the steam turbine for a sustained period of time. It is cooled in passing through the steam generator coils from 565°C (838 K) (hot tank) to 290°C (563 K) (cold tank) and resides in the cold tank ready to be circulated back to the solar heat transfer system to be reheated for another storage cycle.

We will assume that there are negligible heat losses from the steam generator heat exchangers so that the energy balance requires that:

Release rate of stored thermal energy, $\dot{Q}_{dis} = \dot{Q}_{sst}$ = Heat take-up rate by steam, $\dot{Q}_{steam}$, which in terms of specific enthalpies and flow rates becomes

$$\dot{m}_{ms}(h_{bms} - h_{ams}) = \dot{m}_{steam}(h_{3steam} - h_{2steam}). \tag{7.115}$$

We use the values of h_{3steam} (3,500 kJ/kg) and h_{2steam} (135 kJ/kg) employed in Question 6.2.2. So

$$\dot{m}_{ms}c_{pms}(T_2 - T_1) = \dot{m}_{steam}(h_{3steam} - h_{2steam}) \tag{7.116}$$

and so

$$\begin{aligned} [160 \times 1.53 \times (838 - 563)]\ \text{kJ/s} &= \dot{Q}_{sst} = 67{,}320\ \text{kJ/s} \\ &= [\dot{m}_{steam}(3500 - 135)]\ \text{kJ/kg} = 3365\,\dot{m}_{steam}\ \text{kJ/kg}. \end{aligned} \tag{7.117}$$

This gives the steam flow rate in the generator as 67,320/3,365 kg/s = 20 kg/s.

The energy efficiency of this storage discharge stage, $\eta_{th,dis}$, is 100% since we have assumed no heat exchange losses, so $\dot{Q}_{sst} = \dot{Q}_{dis}$, and 67.3 MW has been transferred to drive the power plant. With a superheated Rankine cycle efficiency of 0.33, this will generate 22.2 MW of electricity. The exergy balance is

$$\dot{E}_{sst} = \dot{E}_{steam} + \dot{E}_{l,dis}, \tag{7.118}$$

where $\dot{E}_{l,dis}$ represents the internal losses due to irreversibility. The exergy change absorbed by the heated water/steam, $\dot{E}_{steam}$, is (using steam tables for s_3 and s_2 (Rogers and Mayhew 1994))

$$\dot{E}_{steam} = \dot{m}_{steam}\left[(h_{3steam} - h_{2steam}) - T_0(s_{3steam} - s_{2steam})\right] = 67,320\,\text{kW} - 20 \times 298 \times (7.7586 - 0.464)\,\text{kW}, \quad (7.119)$$

$$\dot{E}_{steam} = [67{,}320 - 43,476]\,\text{kW} = 23,844\,\text{kW} = 23.8\,\text{MW}. \quad (7.120)$$

It can be seen that the "useful" heat that is transferred as exergy to the steam system, though much less than the actual heat energy transferred (67.3 MW), is essentially the same as that estimated from the stored energy $\dot{Q}_{sst}$ once the 33% thermodynamic efficiency of the Rankine cycle has been factored in, 22.2 MW_e. This demonstrates that although the exergy diminishes much faster than the heat energy as we pass through the system from the harvested incident solar energy to convert this into electricity, it does represent the component of the thermal energy that can be transformed into useful mechanical work or electrical energy. In fact, the exergy efficiency of the power cycle will not be 100%, but the two estimates of the final power generation agree within the assumptions and simplifications made. The exergy efficiency for the final discharge stage of the storage process is

$$\eta_{ex,dis} = \dot{E}_{steam}/\dot{E}_{sst} = 23.8/38.3 = 0.62. \quad (7.121)$$

The *overall thermodynamic efficiencies* from harvested solar radiation to electrical power are

For energy: 22%

For exergy: 23.8/95 = 25% (since Equation 7.105 shows that the incident solar radiation already has a reduced exergy compared with its energy)

We can see that to generate ~25 MW_e it is necessary to harvest ~100 MW_{th} of solar energy.

The efficiencies of the energy storage parts of the process (charging, storing and discharging to the steam generator), η_s, are

For energy: $\eta_{th,s} = 0.9 \times 1.0 = 0.9$; this includes a 10% potential inefficiency in transferring the available captured solar power (75 MW) to the molten salt fluid in the receiver heat exchanger. Strictly this may be reduced further by ~2% if we include heat losses during storage and a further ~10% if we include heat exchanger inefficiencies in the steam generator, reducing this to 0.79. This can be used to estimate what is called the "round-trip energy efficiency" of the storage process, η_{RT}, defined as

$$\eta_{RT} = \frac{\text{Energy output with storage}}{\text{Energy output without storage}}. \quad (7.122)$$

For this plant, with storage there are three contributions to energy inefficiency: the input (charging) heat exchanger, heat losses from the storage tanks and the output (discharge) heat exchanger, which reduces the solar energy that reaches the power plant to 75 MW × 0.79 = 59.3 MW. Without storage, the captured solar energy can be fed directly to the steam generator of the power plant via just one heat exchanger, resulting in losses of 10%, meaning that 75 × 0.9 = 67.5 MW is fed into the plant. So, since the efficiency of the steam power plant will be the same in both cases

$$\eta_{RT} = \frac{59.3}{67.5} = 0.88. \tag{7.123}$$

For exergy: $\eta_{ex,s} = 0.54 \times 0.62 = 0.33$, which reflects both internal and non-adiabatic losses in the input receiver heat exchanger. This may be changed slightly by thermal losses from the storage tanks (for which a 2°C drop in temperature actually increases the exergy of the discharge stream by about 0.4%) and in the steam generator of ~10%, giving ~0.31 for the exergy round-trip storage efficiency. However, while there are further large reductions in energy efficiency in the power plant, the reduction in exergy efficiency there is small.

So in energy terms, thermal storage can operate very efficiently, subject to minimizing losses and inefficiencies of process equipment such as pumps and heat exchangers, particularly where, as in this example, the charging and discharging phases only involve heat. The "quality" of the heat does reduce steadily through the process, as measured by the exergy, and this needs to be taken into account in the use of the stored heat. In simple terms, the energy efficiency tells us how effectively the available energy has been transferred to its end application, electricity generation, whereas the exergy efficiency tells us whether this is a sensible use of the energy content of that heat stream. For instance, storing solar-derived heat in a thermal storage device where the storage fluid operated between 100°C and 200°C would not deliver heat of high enough quality to superheat steam significantly, resulting in a significant reduction in power cycle efficiency (cf. Question 3.8).

To illustrate this further, we can consider using the energy stored in the same hot molten salt tank to heat water for a district heating system, rather than generating electricity. If we transfer the heat to a cold water stream at 10°C (T_c = 283 K) and heat this to 40°C (T_h = 313 K) to be used for hot water central heating, then an energy balance on an adiabatic heat exchanger connecting the molten salt and water streams is

$$\dot{m}_{ms} c_{pms} (T_2 - T_1) = \dot{m}_{water} c_{pwater} (T_h - T_c) \tag{7.124}$$

$$160 \times 1.53 \times (838 - 563) \text{ kg/s} = \dot{m}_{water} \times 4.18 \times (313 - 283) \tag{7.125}$$

$$\dot{m}_{water} = 67{,}320/(4.18 \times 30) = 537 \text{ kg/s}. \tag{7.126}$$

The energy efficiency η_{th} is 100%. To calculate the exergy efficiency,

$$\dot{E}_{extracted} = \dot{m}_{water} [c_{pwater} (T_h - T_c) - c_{pwater} T_0 \ln(T_h/T_c)] \tag{7.127}$$

$$= 537 \times [4.18 \times 30 - 4.18 \times 298 \ln(313/283)] \text{ kJ/s} \tag{7.128}$$

$$= 537 \times [125.4 - 123.4] \text{ kJ/s} = 1074 \text{ kJ/s} = 1.1 \text{ MW}. \tag{7.129}$$

Now the available exergy flux in the thermal storage is 38.3 MW (Equation 7.113), so the exergy, Second Law efficiency of this water heating process is

$$\eta_{ex} = 1.074/38.3 = 0.028 = 2.8\%. \tag{7.130}$$

The heat recovered from storage is useful for heating but has been downgraded so that on cooling it is of little further use as an energy source. All the energy has been used, but 97% of the available useful work in the high-quality thermal storage heat has not been utilized. This is vividly illustrated by recognizing that the 23.8 MW_e of electricity that can be generated from the 38.3 MW_{th} in the store could be used to supply electrical heating to 23.8/1.1 = 21.6 times more homes than by choosing to use high-quality heat to heat water by 30°C for domestic heating systems.

A key question about energy storage systems concerning capital and operating costs is their size. We can use the calculations above to estimate the size and capacity of a thermal storage plant, using "solar salt" as the storage fluid, that can service a 100 MW power plant. To deliver 23.8 MW_e of electricity requires a discharge molten salt flow rate $\dot{m}_{ms}$ of 160 kg/s, the typical value assumed in our example calculation. If we target supplying the required heat for t_{dis} = 10 hours, then we can calculate the total mass of molten salt required, m_s, from

$$m_s = \dot{m}_{ms} \times t_{dis} \times \frac{100}{23.8}\,\text{kg} = 160 \times 10 \times 3600 \times \frac{100}{23.8}\,\text{kg} = 24.2 \times 10^6\,\text{kg} = 24{,}200\ \text{t}. \tag{7.131}$$

At a molten salt density of 1.73 t/m³, the capacity of the hot and cold storage tanks needs to be 24,200/1.73 m³ ~ 14,000 m³. This corresponds to a tank which is 15 m tall and 35 m in diameter, indicating the size and cost challenges of using a single molten salt storage system for large power plants.

7.5.4 How Might the Efficiency of a Thermal Energy Storage System Be Improved?

We can gain more insight into how to optimize the design of a thermal energy storage system by expressing the storage phase energy and exergy efficiencies in terms of the key temperature variables. From Equation 7.114 for the exergy efficiency, using Equation 7.105 for $\dot{E}_{sc}$ and Equation 7.112 for $\dot{E}_{sst}$, we have

$$\eta_{ex,st} = \frac{\dot{E}_{sst}}{\dot{E}_{sc}} = \frac{\dot{m}_{ms}\left[c_P(T_2 - T_1) - T_0\, c_P \ln(T_2/T_1)\right]}{\dot{Q}_{sc}(1 - T_0/T_s)}. \tag{7.132}$$

Further, since from Equation 7.108, $\dot{Q}_{sc} = \dot{Q}_{sst}/\eta_{th,st}$, we can replace $\dot{Q}_{sc}$ by using Equation 7.109 for $\dot{Q}_{sst}$:

$$\eta_{ex,st} = \frac{\dot{E}_{sst}}{\dot{E}_{sc}} = \frac{\dot{m}_{ms}\left[c_P(T_2 - T_1) - T_0\, c_P \ln(T_2/T_1)\right]}{\dot{m}_{ms} c_P(T_2 - T_1)(1 - T_0/T_s)/\eta_{th,st}}. \tag{7.133}$$

Rearranging, we obtain

$$\eta_{ex,st} = \eta_{th,st}\frac{1 - T_0 \ln(T_2/T_1)/(T_2 - T_1)}{(1 - T_0/T_s)}. \tag{7.134}$$

For the conditions of the example used, with the 60wt% $NaNO_3$/40wt% KNO_3 molten salt system having a "hot" storage temperature of 565°C (838 K) and the "cold" inlet temperature at 290°C (563 K), Equation 7.134 shows that $\eta_{ex,st}$ is 0.60 $\eta_{th,st}$ i.e 0.54, in agreement with Equation 7.114. Equation 7.134 shows that the exergy efficiency of the storage system is determined by the "hot" and "cold" operating temperatures. Even with no energy losses from the system so that $\eta_{th,st}$ improves from the 0.9 value of the example to 1.0, the maximum exergy efficiency for the operating temperatures used is 60%. There is little scope for operating with this particular molten salt at temperatures much above 600°C. However, using a heat storage fluid with a working range from 400°C to 900°C, which is within the liquid range of Li/Na/K carbonate salt eutectics, would raise the maximum exergy efficiency to 72%. The practical feasibility of operating above 600°C is determined by other factors such as stability and corrosion of the materials used to construct the tanks and pipelines. There are a number of alternative ways to increase the capacity of thermal energy storage systems, including the use of phase-change materials to exploit latent heat of melting, or other transitions, in addition to sensible heat (Fletcher 2015).

7.5.5 How Effective Is a Heat Pump at Using the Solar Energy Stored in the Ground?

A ground-source heat pump extracts thermal energy from the ground, where it is stored as a result of solar radiation incident on the Earth's surface, and uses it for residential or commercial heating. It can also run in reverse to extract heat from the air to provide cooling (i.e. air conditioning) and transfer it back into the ground to replenish the heat source there. The device therefore exploits the ground as a useful energy storage medium, acting as a heat source or heat sink. At a depth of 5–10 m the ground temperature remains almost constant throughout the year (around 11°C in the United Kingdom, for instance) and its sensible heat may be extracted through a buried coiled pipe heat exchanger, or by drilling a borehole within which is fitted a heat exchanger.

The principle of a heat pump was introduced in Question 6.5. Like a refrigerator, it is essentially a heat engine run in reverse where (usually electrical) energy is used to move heat from a cold source to a warmer sink. The working fluid for such a heat or cooling pump is usually a fluid that can readily undergo a vapor compression cycle (see Question 6.5.1) at moderate pressures within the temperature range required for the heat transfer process, where heat is absorbed at a temperature T_0 and rejected at a higher temperature T. Question 6.5.1 showed that we define the efficiency of such a device by a *coefficient of performance* (COP_h), given by Equation 6.105,

$$COP_h = \frac{\text{Heating output}}{\text{Work input}} = \frac{(h_2 - h_3)}{(h_2 - h_1)}, \tag{7.135}$$

where (see Figure 6.25) $(h_2 - h_1)$ is the enthalpy change (work input) in the adiabatic compression of the fluid vapor which raises its temperature from T_1 to T_2 and $(h_2 - h_3)$ the enthalpy change (heat output to the environment) as the vapor cools isobarically from T_2 and condenses to saturated liquid at T_3. The COP_h for an ideal, reversible heat pump operating according to a Carnot cycle between temperatures T_1 and T_2 is given by Equation 6.107

$$COP_h = \frac{T_2}{T_2 - T_1}. \tag{7.136}$$

As an example, we consider a ground-source heat pump (GSHP) system for domestic heating and/or air conditioning, shown schematically in Figure 7.21. Such a system has three main components. First there is a borehole (or buried coiled heat exchanger pipeline) through which is circulated a heat transfer fluid (usually water or brine containing an anti-freeze such as ethylene glycol to prevent freezing in winter) to couple with the ground and transfer stored ground heat to the heat pump itself. In this case, we specify a length of 50 m for the U-tube subsurface heat exchanger (A). Next, there is the heat pump system. The ground fluid transfers its heat via another heat exchanger (B) to the liquid working fluid (a refrigerant, typically a low global warming potential (GWP) mixture of hydrofluorocarbons) which is vaporized at temperature T_1 in the heat pump evaporator (B) as part of the vapor compression cycle. This vapor is compressed adiabatically (in the idealized case isentropically) by compressor C to the upper working temperature T_2, from which it cools isobarically and is condensed back to a liquid at temperature T_3 in the condenser D. The rejected heat is transferred in the condenser to another water stream (5–6) which transports the heat in this third component of the system to the fan-coil unit (F), which releases the heat to warm the air in the space to be heated. Meanwhile, the condensed working fluid is expanded through a throttling valve or capillary tube (E) at constant enthalpy before entering the evaporator to complete the cycle.

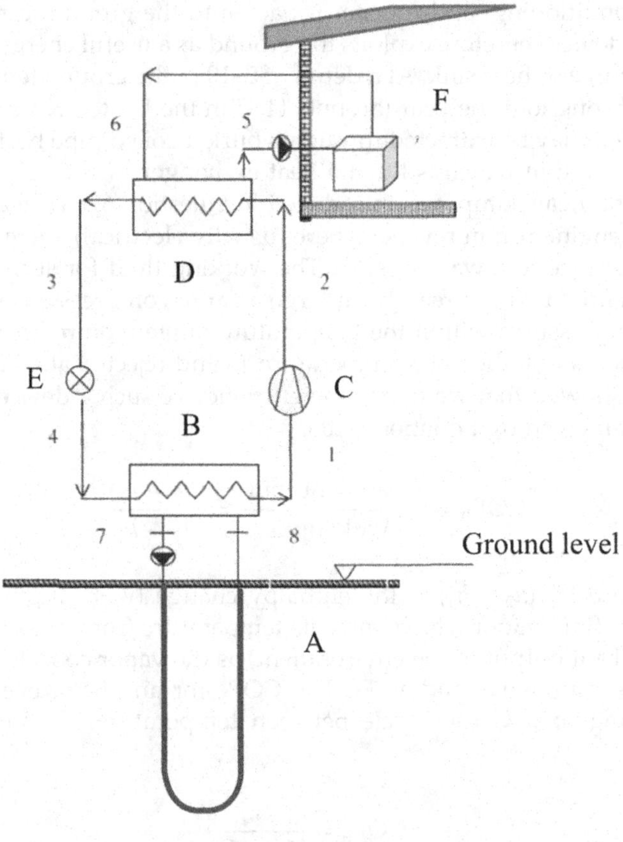

FIGURE 7.21
Schematic diagram of a ground-source heat pump system to provide domestic or district heating. A: ground heat exchanger; B: evaporator; C: compressor; D: condenser; E: throttling valve; F: fan-coil unit (heater or air conditioner).

A full thermodynamic analysis involving both enthalpy and exergy balances can be carried out on this system in much the same way as for the molten salt thermal storage system in Question 7.4.3. The sensible heat stored in the ground replaces that stored in the molten salt and the heat-pump and fan-coil systems replace the steam turbine driven power plant. We will not reproduce the full analysis here but simply give the main outcomes of such calculations. For a typical system, we assume that the subsurface ground temperature is 16°C, the ground heat exchange water enters (7) at 12°C with a flow rate, $\dot{m}_w$, of 0.25 kg/s and exits (8) at 15°C. The working fluid (flow rate $\dot{m}$) leaves the evaporator (1) at −6°C (T_1 = 267 K) and is compressed to superheated vapor (2) at 99 °C (T_2 = 372 K) and 19 bar (1.9 MPa); it cools at this same pressure and emerges as saturated liquid from the condenser (3) at 41°C (T_3 = 314 K). It is throttled down to 3.5 bar (350 kPa) (4) and enters the evaporator from where it emerges as vapor at −6°C, completing the vapor-compression cycle of Figure 6.25. The condenser heats the fan-coil water stream up from 40°C (6) to 45°C (5), from where it ejects its heat for space heating.

Equation 7.136 gives us the Carnot COP_h of the GSHP as (372)/(372 − 267) = 372/105 = 3.54. Using a heat pump is a very efficient way to transfer heat from cool to hotter bodies; the closer the temperature of the ground is to that of the space to be heated (strictly $T_{evaporator}$ to $T_{condenser}$), the higher will be the COP_h. In fact, the heating capacity of the heat pump, the heat released per second in the condenser $\dot{m}(h_3 - h_2)$, is 4.27 kW, and the power consumed by the compressor $\dot{m}$ $(h_2 - h_1)$ is 1.50 kW. So, not surprisingly, Equation 7.135 gives a lower COP_h of 2.85. Once the inefficiencies of the compressor and the work needed to pump the two water streams around the borehole and fan-coil condensers are taken into account this reduces further to 2.65. This is fairly typical of systems with a heating load of a few kW, but GSHPs with a COP_h as high as 4–6 are available commercially.

This demonstrates the key benefit of heat pumps as a highly efficient means not only of transferring heat but also of using electricity. Here, 1.5 kW of direct electrical filament heating is replaced by using that 1.5 kW of electricity to provide 4.3 kW of space or water heating, using solar heat stored in the ground. The actual rate of heat extraction from the ground is

$$\begin{aligned} \Delta\dot{H} = \dot{m}(h_1 - h_4) &= \dot{m}_w(h_8 - h_7) = \dot{m}_w c_P (T_8 - T_7) \\ &= 0.25 \times 4.18 \times (288 - 285) \text{ kJ/s} = 3.1 \text{ kJ/s} = 3.1 \text{ kW}. \end{aligned} \quad (7.137)$$

So the *specific performance* of the ground heat exchanger, a key parameter in determining the overall effectiveness of the ground heat coupling system, is 3,100/50 W/m = 62 W/m of borehole depth, which is typical of heat extraction rates in Europe. In terms of the *heating capacity* of the entire GSHP system, the borehole length requirement is 50/4.27 m/kW_{th} = 11.7 m/kW_{th}, which gives a design guideline for the sub-surface coupling system required to deliver a specified heating capacity.

An exergy analysis can be carried out in exactly the same way as for the molten salt heat storage system in Question 7.5.3 by following equations analogous to Equations 7.102 to 7.121. This gives

$$\text{exergy output from condenser for space heating:} \quad (\dot{E}_3 - \dot{E}_2) = 0.14 \text{ kW} \quad (7.138)$$

$$\text{net exergy input from ground heat storage:} \quad (\dot{E}_8 - \dot{E}_7) = 0.125 \text{ kW} \quad (7.139)$$

$$\text{pure exergy input as electrical energy to drive the compressor:} \quad (\dot{E}_2 - \dot{E}_1) = 1.50 \text{ kW,} \quad (7.140)$$

so that

$$\eta_{ex} = \frac{\dot{E}_3 - \dot{E}_2}{(\dot{E}_8 - \dot{E}_7) + (\dot{E}_2 - \dot{E}_1)} = \frac{0.14}{0.125 + 1.50} = 0.086 \text{ or } 8.6\%. \tag{7.141}$$

The low exergy efficiency reflects the fact that the heat extracted from the ground using a heat pump is low-quality (low-temperature) heat with low exergy content (100 × 0.125/3.1 = 4% of the actual heat extracted from the ground storage) with little potential to do useful work but good heating potential within a heat transfer device efficiently powered by high exergy electrical energy.

7.5.6 How Sustainable Is Using a Ground Source Heat Pump?

The small GSHP system of Question 7.5.5 extracts heat from the ground at the rate of 3.1 kW to deliver a heating capacity of 4.3 kW. Assuming an average annual household heating consumption of 12,500 kWh/a or 33 kWh/d = 33 × 3,600/(24 × 3,600) kW = 1.43 kW, we see that this plant would meet the heating needs of 4.3/1.43 = 3 households. It would need scaling by a factor of 1,000 or so (~1.5 MW) to reach the capacity of a viable district heating system.

Equation 7.137 shows that the heat flux extracted from the ground ($\dot{Q}_{ext}$) is 3.1 kW, which amounts to 3.1 × 3600 × 24 × 365 = 97.8 × 10^6 kJ or ~100 GJ per annum. An important question for the sustainability of GSHP pump use is: how quickly can solar radiation replenish this energy by heating the ground? The heat capacities and thermal conductivities of damp soil or rocks like granite are such that the ground heat exchanger can only effectively recover heat from up to a distance of r_h ~ 6 m from the borehole. So the ground does not act as an infinite heat sink; on the time scales of operation, for a ground heat exchanger of length L, heat recovery is restricted to a ground volume of

$$V_{heat} \sim \pi r_h{}^2 L = 3.142 \times 6^2 \times 50 \text{ m}^3 = 5{,}655 \text{ m}^3. \tag{7.142}$$

Using a "volumetric heat capacity" $C_{V,p}$ (damp soil) ~3 MJ/(m^{3}K), the subsurface ground temperature will be reduced from 16°C by

$$\Delta T = \frac{Q_{ext}}{V_{heat} C_{V,p}} = \frac{100 \times 10^9 \text{ K}}{5{,}655 \times 3 \times 10^6} = 6 \text{ K}. \tag{7.143}$$

Running the heat pump for a full year will therefore reduce the surrounding ground temperature to 10°C or for the six months of autumn and winter to 13°C. This depletion cannot continue indefinitely and needs to be offset by heat replenishment from solar heating. In many parts of the world, there is no shortage of high intensity solar radiation for much of the year. Taking a less favorable situation, northern Europe, the direct normal radiation on the Earth's surface H_e is ~1,000 kWh/m^2 or 3.6 GJ/m^2. The radiation incident on the surface area of the "insulated" volume V_{heat} around the ground heat exchanger is therefore

$$\text{Incident solar energy on } V_{heat},\ H_{e,s} = H_e \times \pi r_h{}^2 = 3.6 \times 113.0 \text{ GJ} = 406.8 \text{ GJ}, \tag{7.144}$$

which is almost ten times the amount of autumn/winter extracted energy we need to replace, $Q_{ext}/2 = 50$ GJ. However, we are not limited by the incident solar intensity but by the rate at which the solar heat can be transported to the cooled zone; the maximum heat flux through soil and rock is about 5 W/m^2. So the amount of heat that can be practically absorbed by the ground volume V_{heat} is

$$Q_{abs} = 5W \times \pi r_h^2 = 5 \times 113 \text{ W} = 565 \text{ W or } 565 \times 3600 \times 24 \times 365/2 \text{ J} = 9 \text{ GJ in 6 months.} \quad (7.145)$$

With this rate of heating it would take 50/9 = 5.5 summers to replenish the heat store, with an increase of about 1 K per annum. So for this system and conditions it is essential to use heat replenishment through injection of the rejected heat from using the heat pump in reverse during spring and summer for air conditioning.

To run the heat pump as an air conditioner in the summer for the building that was heated in the winter, the evaporator heat exchanger is connected to the fan coil system to extract heat from the air in the building to cool it down and the condenser heat exchanger is connected to the ground source system to reject heat to the water circulating in the borehole, which heats the surrounding ground to raise its temperature. If run under the same conditions and in the absence of heat losses, the rate of heat injected to the ground through the buried heat exchanger in air-conditioner mode is the same as the heat released from the fan-coil unit in space heating mode, $\dot{m}_w(h_2 - h_3) = 4.27$ kW, higher than that extracted in the evaporator from the cooled air. So the ground heat reservoir replenishment over six months is

$$\Delta H = 4.27 \times 3600 \times 24 \times 365/2 \text{ GJ} = 67.3 \text{ GJ} \quad (7.146)$$

Assuming the heat take-up is effectively restricted to the volume V_{heat}, then the rise in temperature of the subsurface will be

$$\Delta T = \frac{\Delta H}{V_{heat} C_{V,p}} = \frac{67.3 \times 10^9}{5{,}655 \times 3 \times 10^6} \text{ K} = 4.0 \text{ K}, \quad (7.147)$$

which brings the ground temperature at depth up to 17°C, with some of the power input to the compressor being converted to additional stored heat through the latent heat of the refrigerant released in the condenser. If the air conditioning unit was not part of a GSHP system, the rejected heat from the reverse heat pump cycle could of course be used to heat a domestic hot water system.

Note that the energy requirements for water pumping have not been included in the examples and inefficiencies and heat losses from the compressor, pumps and heat exchangers will mean that these ideal situations will not be realized in practice. Even so, for real systems the combination of GSHPs and air-conditioning units does give an efficient and reasonably sustainable system for heat management in houses and offices. They exploit sources of "free" heat, from the ground or from the air in hot buildings, both arising from natural solar heating. They are not readily or cheaply retrofitted but may perhaps be the system of choice for new builds, depending on the availability of hydrogen, the other major low-carbon heating option.

An alternative, and often cheaper, heat pump option is an air-source heat pump (ASHP), which operates on exactly the same thermodynamic principle as GSHPs. In fact, the air conditioner described previously is an example of such an ASHP. ASHPs are intrinsically less efficient than GSHPs because, whereas the temperature of the ground a few meters down is very stable throughout the year (10–20°C, depending on location and roughly equal to the mean annual air temperature), the air temperature has large daily and seasonal variations and can be quite close to, or even below, 0°C in regions where winter ASHP heating is required. This means that the heat pump evaporator will need to operate at a lower temperature, so that $(T_2 - T_1)$ in Equation 7.136 or $(h_2 - h_1)$ in Equation 7.135 will be higher for an ASHP. On the other hand, an ASHP external heat exchanger unit is readily mounted on the side of a building and is smaller and cheaper to install than a borehole or buried GSHP system. In addition, there is no need to replenish the extracted heat from the air, which is a concern for GHSPs, and ASHPs can continue to operate with COPs around 3 down to air temperatures of –15°C to supply energy for space and water heating. Ice formation from humid outside air can be a problem and here there will then be an additional energy requirement for de-icing.

7.5.7 How Can We Use Air to Store Energy?

An important "mechanical" large-scale energy storage method is to use intermittently generated or off-peak electricity to power a compressor which raises the pressure of a gas, specifically air, in *compressed air energy storage (CAES)*. The stored PV energy can then be released and used to regenerate electricity via gas turbines. The most common storage option is to use solution-mined salt caverns (where underground salt seams have been dissolved in water and pumped out) which provide well-insulated, large-volume underground storage sites. Porous deep saline aquifer formations or depleted gas reservoirs are also an option for air storage. In subsurface gas storage systems, pressures up to 80 bars, 8 MPa, are typically used. The alternative is to operate at constant pressure using a variable volume vessel, for instance positioned underwater at a depth of several hundred meters, where the hydrostatic pressure of the water column maintains a constant pressure. This has some advantages in increasing the efficiency of turbines and compressors working at constant inlet pressure conditions. By displacing water in a deep salt cavern into pipes connected to a surface reservoir, the cavern pressure can be maintained at the constant hydrostatic head and so also operate as a constant pressure system. Salt caverns are also used to store hydrogen to enable it to be used on demand for zero-carbon electricity generation in gas turbines.

7.5.7.1 Adiabatic compression mode

A major issue with CAES is how to manage the heat generated on compression of the air or absorbed on its subsequent expansion. The most straightforward method is to carry out the storage process adiabatically, and to return the heat generated during that compression stage to the surroundings in the subsequent release stage as the air expands and cools through an air turbine to produce electricity. For a perfectly insulated cavern operating reversibly, the theoretical efficiency of CAES is 100%. In practice, heat losses and the irreversibilities of the compression process result in a round trip efficiency of ~70%, although this will decrease for longer storage times due to larger heat losses. We can model this system approximately by assuming it to be a reversible, single-stage adiabatic compression for an open system in steady state. The energy that can be stored at 70 bar (7 MPa) in a salt cavern of volume 5×10^5 m^3 at a depth of 600 m is then given by Equation 1.107 of Question 1.6.5, using $\gamma_{air} = 1.4$:

$$(W_{s12})_{rev}^{adiab.} = \frac{\gamma p_1 V_1}{\gamma - 1}\left(\frac{T_2}{T_1} - 1\right) = \frac{\gamma p_1 V_1}{\gamma - 1}\left[\left(\frac{p_2}{p_1}\right)^{(\gamma-1)/\gamma} - 1\right]. \tag{7.148}$$

Since $p_1 V_1^{\gamma} = p_2 V_2^{\gamma}$ so $V_1 = V_2 (p_2/p_1)^{1/\gamma}$, we can rewrite Equation 7.148 as

$$(W_{s12})_{\text{rev}}^{\text{adiab.}} = \frac{\gamma\, p_1 V_2}{\gamma - 1}\left(\frac{p_2}{p_1}\right)^{1/\gamma}\left[\left(\frac{p_2}{p_1}\right)^{(\gamma-1)/\gamma} - 1\right] \tag{7.149}$$

$$\begin{aligned} &= \frac{1.4 \times 1 \times 10^5 \times 5 \times 10^5}{1.4 - 1}\left(\frac{70}{1}\right)^{1/1.4}\left[\left(\frac{70}{1}\right)^{(1.4-1)/1.4} - 1\right] \text{J} \\ &= 1.4 \times 12.5 \times 10^{10} \times 20.8 \times (3.37 - 1)\ \text{J} \\ &= 8,624\ \text{GJ}. \end{aligned} \tag{7.150}$$

This corresponds to a storage energy density, ρ_e = 8,624/500 MJ/m^3 = 17.3 MJ/m^3. Equation 1.102 tells us that the temperature of the air rises from, say, T_1 = 288 K at ambient to

$$T_2 = T_1\left[\left(\frac{p_2}{p_1}\right)^{(\gamma-1)/\gamma}\right] = 288 \times 3.37\ \text{K} = 970\ \text{K} \cong 697\ {}^\circ\text{C}. \tag{7.151}$$

Assuming air behaves as a perfect gas, the mass of air stored is

$$\begin{aligned} \text{Mass stored} &= nM_{\text{air}} = M_{\text{air}}\frac{PV}{RT} = \frac{29 \times 10^{-3} \times 70 \times 10^5 \times 5 \times 10^5}{8.314 \times 970}\text{kg} \\ &= 1.26 \times 10^7\ \text{kg or } 12600\ \text{t}. \end{aligned} \tag{7.152}$$

Hence, the energy stored per unit mass of air, precisely the specific enthalpy, h, is

$$h = \frac{8624 \times 10^6}{1.26 \times 10^7}\ \text{kJ/kg} = 684\ \text{kJ/kg}. \tag{7.153}$$

Since the real process will be unsteady and not completely open, culminating in the closed storage cavern, and we have not included any flow work effects within the cavern, Equation 7.151 probably gives a lower limit to the temperature rise and Equation 7.152 therefore an upper limit to the mass of air stored. However, the model still gives a reasonable approximation for the energies involved.

If the process is truly adiabatic, the cavern can be charged by intermittent renewable electricity supplies, or during off-peak periods when demand is low by a gas or nuclear generation plant. When power is needed, the hot pressurized gas can be released and expanded through a gas turbine to produce electricity, much as in the gas turbine section of a CCGT gas plant discussed in Question 6.2.2. Here, however, the gas has been heated by adiabatic compression rather than by gas combustion (see Figure 6.4). Flowing the gas through an air turbine at a typical flow rate of 150 kg/s would ideally generate 684 × 150 kJ/s = 103 MW. At this rate, the CAES plant could supply power for 1.26 × 10^7/(150 × 3600) h = 23.3 h, adequate to cover peak demand for several days.

In this ideal process, the cavern storage process is adiabatic and the compressor and turbine are reversible with efficiencies of 100%. In practice, the compressor and the turbine would typically have efficiencies of 0.85. So the round-trip energy storage efficiency η_{RT}, defined as (cf. Equation 7.122)

$$\eta_{RT} = \frac{(w_{el})_{out}}{(w_{el})_{in}}, \tag{7.154}$$

where $(w_{el})_{out}$ is the electrical energy produced by expanding the stored air in the gas turbine and $(w_{el})_{in}$ is the electrical energy used to compress the gas in the storage cavern in the charging phase, will not be as high as 1.0 but ~0.85 × 0.85 = 0.72, which is close to the efficiencies found in practice.

7.5.7.2 Isothermal compression mode

The more common approach is to carry out the charging process approximately isothermally by continuous heat exchange with the environment during the charging stage using, for instance, a multistage reciprocal compressor with interstage cooling or small compressors with a high surface-to-volume compression chamber. However, practical processes can only approximate to isothermal conditions and inevitably have irreversibilities. We can calculate the ideal work of isothermal compression by again assuming perfect gas behavior for the air and using Equation 1.93 for an isothermal expansion from state p_1V_1 to state p_2V_2:

$$(W_{s12})_{rev}^{isoth} = nRT \ln(p_2/p_1) = mR_sT \ln(p_2/p_1). \tag{7.155}$$

Since $nRT = p_1V_1 = p_2V_2$, then

$$(W_{s12})_{rev}^{isoth} = p_2V_2 \ln(p_2/p_1). \tag{7.156}$$

If we carry out the compression for the same CAES system described previously, but this time isothermally, we find

$$(W_{s12})_{rev}^{isoth} = 70 \times 10^5 \times 5 \times 10^5 \times \ln(70/1)\,\text{J} = 1487 \times 10^{10}\text{J} = 14870\,\text{GJ}. \tag{7.157}$$

Although this is 1.7 times the energy required for the adiabatic compression, the mass of gas stored isothermally is significantly greater as the temperature stays close to ambient, here taken as 288 K. The perfect gas estimate of gas stored changes to 42440 t and hence

$$\text{Energy stored per unit mass of air, } h = \frac{14870 \times 10^6\,\text{J}}{4.24 \times 10^7\,\text{kg}} = 350\,\text{kJ/kg}. \tag{7.158}$$

This is half the value for the adiabatic process (Equation 7.153).

The advantage of CAES in terms of subsequent power generation is that the stored gas on discharge to an air turbine is already at high pressure. The compression stage, which accounts for 30–40% of the energy input to the gas turbine section of a CCGT plant (see Table 6.1, Question 6.2.2), is therefore decoupled from the power generation plant and serves the dual function of combining energy storage with the first stage of a power generation cycle, such that on discharge the storage medium becomes the working fluid.

However, the major issue with the near-isothermal CAES process is that in removing heat to keep the temperature almost constant, just above ambient, the compressed gas does not have enough energy to produce sufficient useful work when expanded through the gas turbine in the discharge stage, when it will cool further, causing ice formation problems with humid air. It is therefore necessary to add heat as it discharges from the cavern to raise its temperature to ~500°C prior to expansion. There are several ways to do this in practice:

- The pressurized air can be discharged into a combustion chamber fed with natural gas which is burned to heat the gas. It can then be expanded in a high-pressure turbine with an option to re-heat it on exit at an intermediate pressure and generate more power in a low-pressure turbine before being discharged into the atmosphere.
- Capture the heat released in the near-isothermal multi-stage compression process using high-efficiency heat exchangers, store this energy using, for example, molten salt thermal storage as in Question 7.5.3 and then release it to heat the air on discharge, creating a quasi-adiabatic system without the need for major additional energy input. From a thermodynamic point of view, this is the preferable approach.
- Integrate the air turbine into a CCGT power plant and divert some of the exhaust gas from the gas turbine, which normally provides heat to raise steam for the second cycle in a heat recovery steam generator (see Question 6.2.2), to heat the discharged air before it enters the air turbine.
- If the stored gas is hydrogen, then it can be directly burned in a gas turbine, with the added advantage that the compression energy and costs are largely covered by the storage process which also acts as the first stage of the power generation cycle.

7.5.8 How Can We Use Water Systems to Store Energy?

One of the best-established methods for large-scale energy storage is pumped hydro energy storage (PHES). It is based on the same principles as hydroelectric electricity generation, whereby the vast potential energy stored in large volumes of water in reservoirs and lakes situated at different heights is exploited for energy conversion and, in this case, storage. Electricity from intermittent sources such as wind and solar, or surplus off-peak electricity from continuous base-load facilities waiting to be used when demand or prices are higher, is used to pump water from a low elevation water system to a higher level, converting electrical pumping energy into water potential energy (see Figure 7.22).

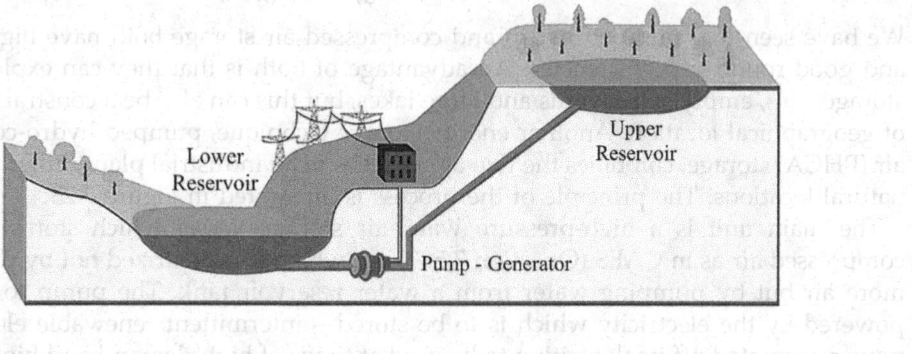

FIGURE 7.22
Schematic diagram of pumped hydro energy storage system with reversible pump-turbine assembly connecting the lower and upper water reservoirs.

When electricity is required, the water is discharged back to the lower level through hydro-turbines to produce electricity. The pump and turbine are usually combined into a single, reversible assembly. In contrast to hydroelectric power systems, the lakes and reservoirs used are relatively small (tens of millions of m^3) and the storage plants are very agile, being able to switch on and off quickly (zero to 1 GW in 10 s is typical) and respond to both demand and supply fluctuations very efficiently.

PHES is the most widespread high-capacity, grid-scale energy storage method in the world. Its challenges are that it requires an appropriate geographical location with large-volume lakes separated by ~500 m in altitude, although artificial twin reservoir systems have been built. A typical energy storage is about 1 GWh per million m^3 of water, which means that pumped storage systems have quite low-energy densities, ~3 kJ/L, compared with lithium ion batteries at ~3 MJ/L. So to serve GW scale power stations and electricity grids PHES needs large water volumes, large height differences between reservoirs and large flow rates. The largest systems can be rated at 2–3 GW. The energy stored in 10 million m^3 of water 500 m above the lower reservoir is

$$\begin{aligned}\text{Stored potential energy, } E_{\text{pot}} &= \rho\, Vz\, g = 10^3 \times 10 \times 10^6 \times 500 \times 9.807\ \text{J} \\ &= 4904 \times 10^{10}\,\text{J} = 49040\ \text{GJ} = 13.6\ \text{GWh},\end{aligned} \tag{7.159}$$

where ρ is the water density, V the water volume, z the height of the reservoir and g the acceleration due to gravity. This energy can be recovered by flowing the elevated water under gravity through a turbine generator; at a flow rate of 400 m^3/s, this gives a potential power generation capacity of

$$\text{Power capacity, } P = 49040 \times 400/10 \times 10^6\ \text{W} = 1.96\ \text{GW}. \tag{7.160}$$

The system involves just hydraulic and mechanical energy with the main losses coming from electrical and mechanical inefficiencies in the pump-turbine system and evaporative losses. They therefore operate with high efficiency, varying between 70% and 80% depending on the installation. They are costly systems to install but are reliable and can remain in service for many decades. Variations on the standard natural reservoir systems include underground reservoirs exploiting both natural systems and abandoned mines, and smaller facilities for distributed energy storage linked to decentralized wind and solar power.

7.5.9 What Is the Potential for Hybrid Energy Storage Systems?

We have seen that pumped hydro and compressed air storage both have high capacity and good round-trip efficiencies. An advantage of both is that they can exploit natural storage sites, empty salt caverns and large lakes, but this can also be a constraint in terms of geographical location. Another energy storage technique, pumped hydro-compressed air (PHCA) storage, combines the two approaches in an industrial plant rather than using natural locations. The principle of the process is illustrated in Figure 7.23.

The main unit is a high-pressure water-air storage vessel which stores energy as compressed air as in CAES (Question 7.5.7), but which is pressurized not by pumping in more air but by pumping water from a water reservoir tank. The pump to do this is powered by the electricity which is to be stored – intermittent renewable electricity or power generated off-peak waiting to be used at times of high-demand and higher price – just as in PHES (Question 7.5.8). When electricity is needed, it is produced by release of

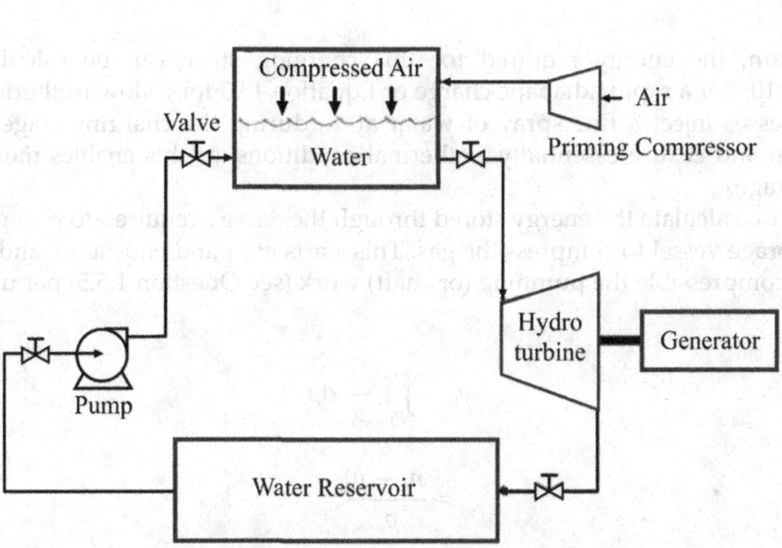

FIGURE 7.23
Schematic diagram of a pumped hydro-compressed air (PHCA) energy storage system.

the pressurized water through a hydro-turbine connected to a generator, as in PHES (exploiting the high efficiency of such a turbine), rather than release of the air, which remains in the storage vessel ready to be re-compressed in the next storage cycle. After passing through the turbine, the water returns to its reservoir, also ready for use in the next cycle. A small compressor is used to prime the storage vessel with air to a pre-set pressure at the start of the process; after this, there is a closed water-air system that acts as an energy pump to take in, store and release electrical energy. The process can be viewed as the pressurized air creating a virtual dam between the water reservoir and the storage vessel; an interior pressure of 6 MPa represents a dam with a virtual height of 600 m.

During charging with the water being pumped into the storage vessel, the air is compressed and its temperature rises in a polytropic process (for which pV^n = constant, a generalization of processes ranging between isothermal ($n = 1$) and reversible adiabatic ($n = \gamma$) conditions; see Question 1.6.5). The compressed air exchanges heat with the water which moderates the temperature rise of the air. If the charging process is very slow, the rate of heat transfer can be high enough to keep the air temperature essentially constant and the process approaches a near-isothermal compression with $n = 1$. On the other hand, if the charging is very rapid, heat cannot be transferred from the heated air to the water fast enough to decrease the air temperature significantly and the charging comes close to an adiabatic, isentropic compression with $n = \gamma$. During discharge involving removal of the water and expansion of the air, similar processes can occur: quasi-isothermal expansion for slow discharge or near-isentropic expansion when the discharge rate is large.

The storage vessel containing air at atmospheric pressure p_0 is initially pressurized to a pre-set pressure p_1 using the small compressor. Once this is done, the compressor is no longer required for normal operation. The work required for this stage can be calculated using Equation 7.149 and the mass of gas contained within the vessel using Equation 7.152. For a storage vessel of 100 m^3 pre-pressurized to 10 bar (1 MPa), the mass stored is 0.63 t with an energy of 120 MJ. The water pump, powered by the electricity to be stored, then compresses this air to the target storage pressure p_2. Depending on the rate of compression and the rate of heat transfer from the compressing air to the incoming water stream, the energy required for this charging stage can be calculated from Equation 1.105 for a rapid adiabatic charge or Equation 1.93 for a slow isothermal charge. Some processes inject a fine spray of water at T_0 during the charging stage to absorb further heat and ensure essentially isothermal conditions, as this enables more efficient energy storage.

We can also calculate the energy stored through the energy required to pump the water into the storage vessel to compress the gas. This starts at p_1 and ends at p_2, and assuming water is incompressible the pumping (or shaft) work (see Question 1.5.5) per unit mass is given by

$$w_p = \int_{p_1}^{p_2} \frac{1}{\rho_w} dp \tag{7.161}$$

$$= \frac{p_2 - p_1}{\rho_w}, \tag{7.162}$$

where ρ_w is the density of water under the conditions of the storage vessel. At the end of the essentially isothermal charging phase, the water volume V_w introduced into the storage vessel, volume V_s, is

$$V_w = V_s - V_{air} = V_s - \frac{p_1 V_s}{p_2} = V_s \frac{(p_2 - p_1)}{p_2}. \tag{7.163}$$

Since the total mass of water charged (m_p) is $V_w \rho_w$, the total water pumping energy consumption, using Equation 7.163 for V_w and assuming perfect efficiency, is

$$W_{p,id} = m_p w_p = V_w \rho_w \frac{(p_2 - p_1)}{\rho_w} = V_s \frac{(p_2 - p_1)^2}{p_2}. \tag{7.164}$$

Neglecting losses in the hydro-turbine/generator system, then the total amount of electricity generated W_{el} will be equal to $W_{p,id}$. The main losses in this essentially mechanical and hydraulic energy system are the result of fluid and mechanical friction within the equipment but it is still very efficient compared with heat engines. Typical values are η_p = 85% for water pumps and in excess of η_t = 90% for hydropower turbines; electrical generators are even more efficient at η_{gen} = 98–99%. Taking these into account, the round-trip efficiency of a PHCA energy storage system of this type, relating the actual electrical output to the actual input pumping energy, is

$$\eta_{PHCA} = \frac{W_{el}}{W_p} = \frac{W_{p,id}\, \eta_t\, \eta_{gen}}{W_{p,id}/\eta_t} = \eta_p\, \eta_t\, \eta_{gen}. \tag{7.165}$$

Using η_p = 0.85, η_t = 0.90 and η_{gen} = 1.0, η_{PHCA} will be ~0.75. Irreversibilities and heat losses in the compression stage are small and so the efficiency of PHCA is largely a result of the efficiency of the hydraulic machinery.

The sensitivity of the storage to the process parameters can be demonstrated by defining a storage parameter ζ, which is the ratio of the storage volumes occupied by water and air. Using Equation 7.163, we find

$$\zeta = \frac{V_w}{V_{air}} = \frac{V_w}{V_s - V_w} = \frac{p_2 - p_1}{p_1}. \tag{7.166}$$

Using this parameter, we can re-write Equation 7.164 as

$$W_p = \frac{V_s\, p_1\, \zeta^2}{\eta_p(1+\zeta)}. \tag{7.167}$$

As an example, consider a system having a storage vessel of volume 100 m^3, with an initial pre-set pressure of 20 bar (2 MPa). If we store the energy at a final pressure of 200 bar (20 MPa), then from Equation 7.166 we see that $\zeta = (200 - 20)/20 = 9$ and using Equation 7.167 we find that the stored energy is

$$W_p = \frac{100 \times 2 \times 10^6 \times 9^2}{0.85 \times (1+9)}\ \text{J} = 190.6 \times 10^7\ \text{J} = 1.91\ \text{GJ}, \tag{7.168}$$

with a storage density ρ_e of 1,906/100 MJ/m^3 = 19.1 MJ/m^3. This is a similar value to the storage density for adiabatic CAES (the example in Question 7.5.7 was 17.3 MJ/m^3). Although PHCA can generally use higher pressures than CAES, which is limited by the fracture stress of the rock in which the underground cavern is formed, CAES uses vastly greater storage volumes (by a factor 1,000 or more) and so the total energy capacity of PHCA will be significantly smaller, but still substantial; for this example using Equation 7.165, $W_{el} = 1.91 \times 0.85 \times 0.9$ GJ = 1.46 GJ = 0.4 MWh of electricity per storage cycle. A 100 m^3 cylindrical storage vessel could be 5 m high and 2.5 m in radius, so has a small footprint.

From Equation 7.164, we see that

$$W_p = \frac{V_s(p_2 - p_1)}{\eta_p}\left(1 - \frac{p_1}{p_2}\right), \tag{7.169}$$

so that for a given difference between the storage and pre-set pressures, maximizing their ratio will lead to greater energy storage. For our example system, decreasing the initial pre-set pressure to 10 bar, 1 MPa will increase the energy storage density by 10% to 21.2 MJ/m^3.

7.5.10 How Do Efficiencies, Scale and Capacities of Different Storage and Conversion Processes Compare?

In Questions 7.5.1 to 7.5.9, we examined a range of methods for storing energy. These differ in scale, capacity and efficiency and it is clear that no single approach can be used for all situations. Some are better suited to storing relatively small amounts of energy to address the needs of portability for devices and transport (e.g. batteries) while others can store extremely large quantities of energy to meet the demands of energy management at the large grid scale. Here we bring together and compare a number of factors that characterize the storage processes and which together enable a choice to be made about which method to use in different situations. We also include in the comparison two major energy conversion processes for electricity generation: hydrogen fuel cells (Question 7.2.3) and solar PV (Questions 7.2.5 and 7.5.3). The comparison measures are those introduced in Question 7.5:

- *Efficiency*: For the energy storage processes we use the round trip efficiency, η_{RT}, whereas for fuel cells and solar PV we use the efficiency of conversion of the input energy content into electrical energy, η_{th}.
- *Energy density, ρ_e*: The amount of energy stored or produced per unit mass, volume or surface area of the system, a measure of the size of device needed for particular applications and its portability.
- *Power delivery*: Overall power (or sometimes total energy capacity) that can be delivered from a single charge-discharge cycle for storage or with continuous supply of energy source for fuel cells and solar PV. This gives an indication of the scale on which the different methods can be deployed.

The comparison is given in Table 7.7; the values given are typical of the device or process and intended to indicate the order of magnitude of the characterizing parameters rather than definitive values for each system. The latter is not possible of course because any precise value will depend on specific design features and process conditions and in many cases can vary over a significant range.

TABLE 7.7

A comparison of the key characteristics of a range of energy storage and conversion methods

Method	Efficiency η	Energy Density ρ_e	Total Energy Delivery (per single charge for storage) and/or Power
Energy Storage			
Batteries			
Lead Acid	0.85	0.3 MJ/L; 150 kJ/kg	2MJ, 100 MW
Lithium Ion	0.95	3 MJ/L; 1.7 MJ/kg	430 kJ (cell); 750 MJ (pack)
Molten Salt	0.88	150 kJ/L	150 MW
Thermal		270 kJ/L	20 GJ p.a. heating
Ground Source Heat Pump	3–4 (COP)	20 MJ/m^3	100 GJ p.a., 3.2 kW with air-conditioning replenishment
Compressed Air	0.72	25 MJ/m^3	8–15 TJ, 1 GW
CCGT integration	0.63	500 kJ/kg	
Pumped Hydro	0.7–0.8	3–5 kJ/L	50 TJ, 2–3 GW
Pumped Hydro Compressed Air	0.75	20 MJ/m^3	2 GJ_{st} → 420 kWh
Energy Conversion			
Solar			
PV panel	0.25	220 W/m^2 (peak) to ~ 20 W/m^2 (mean)	7 GJ/m^2 p.a. (peak) to ~ 0.7 GJ/m^2 p.a. (mean)
CSP	0.75	750 W/m^2	100 MW_{th} → 25 MW_{el}
H_2 Fuel Cells			
SOFC – micro-CHP	0.85	8 MJ/L H_2(l)	2 kW_{el}/1kW_{th}
SOFC – electric	0.60	6 MJ/L H_2(l)	5 kW–10 MW
PEM	0.50	5 MJ/L H_2(l)	300 kW

Battery technology is the area where innovation is improving performance most rapidly and where many of the storage issues may eventually come together all the way from micro-devices to grid-scale power. We can use the data in Table 7.7 to illustrate how the practical viability of batteries for off-grid domestic electricity supply has improved over the last few decades and how they compare with alternatives. We will take the average monthly household electricity usage to be 350 kWh, which is a reasonable global average; some countries such as USA use up to three times this amount, whereas others like China use about a third. If this is all to be supplied by batteries then, assuming recharging at a monthly frequency is acceptable, the battery stack must deliver a total electrical energy E of

$$E = 350 \times 3600 \text{ kJ} = 1.26 \times 10^6 \text{ kJ}. \tag{7.170}$$

In the 1980s, the best battery option was probably the lead acid batteries routinely used for auxiliary power in petroleum powered road vehicles. The volume of battery required would be $E/\rho_e = 4.2 \text{ m}^3$. The battery stack would therefore occupy the space of about eight large domestic freezers.

Battery technology has improved significantly over the past four decades to the point where lithium-ion batteries are available for both electric cars and residential use. Using the typical energy density from Table 7.7 of 3 MJ/L gives the volume of batteries required to power the average home for a month as 0.42 m^3. The energy storage space is now reduced to the size of a medium domestic freezer and can almost be part of the furniture. The major drawback of this energy solution is the enormous cost of these re-chargeable batteries; in 2020, a high-performance lithium-ion electric car battery cost approximately 50% of the cost of the vehicle. The performance of rechargeable batteries will continue to improve and the cost will decrease as technology improves and their use for transport, domestic power and other applications increases.

If we used a SOFC hydrogen fuel cell working at an efficiency of 80% (see Question 7.2.3.2) to power this typical house for a month, then with a liquid hydrogen energy density of 6 MJ/dm^3, the volume of the hydrogen storage tank required would be 0.2 m^3. So the tank would be 200 L, which is roughly one barrel of oil, similar in size to a domestic oven. A promising option that mitigates the safety risk of storing large volumes of hydrogen is the use of liquid organic hydrogen carriers (LOHC) with an energy density similar to those of liquid hydrogen (Haupt and Müller 2017). An alternative to batteries and fuel cells for stand-alone power generation is solar energy. To power our typical house consuming 350 kWh over a month requires $350 \times 3.6 \times 10^6/(30 \times 24 \times 3{,}600)$ W = 486 W, which for a typical 2020 PV Panel from Table 7.7 will require a roof area of $486/20 \text{ m}^2 = 24 \text{ m}^2$, which is comparable with the south-facing roof area of a typical house. In many climates, the best use of the solar electricity is to recharge the batteries to provide low-carbon renewable electricity round the clock.

These examples and Questions 7.5.1–7.5.9 illustrate how important energy storage options are at all scales in order to take best advantage of increasing supplies of intermittent low-carbon electricity. Storage technologies will continue to evolve over the coming decades. However, it is thermodynamics that sets the limits on what is achievable in terms of efficiency and inter-energy conversion and the thermodynamic approaches described previously will underpin the design and optimization of new versions of the methods covered and any new approaches to energy conversion and storage.

7.6 What Does Thermodynamics Tell Us About Wind Energy?

The most common way to extract the kinetic energy from a wind flow is to use a horizontal axis wind turbine, which converts it directly to electrical energy. The shaft of the rotor blades is connected through a gearbox to a generator mounted at the top of the vertical tower structure. To maximize utilization of the wind, the system can rotate on the vertical tower to orientate the rotor blades perpendicular to the prevailing wind. The wind impinging on the rotor blades flows through a certain area upstream of the blades and, as kinetic energy is extracted, the wind slows down so must redistribute to a wider area. This is illustrated in Figure 7.24.

For a constant density fluid, the cross-sectional area of the stream boundary will vary inversely with the fluid speed. If we apply continuity of mass through each of the cross sections of the stream tube shown in Figure 7.24, then the mass flow rate of air $\dot{m}_a$ is given by

$$\dot{m}_a = \rho A_1 u_1 = \rho A_r u_r = \rho A_2 u_2, \tag{7.171}$$

where A_r is the swept area of the rotor ($=\pi L^2$ where L is the rotor length), $A_{1,2}$ are the corresponding cross-sectional areas of the control volume upstream and downstream of the rotor, respectively, with $u_{1,2,r}$ the respective air velocities at these positions and ρ the density of the air. The rate of change of momentum, $\dot{M}$, is this flow rate times the overall change in air velocity:

$$\dot{M} = m_a(u_1 - u_2) = \rho A_r u_r (u_1 - u_2). \tag{7.172}$$

This is equal to the force on the rotor blades so the power P absorbed by the turbine from the wind is

$$P = \rho A_r u_r (u_1 - u_2)\, u_r. \tag{7.173}$$

Now we can also calculate the power from the rate of change of the kinetic energy of the wind moving through the turbine:

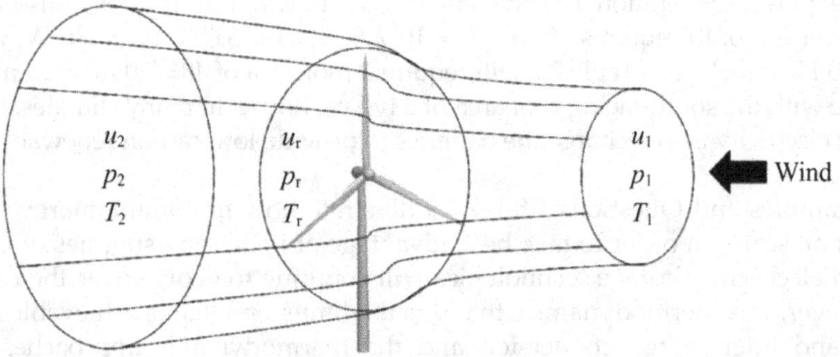

FIGURE 7.24
Schematic diagram of wind stream tube for wind turbine flow. u_1, p_1 and T_1 are the wind speed, pressure and temperature respectively upstream of the turbine; subscripts 2 and r represent the same quantities downstream and at the position of the rotor.

$$P = \frac{1}{2}m_a(u_1^2 - u_2^2) = \frac{1}{2}\rho A_r u_r(u_1^2 - u_2^2), \tag{7.174}$$

which is a form of the Bernoulli Equation 1.79. Equating Equations 7.173 and 7.174, we find that

$$\frac{1}{2}(u_1^2 - u_2^2) = \frac{1}{2}(u_1 - u_2)(u_1 + u_2), \tag{7.175}$$

so that the air speed at the rotor is

$$u_r = \frac{1}{2}(u_1 + u_2) \tag{7.176}$$

and the reductions in the wind speed upstream and downstream of the rotor turn out to be the same. Using this in Equation 7.173, we can re-write the power absorbed from the wind as

$$P = \frac{1}{4}\rho A_r(u_1 + u_2)^2(u_1 - u_2), \tag{7.177}$$

which can be re-expressed in terms of the speed ratio $\omega = u_2/u_1$ as

$$P_{\text{wind}} = \frac{1}{4}\rho A_r u_1^3(1 + \omega)^2(1 - \omega). \tag{7.178}$$

This indicates that the power increases as the cube of the wind speed, so the power generated increases dramatically with wind speed. Differentiating the expression for P with respect to ω and setting the result equal to zero shows that P is a maximum when $\omega = 1/3$. Under these conditions, the outgoing wind leaves at a third of the speed it had approaching the turbine. This maximum power is therefore

$$P_{\text{max}} = \frac{16}{27}\frac{1}{2}\rho A_r u_1^3. \tag{7.179}$$

Defining the *power coefficient* P_w ($=P_{\text{out}}/P_{\text{in}}$, essentially the First Law efficiency for wind power generation) as the ratio of the power that can be extracted from the wind (Equation 7.178) to the kinetic power of the wind in a moving cylinder which has velocity u_1 and a cross-sectional area A_r, which is $\frac{1}{2}\rho A_r u_1^3$, then

$$P = P_w \tfrac{1}{2}\, \rho A_r u_1^3 \text{ with } P_{w,\,\text{max}} = 16/27 = 0.593. \tag{7.180}$$

This is known as *the Lanchester-Betz limit*, which indicates that at best only about 60% of the wind energy can be extracted using a wind turbine, even before other inefficiencies of the wind-turbine energy transfer and of the turbine-generator system are taken into account. A physical explanation of this is that if all the kinetic energy of the wind was converted into useful work through interaction with the rotor blades, then the downstream wind speed would be zero. If it stopped moving at the exit from the turbine, no more wind could enter and the turbine would be blocked. In order to keep the wind moving through the turbine, there has to be some wind movement downstream with residual energy left in it.

The systems are designed to constrain the power input when the inflow wind speed exceeds a rated value, either by altering the pitch of the blades or moving the rotor away from being head-on to the wind, with a consequent drop in the power coefficient, P_w. Taking into account aerodynamic, transmission and generation inefficiencies, modern turbines achieve 70–80% of the Lanchester-Betz limit, giving an actual P_w of 0.45 when generating at the rated wind speed. If we include the power reductions when operating below the rated wind speed and the downtimes at extreme wind speeds, this translates into capacity factors of ~0.25 on land and up to ~0.40 offshore. Applying the above analysis to a wind turbine having rotor blades of length $L = 60$ m operating at a rated wind speed u_1 of 12 m/s, then using Equation 7.180 with $\rho = 1.23$ kg/m^3 (air at sea level) and $P_w = 0.45$, the power produced operating at the rated wind speed, the nameplate capacity, is

$$P = 0.45 \times 0.5 \times 1.23 \times \pi \times 60^2 \times 12^3 \text{ W} = 5.4 \text{ MW}. \quad (7.181)$$

The overall capacity factor of an offshore system is typically only 40% of the nameplate capacity, so the actual power generated would be 0.4 × 5.4 MW = 2.2 MW. The annual electricity production is 2.2 × 3,600 × 24 × 365 MJ/a = 69.4 TJ/a or 19.3 GWh/a, enough to supply about 5,000 households with electricity. This illustrates the high generating capacity of a single wind turbine given the appropriate weather conditions; however replacing traditional large fossil fuel power plants with windfarms results in a much greater land footprint. As part of a windfarm, the turbine spacing has to be at least ten times the rotor length L to avoid wind interference and shadows; our example case would require $(10 \times 60)^2$ m^2 = 360,000 m^2 per turbine, giving a power density for the windfarm of 6 W/m^2 (only about 30% of that delivered by a typical solar panel; see Table 7.7). Wind systems are therefore best suited to locations with a lot of available area and wind (e.g. offshore) but not much sun.

7.7 Can We Use the Kinetic Energy in Water Waves and Tides to Produce Electricity?

Tidal currents also possess significant kinetic energy and it is possible to harness this energy in much the same way as for wind. Although marine turbines are designed using similar principles to the wind turbines discussed in Question 7.6, there are some significant differences in exploiting tidal currents. The major one is that water has a much higher density than air, that of seawater being 1,025 kg/m^3 compared with air at 1.23 kg/m^3, a factor of almost 1,000. This makes the power of water flow almost 1,000 times greater than the blowing wind. The theoretical limit of marine turbines is still the Lanchester-Betz maximum, but they tend to be less efficient than wind turbines, with $P_w \sim 0.35$.

We can examine how these two factors play out in terms of the size of marine turbines by looking at the length of rotor blade required to generate the same power as the wind example in Question 7.6, 5.4 MW, for a typical rated tidal flow speed of 2.5 m/s. Using Equation 7.180, we find

$$\begin{aligned} P &= P_w \tfrac{1}{2} \rho A_r v_1^3 = 0.35 \times 0.5 \times 1025 \times \pi \times L^2 \times 2.5^3 \text{ J/(m}^2\text{s)} \\ &= 8805 \times L^2 \text{ J/(m}^2\text{s)} = 5.4 \times 10^6 \text{ J/s}. \end{aligned} \quad (7.182)$$

So *L* turns out to be 24.8 m, about half the length of the air turbine blade. Assuming that marine turbines can be assembled on the seabed into tidal energy farms in much the same way as wind farms then a 1 MW facility could be constructed by using 100 10 kW turbines, for which the required blade length would be only 1.1 m. If the required spacing for seabed water turbines is similar to that for wind turbines, 10 L, then the seabed area required for the 100 turbines would be 22,222 m^2; i.e. a relatively small plot size of 150 m square, with a power density of 45 W/m^2. This demonstrates the significant potential advantage that the much higher fluid density of water gives marine over wind turbine generation in terms of power density, provided adequate tidal currents can be located.

7.8 How Can We Use Thermodynamics to Compare the Range of Energy Sources and Conversion/Storage Systems to Help Decide Which Systems to Use for Different Applications?

In this chapter, we have seen how a thermodynamic analysis can tell us that, no matter how hard we try, there are limits to the efficiency of energy extraction and interconversion that we cannot go beyond. The data and analysis for different energy options are distributed amongst the seven sets of questions we have posed in order to explore the energy transition landscape, from fossil fuels (see also Chapter 6) to various low/zero carbon forms of energy. We now ask five more critical questions that will enable us to pull together the data for different systems and make direct comparisons of how they compare with each other and for which applications they are likely to be best suited.

7.8.1 How Does the Energy Content of Fossil Fuels Degrade as We Use them for Power and Hydrogen?

Using the thermodynamic data and calculations described in this chapter and Chapter 6, we can construct the energy staircase shown in Figure 7.25.

This diagram starts with 3GJ of solar energy, equivalent to the radiation incident on the area of a tennis court for three hours on a sunny day where the solar radiation intensity is equal to the global average of ~1,000 W/m^2. At the typical efficiency of photosynthesis, ~2%, this is enough energy to create the plant biomass required to produce about 1 kg of fossil fuel with an energy content of ~50 MJ (gas 53.6 MJ/kg, oil 41.9 MJ/kg, coal ~30.0 MJ/kg). The diagram shows how the availability of energy decreases as the fossil fuels are refined and used to produce electrical power and hydrogen.

Points to note from Figure 7.25 are that despite their high specific energies, and energy densities in the case of the liquid and solid fuels, their energy value is considerably degraded by the time they reach the point of use. In the case of petrol and diesel, this is mainly due to the inherent inefficiencies of the internal combustion engine (see Question 6.4). The superior performance of CCGT power generation, even with CCS added, is clear. Despite the environmental credentials of hydrogen, it is seen that using decarbonized fossil fuel electricity to produce it by water electrolysis results in an energy carrier that can deliver only ~40% at most of the energy in the primary fuel used to generate it, natural gas. We will see shortly (Questions 7.8.2 and 7.8.3) that SMR with CCS or using renewable electricity

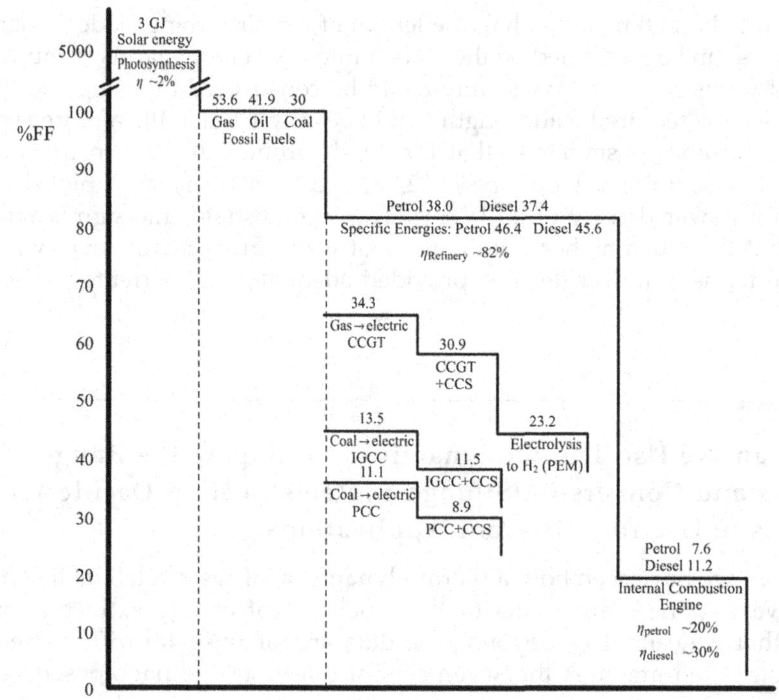

FIGURE 7.25
The fossil fuel power staircase. The numbers on each step are the energy content in MJ/kg (thermal or electric), for the primary fuels taken from Table 7.1 in Question 7.2.1 and for subsequent conversion steps as described in Questions 6.3, 7.2 and their sub-questions. The efficiency η of different processes is taken as a typical value, also given in the answers to those questions. The left-hand axis gives the percentage of the original fossil fuel energy remaining at each stage of the energy conversion process.

electrolysis are much more energy-efficient routes to generate hydrogen. The figure also shows why it is much better to use CCS with gas power plants than to retro-fit coal plants, even for the more efficient IGCC systems.

7.8.2 How Efficiently Can We Use Natural Gas for Power and Conversion to Other Fuels?

In Figure 7.26, we draw on the data from Questions 6.3 and 7.2 to show the energy cascade from natural gas to power and other energy vectors. Here we see that producing hydrogen from natural gas by SMR is energetically the most favorable route to hydrogen, as it is in terms of cost and carbon footprint if CCS is added to the process, shown in more detail in Question 7.2.5. The use of the exothermic water gas shift (WGS) reaction to extract additional hydrogen from water (see Questions 7.2(a) and 7.2.5) increases the yield and thermal efficiency, although adding CCS brings the net specific energy down by 8.8 MJ/kg NG. This step is necessary for both routes to hydrogen using SMR in order that the product can be classified as "blue" (decarbonized route from fossil fuels). The diagram shows how relatively inefficient the use of solar power for PEM electrolysis is, which makes it difficult to compete on cost with decarbonized SMR with the current (2020) PV conversion efficiencies. Solar photocatalysis, using synthetic band gap materials attempting to mimic natural photosynthesis systems (an "artificial leaf") through

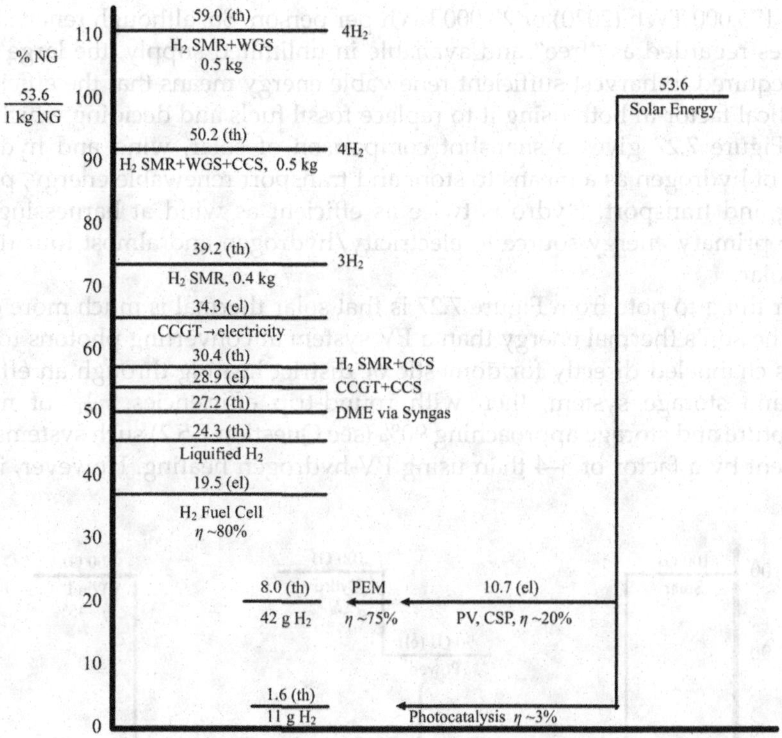

FIGURE 7.26
Gaseous energy vectors: natural gas and hydrogen. The numbers are the net specific energies remaining in the system at each stage (thermal or electric), starting from 1 kg of natural gas (NG) or, on the right, from an equivalent amount of solar energy. The calculations of the net energies can be found in the answers to Questions 6.3 and 7.2. The efficiencies η assumed for different processes are given where relevant.

light-activated direct electron transfer without the need for electricity, is currently only about 3% efficient at best and a long way from competing with electrolysis routes for hydrogen. We can see that when hydrogen produced by the SMR route is used in a fuel cell, less than 40% of the original energy in the primary NG fuel is utilized.

Natural gas can be converted to other liquid energy vectors such as DME (see Question 7.3.4) with a thermal efficiency of about 50%. This fuel has the advantage of having a reasonable volumetric energy density similar to LNG (see Table 7.1 in the answer to Question 7.2.1), while being capable of being stored as a liquid at normal temperatures and moderate pressures. It has a C:H ratio of 1:3 compared to the 1:2 of the alkanes in petrol (gasoline) and diesel, giving a significantly lower carbon footprint. Methanol can also be made from NG (see Question 7.3.4) and has a lower net energy content at 18.7 MJ/kg after synthesis by this route, only 35% of that in the original NG but with a comparable C:H ratio of 1:4. Both DME and methanol have applications as lower carbon substitutes/additives for diesel and petrol, respectively.

7.8.3 How Do Renewable Energy Sources Compare for Different Applications?

The main renewable energy sources, solar, wind and hydro, all have a low energy density compared to fossil fuels, with the energy collection and conversion processes all requiring very large land areas if they are to provide the majority of the world's annual energy demand of 175,000 TWh (2020) or 23,000 kWh per person. So, although renewable energy is sometimes regarded as "free" and available in unlimited supply, the large capital investment required to harvest sufficient renewable energy means that the efficiency of its use is a critical factor in both using it to replace fossil fuels and deciding which source to invest in. Figure 7.27 gives a snapshot comparison of solar, wind and hydro for the generation of hydrogen as a means to store and transport renewable energy, particularly for heating and transport. Hydro is twice as efficient as wind at harnessing and converting the primary energy source to electricity/hydrogen and almost four times as efficient as solar.

The other thing to note from Figure 7.27 is that solar thermal is much more efficient at capturing the sun's thermal energy than a PV system at converting photons to electrons. So if this is channeled directly for domestic or district heating through an efficient heat exchange and storage system, then with round-trip efficiencies, η_{RT}, of molten salt thermal capture and storage approaching 90% (see Question 7.5.2) such systems are much more efficient by a factor of 3–4 than using PV-hydrogen heating. However, if the solar

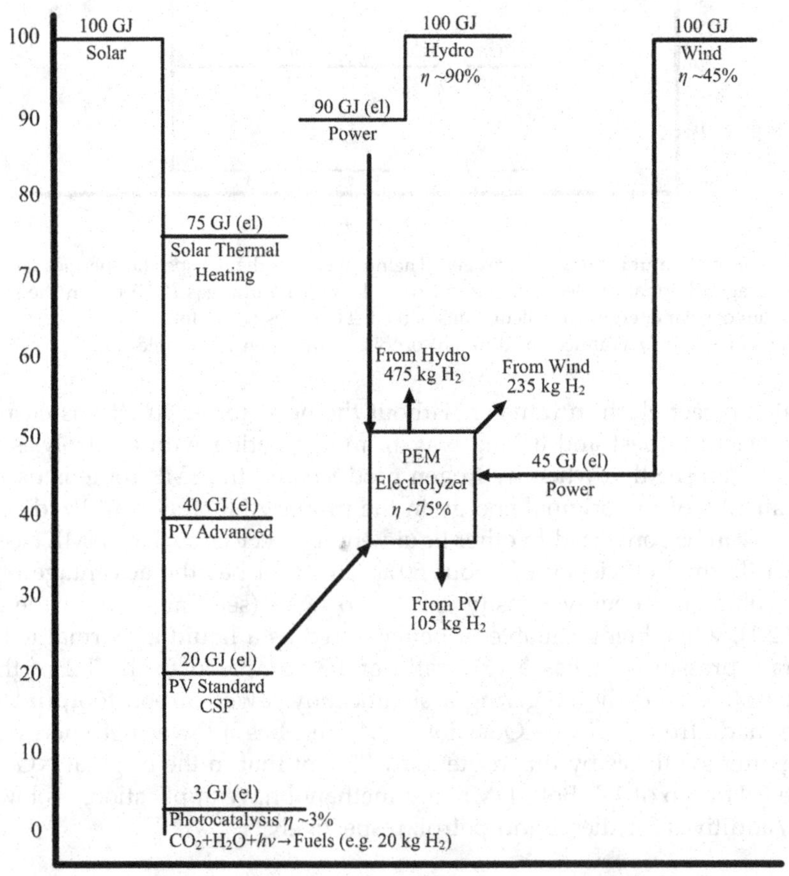

FIGURE 7.27
Renewable energy vectors and production of "green" hydrogen. The figure compares the efficiency of power generation from the renewable source and the amount of renewable hydrogen that can be produced from an initial 100 GJ of each energy source. Efficiencies are based on 2020 typical performance relative to the thermodynamic limits. PV = photovoltaic; CSP = concentrated solar power; PEM = polymer exchange membrane.

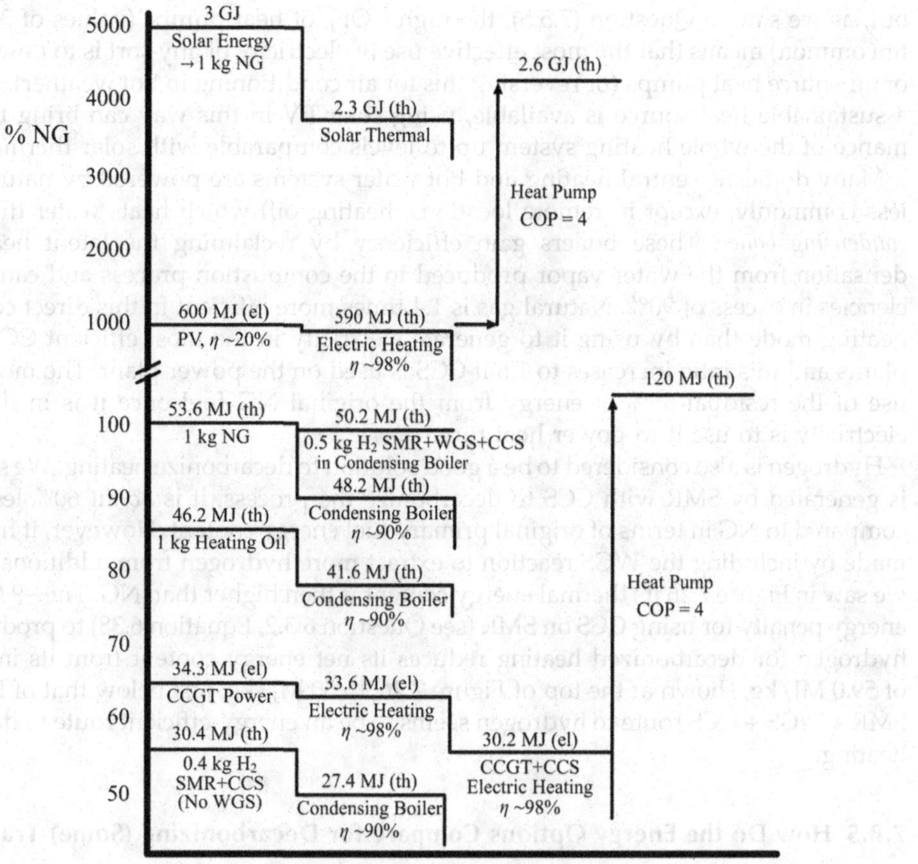

FIGURE 7.28
Energy content of different heating sources. The numbers on each step correspond to the net specific energy at the end of each step in MJ/kg (thermal or electric) when starting from 1 kg of NG, or from 3 GJ of solar energy which was converted over time to 1 kg of NG. The efficiencies η and COPs are given where relevant, based on typical values given earlier.

thermal heat is converted into electricity as in concentrated solar power (CSP) then the Second Law inefficiencies of converting heat to electrical work result in CSP having a similar conversion efficiency as PV at 20–25%.

7.8.4 Which Energy Sources Are Best Suited for Domestic and District Heating?

One advantage for heating systems is that if the primary source of energy is thermal, there is the opportunity to transfer it directly to the space to be heated at high efficiency, without large entropy or availability losses in the conversion of the thermal energy into electricity. There will always be exergy losses in heat exchangers and pumps as these cannot operate completely reversibly, but these losses can be minimized by, for instance, trying to ensure that temperature differences in heat exchanger streams are kept as low as possible (see Question 3.8). Figure 7.28 compares a range of energy sources commonly used for heating applications.

The high efficiency of solar thermal for heating was noted in Question 7.8.3. Solar PV electricity can be used for resistive space and water heating with little drop in efficiency but, as we saw in Question (7.5.5), the high COP_h of heat pumps (values of 3–6 are not uncommon) means that the most effective use of electricity of any sort is to power ground or air source heat pumps (or reversing this for air conditioning in hot weather). As long as a sustainable heat source is available, using solar PV in this way can bring the performance of the whole heating system up to levels comparable with solar thermal.

Many domestic central heating and hot water systems are powered by natural gas (or less commonly, except in remote locations, heating oil) which heats water directly in a *condensing boiler*. These boilers gain efficiency by reclaiming the latent heat of condensation from the water vapor produced in the combustion process and can have efficiencies in excess of 90%. Natural gas is 1.4 times more efficient in this direct combustion heating mode than by using it to generate electricity in the most efficient CCGT power plants and this ratio increases to 1.6 if CCS is used on the power plant. The most efficient use of the residual 56% of energy from the original NG fuel once it is in the form of electricity is to use it to power heat pumps.

Hydrogen is also considered to be a good solution to decarbonize heating. We see that if it is generated by SMR with CCS to decarbonize the process, it is about 60% less efficient compared to NG in terms of original primary fuel energy content. However, if hydrogen is made by including the WGS reaction to extract more hydrogen from additional water, as we saw in Figure 7.26 its thermal energy content is then higher than NG. The ~9 MJ/kg NG energy penalty for using CCS on SMR (see Question 6.3.2, Equation 6.38) to produce "blue" hydrogen for decarbonized heating reduces its net energy content from its initial value of 59.0 MJ/kg, shown at the top of Figure 7.26, to 50 MJ/kg, just below that of NG. So the SMR + WGS + CCS route to hydrogen seems to be an energy-efficient route to decarbonize heating.

7.8.5 How Do the Energy Options Compare for Decarbonizing (Some) Transport?

There are two main routes for decarbonizing road transport. One direction is to use battery-driven vehicles, where a big challenge is to increase the capacity of Li-ion or other portable storage devices to increase the range between charging. The other option is to use liquid, compressed or chemically bound hydrogen as the fuel to power either fuel cells or modified internal combustion engines. Figure 7.29 compares the performance of different routes to hydrogen and portable electricity.

On the left-hand side of Figure 7.29 we have fossil fuel sources and we start with 1 kg of crude oil or 0.78 kg of NG, which have the same specific energy, 42 MJ/kg. The oil is refined to fuels and other products; here we focus on petrol (gasoline) which retains over 80% of the net thermal energy content of the original crude. We see that the use of this in an internal combustion engine (ICE) having a typical efficiency of ~0.25 results in an overall use ("well-to-wheels") of about 20% of the original oil energy to propel the vehicle (option F). The next option is the use of NG in a CCGT power plant, fitted (option D) or not (option C) with CCS. Only option D will contribute to decarbonization and if this "blue" electricity is used to charge the battery of an electric vehicle, of the original 27 MJ(el)/kg energy content of the NG, 18 MJ(el)/kg is available to power the vehicle, with a well-to-wheels efficiency of 0.43. Liquid hydrogen produced from NG by SMR + CCS, with the added liquefaction energy penalty of about 20%, may be used in a fuel cell vehicle, where there are inefficiencies in both the fuel cell and the battery/electric motor system. The overall net energy used by the vehicle in this case (option E) is 12 MJ(el)/kg, representing a (NG) well-to-wheels efficiency of only 0.29.

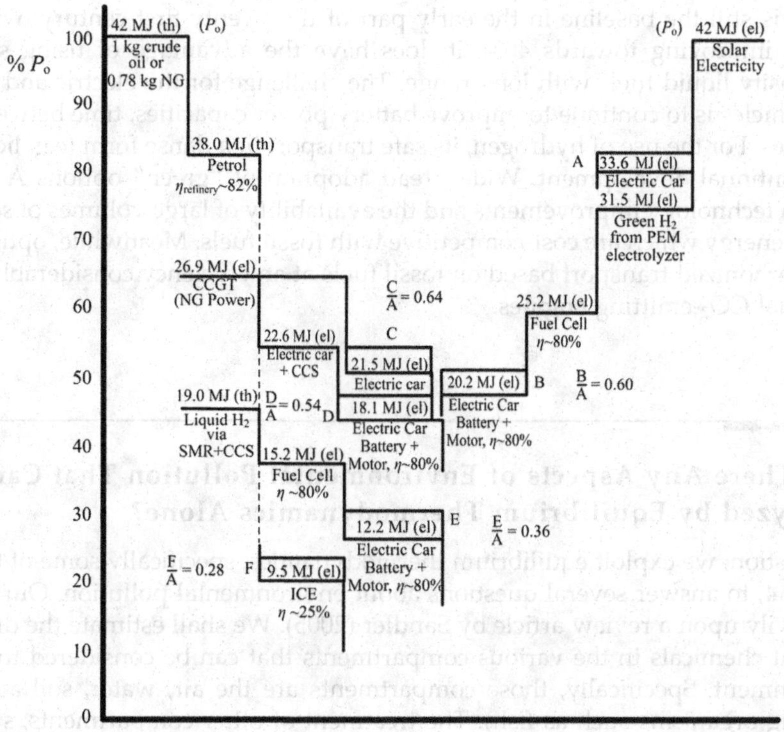

FIGURE 7.29
Energy vectors for transport and their relative energy content. The data are taken mainly from Questions 6.3.2, 6.2.4, 7.2.1-5 and 7.5.10, and from the other figures in Questions 7.8.1-4. The numbers on the steps are net energy content values in MJ/kg, thermal and electric. The left-hand axis shows the percentage of the original source energy (oil, NG or solar) that remains available after the cascade of energy conversion processes and hence gives the "well-to-wheels" or "PV-to wheels" efficiencies for the various transport options. These efficiencies are compared for options B–E with the most efficient option A, a solar-charged electric battery vehicle (see text).

On the right-hand side of Figure 7.29, we start with the same quantity of energy 45 MJ, this time from solar electricity. This can be used in an electric battery vehicle (option A) with a PV-to-wheels efficiency of 0.8 or higher, essentially the efficiency of the battery-motor train. This electricity could also be used to produce "green" hydrogen using a PEM (or other) electrolyzer (efficiency ~0.75), producing 31.5 MJ(th), which powers the fuel cell to produce electricity which in turn feeds the battery-motor unit, delivering 20.2 MJ(el) with an overall PV-to-wheels efficiency of 0.48 (option B).

Figure 7.29 gives the ratio of the net available energy to power each of these options to option A, the solar-powered electric battery vehicle, which has a clear advantage over all the other options. The order is

1. Non-decarbonized fossil fuel electricity: C/A = 0.64
2. Renewable "green" hydrogen fuel cell: B/A = 0.60
3. Decarbonized fossil fuel electricity: D/A = 0.54
4. SMR "blue" hydrogen fuel cell: E/A = 0.36
5. Petrol ICE: F/A = 0.28

Petrol ICE is still the baseline in the early part of the twenty-first century; with engine efficiencies improving towards 40%, it does have the advantage of using safe, high-energy density liquid fuels with long range. The challenge for the electric and hydrogen fuel cell vehicles is to continue to improve battery power capacities, time between charge and lifetimes. For the use of hydrogen, its safe transport in a dense form (e.g. liquid) is an area for continual development. Widespread adoption of "green" options A and B requires both technology improvements and the availability of large volumes of solar, wind and hydro energy which are cost-competitive with fossil fuels. Meanwhile, options D and E give decarbonized transport based on fossil fuels at an efficiency considerably superior to traditional CO_2-emitting engines.

7.9 Are There Any Aspects of Environmental Pollution That Can Be Analyzed by Equilibrium Thermodynamics Alone?

In this question, we exploit equilibrium thermodynamics, specifically some of the results of Chapter 4, to answer several questions about environmental pollution. Our treatment draws heavily upon a review article by Sandler (2005). We shall estimate the distribution of pollutant chemicals in the various compartments that can be considered to comprise the environment. Specifically, those compartments are the air, water, soil and aquatic biota (living organisms such as fish). The treatment of other compartments, such as vegetable biota or soil sediments, is also possible (Sandler 2005), but we shall not deal with them here.

We shall confine our attention to those chemical pollutants that have a long lifetime undegraded within the environmental compartments. Such pollutants are the insecticide dichlorodiphenyltrichloroethane (DDT), polychlorinated biphenyls (PCBs), dioxins and similar chemicals which have lifetimes in the environment of several decades. For these materials, it is reasonable to assume that their distribution among the environmental compartments is determined by phase equilibrium. On the other hand, many chemicals may be more rapidly degraded by hydrolysis, oxidation or by photochemical or microbial processes so that their fate will be determined by the kinetics of those processes rather than any equilibrium concerns.

Since we are concerned with estimates of the levels of pollutants in the various components of the environment, we shall adopt a common model for both the aquatic biota and soil that retains the essential features of those complex systems that matter to phase equilibrium but introduces significant simplification. In particular, we shall use a very simple model for the lipids in the bodies of aquatic and animal biota as well as the organic content of soil and sediments. In each case, we shall represent these systems using *n*-octanol. Of course, lipids and the organic content of soils have more complex molecular features than *n*-octanol but there are many reasons why *n*-octanol is a suitable model for phase equilibrium in these systems. First, when *n*-octanol is mixed with water under ambient conditions, two liquid phases form in which one is almost pure water and the other is about 73 mol percent *n*-octanol. If a small amount of a third material is added and allowed to come to equilibrium its concentration will be very different in the two phases. A hydrophilic material will have a higher concentration in the water phase while a hydrophobic material will be more concentrated in the *n*-octanol phase. These are exactly the characteristics displayed by the lipid systems in animals and the organic material in soil. Secondly, there is a great deal of information available on the n-octanol-water system and the distribution of a variety of materials within the two phases that we can exploit to model biota in equilibrium with water.

7.9.1 How Can We Estimate the Concentration of a Noxious Volatile Material in a Large Body of Water Near a Source?

In Question 4.6.6, we explained how to model phase equilibrium using activity coefficients and fugacity. In order to answer the question posed in this section, we need to consider phase equilibrium between the air and water compartments of the environment.

The general condition for phase equilibrium was expressed as

$$\tilde{p}_{B,\alpha} = \tilde{p}_{B,\beta} = \cdots = \tilde{p}_B(\pi), \tag{7.183}$$

where $\tilde{p}$ is the fugacity of each substance B of the mixture of C components and the phases are indicated by α, β, γ … π.

For the case of equilibrium between the air (A) and water (W) compartments of the environment, we may write this for a single species i to be distributed as

$$\tilde{p}_{i,A}(T, p, x_{i,A}) = \tilde{p}_{i,W}(T, p, x_{i,W}), \tag{7.184}$$

where $x_{i,A}$ is the mole fraction of species i in the air and $x_{i,W}$ its mole fraction in the water.

Now we shall assume that air is composed of a single component and that it behaves as an ideal gas under ambient conditions. In that case the fugacity of species i, which is present in the air in very small concentration, is from Equation 4.72:

$$\tilde{p}_{i,A}(T, p, x_{i,A}) = x_{i,A}\, p = \frac{C_{i,A}}{C_A} p = C_{i,A} RT, \tag{7.185}$$

where $C_{i,A}$ and C_A are the molar concentrations (mol m^{-3}) of the species i and of the air, respectively.

Using Equation 4.95, the fugacity of the species i in water is

$$\tilde{p}_{i,W}(T, p, x_{i,W}) = x_{i,W}\, \gamma_{i,W}(T, p, x_{i,W})\tilde{p}_i^* = x_{i,W}\, \gamma_{i,W}(T, p, x_{i,W})\tilde{p}_i^{sat}. \tag{7.186}$$

The pure component fugacity of the species i as a liquid, $\tilde{p}_i^*$, has been replaced by its saturation pressure at the relevant environmental temperature and $\gamma_{i,W}(T, p, x_{i,W})$ is the activity coefficient of the species i in water (Question 4.6.6). The Poynting factor of Equations 4.95 and 4.178 has been set to unity owing to the closeness of the pressures to the saturation pressure. Generally, the concentration of pollutants in the aqueous phase will be quite small, so that it is legitimate to replace the activity coefficient $\gamma_{i,W}(T, p, x_{i,W})$ by its value at infinite dilution, which we denote by γ_i^∞. Although the activity coefficient at infinite dilution in aqueous solution and the vapor pressure of many substances have been measured, it is not true for all pollutants of interest. So it is sometimes useful also to use as an alternative the Henry's law coefficient of Question 4.11.2, which is shown in Question 5.4.2.1 to be the equivalent infinite dilution activity coefficient, $\gamma_{i,W}^{\infty,m}$, when we use molality, $m_{i,W}$, as the concentration measure rather than mole fraction. So we can write

$$\tilde{p}_{i,\mathrm{W}}(T, p, x_{i,\mathrm{W}}) = x_{i,\mathrm{W}}\, \gamma^{\infty}_{i,\mathrm{w}}(T, p, x_{i,\mathrm{W}}) p_i^{\mathrm{sat}} = m_{i,\mathrm{W}} H_i, \tag{7.187}$$

where $m_{i,\mathrm{w}}$ is the molality of species i in the aqueous solution and H_i is the Henry's law coefficient defined in Equation 4.252.

We can now return to Equation 7.184 and, using Equations 7.185 to 7.187, the equality of the fugacities of species i in the water and air phases becomes

$$\tilde{p}_{i,\mathrm{A}} = C_{i,\mathrm{A}} RT = x_{i,\mathrm{W}}\, \gamma^{\infty}_{i,\mathrm{w}}(T, p) p_i^{\mathrm{sat}} = m_{i,\mathrm{W}} H_i = \tilde{p}_{i,\mathrm{W}}. \tag{7.188}$$

If we now define the air-water concentration ratio or partition coefficient (Question 4.6.9, Equations 4.179 and 4.180) of species i, as $K_{\mathrm{AW},i}$, we obtain

$$K_{\mathrm{AW},i} = \frac{C_{i,\mathrm{A}}}{C_{i,\mathrm{W}}} = \frac{C_{i,\mathrm{A}}}{\rho_{\mathrm{W}} m_{i,\mathrm{w}}} = \frac{C_{i,\mathrm{A}}}{x_{i,\mathrm{w}} C_{\mathrm{W}}} = \frac{\gamma^{\infty}_{i,\mathrm{w}}(T, p) p_i^{\mathrm{sat}}}{\rho_{\mathrm{W}} RT} = \frac{H_i}{\rho_{\mathrm{W}} RT}, \tag{7.189}$$

where C_{W} is the molar density of water. Thus, if we know the equilibrium concentration of the pollutant, species i, in the air and its infinite dilution activity coefficient and saturation vapor pressure at the temperature in question, then we can calculate its concentration in water. Alternatively, we may use its Henry's law constant in water if we have that number available.

As an example, we consider the distribution of the chemical benzo[*a*]pyrene, a toxic polycyclic aromatic hydrocarbon (PAH) that is released into the atmosphere by the partial combustion of coal and coke. Its concentration in the air in Europe has been monitored and values for the levels in several locations published by the European Commission (2001). In London, UK, a figure of 6.3 ng/m^3 was reported as a mean value. The vapor pressure of benzo[*a*]pyrene at 25°C is 2.13×10^{-5} Pa, its activity coefficient at infinite dilution is $\gamma^{\infty}_{i,\mathrm{W}}(T, p) = 3.78 \times 10^{8}$ and its molar mass is 0.252 kg/mol. Thus, using Equation 7.189 at 298 K, we find for this substance

$$\frac{C_{i,\mathrm{A}}}{C_{i,\mathrm{W}}} = 5.8 \times 10^{-5}. \tag{7.190}$$

It follows that the concentration of benzo[*a*]pyrene in natural water systems near to a source of the material will be approximately 110 µg/m^3.

7.9.2 How May the Concentration of a Pollutant in a Fish Be Several Orders Greater Than in the Water in Which the Fish Lives?

Our earlier discussion of the importance of the representation of the lipid content of aquatic biota (fish) by *n*-octanol suggests that an important parameter for characterizing the concentration of a chemical in fish relative to its concentration in water will be the octanol-water partition coefficient of a chemical species i. Denoting the octanol phase by O we can define

$$K_{\mathrm{OW},i} = \frac{C_{i,\mathrm{O}}}{C_{i,\mathrm{W}}}, \tag{7.191}$$

where $C_{i,O}$ and $C_{i,W}$ are the equilibrium concentrations in the octanol-rich and water-rich phases, respectively. For the purposes of estimation, here we use a correlation for this partition coefficient developed by Sandler (1996) that relates it to the infinite dilution activity coefficient of species in the water-rich phase:

$$\log_{10} K_{OW,i} = 0.486 + 0.806 \log_{10} \gamma_{i,W}^{\infty}. \tag{7.192}$$

We also now define a biota-water partition coefficient for species i, $K_{BW,i}$, as

$$K_{BW,i} = \frac{C_{i,B}}{C_{i,W}}, \tag{7.193}$$

and it is usual to approximate this by

$$K_{BW,i} = z_B K_{OW,i}, \tag{7.194}$$

where z_B is the fraction of the biota mass that is lipid. Here we should note that a similar notion may be followed for other forms of biota such as terrestrial animals or vegetable matter using the fraction of each form of biota that is lipid which is, of course, a considerable approximation (Sandler 2005).

We can use our previous results to estimate the mean concentration of benzo[*a*]pyrene in fish on a whole animal basis assuming that for fish the fraction of lipids, z_B, is 5%. First, using the earlier value for $\gamma_{i,W}^{\infty}(T, p) = 3.78 \times 10^8$, we find that $K_{OW,i} = 2.67 \times 10^6$ from Equation 7.192 so that, using the concentration of benzo[*a*]pyrene in water found earlier, we expect the concentration in fish to be

$$C_{i,\text{fish}} = C_{i,B} = 0.05 \times 2.67 \times 10^6 \times 110\ \mu\text{g/m}^3, \tag{7.195}$$

or 14.6 g/m^3. We see that the concentration in fish is much higher than in the water in which it lives, almost entirely because the organic pollutants are hydrophobic and typically have large values of the octanol-water partition coefficient. Typical measured values of PAH concentrations in fish (Eisler 1987) are 5–100 g/m^3, indicating that the octanol lipid proxy gives a reasonable estimate of the solubility of such solutes in living fish. Specifically, $K_{B,w,i}$ is an effective proxy for the so-called "bioconcentration factor" in biota systems. Studies show that lethal concentration 50% (LC_{50}) values can be as low as 200 mg/m^3 for invertebrates such as crustaceans and earthworms (Honda and Suzuki 2020) and for fish are in the region of 50–100 g/m^3 (Oneyemacchi and Okaliwe 2018).

7.9.3 Can We Estimate the Concentration of a PCB in Soil?

In order to estimate the concentration of a chemical in the soil we need to know the fraction of the soil that is organic content, which can be modeled using data for *n*-octanol and a partition coefficient. It is observed empirically that the soil-water partition coefficient, $K_{SW,i}$, can be correlated as (Sandler 2005)

$$K_{SW,i} = \frac{C_{i,S}}{C_{i,W}} = 4.1 \times 10^{-4} z_S \rho_S K_{OW,i}, \tag{7.196}$$

where z_S is the fraction of soil that is organic content (typically 2%) and ρ_s the soil mass density (typically 1,500 kg/m^3). For the estimated concentration of benzo[*a*]pyrene in London water, $C_{i,W} = 110\mu g/m^3$, this leads, through Equation 7.196, to a soil concentration of $C_{i,s} = 3.6\ g/m^3$.

Observations of the concentration of benzo[*a*]pyrene in Ontario, Canada, have found concentrations in fish of 0.14 g/m^3 and in soil of 0.1 g/m^3. The concentration of benzo [*a*]pyrene in Ontario water is a factor of four less than that in the London study and the estimates provided by our simple analysis may therefore be regarded as quite satisfactory. We should note that if ideal solution behavior had been assumed, then the concentrations we would have found would have been six orders of magnitude smaller.

In the context of this analysis, Sandler (2005) has also discussed magnification of the concentration of strongly hydrophobic chemicals as one moves up a food chain but the topic is beyond the scope of this work and the reader should consult Sandler (2005) and references therein.

Some possible sources of data for the quantities needed for this type of calculation are included in Chapter 9.

References

Agarwal A.S., Rode E., Sridhar N., and Hill D., 2015, Conversion of CO_2 to Value-Added Chemicals: Opportunities and Challenges, in *Handbook of Climate Change Mitigation and Adaptation*, eds. Chen W.Y., Suzuki T., and Lackner M., Springer, New York, NY.

Appleby A.J., 1990, "From Sir William Grove to today: Fuel cells and the future", *J. Power Sources* **29**: 3–11.

Arona S., and Prasad R., 2016, "An overview on dry reforming of methane: Strategies to reduce carbonaceous deactivation of catalysts", *RSC Adv.* **6**: 108668–108688.

Atkins P.W., and de Paula J., 2014, *Atkins' Physical Chemistry*, 10th ed., Oxford University Press, Oxford, UK.

Belessiotis V., Kalogirou S., and Delyannis E., 2016, *Thermal Solar Desalination*, Academic Press, US.

Brandani S., 2012, "Carbon dioxide capture from air: A simple analysis", *Energy Environ.*, **23**: 319–328.

Breeze P., 2017, *Combined Heat and Power*, 1st ed., Academic Press, London, UK.

Daggash H.A., Fajardy M., and Mac Dowell N., 2020, *Negative Emission Technologies*, in *Carbon Capture and Storage*, Bui M. and Mac Dowell N., RSC Publishing, Cambridge UK, pp 447–511.

Darensbourg D.J., and Yeung A.D., 2013, "Thermodynamics of the carbon dioxide-epoxide copolymerisation and kinetics of the metal-free degradation: A computational study", *Macromolecules*, **46**: 83–95.

Dicks A.L., and Rand D. A. J., 2018, *Fuel Cell Systems Explained*, 3rd ed., John Wiley and Sons Ltd., Chichester, UK.

Dumont, O., Frate, G. F., Pillai, A., Lecompte, S., and Lemort, V., 2020, "Carnot battery technology: A state-of-the-art review", *J. Energy Storage* **32**: 101756.

Eisler R., 1987, "Polycyclic aromatic hydrocarbon hazards to fish, wildlife and invertebrates; a synoptic review", *Contaminant Hazard Reviews Report No. 11*, U.S. Fish and Wildlife Service, Patuxent Wildlife Research Center, Laurel, MD, U.S.A.

European Commission, 2001, "Ambient Air Pollution by Polycyclic Aromatic Hydrocarbons (PAH)", https://ec.europa.eu/environment/archives/air/pdf/annex_pah.pdf

Fischer J., Lehman T., and Heitz E., 1981, "The production of oxalic acid from CO_2 and H_2O", *J. Appl. Electrochem.* **11**: 743–750.

Fletcher A.S., 2015, *Thermal Energy Storage Using Phase Change Materials – Fundamentals and Applications*, Springer Briefs in Thermal Engineering and Applies Science, Springer International Publishing.

Grove W.R., 1839, "On voltaic series and the combination of gases by platinum", *Phil. Mag.*, **21**: 127–130; 1842, "On a gaseous voltaic battery", *Phil. Mag.*, **24**: 417–420; 1845, "On the gas voltaic battery: Voltaic action of phosphorus, sulphur and hydrocarbons", *Phil. Trans. Roy. Soc.*, **135**: 351–361.

Haupt, A., and Müller, K., 2017, "Integration of a LOHC storage into a heat-controlled CHP system", *Energy* **118**: 1123–1130.

Honda M., and Suzuki N., 2020, " Toxicities of polycyclic aromatic hydrocarbons for aquatic animals", *Int. J. Environ. Res. Public Health* **17**: 1363–1385.

IEAGHG Technical Report 2017-02, 2017, *"Techno-economic evaluation of SMR standalone (merchant) hydrogen plant with CCS,"* International Energy Agency, Paris, France.

Jia J., Seltz L.C., Benck J.D., Huo Y., Chen Y., Ng J.W.D., Bilir T., Harris J.S., and Jaramillo T.F., 2016, "Solar water splitting by photovoltaic-electrolysis with a solar-to-hydrogen efficiency over 30%", *Nat. Commun.* **7**: 13237.

Kalamaras C.M., and Efstathiou A.M., 2013, "Hydrogen production technologies: Current state and future developments", *Conf. Papers Sci.* **2013**: 690627.

Kim Y.S. and Pivovar B., 2007, "Polymer electrolyte membranes for direct methanol fuel cells", *Adv. In Fuel Cells*, **1**, Chapter 4, pp. 187–234.

Kusoglu A., and Weber A.Z., 2017, "New insights into perfluorinated sulfonic acid ionomers", *Chem. Rev.* **117**: 987–1104.

Mahmoudi H., Mahmoudi M., Doustdar O., Jahangiri H., Tsolakis A., Gu S., and LechWyszynski M. 2017, "A review of Fischer Tropsch synthesis process, mechanism, surface chemistry and catalyst formulation", *Biofuels Eng.* **2**: 11–31.

Metz, B., Davidson, O., Coninck, H., Loos, M., and Meyer, L. (Eds), 2005, *IPPC Special Report on Carbon Capture and Storage*, Cambridge University Press, UK.

Mond L. and Langer C., 1889, "A new form of gas battery", *Proc. Roy. Soc., London*, **46**: 296–304.

Müller K., Mokrushina L.,and Arlt W., 2014, "Thermodynamic constraints for the utilization of CO_2", *Chem. Ing. Tech.* **86**: 497–503.

Muñoz C.A., and Fernández A.M., (Eds), 2012, *Urea: Synthesis, Properties and Uses*, Chemical Engineering Methods and Technology Series, Nova Science Publishers, 266 pp.

National Academies of Science, Engineering and Medicine, 2019, *Negative Emissions Technologies and Reliable Sequestration - A Research Agenda*, Chapter 6, Carbon mineralization of CO_2, pp. 247–318.

Nolan, C.J., Field, C.B., and Mach, K.J. 2021, "Constraints and enablers for increasing carbon storage in the terrestrial biosphere", *Nat. Rev. Earth Environ.* **2**: 436–446.

Olwig, R., Hirsch, T., Sattler, C., Glade, H., Schmeken, L., Will, S., and Messalem, R., 2012, "Techno-economic analysis of combined concentrating solar power and desalination plant configurations in Israel and Jordan", *Desalination Water Treat.* **41**: 9–25.

Oneyemacchi E.C., and Okaliwe E.F., 2018, "Polycyclic aromatic hydrocarbons (PAH) in fingerlings of *Clarias gariepinus* (Burchell, 1822) exposed to petroleum", www.preprints.org, January 2018, 10.20944/preprints201801.009.v1.

Parrott J.E., 1992, "Thermodynamics of solar cell efficiency", *Sol Energy Mater. Sol. Cells* **25**: 73–85.

Preuster P., Alekseev A., and Wasserscheid P., 2017, "Hydrogen Storage Technologies for Future Energy Systems", *Annu. Rev. Chem. Biomol. Eng.* **8**: 445–471.

Rogers G. F. C. and Mayhew Y. R., 1994, *Thermodynamic and Transport Properties of Fluids, 5th Edition*, Wiley-Blackwell, Oxford, UK, 32 pp.

Sandler S.I., 1996, "Infinite dilution activity coefficients in chemical, environmental and biochemical engineering", *Fluid Phase Equilib.***116**: 343–353.

Sandler S.I., 2005, *Environmental Pollution*, Ch.2, *Chemical Thermodynamics for Industry*, ed. Letcher, T., for IUPAC, Royal Society of Chemistry, Cambridge UK.

Sella A, 2020, "Grove's gaseous voltaic battery", *Chemistry World*, June 2020, Article 4011756.

Sievi, G., Geburtig, D., Skeledzic, T., Bösmann, A., Preuster, P., Brummel, O., and Wasserscheid, P., 2019, "Towards an efficient liquid organic hydrogen carrier fuel cell concept", *Energy Environ. Sci.* **12**: 2305–2314.

Stanescu I., Gupta R.R., and Achenie L.E.K., 2006, "An *in-silico* study of solvent effects on the Kolbe-Schmitt reaction using a DFT method", *Mol. Simul.* **32**: 279–290.

The Chemical Company, Products: Urea, 2021: https://thechemco.com/chemical/urea.

Wang T., Lackner K.S., and Wright A., 2011, "Moisture swing sorbent for carbon dioxide capture from ambient air", *Environ. Sci. Tech.* **45**: 6670–6675.

8

Biothermodynamics

8.1 Are Large Biological Molecules in Aqueous Solutions Any Different from Familiar Molecules in Organic Mixtures?

Large molecules with significance in life processes differ in three main aspects from familiar molecules in conventional chemical reactions. These are, first, the special significance of the three-dimensional arrangement of atoms, second, the crowded environment in which the molecules typically occur and third, their charge (Figure 8.1).

The three-dimensional arrangement of proteins is an example of the primary structure of large biomolecules defined by the sequence of amino acids in the polypeptide chain. Interactions between the amino acid side chains determine additionally the conformation and behavior of the proteins. These interactions define the secondary and tertiary structure (Figure 8.1a). The quaternary structure (not presented) considers additionally the spatial structure of the entire protein complex with all of its subunits. All of these fine structures are responsible for the proper functioning of the macromolecules.

Living biological cells are crowded with macromolecules (Figure 8.1c). For instance, the cytosol (the liquid inside a cell) of *E. coli* contains about 300–400 mg cm^{-3} macromolecules (Zimmerman et al. 1991). The crowding reduces the volume available for the solvent molecules and affects drastically their effective concentration. Therefore, biological macromolecules behave differently in the cytosol than in the usual buffered test assays in a laboratory.

Almost all molecules occurring in living systems (e.g. metabolites, proteins, nucleic acids, lipids and functionalized carbohydrates) carry discrete fixed charges (shown schematically for carboxylic acid and amino groups in Figure 8.1b). Furthermore, resins used for chromatography and biomolecule separation or clay interacting with microorganisms in ecosystems are also charged species. The specifics of thermodynamic considerations for charged (bio-)molecules are thus highly important in life sciences and biotechnology.

8.1.1 How Do Charges on Biological Molecules Affect Phase Equilibrium and Reaction Equilibrium?

Charges on biological molecules influence phase and reaction equilibria by additional electrostatic attractive (between unlike charges) or repulsive (between like charges) intra- or intermolecular forces. The forces F_{ij} between two charges (q_i and q_j) as a function of the distance r, are described by Coulomb's law

$$\mathbf{F}_{ij}(r) = \frac{q_i\, q_j}{4\,\pi\,\varepsilon_o\,\varepsilon_r\,r^2}\mathbf{i}_r = \frac{z_i\, z_j\, e^2}{4\,\pi\,\varepsilon_o\,\varepsilon_r\,r^2}\mathbf{i}_r, \tag{8.1}$$

DOI: 10.1201/9780429329524-8

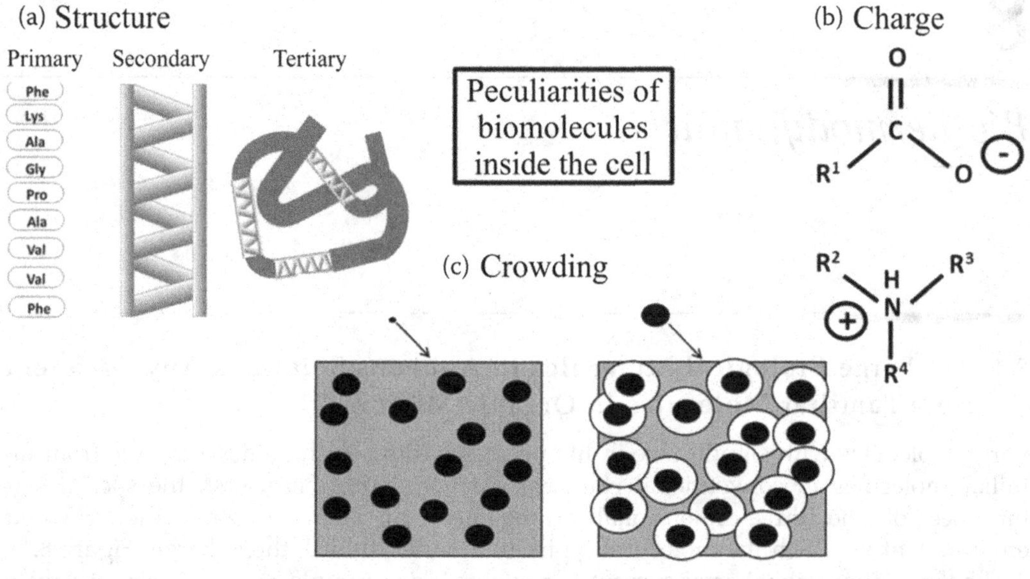

FIGURE 8.1
Causes of the three peculiarities in the thermodynamic treatment of biomolecules under cytosolic conditions (intracellular liquid). (a) the categories of the three-dimensional structure; (b) fixed charges (R1-R4 are functional organic groups such as alkyl (hydrocarbon) groups); (c) the macromolecular crowding (the grey area is available for reactions: left-small metabolites, right: large molecules).

where ε_o, ε_r, z_i, and e are the vacuum permittivity (8.8541878128(13) × 10^{-12} J^{-1} C^2 m^{-1}), the relative permittivity of the medium, the number of charges and the charge of a proton (1.602176634 × 10^{-19} C), respectively and $\mathbf{i}_r$ is a unit vector in the direction of r. For the calculation of phase and reaction equilibria, these electrostatic interactions have to be considered. Instead of the well-known equilibrium conditions ($\mu_B^\beta = \mu_B^\alpha$) requiring equality of chemical potentials of the uncharged species, see Question 4.4.1, equality of electrochemical potentials $\tilde{\mu}_B^\alpha = \tilde{\mu}_B^\beta$ (see Question 5.5.1) is now required. The chemical potential is thus extended by an electrostatic contribution (zFV_B)

$$\tilde{\mu}_B = \mu_B + z_B FV_B. \tag{8.2}$$

Here, $F = Le$ is the Faraday's constant and V_B is the electrostatic potential acting on species B. In non-electrolytic systems, it is often useful to separate the chemical potential into a first part with an ideal concentration dependence μ_B^{id} and a second, excess part μ_B^E

$$\mu_B = \mu_B^{id} + \mu_B^E = \mu_B^\ominus + RT \ln x_B + \mu_B^E = \mu_B^\ominus + RT \ln x_B + RT \ln \gamma_{B,l}. \tag{8.3}$$

Excess functions are defined in Question 4.5.1. Here, $\mu_B^\ominus$ represents the chemical potential in the standard state, and the remaining terms express the result for an ideal solution and a non-ideal solution respectively. As usual, R is the universal gas constant, x_B is the mole fraction and $\gamma_{B,l}$ is the activity coefficient of species B in the liquid phase mixture (see Question 4.6). If there are charges present, Equation 8.3 must be extended and the electrochemical potential reads

$$\tilde{\mu}_B = \mu_B^{id} + \mu_B^E + z_B F \phi_B = \mu_B^{\ominus} + RT \ln x_B + RT \ln \gamma_{B,1} + z_B F V_B. \quad (8.4)$$

From Equation 8.4, three important conclusions can be drawn:

1. Charge is a quantity that makes solutions non-ideal.
2. Non-idealities resulting from the introduction of charged species B possess both an electrostatic ($z_B F V_B$) and a non-electrostatic component μ_B^E. The first is due to "long-range" electrostatic interactions, whereas the second is due to "short-range" interactions (see Question 2.5.2).
3. The contribution of the ion charge to $\tilde{\mu}_B$ is directly proportional to valency z_B. Thus, species with a higher valency tend to show stronger deviations from ideality.

It is worthy of note that biochemical solutes are often only partially soluble in liquid mixtures and that the mixtures of relevance often consist of many substances. For that reason, it is useful to apply the infinite dilution standard states (see Question 5.4.2) and to define the concentration through the molality (Question 5.4.2.1) or molarity (Question 5.4.2.2).

8.1.2 How Do We Calculate the Activity Coefficients for Long-Range Electrostatic Interactions?

The last unknown quantity required for the application of Equation 8.4 for phase and reaction equilibria is the activity coefficient. Activity coefficients enable us to correct for non-idealities using either the pure liquid compound or the compound in infinite dilution as standard state (see Question 5.4.2). In biological systems, it is more convenient to use the first approach for the solvent (in our case water) and the latter for the highly diluted (bio-)molecules or salts. In 1923, Peter Debye and Erich Hückel succeeded in developing a basic model that takes long-range interionic interactions into account (Debye and Hückel 1923). In highly dilute solutions of charged molecules, long-range interactions dominate. They expressed the long-range interaction effect in terms of a mean activity coefficient, $\gamma_\pm^{\infty,m}$, instead of the individual activity coefficients for the cation, $\gamma_+^{\infty,m}$, and the anion, $\gamma_-^{\infty,m}$ (see Question 5.6). The mean activity coefficient is given by

$$\ln \gamma_\pm^{\infty,m} = \frac{\nu_+\gamma_+^{\infty,m} + \nu_-\gamma_-^{\infty,m}}{\nu_+ + \nu_-}. \quad (8.5)$$

Their model allows a good estimation of the mean activity coefficient for highly dilute solutions as

$$\ln \gamma_\pm^{\infty,m} = -\alpha \,| z_+ z_- |\, \sqrt{I}, \quad (8.6)$$

with

$$I = 0.5 \sum_i z_i^2 m_i, \quad (8.7)$$

and

$$\alpha = \frac{e^3 L^{1/2}}{4\sqrt{2}\pi\rho_A^{1/2}(\varepsilon\, k_B T)^{3/2}}. \tag{8.8}$$

Here, m_i, L, ρ_A, and k_B are the molality of the ion i, the Avogadro constant, the density of the solvent A and the Boltzmann constant, respectively. For water at 25°C, the parameter $\alpha = 1.172\ \text{mol}^{1/2}\ \text{kg}^{-1/2}$. Note that this activity coefficient refers to the molality and not the mole fraction (as in Question 4.6) for reasons that are explained in Question 8.4. Details on the use of molality are provided in Question 5.4.2.1. Debye and Hückel used the following approximations to derive Equation 8.6:

- In highly dilute solutions strong electrolytes dissociate completely into ions.
- Electrostatic interactions impose some degree of order over random thermal motions.
- Non-ideality is only a result of electrostatic interactions.
- Ions are considered as non-polarizable point charges.
- The solvent is considered to be a structureless, continuous medium characterized by a bulk macroscopic property with relative permittivity, ε.
- No electrostriction (i.e. no shape changes under the application of an electric field) is allowed.
- Each ion is assumed to have an ionic atmosphere owing to all other ions in the electrolyte solution. The charge of each ion has to be balanced by the charge of the ionic atmosphere owing to the electro-neutrality condition. Although the ions of the ionic atmosphere are discrete charges, the ionic atmosphere itself is described as though it would be a smeared-out cloud of charge whose density varies continuously throughout the solution.

Considering all of these simplifications, it is astonishing that predictions using the Debye-Hückel theory are quite accurate up to an ionic strength of 10^{-2} mol kg^{-1}, but it becomes progressively unreliable at higher electrolyte concentrations (Figure 8.2). Equation 8.6 is therefore often called the *Debye-Hückel limiting law*. The simplest extension (called the *extended Debye-Hückel limiting law*) takes the incompressible radius of the ion, d, into account and increases the validity range by a power of ten ($I < 10^{-1}$ mol kg^{-1}). The result is

$$\ln \gamma_{\pm}^{\infty,\text{m}} = -\frac{\alpha\,|z_+ z_-|\sqrt{I}}{1 + \beta\, d\sqrt{I}}, \tag{8.9}$$

with

$$\beta = \sqrt{\frac{2\rho_A L e^2}{\varepsilon\, k_B T}}. \tag{8.10}$$

For water at 25°C, the parameter $\beta = 3.287\ \text{nm}^{-1}\ \text{kg}^{1/2}\ \text{mol}^{-1/2}$. A more detailed explanation of the terms in these equations, together with forms that extend to even higher molalities, is given in Question 5.6.

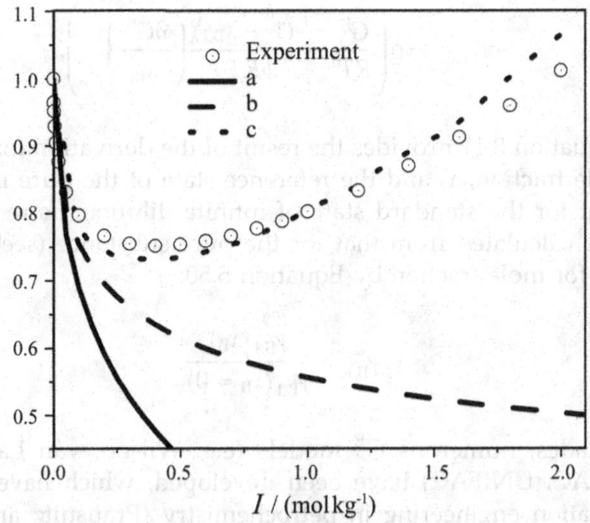

FIGURE 8.2
Mean ion activity coefficient (MIAC) of lithium bromide dissolved in water as function of the ionic strength. (a) Debye-Hückel limiting law (Equation 8.6); (b) extension considering an effective ion radius d = 0.3 nm (Equation 8.9); (c) ePC-SAFT. The experimental data were taken from Hamer and Wu (1972) and the ePC-SAFT data from Held et al. (2008).

8.1.3 Why Should We Also Consider Short-Range Interactions of Charged Biomolecules?

Especially for biomolecules with uncharged parts as well as for highly concentrated solutions, in addition to the long-range electrostatic interactions, the short-range interactions play an important role. Short-range interactions are caused by (i) the force between two permanent dipoles (Keesom force), (ii) the force between a permanent dipole and a corresponding induced dipole (Debye force) and (iii) the force between two instantaneously induced dipoles (London dispersion force) and often summarized as van der Waals interactions (see Chapter 2, Question 2.5.2). In general, at long intermolecular or interionic distances, Coulomb-interactions dominate and that is especially true at low concentrations; on the other hand, at short distances, corresponding to high concentrations, the van der Waals interactions govern the real behavior; they are the only forces present for neutral molecules. In the crowded molecular conditions of the cytosol, ionic strengths are often greater than 0.1 mol kg^{-1} so that short-range interactions are always important there.

8.1.4 What Are Excess Energy Models, and What Are Equation-of-State Models?

In the context of the cytosol, there are two methods in use that consider both short-range and long-range interactions. These are excess Gibbs energy (G^E) models and equations of state (EOS). Conventional G^E and EOS models require extension to allow for Coulomb interactions in electrolyte systems. The first category of models (G^E) describes the difference between the real Gibbs energy and the ideal Gibbs energy. Activity coefficients for non-ideal liquid mixtures γ_i can then be calculated for binary mixtures by differentiation (see Equation 4.137 in Question 4.6.1)

$$\gamma_{B,1} = \exp\left(\frac{G^E}{RT} - \frac{(1 - x_{B,1})}{RT}\left(\frac{\partial G^E}{\partial x_{B,1}}\right)_{T,p}\right). \tag{8.11}$$

For simplicity, Equation 8.11 provides the result of the derivation for activity coefficients based on the mole fraction, x, and the reference state of the pure liquid species B. The activity coefficient for the standard state of infinite dilution, as required for most biomolecules, can be calculated from that for the pure substance (see Question 5.4.2), exemplarily shown for mole fraction by Equation 5.50:

$$\gamma_{B,1}^{\infty} = \frac{\gamma_{B,1}(x_B)}{\gamma_{B,1}(x_B = 0)}. \tag{8.12}$$

Over several decades, numerous G^E models (e.g. Wilson, van Laar, regular solution theories, UNIQUAC, UNIFAC) have been developed, which have had a tremendous success for separation engineering in petrochemistry (Prausnitz and Tavares 2004). A frequently used G^E model for charged molecules is the Pitzer model (Pitzer 1973). The weak point of G^E models is that they can only be used within a limited temperature and pressure range.

The second category of models (EOS) is designed for the prediction of vapor-liquid equilibria (VLE) over wide pressure and temperature ranges (see Question 4.7.3). An EOS is an equation relating state variables (e.g. pressure, volume, temperature, composition), which describe the state of matter under a given set of physical conditions. For instance, the ePC-SAFT EOS (electrolyte perturbated-chain statistical associating fluid theory) is already proven to account for both long-range and short-range interactions and to describe phase and reaction equilibria of charged biomolecules (Held 2008). For a brief description of the ePC-SAFT EOS, the reader is referred to Question 2.9.2. Figure 8.2 compares the predictions of the two Debye-Hückel limiting laws and the EOS ePC-SAFT.

8.2 How Can Thermodynamics Be Applied to Cultures of Living Cells?

Biotechnology often works with cultures of living cells, such as microorganisms, animal or human cells, which are cultivated by being suspended in aqueous solutions of nutritional compounds contained in large vats ("bioreactors"). Starting from a laboratory culture of the target cells, which is called the inoculum, the cells are intended to proliferate and grow in number in the bioreactor. At the same time, they are expected to produce the desired product, i.e. to serve as a catalyst for its synthesis. This section explores how thermodynamics can be applied to such cultures.

8.2.1 Why Is the Application of Thermodynamics to Cultures of Live Cells Important?

One of the by-products of most biological processes, but as we shall see in Question 8.3.3, not all, is the generation of heat. This links the cellular cultures directly to the topic of thermodynamics, as this word literally means "force of heat." In the laboratory, biological heat effects often go unnoticed because most of the heat is lost to the environment too quickly to give rise to a measurable heating effect. But this is completely different at large scale (von Stockar and Marison 1991). Owing to their dramatically smaller surface-to-volume ratio compared to laboratory equipment, large-scale bioreactors work essentially adiabatically, and because of the small temperature difference between the cultivation temperature and that of the surroundings, the design of appropriate cooling facilities is often a severe technical challenge.

The continuous generation of heat by microbial cultures can also be used as a basis for online monitoring of the microbial activity and metabolism. Measuring the heat generation rate continuously can be used with other online data in order to optimize the bioprocess and for online process control (Voisard et al. 2002).

A further important topic is the link between heat generation and Gibbs energy dissipation. A thermodynamic understanding of Gibbs energy dissipation helps to understand what determines the most important performance parameters of cultures, such as biomass and product yields, growth rates and maintenance requirements and how these can be optimized.

8.2.2 How Can the First Law Be Applied to a Growing Cellular Culture?

Biological growth processes and metabolic reaction cascades are subject to material and energy balances, just like conventional chemical reactions or physical processes. While material balances are frequently used in biotechnology, this is rarely the case with energetic balances. In the following, the application of the First Law of Thermodynamics to growing cultures and its enormous potential for generating understanding and optimizing bioprocesses will be explained.

8.2.2.1 How Do We Formulate Energy and Molar Balances for Growing Cellular Cultures?

A typical growth process of microbial or other cells is shown in Figure 8.3. The cells absorb a number of nutritional compounds, termed "substrates." From these, they synthesize new cells, shown as new biomass. The growth process requires energy which, in the case of so-called "chemotrophic" organisms, is obtained by oxidizing or otherwise degrading some of the substrates into catabolic products (CO_2, ethanol, water or similar compounds). The whole metabolic activity of the growing cell is thus subdivided into anabolism, comprising all the biochemical reactions needed to synthesize the new cells, and catabolism, designating the oxidation or degradation processes required to obtain energy.

Before formulating an energy balance around a growing cellular culture, the stoichiometry of the process shown in Figure 8.3 ought to be known. In simple cases, this

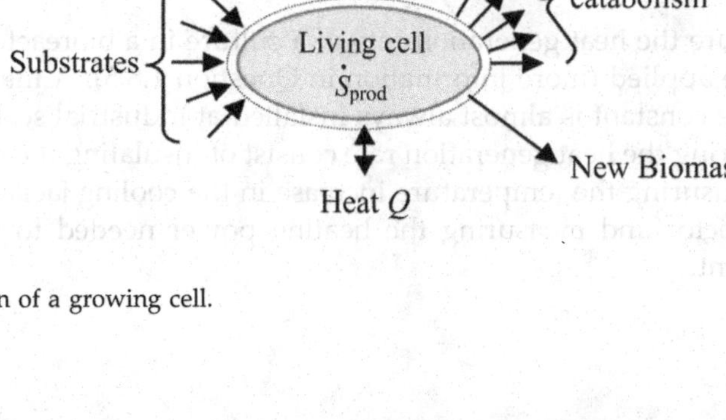

FIGURE 8.3
Schematic description of a growing cell.

stoichiometry may be approximated by a so-called "macrochemical equation" and assumed to be constant during at least part of the growth process. In general, such equations have the following structure

$$Y_{S/X}S + Y_{A/X}A + Y_{N/X}NS \rightarrow X + Y_{P/X}P + Y_{C/X}CO_2 + Y_{W/X}H_2O, \tag{8.13}$$

where S represents the chemical formula of the so-called carbon and energy source, i.e. the main nutritional compound (often glucose, $C_6H_{12}O_6$). A, NS, X and P stand for the chemical formulae of, respectively, an electron acceptor (e.g. oxygen), a nitrogen source (e.g. NH_3), the newly synthesized dry biomass and a catabolic waste product (e.g. ethanol). The stoichiometric coefficients are represented by the yield coefficients $Y_{B/X}$.

A typical example of an experimentally observed macro-chemical equation for aerobic growth of the yeast *Klyveromyces fragilis* is (von Stockar and Liu 1999)

$$1.758\ CH_2O + 0.695O_2 + 0.15\ NH_3 \rightarrow CH_{1.75}O_{0.52}N_{0.15} + 0.758\ CO_2 + 1.11\ H_2O. \tag{8.14}$$

As can be seen, all compounds are written in C-molar notation, which simplifies balance calculation considerably. For instance, the carbon and energy substrate glucose ($C_6H_{12}O_6$) is represented as CH_2O. One mole of glucose is therefore equal to six C-moles. The newly grown dry biomass is included in this reaction as a C-mole of a compound containing the elements C, H, O and N. All other elements can be neglected and allowed for in the estimation of the C-molar mass of dry biomass together as "ash". No catabolic product P other than CO_2 and water was formed in this case. The stoichiometric coefficients $Y_{B/X}$ must be expressed in C-moles of B per C-mole of X. In the biological literature, $Y_{S/X}$ is often given as the inverse and is called the biomass yield $Y_{X/S}$, which indicates the number of C-moles of X obtained per C-mole S consumed (see Question 8.3.1).

For the energy balance, the answer to Question 1.5.6 can be used. The thermodynamic system is typically taken as the culture suspension of the microbial cells but without the gas phase. For only a single growth reaction one obtains from Equation 1.87

$$\rho V\, \bar{c}_p \frac{\mathrm{d}T}{\mathrm{d}t} = \dot{Q} + \dot{W}_s + \sum_k \left\{\bar{c}_{p,k}(T_k - T)\, \dot{V}_k \rho\right\} - \Delta_r H_X\, V\, r_X, \tag{8.15}$$

where $\Delta_r H_X$ stands for the heat of the growth reaction, i.e. the enthalpy change of the macro-chemical reaction given previously (Equations 8.13 and 8.14) in units of kJ per C-mole of freshly grown dry biomass. The rate of the growth reaction is designated as r_X (C-mol h^{-1} m^{-3}), while ρ and V denote the density and the volume of the culture broth, respectively. It is worthy of note here that for practical reasons the unit of time used for rate constants in biochemical reactions is commonly the hour (h), rather than the SI unit of the second.

In order to measure the heat generation rate of a culture in a bioreactor, a calorimetric technique has to be applied (more information in Question 1.8.5). A method to keep the culture temperature constant is almost always installed at industrial scale. Two common methods for measuring the heat generation rate consist of insulating the reactor thermally and then either measuring the temperature increase in the cooling jacket, or deliberately overcooling the reactor and measuring the heating power needed to keep the reactor temperature constant.

8.2.2.2 How Can We Monitor a Growing Cellular Culture Based on the First Law?

Based on the energy balance (Equation 8.15), one can either (i) predict the heat evolution rate $\dot{Q}$ if all other elements of the balance are known, or (ii) observe the biological activity in terms of the measured $\dot{Q}$ and monitor the growth rate r_X online if $\Delta_r H_X$ is known or (iii) observe the biological activity in terms of $\dot{Q}$ and measure the heat of growth, $\Delta_r H_X$, if r_X is known. A common procedure in laboratory bioreactors consists of measuring heat evolution rates per unit volume, $\dot{q}_V$, as a function of time during growth, where we now define the thermodynamic system as one unit volume of culture broth. Integrated forms of such curves ($\int \dot{q}_V \, dt = Q/V$) often follow nicely the concentration of live cells (X) as a function of time. As a consequence, a plot of Q as a function of X yields then a straight line. The slope of this line is the enthalpy of the growth reaction per C-mole of biomass grown $\Delta_r H_X$ (von Stockar 2010).

8.2.2.3 What Is Indirect Calorimetry?

It is also possible to monitor the effect of heat release indirectly by following, for example, the oxygen uptake during growth. For many aerobically growing microorganisms, studies show that heats of the growth reaction, expressed per mole of oxygen, scatter around an average value of about –450 kJ ± 10% per mole of oxygen respired (von Stockar 2010; Birou et al. 1987). This value is called the *oxycaloric equivalent*.

Oxygen uptake always reflects the oxidation of some organic molecules into CO_2 and water. Since this reaction releases much more heat than do transformations of various organic molecules into one another, it dominates the heat of the growth reaction, which will therefore be close to the heat of combustion of organic compounds. For the same reason, the latter is about –440 kJ per mole of oxygen for any organic compound (Sandler and Orbey 1991). This finding permits us to estimate $\dot{Q}$ by measuring the oxygen up-take rate, which is the basis of what is known as *indirect calorimetry*.

For anaerobic life processes, indirect calorimetry does not work because these processes do not consume oxygen. Interpreting the absence of oxygen up-take by a growing culture according to the philosophy of indirect calorimetry would either mean there is no growth or that the culture grows without producing heat; both are incorrect.

8.2.3 How Can the Second Law Be Applied to a Growing Cellular Culture?

From the Second Law of Thermodynamics, statements can be derived about the direction of naturally occurring bioprocesses, their feasibility and their velocity as a function of thermodynamic driving forces. To what extent this can be applicable to the complex processes in growing cellular cultures and what advantages this has for life sciences and the related technologies will be discussed in the following questions.

8.2.3.1 What Is the Driving Force for Growth?

Microbial growth occurs spontaneously and is obviously a highly irreversible phenomenon. As with any spontaneous process, the driving force must therefore be the production of entropy. In relation to growth reactions, this seems contradictory, because growth reactions produce matter in a highly organized form from a set of very simple small molecules. One intuitively gets the impression that microbial growth decreases the entropy rather than producing it.

8.2.3.2 How Can Growth Be Reconciled with the Need for Entropy Production?

The apparent contradiction may be resolved by writing an entropy balance around a growing cell. The system for this balance will be one of the cells (or all the cells in the culture). We will assume that the cells multiply by budding, as shown in Figure 8.4.

The freshly grown biomass, i.e. the buds, can thus be treated as a product of the cell, and do not belong to the system. The mother cells, on the other hand, can be assumed to represent a system at the steady state that acts as a catalyst for producing the new biomass from the substrates dissolved in the culture broth, as shown in Figure 8.3.

The application of an entropy balance to a growing microbial cell yields (see Equation 3.16)

$$\frac{dS}{dt} = \frac{\dot{Q}}{T} + \sum_{B} S_B\, \dot{n}_B - S_X\, |\dot{n}_X| + \dot{S}_{gen} = 0. \tag{8.16}$$

Equation 3.16 has been simplified for this case assuming the whole surface of the cell operates just as a single material exchange port. According to this balance, the variation of entropy in the cell with time is given by the sum of all entropy fluxes exchanged with the environment plus the rate at which entropy is produced by irreversible processes. Entropy may be exchanged with the environment owing to heat transfer to or from the cell, denoted by $\dot{Q}/T$. In open systems, entropy is also imported or exported through metabolites entering or leaving the cell, where S_B denotes the partial molar entropy carried by the metabolite B and $\dot{n}_B$ its molar rate of exchange, whereby positive values indicate entry into the cell. Newly formed biomass from the growth reaction is treated as a product of the cell leaving it at a C-molar rate of $\dot{n}_X$. Its partial molar entropy S_X is expected to be rather low owing to the high degree of organization of matter. The rate of entropy production by irreversible processes $\dot{S}_{gen}$ can only be positive, according to the Second Law of Thermodynamics and represents the real driving force for the process.

Owing to the constant entropy production at the rate $\dot{S}_{gen}$ and because newly formed cells of low entropy content leave the cell (but have been synthesized by importing high-entropy metabolites), entropy could in principle accumulate in the cell and lead to thermal cell death or to structural disorganization. In order to avoid this, the cell must constantly export the excess entropy, i.e. it must keep dS/dt at zero by making the sum of the first two terms on the right-hand side of Equation 8.16 negative. This is precisely the role of catabolism. There are two ways in which catabolism can export excess entropy:

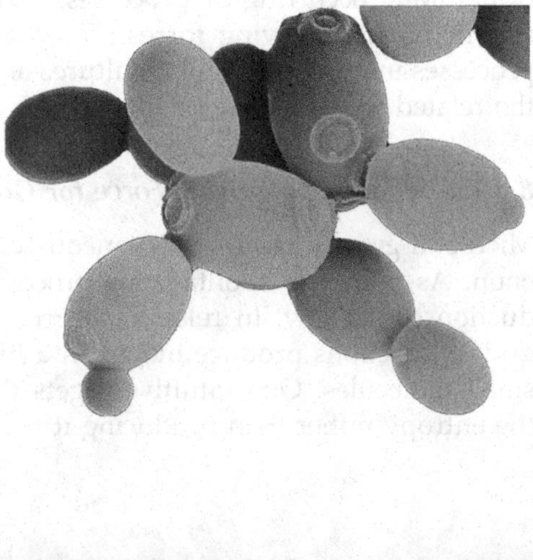

FIGURE 8.4
Growing yeast cells. These are propagating according to a mechanism known as budding. First, a bud appears on their surface as at the bottom left of the figure. Once the bud has increased in size to an adult cell, it breaks off leaving a "budding scar" on the mother cell (see cell shown near the right border at mid-height of the figure.

First, by creating a large flux of small waste molecules from the substrate, thereby exporting it in the form of chemical entropy and making $\Sigma_B S_B \dot{n}_B$ strongly negative, and second, by releasing considerable amounts of heat, thereby making $\dot{Q}/T$ strongly negative. This is the fundamental reason why virtually all living chemotrophic organisms constantly generate heat. Indeed, the heat released by living cells serves much the same purpose as the heat "wasted" in heat engines (see Chapter 3), namely to avoid accumulating excess entropy by exporting it.

8.2.3.3 What Is Schrödinger's Negentropy?

In his book *What is Life?*, Schrödinger (1944) hypothesized that "Life feeds on negative entropy." The term was later shortened into "negentropy." Although Schrödinger later added a note saying that a more precise concept would be Gibbs energy, this famous statement created quite a lot of confusion and discussion. Today, we would interpret it as meaning that living cells must feed on substrates (nutritional products) that have a lower entropy content than the average of the waste products and the biomass synthesized during growth in order to maintain a low level of entropy.

8.2.3.4 How Can Gibbs Energy Balances Be Applied to Growing Cells?

Adapting the Gibbs energy balance of Equation 3.22 for a single cell yields

$$\frac{dG}{dt} = \dot{W}_o + \sum_B \mu_B \dot{n}_B - \mu_X \,|\, \dot{n}_X \,| \; - T\dot{S}_{gen}, \tag{8.17}$$

where W_o represents any the work the environment does on the cell or that the latter does on the former, and G is the Gibbs energy of one cell, the derivative of which can be assumed zero because the cell operates at steady state. (The same result is obtained if the entropy balance for a living cell, Equation 8.16, is multiplied by T and subtracted from an enthalpy balance such as Equation 1.71, but formulated for a single cell, and then applying the definitions of G and μ to simplify the resulting differential as well as the sum of partial molar quantities.)

If we do assume that the cell operates at steady state, the left-hand side of the Equation 8.17 is zero. Moreover, the rates $\dot{n}_B$ at which the metabolites are consumed or excreted by the cell may be linked to the growth rate by means of molar balances. For a steady-state system, where the accumulation rate dn_B/dt can be set to zero, the molar balance for B (Equation 1.40) reads

$$\dot{n}_B = -\upsilon_B \, \dot{\xi}_X. \tag{8.18}$$

The term $\dot{\xi}_X$ represents the rate of synthesis of fresh live biomass $\dot{n}_X$ by the cell in C-mol/h. It replaces the term rV in Equation 1.40 according to Equation 1.38. Substituting Equation 8.18 into Equation 8.17 and replacing the stoichiometric coefficients by the yields $Y_{B/X}$ of the growth stoichiometry of Equation 8.13 gives

$$\frac{dG}{dt} = \dot{W}_o - \left(\sum_B Y_{B/X} \mu_B - \mu_X \right) \dot{\xi}_X - T\dot{S}_{gen} = 0. \tag{8.19}$$

The expression $\Sigma_B Y_{B/X}\mu_B - \mu_X$ is equal to $\Delta_r G_X$, the Gibbs energy change of the growth reaction. Hence, the previous equation becomes identical to Equation 5.16

$$\Delta_r G_X \dot{\xi}_X = \dot{W}_o - T\dot{S}_{gen}. \tag{8.20}$$

Formally the same equation is obtained if one performs a Gibbs energy balance over the biomass in the whole culture, with the exception that $\dot{\xi}_X$ would be replaced by $r_X V$. The system would not be at a steady state anymore, but it could be treated as closed.

In chemotrophic growth, the environment does no work on the cell and neither does the cell do work on the environment, so that $\dot{W}_o$ in Equation 8.20 is zero. The right-hand side of the remaining equation then represents the rate of Gibbs energy dissipation and can only be negative. For a positive growth rate $\dot{\xi}_X$, the Gibbs energy of growth $\Delta_r G_X$ must therefore be negative, i.e. Gibbs energy must be dissipated. The entropy generation is the actual driving force for spontaneous processes, but the same can also be said for the negative Gibbs energy of chemotrophic growth reactions. The negative Gibbs energy of the growth reaction, which is identical with the affinity, defined in Equation 5.105, is sometimes also described as "Gibbs energy dissipation." According to irreversible thermodynamics, the negative Gibbs energy of reaction is the thermodynamic driving force, and the growth rate is the "conjugate" flux (Westerhoff and van Dam 1987). The reaction will proceed the faster the more negative $\Delta_r G_X$ is.

In phototrophic organisms (that require light to grow), biomass is synthesized from carbon dioxide and water, and $\Delta_r G_X$ is clearly positive. But, in this case, a flux of photons provides the necessary additional work at a rate, $\dot{W}_o$, to permit a positive $\Delta_r G_X$ in Equation 8.18, and still to drive the synthesis by allowing for a sufficiently negative $- T \dot{S}_{gen}$ term, or a sufficiently positive entropy production $\dot{S}_{gen}$. Each photon of frequency f contains the energy hf with h being the Planck constant. As an example, 1 mole of red photons (wavelength 700 nm, frequency $f = 4.28 \times 10^{14}$ Hz) absorbed by a photosystem provides an energy of 171 kJ.

8.3 How Can a Gibbs Energy Analysis Be Used to Predict the Growth Yield of Cellular Cultures?

In order to demonstrate the potential of thermodynamics to predict, or at least to correlate, important cell-culture parameters, we choose the biomass or growth yield, $Y_{X/S}$, as an example because of its dominant importance both for scientific research and for industrial manufacture.

8.3.1 Why Is the Growth Yield Related to the Gibbs Energy Dissipation?

The growth yield is a stoichiometric coefficient and indicates how much biomass can be obtained from a given amount of carbon and energy substrate (e.g. glucose) in C-mole per C-mole

$$Y_{X/S} \equiv -\frac{\Delta n_X}{\Delta n_S} = -\frac{r_X}{r_S}. \tag{8.21}$$

8.3.1.1 What Is the Growth Yield and Why Is it Important?

The importance of the biomass yield resides in the fact that it determines the number of cells and thus the biomass concentration in the culture, once all the carbon and energy substrate has been exhausted. Since the cells are catalysts for product synthesis, this will often also determine the final product concentration. In research, it is much easier to work with large cell and product concentrations than with dilute cultures. In industrial production, the final product concentration is, arguably, one of the most important key factors determining the final production cost.

Growth yields in different microbial strains can vary from just a few percent to values close to 90% in C-mol/C-mol. A way to estimate the value of growth yields before even embarking on costly experimental trials is therefore of considerable practical value.

8.3.1.2 How Do Growing Organisms Reconcile the Need to Incorporate Gibbs Energy Dissipation with the Need to Incorporate Gibbs Energy into the Fresh Biomass?

We have seen in Question 8.2 that cellular growth, as a highly spontaneous process, must dissipate Gibbs energy. As with the associated entropy change, this seems counter-intuitive because the highly structured biomass produced by the process is generally expected to have a higher Gibbs energy than the simple molecules used as nutritive substrates. Therefore, growth might be expected to increase Gibbs energy rather than dissipating it.

If the biosynthetic reactions (anabolism) are taken by themselves, they would indeed often have a positive overall reaction Gibbs energy, denoted as $\Delta_{an}G$. The process thus would occur in the "wrong" direction, against its own driving force. In chemotrophs this is, however, coupled to energy yielding reactions (catabolism), with highly negative Gibbs energies of reaction $\Delta_{cat}G$, such that the Gibbs energy of the overall growth reaction remains negative. In some cases, the anabolic Gibbs energy change $\Delta_{an}G$ can be close to zero or even very slightly negative. However, with these small negative values of $\Delta_{an}G$, growth could not occur spontaneously without being coupled to a catabolic process.

A quantitative analysis of this situation is easily performed if we formally assume that the cells first catabolize all the carbon and energy substrate to the products of catabolism and that they synthesize the new biomass starting from there (see Figure 8.5). This split of the growth reaction into a catabolic and an anabolic reaction is obviously not realistic, and $\Delta_{an}G$ will assume a very large positive value. On the other hand, the amount of substrate catabolized will also release a large amount of Gibbs energy if we assume that even the substrate molecules that end up in the biomass are first catabolized. Therefore, we end up with the correct final state and so the result for the overall energy state remains valid (Hess's law) (von Stockar 2010).

Based on this definition of $\Delta_{cat}G_c$ and $\Delta_{an}G$, the overall Gibbs energy change of the growth reaction per 1 C-mole of biomass grown can be calculated simply by

$$\Delta_r G_X^{\ominus} = \frac{1}{Y_{X/S}} \Delta_{cat}G^{\ominus} + \Delta_{an}G^{\ominus}, \tag{8.22}$$

where $Y_{X/S}$ is the biomass yield in C-mol/C-mol, i.e. the number of C-moles of dry biomass synthesized per C-mole of energy substrate consumed, $\Delta_{cat}G^{\ominus}$ represents the strongly negative standard Gibbs energy change of the energy yielding reaction, $\Delta_{an}G^{\ominus}$ is the

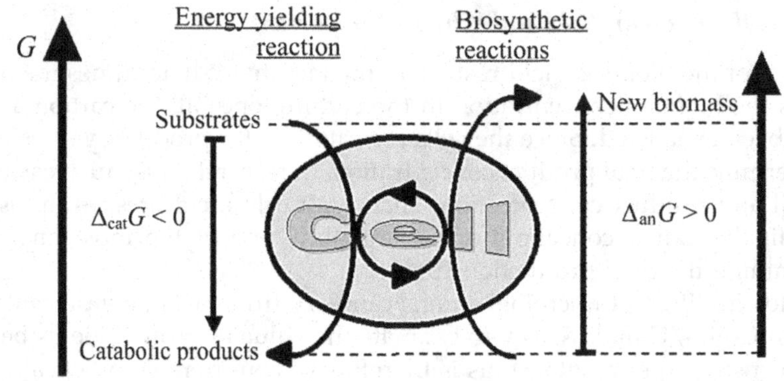

FIGURE 8.5
Splitting the overall growth stoichiometry into an energy yielding catabolic reaction that first catabolizes all the carbon and energy substrate and then uses part of the catabolic waste products for the synthesis of the new biomass, in a hypothetical anabolic reaction. The figure shows schematically the Gibbs energy level of the various substances. (Adapted from von Stockar 2013.)

positive standard Gibbs energy change of the biosynthetic reaction, and $\Delta_r G_X^\ominus$ denotes the standard Gibbs energy change of the combined process per C-mole of dry biomass formed.

Equation 8.22 represents an application of Hess's law (see Question 5.2.1). The stoichiometry expressed in Equation 8.13 is split into first catabolizing the whole of the $Y_{S/X}$ C-moles of carbon and other substrates into the products of catabolism, liberating $Y_{S/X} = 1/Y_{X/S}$ times $\Delta_{cat}G$ of catabolic energy. In a second step, a C-mole of carbon compound and other catabolic products are used for synthesizing 1 C-mole of new biomass, absorbing $\Delta_{an}G$ of Gibbs energy. For a numerical example, see von Stockar (2010), Appendix B. (Equation 8.22 uses standard values, $\Delta G^\ominus$, because in many cases these are so large that the concentration dependencies may be neglected in an approximate calculation). The equation shows that the higher the growth yield is, the lower the (negative) dissipation will be. This means that for large biomass yields $\Delta_r G_X^\ominus$ will be less negative, and the overall driving force will be smaller.

As pointed out previously, the anabolic reaction must proceed from a lower to a higher level of Gibbs energy, that means against its own driving force (Figure 8.5). To understand how this occurs, it is useful to consider the analogy with a mountain cable car used to bring tourists to the top of a mountain (see Figure 8.6). Traditionally, such cable cars were operated by filling a tank on the car at the top of the mountain with water, making it heavier

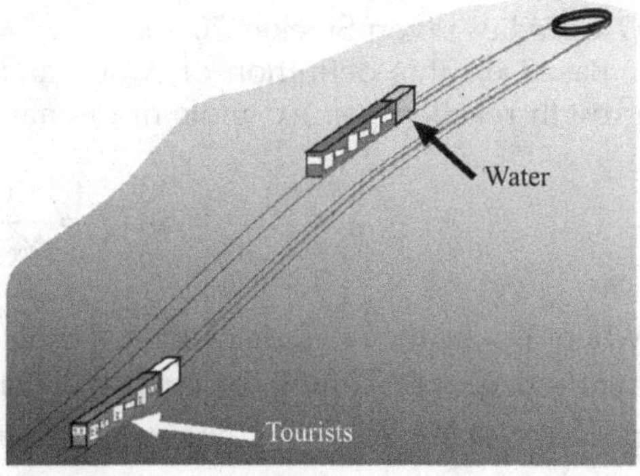

FIGURE 8.6
Mechanical analogy for driving a endergonic reaction "uphill" by coupling it to an energy yielding, or a catabolic, strongly exergonic reaction (see Question 7.3.4). (Adapted from von Stockar U. 2018.)

than the car at the bottom containing the tourists. The car with the water tank therefore had a larger driving force for rolling down the slope than the car with the tourists and because the two cars were coupled by a rope, it could thus force the car with its payload of tourists to roll uphill against its own (downward) driving force. The driving force for the combined system was obviously proportional to the weight difference between the two cars. It decreased with the increase of payload, i.e. with an increasing number of tourists transported uphill, and this obviously slowed the rate of travel down.

In the biological case, the payload is the biomass yield, $Y_{X/S}$, that indicates the amount of dry biomass X that the cell can synthesize per amount of substrate S consumed. In both the cellular metabolism and the cable car, the energy efficiency may be increased by increasing the payload (less energy or water spent per payload), but as the combined driving force decreases, the whole process is slowed down. This may be seen quantitatively from Equation 8.22.

Many textbooks state that the dissipated energy in growing organisms is wasted, but this is inexact: the price for high growth efficiency is slow growth. For a reasonable growth rate, a certain driving force is needed and thus some Gibbs energy must necessarily be dissipated.

8.3.1.3 What Is the Thermodynamically Highest Possible Growth Yield?

Equation 8.22 can be used to estimate the thermodynamically maximum possible biomass yield, which is reached when $Y_{X/S}$ becomes so large that $\Delta_r G_X^{\ominus}$ goes to zero. The biomass yield then becomes

$$Y_{X/S}^{max} = -\frac{\Delta_{cat}G^{\ominus}}{\Delta_{an}G^{\ominus}}. \tag{8.23}$$

Without a driving force, the growth would then proceed infinitely slowly because the system would be locked in a thermodynamic equilibrium. In the cable car analog, this would correspond to a transport of so many tourists that the two cars have the same weight.

8.3.1.4 How Can the Standard Gibbs Energy of the Anabolic, the Catabolic and Consequently the Whole Growth Reaction Be Estimated?

The standard Gibbs energy of the growth reaction can be estimated using Equation 8.22, for which $\Delta_{cat}G^{\ominus}$ and $\Delta_{an}G^{\ominus}$ are the standard Gibbs energies of the formal catabolic and anabolic reactions as defined by Figure 8.5. The chemical equations for these formal reactions are developed in Question 9.4.2. Typical values for the standard Gibbs energies of combustion appearing in these equations are listed in Question 9.4.5.

8.3.2 What Is the Actual Relationship Between the Gibbs Energy Change of Growth and Biomass Yield and How Can We Use It to Predict $Y_{X/S}$?

This question can be separated into a number of smaller questions and so we first explore the nature of the $\Delta_r G_X - Y_{X/S}$ relationship for aerobic growth (i.e. in the presence of oxygen) and then consider the special features of anaerobic growth, where oxygen is absent from the system. Subsequently, we briefly examine methods to predict or at least correlate, growth and biomass yield using known data on the system.

8.3.2.1 What Is the $\Delta_r G_X$–$Y_{X/S}$ Relationship for Aerobic Growth?

In order to illustrate this relationship, Equation 8.22 is plotted as the solid curve in Figure 8.7 for aerobic growth of microorganisms on glucose as a carbon and energy substrat. At any given biomass yield, the standard Gibbs energy of growth, i.e. the driving force for growth, can be seen as the solid downward-pointing arrow. The black line shows clearly the decrease of the driving force as a function of increasing assumed biomass yield. It reaches zero at the maximal possible biomass yield of about 1.04 C-mol/C-mol, as predicted by Equation 8.23.

Also shown as the broken curve and the broken downward-pointing arrow is the enthalpy of the growth reaction, calculated by an equation for $\Delta_r H_X^{\ominus}$, the heat of the growth reaction, analogous to Equation 8.22. The heat and the Gibbs energy of growth are related by applying the definition of Gibbs energy, Equation 3.20, to the reaction variables of Gibbs energy, enthalpy and entropy so that

$$\Delta_r G_X^{\ominus} = \Delta_r H_X^{\ominus} - T\,\Delta_r S_X^{\ominus}. \tag{8.24}$$

The driving force may thus be separated into two parts: an enthalpic one ($\Delta_r H_X^{\ominus}$) corresponding to the entropy exportation in the form of heat, and an entropic one ($T\ \Delta_r S_X^{\ominus}$), corresponding to the entropy exportation in the form of small, high-entropy waste molecules (see Question 8.2.3.2 and the remarks concerning Equation 8.16).

The fact that the lines representing $\Delta_r G_X^{\ominus}$ and $\Delta_r H_X^{\ominus}$ lie very close together indicates that $T\ \Delta_r S_X^{\ominus}$, which separates the two lines, is close to zero. It may be concluded that aerobic growth on glucose is driven almost entirely by an enthalpy change, or that the entropy produced by irreversible processes is exported practically exclusively in the form of heat, and not in the form of high entropy metabolites.

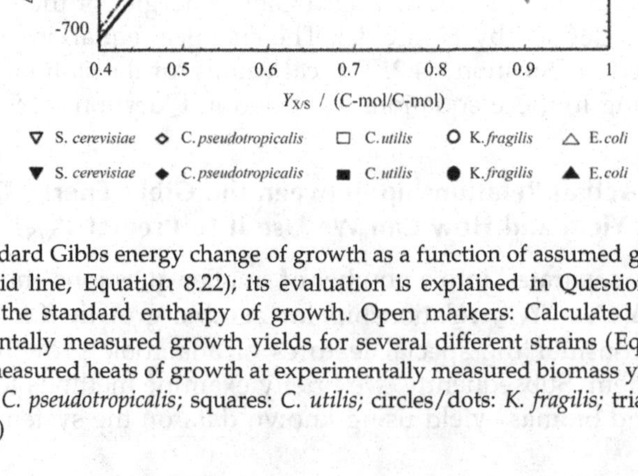

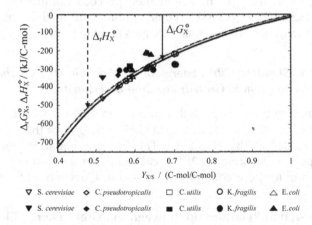

FIGURE 8.7
Relationship of the standard Gibbs energy change of growth as a function of assumed growth yields for aerobic growth on glucose (solid line, Equation 8.22); its evaluation is explained in Question 8.2.3.2. Broken curve: Similar calculation for the standard enthalpy of growth. Open markers: Calculated standard Gibbs energy dissipation at experimentally measured growth yields for several different strains (Equation 8.22). Solid markers: Calorimetrically measured heats of growth at experimentally measured biomass yields. Inverted triangles: *S. cerevisiae*; diamonds: *C. pseudotropicalis*; squares: *C. utilis*; circles/dots: *K. fragilis*; triangles: *E. coli*. (Adapted from von Stockar 2013.)

The points in Figure 8.7 represent the results of a number of aerobic growth experiments on glucose, for which the biomass yields have been measured. The values for $\Delta_r G_X^{\ominus}$ (open symbols) lie on the solid curve because they have been calculated by Equation 8.22, which also gave the curve. While the Gibbs energy dissipation cannot be measured, the heat "dissipation" may be experimentally observed by calorimetry (solid symbols). There is a considerable scatter of the experimental points around the broken curve but, overall, they confirm the trend predicted by it.

As can be seen in Figure 8.7, the experimentally observed biomass yields lie between 0.5 and 0.7. They are thus considerably lower than the theoretical maximum of 1.04 at which the Gibbs energy of the growth reaction would become zero. The reason is obviously the need to have a sufficiently large driving force for assuring growth at a reasonable rate. This forces the cultures to dissipate between –250 and –500 kJ of Gibbs energy per C-mole of freshly-grown dry biomass.

8.3.2.2 What Special Features Arise in Anaerobic Growth?

The catabolic reactions of many anaerobic growth systems do not use an external electron acceptor. In these cases, the energy substrates are not oxidized as they are in Equation 8.14 but are subject to a disproportionation into more oxidized and more reduced molecules. Fermentation of glucose into ethanol and carbon dioxide is a typical example for which the catabolic equation reads

$$C_6H_{12}O_6 \Rightarrow 2\,CH_3CH_2OH + 2\,CO_2. \quad (8.25)$$

Such catabolic reactions yield much less energy; their $\Delta_{cat}G$ is much less negative than the one in aerobic growth. The relationship between the biomass yield and the Gibbs energy dissipation is therefore displaced dramatically towards smaller yields, as shown in Figure 8.8.

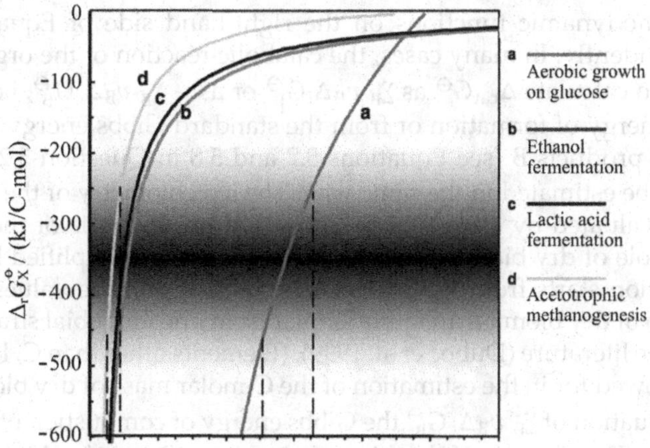

FIGURE 8.8
Relationship between Gibbs energy change of growth and assumed biomass yields (calculated using Equation 8.22 as explained in Question 8.2.3.2) for different types of catabolism. If one assumes that the Gibbs energy dissipation of most organisms fall in the shaded range, it is clear that biomass yields for anaerobic strains (b-d) should fall between 0.05 and 0.15, but those for aerobic growth between 0.5 and 0.65. (Adapted from von Stockar 2013.)

It can be seen that the Gibbs energy of the growth reaction for anaerobic systems approaches zero at much smaller biomass yields than is the case for aerobic growth, typically reducing to about –150 kJ/C-mol by $Y_{X/S}$= 0.3 C-mol/C-mol. The Gibbs energies of growth observed in practice, on the basis of measured biomass yields, still vary between –250 to –500 kJ/C-mol, so that they lie in a similar range as for aerobic growth (shaded region in Figure 8.8). This range appears to be a reasonable compromise between excessively efficient growth, affording very high biomass yields but low driving force, entailing very slow growth, and large driving forces affording very rapid metabolism but only minimal growth yields. Figure 8.8 shows how the different shape of the anaerobic $\Delta_r G_X - Y_{X/S}$ curves compared to that for aerobic growth explains the much smaller biomass yields observed in anaerobic growth.

Since in anaerobic growth there is no electron acceptor, the absolute values of the heats of growth $\Delta_r H_X^\ominus$ are much smaller than $\Delta_r G_X^\ominus$. The origin of the driving force for growth in this case is therefore not so much a result of a negative enthalpy of reaction but rather because the energy substrate is ripped apart to produce smaller molecules, thereby generating a large negative value of $-T\Delta_r S_X^\ominus$. Growth is thus less driven by an enthalpy change than by an increase of chemical entropy. In other words, while aerobically growing organisms export their excess entropy almost exclusively by heat, anaerobic organisms export it in large part in the form of high entropy waste molecules.

8.3.2.3 How Can the Relationship Between Gibbs Energy Dissipation and Growth Yield Be Used to Predict the Latter?

By solving Equation 8.22 for the biomass yield, we find

$$Y_{X/S} = \frac{\Delta_{cat} G^\ominus}{\Delta_r G_X^\ominus - \Delta_{an} G^\ominus}. \tag{8.26}$$

Two of the thermodynamic functions on the right-hand side of Equation 8.26 can be estimated independently. In many cases, the catabolic reaction of the organism is known. It is thus simple to calculate $\Delta_{cat}G^\ominus$ as $\Sigma_B\, v_B \Delta_f G_B^\ominus$ or as $-\Sigma_B\, v_B \Delta_c G_B^\ominus$, i.e. from either the standard Gibbs energy of formation or from the standard Gibbs energy of combustion of the reactants and products B (see Equations 5.7 and 5.8 in Question 5.2.1).

$\Delta_{an}G^\ominus$ can also be estimated in the same way. The stoichiometry of the anabolic reaction will have to be evaluated by elemental balances and on the basis of the elemental composition of a C-mole of dry biomass ($CH_hO_oN_n$). This task is simplified by assuming that the anabolic reaction starts from the products of catabolism. The values of h, o, n in the elemental formula of dry biomass are quite similar from one microbial strain to another and are tabulated in the literature (Duboc et al. 1999). (Elements other than C, H, O and N can be neglected and allowed for in the estimation of the C-molar mass of dry biomass together as "ash"). In the evaluation of $\Sigma_B\, v_B \Delta_c G_B^\ominus$ the Gibbs energy of combustion of dry biomass will have to be known. The entropy of dry biomass has been determined by low-temperature calorimetry by Battley et al. (1997) and using this and enthalpies of combustion an average estimation of $\Delta_c G_X^\ominus$ has been found as –515 kJ/C-mol.

The only value that is not independently known is the Gibbs dissipation energy of the growth reaction as a whole ($\Delta_r G_X^\ominus$). A number of studies exist in which experimentally measured biomass yields of large numbers of different microbial strains were used to calculate the Gibbs energies of growth, essentially by applying equations similar to

Equation 8.22. As already pointed out in the previous question, many microbial strains dissipate between –200 and –500 kJ of Gibbs energy per C-mole of freshly synthesized dry biomass (see Figures 8.7 and 8.8). A few strains, however, show Gibbs dissipation energies as high as minus several thousand kJ/C-mol. Thus, attempts have been made to develop correlations of $\Delta_r G_X^{\ominus}$ for practical use.

According to Heijnen and van Dijken (1992), the Gibbs energy dissipation depends, above all, on the characteristics of the carbon source, and they correlated the $\Delta_r G_X^{\ominus}$ values to the degree of reduction and the number of carbon atoms of the carbon substrate. Using this correlation in an equation such as Equation 8.26 permitted them to predict the percentage growth yield of the materials in their database to within ±10%. More recently, Liu et al. (2007) proposed a considerably simpler correlation based on the degree of reduction of the energy substrate, which gave the yields within ±9%.

8.3.2.4 Is the Concept of Energetic Growth Efficiency Useful?

When developing a system to predict biomass yields, it would seem natural to translate a large number of experimentally measured growth yields into thermodynamic efficiency factors and to search for a way to correlate these. To this end, a so-called *energy-transducer efficiency*, characterizing the fraction of the Gibbs energy released by the catabolic process that can be recovered in the form of Gibbs energy stored within the newly grown biomass by anabolism (von Stockar and Marison 1993), has been defined as

$$\eta = -\frac{Y_{X/S}\,\Delta_{an}G^{\ominus}}{\Delta_{cat}G^{\ominus}}. \tag{8.27}$$

While this concept allowed a reasonable correlation of data for aerobic growth, it failed completely for anaerobic processes (von Stockar et al. 2008). When plotted as a function of the degree of reduction of the energy substrate, for instance, the efficiency values scattered in an unpredictable manner about the correlation by amounts varying from a few percent to 60%.

The reason lies mostly in the fact that the values of the efficiencies, η, depend on how one splits the growth stoichiometry into a catabolic and an anaerobic reaction (Heijnen and van Dijken 1993; see Question 8.3.1.2). The choice of formulation of the anabolic reactions is entirely arbitrary. For instance, if one formulates the anabolic reaction as starting not from the products of catabolism, but directly from the carbon and energy substrates, the Gibbs energy increase of the anabolic reaction would become very much smaller and, in some cases even negative, so that some of the efficiency values would also become negative. In real live cells, hundreds of different constituent components of biomass are derived for growth from different points along the catabolic series of reactions. Hence, trying to use the real anabolism for defining the efficiency is impractical owing to the complexity of its details, which are often unknown. This example shows that a thermodynamic analysis is often more useful than empirical concepts and correlations.

8.3.3 Does Endothermic Growth Exist?

Question 8.3.2.2 points out that the driving force for anaerobic growth is often a large $T\Delta_r S_X^{\ominus}$ term in Equation 8.24, so that the heat of growth is much smaller than in aerobic growth. The question arises whether the entropic part of the driving force can become so large that a nonphotosynthetic organism can grow despite the fact that it uses substrates that contain less enthalpy than the catabolic waste products, thereby making growth endothermic?

The answer to this question is yes. An example of an organism that not only does not produce heat while growing, but actually cools its environment down, is *Methanosarcina barkeri*, which obtains its energy from the following catabolic reaction involving the conversion of acetic acid to methane and carbon dioxide

$$CH_3COOH \Rightarrow CH_4 + CO_2. \tag{8.28}$$

This is called "acetoclastic methanogenesis". The enthalpy change of this catabolic reaction is positive ($\Delta_{cat}H = +7.5$ kJ/C-mol; Stockar and Liu 1999). As would be expected, the anabolic reactions are also slightly endothermic ($\Delta_{an}H \sim 0$) so that the growth reaction as a whole is endothermic. The endothermic nature of this fermentation has been proven calorimetrically with a heat of growth of $\Delta_r H_X = 145$ kJ/C-mol (Liu et al. 2001). The driving force comes from the very large entropy increase of 1,650 J K^{-1} C-mol^{-1} when a liquid molecule of acetic acid is split into two gaseous molecules, so that a driving force of $\Delta_r G_X = -366$ kJ/C-mol results, despite the positive heat of growth.

Another case of endothermic growth is photosynthesis. In phototrophic growth or mixotrophic growth (a mixture of phototrophic and chemotrophic metabolism), the Gibbs energy change of growth does not represent the driving force any longer. Phototrophic growth is driven by a constant influx of Gibbs energy in the form of photons ($\dot{W}_o > 0$ in Equation 8.20). This allows the cells to synthesize new biomass from CO_2 as a carbon source, which represents a process accompanied by a large increase of Gibbs energy. As this process resembles the inverse of the catabolic reaction in aerobic chemotrophic growth, such as in Equation 8.14, it is also characterized by a negligible term $T\,\Delta S$ (see Question 8.3.2.1), making the enthalpy change close to the Gibbs energy increase. The growth process per se is therefore endothermic, even though the whole experiment will liberate large amounts of heat due to the physical absorption of light. The endothermic nature of phototrophic growth has been demonstrated in calorimeters (von Stockar et al. 2011).

8.3.4 What Important Culture Performance Parameters Other Than Growth Yield Can Be Predicted Based on Thermodynamics?

The methods presented previously for estimating growth yields may also be used to compute the yields of products from catabolism, which is important in biorefinery design (see Section 8.4).

Other thermodynamic considerations have led to the proposal of methods for estimating microbial culture performance parameters other than optimal biomass yields. Tijhuis et al. (1993) have shown how to estimate growth yields in situations where the growth rate is maintained at an artificially low level, for instance in chemostats operating at low dilution rates. In such experiments, one needs to allow for additional Gibbs energy dissipation for maintenance reactions, which reduces the biomass yield. Maintenance reactions consist of catabolism in order to provide for example, additional energy to drive repair reactions for deactivated enzymes and other proteins, leaky membranes and damage to the DNA molecules. If the dilution rate becomes so low that it permits only the satisfaction of the maintenance reactions, growth ceases and the culture "washes out".

Assuming that the liberation of Gibbs energy in the electron transport chain is the growth limiting phenomenon, Heijnen (1999) developed an expression for predicting the maximum growth rate. He also presented thermodynamic arguments for estimating the minimum substrate concentration at which growth ceases and the threshold concentration at which no further substrate can be taken up.

8.4 Can Thermodynamics Be Used to Develop Biorefineries?

A biorefinery is defined as "the sustainable processing of biomass into a spectrum of bio-based products (food, feed, chemicals, materials) and bioenergy (biofuels, power and/or heat)" (IEA Bioenergy, Task 42 Biorefinery 2019). Here we consider how thermodynamics can contribute to the development of such processes.

8.4.1 Why Are We Interested in Biorefineries?

The need for an environmentally, economically and socially sustainable global economy, the peak production of oil, gas and coal in the not-too-distant future, the dependency on oil-exporting countries and, most importantly, the requirement to reduce atmospheric greenhouse gases mean that there are strong drivers to move from today's fossil-based economy to a more sustainable economy based on sustainably grown biomass. A typical example for such a biomass source would be lignocellulosic materials, which will have to be degraded to glucose and other basic bioorganic biomolecules to be useful. Then, instead of conventional petrochemical refineries, we will need processes that make all industrial chemical building blocks required for the fuel, polymer, pharmaceutical and the general chemical industries from such basic biomolecules. Given the fact that much of the chemistry involved is bioorganic in nature, it is generally believed that a wide variety of catalysts for such processes are to be found in nature in the form of bacteria, yeasts, fungi and microalgae. New types of refineries, known collectively as *biorefineries*, based on cultures of such microorganisms, will thus have to be developed.

8.4.2 What Are Metabolic Engineering and Systems Biology?

Unfortunately, natural evolution optimized biological organisms for reproduction and maintenance, not to produce the chemicals we desire. Therefore, the organisms need to be metabolically engineered. *Metabolic engineering* is the optimization of genetic and regulatory processes within different strains to increase a biological cell's production of desired products. The final goal of such metabolic engineering consists of predicting and improving the production of target molecules in engineered production strains or even microbial communities. Systems biology is a very important tool here and encompasses the computational and mathematical analysis and modeling of complex biological systems. Thermodynamics can potentially make important contributions to this endeavor.

8.4.3 Is it Possible to Assess the Thermodynamic Feasibility of Genetically Engineered Novel Metabolic Pathways?

One of the main goals of systems biology is the prediction of all enzymatic reaction rates occurring inside live cells as well the substance consumption and excretion rates between the cell and its environment. This would yield a complete overview of the product distribution and provide the necessary insight into the metabolism needed to genetically modify it as a basis for designing and tailor-making novel biocatalysts for biorefineries.

One of the core tools of systems biology is metabolic flux analysis. This consists of the construction of a model of the whole cellular network based on a knowledge of the important enzymes present in the cell from genomics (studying an organism's complete set of DNA) and proteomics (considering the complete set of proteins) as well as of the important metabolites from metabolomics (analyzing the chemical processes of organisms involving small molecules, intermediates and products of metabolism). For each metabolite with the concentration c_j, a molar balance is formulated, as shown in the following equation (see also Question 1.4.3.)

$$\begin{aligned} \nu_{1,1} r_1 + \nu_{1,2} r_2 + \nu_{1,3} r_3 + \cdots &= \frac{dc_1}{dt} \\ \nu_{2,1} r_1 + \nu_{1,2} r_2 + \nu_{1,3} r_3 + \cdots &= \frac{dc_2}{dt} \\ \nu_{3,1} r_1 + \nu_{1,2} r_2 + \nu_{1,3} r_3 + \cdots &= \frac{dc_3}{dt} \\ \vdots \quad\quad \vdots \quad\quad \vdots \quad + \cdots &= \vdots \end{aligned} \tag{8.29}$$

Each line represents the balance of a particular metabolite, whereby $\nu_{i,j}$ stands for the stoichiometric coefficient of the i-th metabolite in the j-th reaction. The vector r_j represents the vector of the rates of all the enzymatic reactions. The solution of the system of Equations 8.29 can be simplified by using a quasi-steady-state approximation of the metabolism, rendering the vector of the time derivatives of the concentration to zero. The approximation is justified by the fact that the metabolism is much faster than other cellular processes such as gene expression and cellular adaptations. The entirety of these equations can be written as a matrix equation with $\underline{\nu}$ representing the stoichiometry matrix with the number of rows equal to the number of metabolites and the number of columns equal to the number of enzymatic reactions

$$\underline{\nu}\,\vec{r} = 0. \tag{8.30}$$

In practice, the number of metabolites in networks is always much smaller than the number of enzymatic reactions. As a result, there are rather more unknowns to be determined than can be found from a smaller number of equations. It is an underdetermined system of equations. It is then impossible to find a single solution, but only a solution space which bounds all possible solutions. Many different methods have been proposed to reduce the size of this solution space. One proposition is to use constraints from the Second Law of Thermodynamics (see Question 3.3) so that each and every feasible enzyme reaction must obey the following constraint:

$$\Delta_{r,j} G < 0. \tag{8.31}$$

8.4.4 How Does a Thermodynamic Feasibility Analysis Work?

The inequality, $\Delta_r G < 0$, is the basis of so-called thermodynamic feasibility analyses (TFA) (Mavrovouniotis 1993). The TFA examines, for a measured or given distribution of metabolite concentrations, whether each individual reaction or each conceivable reaction cascade fulfills the inequality of Equation 8.31. That inequality can be written in terms of the activity coefficients for the solutes (metabolites), $\gamma_{B,1}^{\infty,c}$, and their concentrations, c_B, for a metabolic reaction with the stoichiometric coefficients ν_B as (Question 5.4.2.3)

$$\begin{aligned} \Delta_r G &= \Delta_r G^{\ominus} + RT \ln \prod_B (a_B^{\infty,c})^{\nu_B} \\ &= \Delta_r G^{\ominus} + RT \ln \prod_B (\gamma_B^{\infty,c})^{\nu_B} + RT \ln \prod_B c_B^{\nu_B}. \end{aligned} \tag{8.32}$$

Here, we have made use of the definitions of the Gibbs energy of reaction, $\Delta_r G = \sum_B \nu_B \mu_B$ (from Equation 5.14 in Question 5.2.2), and the infinite-dilution standard state using molar concentrations discussed in Question 5.4.2.2. Equation 8.32 shows that $\Delta_r G$ consists of three terms. The first term is the standard Gibbs energy of reaction, and is concentration independent. The second term, $RT \ln \prod_B (\gamma_B^{\infty,c})^{\nu_B}$, containing the activity coefficients of all reacting metabolites, depends on all material in the cytosol (macromolecules, salts, metabolites etc.). The activity coefficients are usually estimated using the Debye-Hückel limiting law (see Question 8.1.2) or can be calculated more accurately using G^E models (see Question 8.1.4) or equations of state such as the ePC-SAFT equation (see Question 2.9.2). The third term, $RT \ln \prod_B c_B^{\nu_B}$, depends strongly on the metabolically controlled concentrations, c_B. An example of a reaction susceptible to this type of analysis would be the glycolysis reaction, where D-2-phosphoglycerate reacts to phophoenolpyruvat and water (Greinert et al. 2020).

In some work, the relative activity, a_B, is often substituted by the measured metabolite concentration c_B and the influence of the real behavior of the metabolites in the cytosol is neglected by setting the activity coefficient $\gamma_{B,1}^{\infty,c}$ to 1. However, as explained in Question 8.1, illustrated in Figure 8.2 and discussed for the example of glycolysis by Vojinovic and von Stockar (2009), von Stockar et al. (2013) and Greinert et al. (2021), this simplification can lead to considerable errors.

To understand the *thermodynamic feasibility analysis* (TFA), let us consider the following metabolic reaction cascade

$$A \leftrightarrow B \leftrightarrow C \leftrightarrow D \leftrightarrow E \leftrightarrow F. \tag{8.33}$$

The TFA tests whether a distribution of relative metabolite activities over the metabolic sequence exists that makes $\Delta_r G$ of each single reaction (e.g. A↔B) or of arbitrary reaction sequences (e.g. C↔E) positive. If at least one single reaction (called a localized bottleneck) or one combination of reactions (called a distributed bottleneck) exists that does not fulfill inequality of Equation 8.31, the reaction cascade is considered to be thermodynamically unfeasible. This type of analysis is illustrated for the first two enzymatic reactions of the imaginary metabolic reaction cascade

$$A \rightarrow B \rightarrow C, \tag{8.34}$$

$$\Delta_{r1} G = \Delta_{r1} G^{\ominus} + RT \ln \frac{a_B}{a_A} \text{ and } \Delta_{r2} G = \Delta_{r2} G^{\ominus} + RT \ln \frac{a_C}{a_B}. \tag{8.35}$$

Figure 8.9 shows schematically a possible outcome of a TFA to identify relative activity distributions which will satisfy the inequality of Equation 8.31 for the reaction cascade of Equation 8.34.

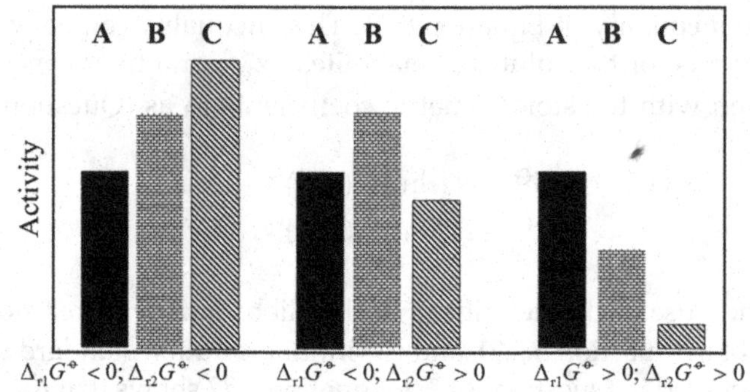

FIGURE 8.9
Possible relative activity distributions of the metabolites A, B and C fulfilling the inequality of Equation 8.31 and making the metabolic reaction cascade A↔B↔C thermodynamically feasible for three different distributions of the standard Gibbs energy of reaction $\Delta_r G^{\ominus}$.

If $\Delta_{r1}G^{\ominus}$ and $\Delta_{r2}G^{\ominus}$ are both negative (first case on the left in Figure 8.9), Equation 8.35 shows many different activity profiles, including some with increasing activities along the reaction path, will yield negative $\Delta_r G$'s for both steps and will thus yield a feasible pathway. If, however, $\Delta_{r2}G^{\ominus}$ is positive, a_C must be considerably smaller than a_B in order to override the positive standard Gibbs energy of reaction and to make $\Delta_{r2}G$ negative (second case, in the center of Figure 8.9) to enable a feasible pathway from A to C. If both $\Delta_r G^{\ominus}$ are positive, the pathway will only be feasible if the two positive standard terms are overridden by a large activity drop over both the first and the second reaction (third case on the right of Figure 8.9).

8.4.5 What Is Required for a Thermodynamic Feasibility Analysis?

As Equation 8.32 shows, we need three categories of data to carry out a full thermodynamic feasibility analysis, namely the standard Gibbs energies $\Delta_r G^{\ominus}$, the activity coefficients $\gamma_{B,1}^{\infty,c}$ and the concentration range of the considered metabolites c_B. The first category of data, $\Delta_r G^{\ominus}$, is independent of the concentration of the metabolites and the cytosolic conditions and can be calculated using the equation

$$\Delta_r G^{\ominus} = \Delta_r H^{\ominus} - T\,\Delta_r S^{\ominus}. \tag{8.36}$$

The problem with this approach is the difficulty of obtaining reliable data for $\Delta_r S^{\ominus}$. Alternatively, $\Delta_r G^{\ominus}$can be calculated using The Law of Hess (see Question 5.2.1)

$$\Delta_r G^{\ominus} = \sum_B \nu_B \Delta_f G_B^{\ominus} = -\sum_B \nu_B \Delta_c G_B^{\ominus}, \tag{8.37}$$

where $\Delta_f G_B^{\ominus}$, $\Delta_c G_B^{\ominus}$ stand for the standard Gibbs energy of formation and combustion of component B, respectively. A third option is to use group contribution methods based on the molecular structure of each metabolite. These group contribution methods assign an effective thermodynamic property contribution to each functional group of any chemical compound. The overall thermodynamic property of the respective compound is then derived from a base term and the summation over the frequency of occurrence of the functional groups multiplied by the property contribution of the respective group. For more details, the reader is referred for instance to Du et al. (2018). Sources for the Gibbs energies and their uncertainties are given in Chapter 9.

The second category of data required is related to the activity coefficient $\gamma_{B,1}^{\infty,c}$. If only long-range interactions are considered such as those represented by the extended Debye-Hückel limiting law (Question 8.1.2), no further parameters are required. However, at high metabolite concentrations, short-range interactions become crucial and the extended Debye-Hückel limiting law is no longer accurate enough. Furthermore, the crowding of the cell interior by macromolecules influences $\gamma_{B,1}^{\infty,c}$. In the simplest approach, this is allowed for means of excluded volume effects as discussed in Question 8.1. The test of the TFA on the example of the glycolysis pathway (Vojinovic and von Stockar 2009; von Stockar, et al., 2013; Greinert, et al. 2021) indicates that short-range interactions may also have an influence on the accuracy of the analysis and they must be considered using activity coefficient models for $\gamma_{B,1}^{\infty,c}$, as indicated in Question 5.4.2.3. However, each model, whether excess Gibbs energy (G^E) or equation of state (EOS), requires specific parameters. The EOS ePC-SAFT seems to be well suited to predict activity coefficients in complex mixtures such as those inside of cells (for instance see Vogel et al. 2020). More details on the SAFT theory are given in Question 2.9.2.

The third category of data required to use Equation 8.32 (i.e. the concentration of all individual metabolites c_B called the metabolome) covers two issues. The first issue is the total concentration of all metabolites inside the cell. It can either be measured using nuclear magnetic resonance (NMR) or liquid or gas chromatography in conjunction with mass spectrometry (LC-MS or GC-MS) or taken from databases such as those discussed in Chapter 9 (e.g. Wishart et al. 2007; Jewison et al. 2012; Guo et al. 2013).

The second issue arises from the fact that metabolites are often weak acids and are sometimes able to form complexes with magnesium. Consequently, instead of simply considering a single metabolite B, a diversity of differently protonated and complexed species of B must be considered. The species distribution depends on the pH and the magnesium ion concentration and can be calculated from the respective equilibrium constants (see Question 5.4). As an example, we consider the protonation equilibria of fructose-6-phosphate (F6P) and the complexation of adenosine triphosphate (ATP) with magnesium

$$H_2F6P \rightleftharpoons HF6P^- + H^+ \rightleftharpoons F6P^{2-} + 2H^+, \tag{8.38}$$

$$ATP^{4-} + Mg^{2+} \rightleftharpoons MgATP^{2-}. \tag{8.39}$$

Alberty (2003) showed there are two possible approaches to calculate the complex interplay of protonations and complexations: Binding polynomials or Legendre transformations. For more details, the reader is referred to the textbook of Alberty (2003).

References

Alberty R.B., 2003. *Thermodynamics of Biochemical Reactions*, New Jersey, Wiley Interscience.
Battley, E.H., Putnam, R.L., and Borio-Gates, J., 1997, "Heat capacity measurements from 10 to 300 K

and derived thermodynamic functions of lyophilized cells of *Saccharomyces cerevisiae* including the absolute entropy and the entropy of formation at 298.15 K", *Thermochim. Acta* **298**:37–46.

Birou, B., Marison, I. W., and von Stockar, U., 1987, "Calorimetric investigation of aerobic fermentations", *Biotechnol. Bioeng.* **30**: 650–660.

Debye P., and Hückel, E., 1923, "Zur Theorie der Elektrolyte", *Physikalische Zeitschrift* **24**: 185–206.

Du, B., Zhang, Z., Grubner, S., Yurkovich, J.T., Palsson, B.O., and Zielinski D.C., 2018, "Temperature-dependent estimation of Gibbs energies using an updated group-contribution method", *Biophys. J.* **114**: 2691–2702.

Duboc P., Marison, i.W., von Stockar, U., 1999, "Quantitative Calorimetry and Biochemical Engineering", *Handbook of Thermal Analysis and Calorimetry 4; From Macromolecules to Man*, ed. R.B. Kemp, Elsevier, Amsterdam, pp. 267 – 365.

Guo, A., Jewison, T., Wilson, M., Liu, Y., Knox, C., Djoumbou, Y., Lo, P., Mandal, R., Krishnamurthy, R., and Wishart, D.S., 2013, "ECMDB: the *E. coli* metabolome database", *Nucleic Acids Res.* **41**: D625–D630.

Greinert, T., Vogel, K., Seifert, A.I., Siewert, R., Andreeva, S.P., Maskow, T., Sadowski, G., and Held, C., 2020, "Standard Gibbs energy of metabolic reactions: V. Enolase reaction", *Biochim. et Biophys.* Acta 1868: 140365.

Greinert, T., Vogel, K., Maskow, T., and Held, C., 2021, "New thermodynamic activity-based approach allows predicting the feasibility of glycolysis", *Sci. Rep.* **11**: 6125.

Hamer, W.J., and Wu,Y.-C., 1972, "Osmotic coefficients and mean activity coefficients of uni-univalent electrolytes in water at 25°C", *J. Phys. Chem. Ref. Data* **1(4)**: 1047–1099.

Heijnen, J. J., and van Dijken, J. A., 1992, "In search of thermodynamic description of biomass yields for the chemotrophic growth of micro-organisms," *Biotechnol. Bioeng.* **39**: 833–858.

Heijnen, J. J., and van Dijken, J. P., 1993, "Response to comments on: In search of a thermodynamic description of biomass yields for the chemotrophic growth of micro-organisms", *Biotechnol. Bioeng.* **42**: 1127–1130.

Heijnen, J. J., 1999, "Bioenergetics of microbial growth", in: *Bioprocess Technology: Fermentation, Biocatalysis and Bioseparation*, Eds.M. C. Flickiger and S.W. Drew, J. Wiley & Sons, Chicester UK, pp. 267–291.

Held, C., Cameretti, L.F., and Sadowski, G., 2008, "Modeling aqueous electrolyte solutions. Part1. Fully dissociated electrolytes," *Fluid Phase Equilib.* **270**: 87–96.

IEA Bioenergy, Task 42 Biorefinery, 2019, *Bio-based chemicals – Value added products from biorefineries.*

Jewison T., Neveu V., Lee J., Knox C., Liu P., Mandal R., Murthy R.K., Sinelnikov I., Guo A.C., Wilson M., Djoumbou Y., and Wishart, D.S., 2012, "The yeast metabolome database", *Nucleic Acids Res.* **40**: D815–D820.

Liu, J.-S., Vojinovic, V., Patino, R., Maskow, Th., and von Stockar, U., 2007, "A comparison of various Gibbs energy dissipation correlations for predicting microbial growth yields", *Thermochim. Acta* **458**: 38–46.

Liu, J. S., Marison, I.W., and von Stockar, U., 2001, "Microbial growth by a net heat up-take: A calorimetric and thermodynamic study on acetotrophic methanogenesis by *Methanosarcina barkeri*", *Biotechnol. Bioeng.* **75**: 170–180.

Mavrovouniotis, M. L., 1993, *Identification of Qualitatively Feasible Metabolic Pathways.* In *Artificial Intelligence and Molecular Biology*, ed. L. Hunter, 325–364, AAAI Press/MIT Press.Menlo Park, California:

Pitzer, K.S., 1973, "Thermodynamics of electrolytes. I. Theoretical basis and general equations", *J. Phys. Chem.* **77(2)**: 268–277.

Prausnitz, J.M., and Tavares, F.W., 2004, "Thermodynamics of fluid-phase equilibria for standard chemical engineering operations," *AIChE J.* **50**: 739–761.

Sandler S. I., and Orbey, H., 1991, "On the thermodynamics of microbial growth processes", *Biotechnol. Bioeng.* **38**: 697–718.

Schrödinger, E., 1944, *What is Life – the Physical Aspect of the Living Cell,* Cambridge University Press, Cambridge, UK.

Tijhuis, L., van Loosdrecht, M., and Heijnen, J. J., 1993, "A Thermodynamically based correlation for maintenance Gibbs energy requirements in aerobic and anaerobic chemotrophic growth", *Biotechnol. Bioeng.* **42**: 509–519.

Vogel, K., Greinert, T., Reichard, M., Held, C., Harms, H., and Maskow, T., 2020, "Thermodynamics and kinetics of glycolytic reactions. Part II: Influence of cytosolic conditions on thermodynamic state variables and kinetic parameters", *Int. J. Mol. Sci.* **21**: 7921

Voisard, D., Pugeaud, P., Kumar, A. R., Jenny, K., Jayaraman, K., Marison, I.W., and von Stockar, U., 2002, "Development of a large-scale biocalorimeter to monitor and control bioprocesses," *Biotechnol. Bioeng.* **80**: 125–138.

Vojinovic, V., and von Stockar, U., 2009, "Influences of uncertainties in pH, pMg, Activity coefficients, Metabolite concentrations, and other factors on the analysis of the thermodynamic feasibility of metabolic pathways", *Biotechnol. Bioeng.* 103: 780 –795

von Stockar, U., and Marison, I.W., 1991, "Large scale calorimetry in biotechnology," *Thermochim. Acta* **193**: 215–242.

von Stockar, U., and Marison, I. W., 1993, "The definition of energetic growth efficiencies for aerobic and anaerobic microbial growth and their determination by calorimetry and other means", *Thermochimica Acta* **229**: 157–172.

von Stockar, U., and Liu, J.-S., 1999, "Does microbial life always feed on negative entropy? Thermodynamic analysis of microbial growth", *Biochim. Biophys. Acta* **1412**: 191–211.

von Stockar, U., Vojinovic, V., Maskow, Th., and Liu, J.-S., 2008, "Can microbial growth yield be estimated using simple thermodynamic analogies to technical processes?" *Chem. Eng. Process.* **47**: 980–990.

von Stockar, U., 2010, "Biothermodynamics of live cells: A tool for Biotechnology and biochemical engineering", *J. Non-Equilib. Thermodyn.***35**: 415–475.

von Stockar, U., Marison, I. W., Janssen, M., and Patiño R., 2011, "Calorimetry and thermodynamic aspects of heterotrophic, mixotrophic, and phototrophic growth", *J. Therm. Anal. Calorim.* **104**: 45–52.

von Stockar, U., 2013, "Biothermodynamics of live cells: Energy dissipation and heat generation in cellular cultures", in: Biothermodynamics, The role of thermodynamics in biochemical engineering, Ed. U. von Stockar, EPFL Press, CRC Press, Taylor & Francis Group, Lausanne and Boca Raton FL 33487, pp. 475–534.

von Stockar U., 2018, "Biothermodynamics: Bridging thermodynamics with biochemical engineering", *Adv. Material Sci. Eng.* **2**:2–7.

Westerhoff H.K., and van Dam K., 1987, "Thermodynamics and control of biological free-energy transduction," Elsevier, Amsterdam.

Wishart D.S., and Querengesser, L., 2007, "HMDB: The human Metabolome Database", *Nucleic Acids Res.* **35**: D521–D526

Zimmerman, S. B., and Trach, S.O., 1991, "Estimation of macromolecular concentration and excluded volume effects for the cytosol of *Escherichia coli*", *J. Mol. Biol.* **222**: 599–620.

9

Sources of Data

9.1 Introduction

The practical application of the principles of thermodynamics to any field of science and engineering ultimately depend upon the physical properties of the materials that make up the thermodynamic system discussed in Chapter 1. Some general notions such as the idea of equilibrium, the ultimate efficiency that can be achieved in a heat engine and the description of phase behavior in multi-component systems can be accommodated without recourse to particular materials but, if one wants to build a real machine or design a real separation process, properties such a density, enthalpy and entropy of components and mixtures really matter. Those properties of materials that are of concern are collectively known as thermophysical properties and include those characteristic of the equilibrium state (thermodynamic properties) and of the nonuniform state (transport properties) for gases, liquids and solids. These properties are the subject of considerable international research (e.g. *Experimental Thermodynamics* 1968, 1975, 1991, 1994, 2000, 2003, 2005, 2010, 2014 and 2016) involving both experimental effort to measure them directly and theoretical effort to provide a sound physical basis for their prediction from first principles or to at least supplement the available experimental information.

In this chapter, we seek to set out some of the issues that surround the supply and use of such thermophysical properties; in particular, where the numerical values of material properties are best found and how one can assess the reliability of such numbers, their pedigree and how one should proceed if there are no sources of the particular information sought. This chapter is again aimed at a general audience encompassing students engaged in projects to design engineers who are not specialists in the field of thermophysical properties.

9.2 What Kind of Numbers Are We Searching For?

Before proceeding to the main question of where to find the value of a required property, in this section we will try to specify the characteristics of the value we are interested in. Specifically, we consider what degree of uncertainty we can tolerate in the value assigned to a particular property and whether it should be an internationally agreed value, an experimental value or if it can be an estimated value.

9.2.1 How Uncertain Should the Values Be?

In any scientific or engineering calculation, design or simulation, thermophysical properties of the materials involved are required to complete the computation. Inevitably, the

DOI: 10.1201/9780429329524-9

values obtained and tabulated for any such property are burdened with an uncertainty whether than have been obtained via experiment, theory or an estimation method. That uncertainty must propagate to the final result of the calculation or design or simulation of industrial or scientific plant and thereby affect its cost or operation. It is therefore quite reasonable to ask the question of this section with respect to the uncertainty in the property values that can be tolerated in the design. Of course, in some cases, it can be seen by students as a triumph to find any value for the required property of a material, but that does not negate the need for a consideration of its uncertainty.

In the first edition of this book (Assael et al. 2011), we presented a number of examples of the effects of property uncertainty upon processes of various kinds that remain valid. However, they are less relevant a decade later and so we use here different examples drawn from the processes considered earlier in this edition to illustrate the point.

As we have seen in Chapters 6, 7 and 8, a thermodynamic analysis can often be used to evaluate the energy requirements of a process or its thermodynamic efficiency. Usually these are related to the enthalpy change, $\Delta_r H$, or Gibbs energy change, $\Delta_r G$, for a reaction or process and the energy requirement is related to these changes in a linear fashion. For example, the energy generated by combustion of methane in a CCGT power plant is evaluated in Equation 6.40, which requires a knowledge of the standard heat of combustion of methane, $\Delta_c H^\ominus$, and the heat capacities of the compounds involved in the reaction, C_p, over a range of temperatures from 298 K to 700 K. If we assume that each value of C_p is about 50 J mol^{-1}K^{-1} with an uncertainty of 1% i.e. 0.5 J mol^{-1}K^{-1}, and $\Delta_c H^\ominus$ (= −890.8 kJ/mol) has an uncertainty of 1 kJ/mol, then the uncertainty in $\Delta_{r1} H$ is approximately

$$\begin{aligned}\delta(\Delta_{r1}H)^2 &= (700-298)^2[0.5^2+1.0^2+4.0^2]+(10^3)^2+(353-298)^2[1.0^2+0.5^2+4.0^2]\\ &= 161{,}604[0.25+1.0+16.0]+10^6+3,025[1.0+0.25+16.0]\\ &= 2.788\times10^6+10^6+52{,}181\\ &= 3.8402\times10^6(\mathrm{J/mol})^2\end{aligned} \tag{9.1}$$

$$\delta(\Delta_{r1} H) = 1.96\times10^3(\mathrm{J/mol})\sim 2\mathrm{kJ/mol} \tag{9.2}$$

and

$$\Delta_{r1} H \sim 776.7 \pm 2.0\mathrm{kJ/mol},\ \text{an uncertainty of} \sim 0.25\%. \tag{9.3}$$

This quantity, along with the energy conversion efficiency, will determine the amount of product (in this case electricity) delivered per unit mass of feedstock and hence the size and/or throughput of the plant. Although, in cases like this, the uncertainty depends on the relative values of the different contributions to the evaluated energy or function, the overall relative uncertainty is comparable with that of the contributing thermodynamic quantities, particularly that of the larger terms where these dominate.

We are often concerned with the efficiency of a device or process, such as that of a hydrogen fuel cell in Equations 7.34 and 7.35, a CCGT power plant (Questions 6.2.2 and 6.3.4) or the coefficient of performance of a refrigeration cycle or a heat pump as in Equations 6.104 and 7.135. These involve the ratio of energy terms and any relative errors in the enthalpies or Gibbs energies involved accumulate essentially additively, or strictly as the root mean square value. Hence, a 10% error in the $\Delta_r H$ and $\Delta_r G$ in Equation 7.34

results in an uncertainty of $(0.1^2 + 0.1^2)^{0.5} = 0.14$ or 14% in efficiency. Since enhancements in efficiency of as little as 1% can have significant implications for the size (to meet a target output), capital and operating costs of a power plant, an energy storage process/device or a chemical conversion process, determining the relevant changes in thermodynamic state functions as accurately as possible is a key aspect of designing and modelling such operations.

However, it is for thermodynamic quantities related to each other through exponential relationships where the greatest sensitivities to uncertainty arise. In particular, the relationships of equilibrium constants to the standard Gibbs energy of reaction (Equation 5.26, the van't Hoff isotherm; Equations 5.60 and 5.64), and to the enthalpy of reaction (through Equation 5.33, the van't Hoff isochore, and its integrated form in Equation 5.35) to give their variation with temperature, are very sensitive to errors in the thermodynamic quantities involved. We can illustrate this by using the example of the Haber-Bosch process, Equation 5.53, described in detail in Question 5.3.5.

Let us examine the implications of a 10% error in the determination of the standard Gibbs energy for this reaction, 7.4 ± 0.74 kJ mol^{-1} at 500 K and 0.5 MPa. Within these bounds (6.66 to 8.14 kJ mol^{-1}), Equation 5.35 shows that the thermodynamic equilibrium constant $K^{\ominus}$ varies from 0.202 through 0.169 to 0.141, a spread of ±18%. The equilibrium extent of reaction, ξ_{eq}, varies from 0.50 to 0.46 and the equilibrium yield of ammonia, y_{NH_3}, from 0.30 to 0.33. When the reactor pressure is raised to 20 MPa, use of Equation 5.58 shows that K_y ranges from 8,050 to 5,650 but that now the spread of ξ_{eq} (0.91–0.89) and y_{NH_3} (0.83–0.80) is smaller. These ranges are very significant.

In Question 5.3.2, we saw that increasing the temperature to 650 K, using the van't Hoff isochore in the form given in Equation 5.35 with the first Ulich approximation and $\Delta_r H^{\ominus}(573\,K) = -105$ kJ/mol, gave $K^{\ominus}(T = 650\text{ K}) = 4.96 \times 10^{-4}$, a dramatic decrease of almost three orders of magnitude. If the uncertainty in $\Delta_r H^{\ominus}(573\text{ K})$ is ±5%, then $K^{\ominus}(T = 650\text{ K})$ ranges from 3.76×10^{-4} to 6.55×10^{-4}, a spread of about ±30%. At p = 20 MPa, this leads to an uncertainty in ξ_{eq} of 0.62 ± 0.03, with $y_{NH_3} = 0.45 \pm 0.03$. Thus, the uncertainty in $\Delta_r H^{\ominus}$ here leads to a very significant uncertainty in the ammonia yield of about 7%, leading to large variations in the selection of reactor size and optimal operating conditions. Again, being able to measure or predict the thermodynamic parameters with an accuracy of ±1% or better enables much tighter specification of the optimal reactor size and process conditions and has direct benefits for other aspects such as cost and process safety.

A further example of the sensitivity of process outputs to the accuracy of the controlling thermodynamic properties is the use of the different forms of Equation 5.35, the van't Hoff isochore. The previous example used the first Ulich approximation, assuming that the enthalpy of reaction is constant over the temperature range of integration to calculate the change in value of an equilibrium constant from one temperature to another. In reality, $\Delta_r H^{\ominus}(T)$ is a continuous function of temperature, varying according to Equation 5.38 (Kirchoff's law) to an extent determined by the heat capacity C_p, which is itself a function of temperature. The second Ulich approximation takes this into account by including the change in C_p over the temperature range. Question 5.4.1, which concerns the $CaCO_3$ decomposition reaction, Equation 5.76, showed that, including the second Ulich approximation, resulted in a decrease in the equilibrium constant $K^{\ominus}(=p_{CO_2}/p^{\ominus})$ at 1,273 K from 16.1 to 9.2, a decrease of 43%. We see therefore that values of $C_p(T)$ are essential for the calculation of the precise temperature dependence of equilibrium constants and accurate reacting mixture compositions, especially over extended temperature ranges and just how errors in them can affect design.

Similar considerations can be applied to other thermodynamic quantities that are related through logarithmic relationships. These include the chemical potential (to activities, activity coefficients and measures of concentration; e.g. Equations 3.53, 3.162, 3.165, 5.57 and 5.69), galvanic cell emfs (to Gibbs energies of reaction and, hence, equilibrium constants; e.g. Equations 5.97, 5.98 and Section 5.5.1) and exergy and entropy changes (to heat capacities and temperature; e.g. Equations 3.177 and 3.178).

These examples illustrate that it is important to be able to estimate the uncertainty in all the property values that enter a calculation as well as finding the property value itself.

9.2.2 Why Should the Numbers Be Internationally Agreed-Upon Values?

In some cases, the prescribed uncertainty of the required property may not be sufficient for a commercial purpose. As an example, we consider custody transfer of material across an international border, as illustrated in Figure 9.1. We assume that Company A in Country X sells a fluid to Company B in Country Y. In both cases, the quantity delivered by Company A and the quantity received by Company B is measured by volume but is paid for by mass. Since the same volume crosses the border, the options are as follows:

a. If both countries employ the same density tables, then the mass calculated in both countries is the same and hence payments requested will be equal to payments to be paid
b. If, however, different density tables are employed, different masses will be calculated, and clearly payments requested and paid will not agree, resulting in a payment dispute

It is thus evident that in the case of custody transfer, the uncertainty of the density and any correlation used to determine it is not of primary importance but what is important is whether the values are accepted internationally.

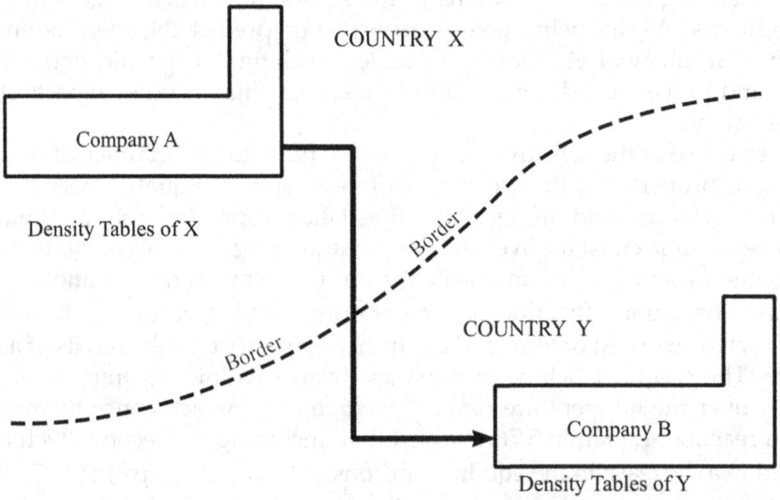

FIGURE 9.1
Custody transfer.

Consider, for example, the pipeline from Country X to Country Y, which will annually transport about 35 Mt of crude oil that originates in Country X. This mass is similar to that transferred through the so-called Trans-Alaska Pipeline. If the density of the crude oil originating in Country X is assumed to be $\rho = 900\ \text{kg m}^{-3}$, the volume of oil transferred annually will be about 38 Mm^3 (about 244×10^6 barrels). At an oil price of \$629 m^{-3} (\$100 bbl^{-1}), this is equivalent to \$24.4 G (24 billion USD). Hence, a difference of 1% in density of the crude oil used by Company A and Company B, shown in Figure 9.1, will result in a \$0.24 G difference, and presumably a dispute with potentially at least legal ramifications if not more! Based solely on this one example, it is not difficult to see the importance of internationally accepted values.

A similar argument can be put forward in the case of technology transfer. A process or plant developed in one country and sold to another must meet detail specifications and design methodology that will also include the data used for the engineering calculations. An example of this fact is provided by considering a nuclear power reactor plant commissioned in a different country than where the original company is. In this case, internationalization of the thermophysical properties of water and steam employed is recognized as highly significant to the generation of electricity from steam-driven power plants; the properties of steam are an essential part of the design as well as form the basis for estimating the energy efficiency of the system that will ultimately be compared with measurements, albeit too late by that stage for major changes because the generator has been designed, constructed, and commissioned. It was usual for each country to have their own values for the thermophysical properties of steam that are often referred to as steam tables. The measurements that underpinned these steam tables were combined, in some cases complimented with new measurements and then fit by a correlation all under the auspices of the IAPWS. This organization has spent more than 40 years developing what are now internationally accepted steam tables and correlations otherwise known as formulations.

A final point an engineer will almost certainly be called upon to consider arises from quality assurance; that is, the requirement to satisfy regulatory requirements imposed for safety and environmental reasons. These may be imposed by a national regulatory body or an international organization. The requirements of these organizations must be satisfied; in some cases, national regulatory bodies, perhaps for the purpose of trade, comply with regulations of other nations. Quality assurance of a plant or a process can often require a demonstrable pedigree for each number used in the design calculations; one example is the calculation of the energy (heat) transfer that would be required during a meltdown of a nuclear reactor.

The previous discussion clearly demonstrates that in such cases the user must search for internationally accepted thermophysical data, which is data that are used by the majority of the world as a basis for trade, regulation or standardization. This refers to supranational bodies that propose such standards. Such bodies include the following:

- International Association for the Properties of Water and Steam (IAPWS)
- International Association for Chemical Thermodynamic (IACT)
- International Association for Transport Properties (IATP)
- International Union of Pure and Applied Chemistry (IUPAC)

International accepted values or standard or reference values can be found in reference journals or textbooks concerned with reference data, for example, the *Journal of Physical and Chemical Reference Data* to name but one.

9.2.3 Should I Prefer Experimental or Predicted (Estimated) Values?

Having discussed the uncertainty associated with property values as well as the international dimension, one obvious question that can arise is whether the reader should be looking specifically for experimental values or for predicted ones. The answer to this question is relatively easy.

Let us assume that we have a need to measure only 10 properties at just 10 temperatures and 10 pressures, for 15 pure fluids and all their mixtures, at 5 compositions in the liquid and gas phases; we will assume there are no values reported in primary tables of the standard equilibrium constant and molar enthalpy of formation that would provide a means of determining the required properties. The total number of measurements required is 3.3×10^8 ($10 \times 10 \times 10 \times 32{,}766 \times 5 \times 2$). If one further assumes that three measurements can be obtained for each normal eight-hour working day and that a person works for 48 weeks (or 240 days per year), then the number of years the task of measurements requires is about 457,000; alternatively, one might employ 457,000 people working for one year. In view of this estimate, it is rather obvious that we cannot rely solely on measurements. In reality, some of the required values can be reliably estimated at least for most purposes from primary tables of standard thermodynamic properties, perhaps when combined with data from secondary tables. Such tables are maintained, for example, by the Thermodynamic Research Center now located at the National Institute of Science and Technology, USA.

If one ends up searching the archival literature and is indeed fortunate to find measured values of the required property then the question arises, should we trust it? Unfortunately, nothing is that simple. There are, of course, just as with every human endeavor, good and bad measurements, and the fact that a measurement exists does not imply that the value is correct. If the measurement satisfies the criteria of quality laid down for the experimental technique, then the measurement results are preferred. In the absence of such measurements, predicted values should be sought, but that does not absolve the user from the obligation to conduct an assessment of the uncertainty of the values so obtained.

9.3 What Software Packages Exist for the Calculation of Thermophysical Properties?

A number of software packages claim to calculate or predict the thermophysical properties of fluids and much of this work has been conducted by the National Institute of Standards and Technology (NIST), USA, and in the subquestions of this question we list a few examples. We note that some are free, and some require a fee. We also note that many of these packages are incorporated in design packages (e.g. Aspen© incorporates NIST REFPROP), or are employed in various handbooks (e.g. CRC *Handbook of Chemistry and Physics*, *Perry's Chemical Engineers' Handbook*, or the *ASHRAE Handbook of Fundamentals*); they also incorporate among others, tables calculated using NIST REFPROP.

9.3.1 What Is the NIST Thermo Data Engine (TDE)?

NIST Standard Reference Database 103a (pure fluids) and NIST Standard Reference Database 103b (pure fluids and mixtures) (https://www.nist.gov/mml/acmd/trc/thermodata-engine/srd-nist-tde-103a and https://www.nist.gov/mml/acmd/trc/

thermodata-engine/srd-nist-tde-103b), provide about 50 properties for pure fluids (Database 103a) and about 120 properties for mixtures (Database 103b), including density, vapor pressure, heat capacity, enthalpies of phase transitions, critical properties, melting and boiling points and so on. It fills the gaps in experimental data by deployment of automated group-contribution and corresponding-states prediction schemes and most of all emphasizes the consistency between properties (including those obtained from predictions), and provides for flexibility in selection of default data models depending on the particular data scenario. The Thermo Data Engine supports several equations of state for pure compounds (original and modified volume-translated Peng-Robinson, Sanchez-Lacombe, PC-SAFT and Span-Wagner) and allows the user to fit parameters to experimental and predicted data. Enthalpies of formation are evaluated on the basis of stored experimental enthalpies of combustion and the modified Benson group-contribution method.

9.3.2 What Is the NIST REFPROP?

NIST REFPROP, NIST Reference Fluid Thermodynamic and Transport Properties Database provide estimates of the thermophysical properties of pure fluids and mixtures and are available from https://www.nist.gov/srd/refprop. REFPROP Version 10 employs correlations or models that represent experimental data. It includes 147 pure fluids, 5 pseudo-pure fluids (such as air) and mixtures with up to 20 components (natural gas, hydrocarbons, refrigerants, alternative and natural refrigerants, air, noble elements and many predefined mixtures). The properties calculated are as follows: Density, energy, enthalpy, entropy, C_V, C_p, sound speed, compressibility factor, Joule-Thomson coefficient, quality, second and third virial coefficients, Helmholtz function, Gibbs function, heat of vaporization, fugacity, fugacity coefficient, K value, molar mass, thermal conductivity, viscosity, kinematic viscosity, thermal diffusivity, Prandtl number, surface tension, dielectric constant, isothermal compressibility, volume expansivity, isentropic coefficient, adiabatic compressibility, specific heat input, exergy and many others. REFPROP incorporates "high accuracy" Helmholtz function and MBWR equations of state, including many international standard equations, the Bender equation of state for several of the refrigerants, an extended corresponding-states model for fluids with limited data, an excess Helmholtz function model for mixture properties, while experimentally based values of the mixture parameters are available for hundreds of mixtures. Finally, predictions of both viscosity and thermal conductivity are provided by fluid-specific correlations (where available): A modification of the extended corresponding-states model, or the friction theory model. Because the compilation was created by NIST, which is a governmental agency, and full reference to the original scientific journals are given, this compilation should be an excellent source for data for the purposes of both science and engineering.

9.3.3 What Is the NIST Chemistry Webbook?

The NIST Chemistry webbook (available from https://webbook.nist.gov/) is free and includes thermochemical data for 7,000 organic and inorganic compounds (enthalpy of formation, enthalpy of combustion, heat capacity, entropy, phase transition enthalpies and temperatures, vapor pressure), reaction thermochemistry data for over 8,000 reactions, infrared spectra for over 1,000 compounds, mass spectra for over 33,000 compounds, ultraviolet and visible spectra for over 1,600 compounds, gas chromatography data for over 27,000 compounds, electronic and vibrational spectra for over 5,000 compounds, constants of diatomic molecules (spectroscopic data) for over 600 compounds, ion energetics data for over 16,000 compounds and thermophysical property data for 74 fluids at the time of writing this.

9.3.4 What Is the DIPPR Database?

The Design Institute for Physical Property Data (DIPPR) provides a database (available from https://dippr.aiche.org/) that contains evaluated thermodynamic and physical property data for process engineering. It is supported by the American Institute of Chemical Engineers (AIChE) and is run by Brigham Young University, USA. DIPPR contains 49 thermophysical properties for 2,332 industrially relevant compounds. It also includes 15 temperature-dependent properties; contains raw data from the literature; contains critically evaluated, recommended thermophysical values and predicts appropriate values when experimental chemical data are not available.

9.3.5 What Is the Landolt-Börnstein Database?

The Landolt-Börnstein database for pure substances incorporates the 400 Landolt-Börnstein volumes that include 250,000 substances and 1,200,000 citations available with a single keystroke. Marketed as "the world's largest resource for physical and chemical data," SpringerMaterials—The Landolt-Börnstein Database (https://materials.springer.com/) brings the print collection's content into one easy-to-access online platform (with 91,000 online documents and 3,000 properties). The core of the database is twofold; first, it employs a user interface with a search engine, and, second, it makes the content findable. Users can search in several ways: With a Google-like search box, an advanced search tab that creates a Boolean search term automatically as the user sets up the parameters or a color-coded periodic table.

9.3.6 What Is the NIST STEAM?

NIST/ASME STEAM PROPERTIES DATABASE, NIST Standard Reference Database 10 (https://www.nist.gov/srd/nist-standard-reference-database-10), is a computer package for the calculation of the properties of water and steam. The STEAM package employs the latest correlations developed by IAPWS (http://www.iapws.org/) for water and steam. As such, they are standard values and their uncertainty are those quoted by IAPWS.

9.3.7 What Is CoolProp?

CoolProp (http://www.coolprop.org/) is a C++ library that implements pure and pseudo-pure fluid equations of state and transport properties for 122 components, mixture properties using high-accuracy Helmholtz energy formulations and correlations of properties of incompressible fluids and brines. It also incorporates fast IAPWS-IF97 (Industrial Formulation) for water/steam.

9.3.8 What Is ILThermo?

ILThermo (NIST Standard Reference Database 147, https://ilthermo.boulder.nist.gov/) is a web-based ionic liquids database available free to the public. It aims to provide users worldwide with up-to-date information on publications of experimental investigation on ionic liquids, including numerical values of chemical and physical properties, measurement methods, sample purity, uncertainty of property values, as well as many other significant measurement details. The database can be searched by means of the ions constituting the ionic liquids, the ionic liquids themselves, their properties and references.

9.3.9 What Is the Clathrate Hydrate Physical Property Database?

The Clathrate Hydrate Physical Property Database (NIST Standard Reference Database 156; https://gashydrates.nist.gov) is a free database that presently contains two main types of data for clathrate hydrates: A comprehensive database of thermophysical properties of gas hydrate systems and the database for the Mallik 2002 drilling project (https://netl.doe.gov/node/7489).

9.3.10 What Is KDB?

KDB (https://www.cheric.org/research/kdb/) is a free website that provides useful information about thermophysical properties and calculation methods for 2,000 hydrocarbons and light gases, 200 polymers and solvents and 2,000 electrolyte solutions commonly encountered in chemical engineering practices.

9.3.11 What Is DETHERM?

DETHERM (https://dechema.de/en/detherm.html) is a numerical database that provides thermophysical property data. The DETHERM database provides thermophysical property data for about 76,000 pure compounds and 253,000 mixtures, literature values, together with bibliographical information, descriptors and abstracts. The DETHERM database is produced from the DECHEMA e.V. in cooperation with the DDBST GmbH, Oldenburg. The complete database consists of sets of property orientated packages, which are maintained and produced by external experts. This guarantees high quality and checked data. Examples for such packages are the Dortmunder Datenbank DDB from the DDBST GmbH, the Wiley/VCH database INFOTHERM (former FIZ-Chemie Berlin) or the electrolyte database ELDAR from the University of Regensburg.

9.4 How Do We Find Enthalpies, Gibbs Energies and Entropies for Biochemical Reactions?

Thermophysical data of very many biomolecules can be found in the software packages and data sources described in Question 9.3. However, that is not always the case, so that here we provide information on some estimation methods that can be employed when no other source is available. A complete overview of all data source is beyond the scope of this chapter and therefore we will limit ourselves to $\Delta_c H^{\ominus}$, $\Delta_c G^{\ominus}$, $\Delta_c S^{\ominus}$ in the following. For the estimation of thermodynamic data for biomolecules, it makes a significant difference if we know a structural formula for the molecule, as it is the case for most metabolites, or whether we only know the elemental composition, as is often the case for enzymes, viruses, bacteria, algae or human cell lines.

9.4.1 How Are $\Delta_c H^\ominus$, $\Delta_c G^\ominus$, $\Delta_c S^\ominus$ Estimated When Only the Elemental Composition of Species B Is Available?

If only the elemental composition is available for a biomolecular species B, then the standard enthalpy of combustion $\Delta_c H^\ominus$ and the standard Gibbs energy of combustion $\Delta_c G^\ominus$ are approximately linearly related to what is called the degree of reduction, $j_{r,B}$, by the equations

$$\begin{aligned} \Delta_c H^\ominus &= Q_H j_{rB} \\ \Delta_c G^\ominus &= Q_G j_{rB} \\ j_{rB} &= 4n_C + n_H - 2n_O + 6n_S + 5n_P. \end{aligned} \tag{9.4}$$

Here, n_C, n_H, n_O, n_S, n_P indicate the total number of atoms of carbon, hydrogen, oxygen, sulfur and phosphorus, respectively, in the molecule and the multiplicative factors are the number of the atoms. The proportionality coefficients, Q_H and, Q_G, are nearly the same for most organic compounds and are therefore assumed to be applicable to any biomolecules or even to biological cells. For instance, Sandler and Orbey (1991) provided $Q_H = -109.0$ kJ/degree of reduction and $Q_G = -110.23$ kJ/degree of reduction. Despite the approximate nature of these relationships, they often provide very good predictions. The entropy of combustion can be calculated using the equation

$$\Delta_c S_B^\ominus = \frac{\Delta_c H_B^\ominus - \Delta_c G_B^\ominus}{T}. \tag{9.5}$$

The manner in which the enthalpies, entropies or Gibbs energies of formation of compounds can be calculated from the respective combustion data is described in Question 5.2.1.

9.4.2 How Are Estimates of Gibbs Energies of Catabolic and Anabolic Reactions Calculated?

In order to illustrate how one can estimate Gibbs energies of catabolic and anabolic reactions, we consider the reaction discussed in Question 8.3. The split of the growth reaction into a catabolic and an anabolic reaction proposed in Figure 8.5 assumes that the carbon and energy substrate is first completely catabolized and that, then, parts of the catabolic products are utilized in order to synthesize the new cells. These two partial reactions can be formalized as follows

$$\text{Catabolism: } \mathrm{S} + Y_A^{cat}\mathrm{A} \rightarrow Y_P^{cat}\mathrm{P} + Y_{SOX}^{cat}\mathrm{SOX} \tag{9.6}$$

$$\text{Anabolism: } Y_P^{an}\mathrm{P} + Y_{SOX}^{an}\mathrm{SOX} + Y_N^{an}\mathrm{NH_3} \rightarrow \mathrm{X} + Y_A^{an}\mathrm{A}, \tag{9.7}$$

where S is the energy and carbon substrate, A is an electron acceptor (e.g. O_2), P the reduced form of A (e.g. H_2O) and SOX the oxidized form of S (e.g. CO_2).

Now, from Question 5.2.2, using the definitions of Equations 5.11 and 5.13, we have

$$\Delta_r G \equiv \sum \nu_B \mu_B = -\sum \nu_B \Delta_c G_B. \tag{9.8}$$

In fermentations, only S, P, NH_3 and X can have a combustion Gibbs energy different from zero. Therefore, the Gibbs energies of Reactions 9.6 and 9.7 are given by

$$\Delta_{cat}G = \Delta_c G_S - Y_P^{cat}\Delta_c G_P \tag{9.9}$$

$$\Delta_{an}G = Y_P^{an}\Delta_c G_P + Y_N^{an}\Delta_c G_N - \Delta_c G_X. \tag{9.10}$$

The stoichiometric coefficient for the nitrogen source, Y_N^{an}, can be found from a nitrogen balance as the coefficient of nitrogen in the elemental formula for dry biomass, x_N. In the example in Question 8.2, the stoichiometric coefficient for the nitrogen source is 0.15, as shown in Equation 8.14.

In *aerobic growth*, the product P is water and thus $\Delta_c G_P$ is equal to zero. Therefore

$$\Delta_{cat}G = \Delta_c G_S \tag{9.11}$$

$$\Delta_{an}G = x_N\Delta_c G_N - \Delta_c G_X. \tag{9.12}$$

In *anaerobic growth*, the stoichiometric coefficients for P, Y_P^{cat} and Y_P^{an}, may be calculated based on a degree of reduction balance on Equations 9.6 and 9.7 as follows

$$j_{rS} = Y_P^{cat} j_{rP} \text{ or } Y_P^{cat} = \frac{j_{rS}}{j_{rP}} \tag{9.13}$$

and

$$j_{rP} Y_P^{an} + j_{rN} x_N = j_{rX} \text{ or } Y_P^{an} = \frac{j_{rX} - j_{rN} x_N}{j_{rP}}. \tag{9.14}$$

Hence, Equations 9.9 and 9.10 become, for anaerobic growth,

$$\Delta_{cat}G = \Delta_c G_S - \frac{j_{rS}}{j_{rP}}\Delta_c G_P \tag{9.15}$$

$$\Delta_{an}G = \frac{j_{rX} - x_N j_{rN}}{j_{rP}}\Delta_c G_P + x_N\Delta_c G_N - \Delta_c G_X. \tag{9.16}$$

The heats of reaction of the two partial growth reactions may be computed by using strictly analogous equations for enthalpy. That is, in the previous equations, all the G's would have to be replaced by H's.

The Gibbs combustion energies and the heats of combustion of the various compounds have to be taken from thermodynamic tables (see Question 9.4.5). Because the concentration dependent parts of the Gibbs reaction and combustion energies and of the heats of reaction and of combustion are often much smaller than their standard values in such calculations, they can usually safely be neglected and the calculation performed only with the standard values.

9.4.3 How Is $\Delta_f G^{\ominus}$ Estimated If the Structural Formula of the Biomolecule B Is Available?

If the structural formula of the species B is known, such as for many metabolites, then group-contribution models can be applied. The concept of group-contribution models is explained in Question 4.6.8 in the case of the prediction of activity coefficients. Group-contribution models for the standard Gibbs energy of formation, $\Delta_f G^{\ominus}$ of biochemically relevant molecules can be found in literature (Jankowski et al. 2008; Noor et al. 2013 and Du et al. 2018).

9.4.4 How Can Metabolic Reaction Networks and Conditions in Cells Be Taken into Account for Thermodynamic Feasibility Analysis?

As discussed in Question 8.4.5, cytosolic conditions such as pH, pMg and ionic strength have a considerable influence on reaction equilibria. To analyzie the influence of the cytosolic conditions on equilibria, the Milo Lab at the Weizmann Institute in Rehovot, Israel, developed a program called "eQuilibrator" (https://equilibrator.weizmann.ac.il). Even Gibbs potentials of transport reactions (i.e. between cellular compartments) can be estimated using the eQuilibrator.

If thermodynamic data are to be incorporated into a thermodynamic feasibility analysis (explained in Question 8.4.4), metabolic reaction routes need to be known, which can be found at KEGG (Kyoto Encyclopedia of Genes and Genomes; https://www.genome.jp/kegg/) or MetaNetX (Automated Model Construction and Genome Annotation for Large-Scale Metabolic Networks from the ETH Zürich and the EPFL Ecole Polytechnique Federale Lausanne; https://www.metanetx.org).

9.4.5 Where Can Reliable, Experimentally Determined Standard Enthalpies and Gibbs Energies for Biochemical Reactions Be Found?

A comprehensive collection of measured thermodynamic data (e.g. equilibrium constants, standard enthalpies of reaction), for more than 400 reactions, has been published by the National Institute of Standards and Technology (NIST) in the TECRDB (Thermodynamics of Enzyme-Catalyzed Reactions Database; https://randr.nist.gov/enzyme/Default.aspx). Standard Gibbs energies of reaction can be calculated from the equilibrium constants using Equation 5.25 (Question 5.3.1). If equilibrium constants for certain reactions cannot be found in the database, combinations can be used to calculate those that are unknown or difficult to measure. An illustrative example would be the reaction (ATP+ $H_2O \rightleftharpoons$ ADP + P_i). The more easily measurable reaction of glucose isomerase (ATP + glucose $\rightleftharpoons$ ADP + glucose-6P; $\Delta_r G_1^{\ominus} = -25$ kJ/mol) and of glucose-6P phosphatase (glucose-6P + $H_2O \rightleftharpoons$ glucose + P_i; $\Delta_r G_2^{\ominus} = -12$ kJ/mol) can be combined. The sum of both reactions yields the standard Gibbs energy for ATP hydrolysis $\Delta_r G^{\ominus} = \Delta_r G_1^{\ominus} + \Delta_r G_2^{\ominus} (=(-25-12)$ kJ/mol $= -37$kJ/mol).

Table 9.1 provides examples of standard values of Gibbs combustion energies and heats of combustion for a number of important biochemical compounds. Question 5.2.1 explains how to calculate standard values of Gibbs reaction energies and heats of reaction from the respective combustion data.

9.5 What About Searching in Scientific and Engineering Journals?

The most serious source for property values is the scientific journals where those values are first published. Today, the retrieval of information from scientific journals is very easy. The most commonly used such search engines are as follows

TABLE 9.1

Heats and Gibbs energies of combustion of selected chemical compounds. Standard state for aqueous products (aq): 1 M, 10^5 Pa and 298.15 K. Standard state for gases (g): 10^5 Pa and 298.2 K. $\Delta_c G^{\ominus}$ uses as reference state CO_2 (g), N_2 (g), and liquid water at 10^5 Pa and 298.2 K. $\Delta_c G^{\ominus\prime}$ describes a combustion at pH=7: Standard state is for all aqueous compounds including HCO_3^-(aq), 1 M, 10^5 Pa except for hydrogen ions, for which $c = 10^{-7}$ M. Data sources: a) Heijnen (1999); b) Battley et al. (1997); c) Roels (1983); d) Taymaz-Nikerel et al. (2013); e) von Stockar and Liu (1999). Reprinted with permission from von Stockar (2013).

		$j_{r,B}$	$\Delta_c H^{\ominus}$ kJ/ C-mol	$\Delta_c H^{\ominus}$ kJ/mol	$\Delta_c G^{\ominus}$ kJ/ C-mol	$\Delta_c G^{\ominus}$ kJ/mol	$\Delta_c G^{\ominus\prime}$ kJ/ C-mol	$\Delta_c G^{\ominus\prime}$ kJ/mol	Source
Inorganic Chemicals									
Water	H_2O	0.00		0.00	0.00	0.00	0.00	0.00	a)
Bicarbonate	$HC0_3^-$	0.00	−286.00	−286.00	−4 .82	−4.82	0.00	0.00	a)
CO (g)		0.00		0.00	0.00	0.00	0.00	0.00	a)
Proton	H^+	0.00		0.00		0.00		0.00	a), c)
Oxygen	O_2	4.00		0.00		0.00		0.00	a)
Hydrogen	H_2	2.00		−286.00		−237.18		−237.18	a)
Methane	CH_4	8.00	−891.10	−891.10	−817.97	−817.97	−817.97	−817.97	a)
Ammonium	NH_4^+	3.00	−383.00	−383.00	−356.00	−356.00	−316.00	−316.00	c)
Ammonia	NH_3	3.00	−383.00	−383.00	−329.00	−329.00	−329.00	−329.00	c)
Biochemicals									
Biomass	$CH_{1.66}O_{0.46}N_{0.14}$	4.74	−526.00		−515.00				b), e)
Formate	CHO_2^-	2.00	−270.10	−270.10	−296.54	−296.54	−291.72	−291.72	a)
Formaldehyde	CH_20	4.00	−572.00	−572.00	−501.00	−501.00	−496.18	−496.18	a), c)
Methanol	CH_4O	6.00	−720 .10	−720.10	−693.33	−693.33	−688.51	−688.51	a)
Acetate	$C_2H_3O_3^-$	4.00	−437.10	−874.20	−446.84	−893.67	−422.08	−844.16	a)
Ethanol	C_2H_6O	6.00	−679.10	−1358.20	−659.26	−1318.51	−654.44	1308.87	a)
Lactate	$C_3H_5O_4^-$	4.00	−451.10	−1353.30	−459.15	−1377.44	−441 .04	1323.11	a)
Propionate	$C_3H_5O_3^-$	4.67	−509.67	−1529.00	−511 .18	−1533.54	−493.07	1479.21	a)
Glycerol	$C_3H_8O_3$	4.67	−550.10	−1650.30	−547.76	−1643.28	−542.94	1628.82	a)
Propanediol	$C_3H_8O_2$	5.33			−601.60	−1804.80	−596.78	1790.34	a)
Propanol	C_3H_8O	6.00	−663.00	−1989.00	−648.67	−1946.00	−644.00	1932.00	c)
Fumaric	$C_4H_4O_4$	3.00	−334.25	−1337.00	−351.50	−1406.00	−346.68	1386.72	d), c)
Fumarate	$C_4H_3O_4^-$	3.00			−355.92	−1423.66	−341.13	1364.51	d)
Fumarate	$C_4H_2O_4^{2-}$	3.00	−386.05	−1544.19	−362.48	−1449.93	−337.73	1350.91	d)
Succinic	$C_4H_6O_4$	3.50	−373.25	−1493.00	−385.59	−1542.34	−380.77	1523.06	d), c)
Succinate	$C_4H_5O_4^-$	3.50			−391.59	−1566.36	−376.80	1507.21	d)
Succinate	$C_4H_4O_4^{2-}$	3.50	−381.35	−1525.40	−399.64	−1598.54	−374.88	1499.52	d)
Butyric	$C_4H_8O_2$	5.00	−548 .50	−2194.00	543.25	2173.00	528.50	2114.00	c)
Butyrate	$C_4H_7O_3^-$	5.00	−546.35	−2185.40	−543.38	−2173.53	−567.96	2271.82	a)
Acetoin	$C_4H_8O_2$	5.00			−561.54	−2246.16	−556.72	2226.88	a)
Acetone	C_3H_6O	5.33	−448.25	−1793.00	−433.50	1734.00	−429.75	1719.00	c)
Butanediol	$C_4H_{10}O_2$	5.50			−610.34	−2441.34	−605.52	2422.06	a)
Butanol	(4 H 10O	6.00	−670.00	−2680.00	648.00	2592.00	643.00	2572.00	c)
Malate	C4H40 s·	2.75	−332.25	−1329.00	361 .00	1444.00	336.25	1345.00	c)
Glucose	C6H 120 6	4.00	−469.43	−2816.60	−478.67	−2872.02	−473.85	2843.10	a)

1. *Scopus*, an abstract and citation database of peer-reviewed literature
2. *Web of Science*, a Thomson Reuters citation database
3. *Google Scholar*

These can be easily used provided the user's institution is registered, the paper can be made to appear directly on the screen. Many papers today are open access, which means that they can be downloaded even if the institution is not registered with the particular journal.

A further source of papers is the archives. In addition to universities' archives, papers can also be found in

1. *ResearchGate.com*
2. *Academia.edu*

In these websites, the title and the abstract of the paper can be found (unless it is open access), and the reader can ask the author for the actual paper.

9.6 How Can I Evaluate Reported Experimental Values?

To evaluate the experimental data that one finds in literature, it is imperative to recognize that not all experimental values are of equal worth. The field of thermophysical properties, and particularly transport properties, is littered with examples of quite erroneous measurements made, in good faith, with instruments whose theory was not completely understood. It is therefore always necessary to separate all of the experimental data collected during a literature search into primary and secondary data by means of a thorough study of each paper.

Data with the lowest attainable uncertainty can be used in developing correlations. These data must satisfy the following conditions

1. The measurements will have been carried out in an instrument for which a complete working equation is available together with a complete set of corrections
2. The instrument will have had a high sensitivity to the property to be measured
3. The primary, measured variables will have been determined with high precision

Occasionally, experimental data that fail to satisfy these conditions may be included in the primary data set if they are unique in their coverage of a particular region of state and cannot be shown to be inconsistent within theoretical constraints. Their inclusion is encouraged if other measurements made in the same instrument are consistent with independent, nominally lower uncertainty data. Secondary data, excluded by the above conditions, are used for comparison only.

In certain occasions, primary data can consequently be employed to develop "reference values" or "reference correlations," as

1. Internationally accepted "**reference values**" (known also as "standard reference values") serve two primary purposes. First, they can provide a means of confirming the operation and experimental uncertainty of any new absolute apparatus

TABLE 9.2

Experimental Thermodynamics. Volumes I–X

Volume I (*Calorimetry of Non-Reacting Systems*) covers the heat capacity determinations for chemical substances in the solid, liquid, solution and vapor states, at temperatures ranging from near the absolute zero to the highest at which calorimetry is feasible. The book provides a historical perspective.
Volume II (*Experimental Thermodynamics of Non-Reacting Fluids*) provides historical information on methods used in measuring thermodynamic properties and tests, including physical quantities and symbols for physical quantities, thermodynamic definitions and definitions of activities and related quantities, up to the time of publication. The text also describes reference materials for thermometric fixed points, temperature measurement under pressures and pressure measurements. The publication takes a look at absolute measurement of volume and equation of state of gases at high temperatures and low or moderate temperatures. Discussions focus on volumes of cubes of fused silica, density of water and methods of measuring pressure. The text also examines the compression of liquids and thermodynamic properties and velocity of sound, including thermodynamics of volume changes, weight methods and adiabatic compression.
Volume III (*Measurement of the Transport Properties of Fluids*) describes the development of the techniques of measurement of the transport properties of fluids and their mixtures including viscosity, thermal conductivity, diffusion coefficients, thermal diffusion factors and thermal transpiration coefficients. A summary of a wide variety of experimental techniques applicable at that time, over a wide range of thermodynamic states, is also included. Many of the techniques described are still in regular use.
Volume IV (*Solution Calorimetry*) is concerned with solution calorimetry and considers the calorimetry of both reacting and non-reacting systems. The contributions reflect and summarize the current state of development of techniques for the measurement of the energy effects involved in reaction and mixing in solution over a range of temperature and pressure. These techniques are also considered in the light of their application, at that time, in chemical research and the chemical industry.
Volume V (*Equations of State for Fluids and Fluid Mixtures*) covers all the major approaches for developing equations of state. The theoretical basis and practical use of each type of equation is discussed and the strength and weaknesses of each is addressed. Topics addressed include the virial equation of state, cubic equations and generalized van der Waals equations, perturbation theory, integral equations and corresponding stated and mixing rules. Special attention is also devoted to associating fluids, polydisperse fluids, polymer systems, self-assembled systems, ionic fluids and fluids near critical points.
Volume VI (*Measurement of the Thermodynamic Properties of Single Phases*) presents the current state of development of the techniques of measurement of the thermodynamic quantities of single phases. It contains a valuable summary of a large variety of experimental techniques applicable over a wide range of thermodynamic states with an emphasis on the precision and accuracy of the results obtained.
Volume VII (*Measurement of the Thermodynamic Properties of Multiple Phases*) summarizes the state of knowledge with respect to experimental techniques in thermochemistry and thermodynamics. It further contains descriptions of recent developments in the techniques for measurement of thermodynamic quantities for multiple phases of pure fluids as well mixtures over a wide range of conditions, while emphasis is given in the precision and accuracy of results obtained from each method.
Volume VIII (*Applied Thermodynamics of Fluids*) serves as a guide to scientists or technicians who use equations of state for fluids, SAFT and corresponding states. Concentrating on the application of theory, the practical use of each type of equation is discussed and the strengths and weaknesses of each are addressed. It includes material on the equations of state for chemically reacting and non-equilibrium fluids that have undergone significant developments and brings up to date the equations of state for fluids and fluid mixtures.
Volume IX (*Advances in Transport Properties of Fluids*) provides a valuable account of the advances in the measurement and prediction of transport properties that have occurred over the last 20 years. It presents the experimental and theoretical background of transport properties, including new experimental techniques and how existing techniques have developed, new fluids e.g. molten metals, dense fluids and critical enhancements of transport properties of pure substances.
Volume X (*Non-equilibrium Thermodynamics with Applications*) covers recent developments in the theory of non-equilibrium thermodynamics and its applications. Four chapters are devoted to the foundations; an overview chapter is followed by recent results addressing the underlying principles of the theory. The applications are concerned with bulk systems, with heterogeneous systems where interfaces are central and with process units in industry where entropy production minimization is useful. There is also a collection of chapters under the heading mesoscopic non-equilibrium thermodynamics, giving in the end an overview of extensions of the theory into the non-linear regime.

and the stability and reproducibility of existing absolute measurement equipment. Second, in the case of instruments operating in a relative way, they provide the basis to calibrate one or more unknown constants in the working equation. Reference values refer to the properties specified at a fixed state condition (specific temperature, pressure and composition) or at a small number of such states. These values are often characterized by the lowest uncertainty possible at the time of their acceptance.

2. "**Reference correlations**" for pure-fluid properties often cover a wide range of conditions – typically from the triple-point temperature to 1,000 K and up to 100 MPa pressure – and are developed to achieve the lowest possible uncertainties (although perhaps higher than those of reference values).

We should also mention that many journals today have cooperative agreement with the Thermodynamics Research Center (TRC) at the National Institute of Science and Technology (NIST). Under this agreement, and for manuscripts that are in scope for NIST, a statistical editor based inside the TRC checks the data in the manuscripts, comparing them with previously published literature results and checking the trends. The reports are provided to the authors for manuscripts containing problems, and the journal's editor handling the manuscript may request corrections based on that report. Where such an agreement is in place, the readers may place great confidence in the data published in that journal.

9.7 What Are the Preferred Methods for the Measurement of Thermodynamic Properties?

An excellent selection of techniques for which complete working equations are available, as well as the best possible ways of prediction or calculation of thermophysical properties, are discussed elsewhere. The series, *Experimental Thermodynamics* (Vol. I 1968, Vol. II 1975, Vol. III 1991, Vol. IV 1994, Vol. V2000, Vol. VI 2003, Vol. VII 2005, Vol. VIII 2010, Vol. IX 2014, and Vol. X 2016), summarizes the historical evolution of methods for the measurement and prediction of thermophysical properties. The early volumes are sometimes superseeded by later material in the series but the most recent volumes give a coherent and modern description of experimental methods although, of course, they continue to evolve (Table 9.2).

References

ASHRAE Handbook - Fundamentals, 2021, ASHRAE, GA, USA.

Assael M.J., Wakeham W.A., Goodwin A.R.H., Will S., and Stamatoudis M., 2011, *Commonly Asked Questions in Thermodynamics*, CRC Press, Boca Raton.

Battley E.H., Putnam R.L., and Boerio-Gates J., 1997, "Heat capacity measurements from 10 to 300 K and derived thermodynamic functions of lyophilized cells of Saccharomyces cerevisiae, including the absolute entropy and the entropy of formation at 298.15 K", *Thermochimica Acta* **298**:37–46.

CRC Handbook of Chemistry and Physics, 2020, 102nd ed., Editor-in-Chief J.R. Rumble, CRC Press, Boca Raton.

Du B., Zhang Z., Grubner S., Yurkovich J.T., Palsson B.O., and Zielinski D.C. 2018, "Temperature-dependent estimation of Gibbs energies using an updated group-contribution method", *Biophys J.* **114**(11): 2691–2702.

Experimental Thermodynamics, Volume I, Calorimetry of Non-Reacting Systems, 1968, eds. McCullough J.P., and Scott D.W., for IUPAC, Butterworths, London.

Experimental Thermodynamics, Volume II, Experimental Thermodynamics of Non-Reacting Fluids, 1975, eds. Le Neindre B., and Vodar B., for IUPAC, Butterworths, London.

Experimental Thermodynamics, Volume III, Measurement of the Transport Properties of Fluids, 1991, eds. Wakeham W.A., Nagashima A., and Sengers J.V., for IUPAC, Blackwell Scientific Publications, Oxford.

Experimental Thermodynamics, Volume IV, Solution Calorimetry, 1994, eds. Marsh K.N., and O'Hare P.A.G., for IUPAC, Blackwell Scientific Publications, Oxford.

Experimental Thermodynamics, Volume V, Equations of State for Fluids and Fluid Mixtures, Parts I and II, 2000, eds. Sengers J.V., Kayser R.F., Peters C.J., and White H.J. Jr., for IUPAC, Elsevier, Amsterdam.

Experimental Thermodynamics, Volume VI, Measurement of the Thermodynamic Properties of Single Phases, 2003, eds. Goodwin A.R.H., Marsh K.N., and Wakeham W.A., for IUPAC, Elsevier, Amsterdam.

Experimental Thermodynamics, Volume VII, Measurement of the Thermodynamic Properties of Multiple Phases, 2005, eds. Weir R.D., and de Loos T.W., for IUPAC, Elsevier, Amsterdam.

Experimental Thermodynamics, Volume VIII, Applied Thermodynamics of Fluids, 2010, eds. Goodwin A.R.H., Sengers J.V., and Peters C.J., for IUPAC, RSC Publishing, Cambridge.

Experimental Thermodynamics Volume IX, Advances in Transport Properties of Fluids, 2014, eds. Assael M.J., Goodwin A.R.H., Vesovic V., and Wakeham W.A., for IUPAC, RSC Publishing, London.

Experimental Thermodynamics Volume X, Non-equilibrium Thermodynamics with Applications, 2016, eds. Bedeaux D., Kjelstrup S., and Sengers J.V., for IUPAC, RSC Publishing, London.

Heijnen, J.J., 1999, "Bioenergetics of microbial growth", in Bioprocess Technology: Fermentation, *Biocatalysis and Bioseparation*, Eds.M.C. Flickiger, S.W. Drew, Wiley & Sons, Inc., 267–291.

Jankowski M.D., Henry C.S., Broadbelt L.J., and Hatzimanikatis V., 2008, "Group contribution method for thermodynamic analysis of complex metabolic networks," *Biophysical J.* **95**: 1487–1499.

Noor, E., Haraldsdottir, H.S., Milo, R., and Fleming, R.M., 2013, "Consistent estimation of Gibbs energy using component contributions", *PLoS Comput. Biol.* **9**: e1003098.

Perry's Chemical Engineers' Handbook, 2018, 9th ed., eds. Green D.W., and Southard M.Z., McGraw-Hill.

Roels, J.A., 1983, *Energetics and Kinetics in Biotechnology*, Elsevier Biomedical Press, Amsterdam.

Sandler, S.I., and Orbey H., 1991, "On the thermodynamics of microbial growth processes", *Biotechnol. Bioeng.* **38**(7): 697–718.

Taymaz-Nikerel, H., Jamalzadeh, A., Espah Borujeni, A., Verhejen, P.J.T., van Gulik, V.M., and Heijnen, J.J., 2013, in Biothermodynamics: *The Role of Thermodynamics in Biochemical Engineering*, Ed.U. von Stockar, E.P.F.L. Press, Lausanne, distributed by CRC Press, Taylor & Francis Group, 547–579.

Thermodynamic Research Center (TRC), 1942–2007, *Thermodynamic Tables Hydrocarbons*, ed. Frenkel M., National Institute of Standards and Technology Boulder, CO, Standard Reference Data Program Publication Series NSRDS-NIST-75, Gaithersburg, MD.

Thermodynamic Research Center (TRC), 1955–2007, *Thermodynamic Tables Non-Hydrocarbons*, ed. Frenkel M., National Institute of Standards and Technology Boulder, CO, Standard Reference Data Program Publication Series NSRDS-NIST- 74, Gaithersburg, MD.

von Stockar, U., and Liu, J.-S., 1999, "Does microbial life always feed on negative entropy? Thermodynamic analysis of microbial growth", *Biochim. Biophys. Acta- Bioenrg.* **1412**:191–211.

von Stockar U., 2013, Biothermodynamics of Live Cells, in Biothermodynamics: The Role of Thermodynamics in Biochemical Engineering, EPFL Press, distributed by CRC Press, Taylor and Francis Group, Lausanne, 475–534.

Index